Nanoparticles Technology Handbook

Nanoparticles Technology Handbook

Edited by **Mindy Adams**

NY RESEARCH
P R E S S

New York

Published by NY Research Press,
23 West, 55th Street, Suite 816,
New York, NY 10019, USA
www.nyresearchpress.com

Nanoparticles Technology Handbook
Edited by Mindy Adams

International Standard Book Number: 978-1-63238-337-2 (Hardback)

Printed in the United States of America.

Contents

Preface

Over the recent decade, advancements and applications have progressed exponentially. This has led to the increased interest in this field and projects are being conducted to enhance knowledge. The main objective of this book is to present some of the critical challenges and provide insights into possible solutions. This book will answer the varied questions that arise in the field and also provide an increased scope for furthering studies.

This book aims to prove helpful for all researching in the field of nanoparticles. Nanoparticles have made a significant impact in the scientific domain in the past few years. The properties of numerous conventional materials changed when shaped from nanoparticles. Nanoparticles have a larger surface area per weight as compared to larger molecules which makes it more reactive and effective than other molecules. This book covers various aspects related to metallic nanoparticles and their testing technology. It includes a number of well researched studies which will help students and experts interested in the field of nanotechnology.

I hope that this book, with its visionary approach, will be a valuable addition and will promote interest among readers. Each of the authors has provided their extraordinary competence in their specific fields by providing different perspectives as they come from diverse nations and regions. I thank them for their contributions.

Editor

Part 1

Metallic Nanoparticles

1

Metallic Nanoparticles Coupled with Photosynthetic Complexes

Sebastian Mackowski
Optics of Hybrid Nanostructures Group
Institute of Physics, Nicolaus Copernicus University, Torun
Poland

1. Introduction

Plasmon excitations in metallic nanoparticles provide an efficient way to manipulate electromagnetic fields at the nanoscale (Maier, 2004). While the interactions between plasmons and simple nanostructures such as organic dyes or semiconductor nanocrystals is relatively well described and understood, application of metallic nanoparticles to multi-pigment structures has started just recently (Carmeli, 2010; Govorov, 2008; Kim, 2011; Mackowski, 2008; Nieder, 2010). Light-harvesting complexes, or more generally, photosynthetic complexes, are quite appealing in this regard as they not only provide an interesting biomolecular system for studying plasmon effect on both the optical properties of pigments and the energy transfer between them, but also they could offer attractive potential application route in photovoltaics (Atwater & Polman, 2010; Mackowski, 2010).

Extending concepts and methods that have been developed for describing the coupling of single organic chromophores with plasmon excitations in metallic nanoparticles (Anger, 2006; Chettiar, 2010; Govorov, 2006) to multi-chromophoric biological systems has not been completely straightforward from both theoretical and experimental points of view. On the one hand, organic molecules or semiconductor quantum dots are much more robust nanostructures than pigment-protein complexes, therefore, the sample preparation in the latter case should be more gentle, so that the protein itself maintains its structure. Preserving protein structure implies that the function of the complex as a whole also remains intact. This assures conservation of the energy transfer pathways between various chromophores comprising the complex as well as identical optical properties, including absorption and fluorescence, to that of the isolated (decoupled from a metallic nanoparticle) biomolecule. On the other hand, from the theory standpoint, biomolecules, and in particular light-harvesting complexes, render themselves a real challenging system to model due to multitude of interactions between chromophores such as chlorophylls and carotenoids (Blankenship 2002), which results in many energy transfer pathways and formation of strongly coupled excitonic systems, as well as conformational changes of the protein itself. Nevertheless, driven by the continuous development of optical spectroscopy and microscopy techniques (Polivka & Sundstom 2004) as well as more efficient modeling tools, significant progress has been achieved in understanding interactions and functions of light-harvesting complexes. It has also been helped by high-resolution crystal structures of the complexes (Hofmann 1996; McDermott, 1995), enabling thus direct association of the pigments as well as their interactions both with themselves and the protein with the actual structure and spatial arrangement of the pigments in these systems.

The purpose of this chapter is to review recent research carried out on hybrid nanostructures composed of metallic nanostructures and light-harvesting complexes. In general, the research is focused on improving the light absorption of the light-harvesting complexes through properly designed plasmonic nanostructures. However, before we start discussing particular hybrid nanostructures fabricated in the context of plasmonically enhanced absorption of light-harvesting complexes, we describe two basic concepts of metal-enhanced fluorescence: distance dependence of the fluorescence intensity and the influence of spectral properties of metallic nanoparticles and placed nearby molecules (Anger, 2006). This brief introductory discussion is essential for understanding the rationale behind designing hybrid nanostructures that involve biological fluorescing complexes.

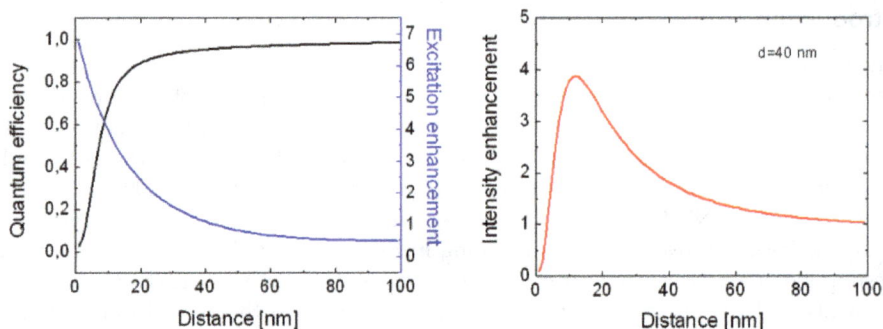

Fig. 1. Dependence of the quantum yield and non-radiative fluorescence quenching upon the distance. Fluorescence intensity of a chromophore displayed as a function of the distance between the chromophore and spherical metallic nanoparticle with 40 nm diameter.

The optical properties of a fluorophore placed in the vicinity of a metallic nanoparticle are strongly affected by plasmon excitations induced in the latter by a laser light. Without a metallic nanoparticle, a fluorophore is characterized with three rates: absorption rate, radiative rate, and non-radiative rate. Since the oscillation of electrons in the metallic nanoparticle results in creation of local electromagnetic field, in principle all three rates can be changed (Lakowicz, 2006). In addition, another process related to non-radiative energy transfer from the fluorophore to the metallic nanoparticle could also take place in such a hybrid nanostructure. The influence of plasmon excitations upon the quantum yield of a fluorophore and non-radiative energy transfer between the fluorophore and metallic nanoparticle has been recently studied theoretically. In particular, the dependence on the separation distance between the two nanostructures has been analyzed in detail. It turns out that the distance between the fluorophore and metallic nanoparticle is of critical importance in regard to the process that plays dominant role in such a system. In Fig. 1 we show the dependence of the excitation rate and quantum efficiency of a fluorophore upon the distance to the metallic nanoparticle. In this example we consider a metallic nanoparticle with diameter of 40 nm. The excitation efficiency increases exponentially with reducing the distance, which is a clear manifestation of stronger electromagnetic field felt by the fluorophore due to plasmon excitation in the metallic nanoparticle. On the other hand, the quantum efficiency, when approaching the range of distances shorter than 20 nm, starts to drop significantly, due to the non-radiative energy transfer from the fluorophore to the nanoparticle. The net result of these two processes in displayed in Fig. 1, where a clear non-monotonic dependence of the intensity of fluorescence emitted by the fluorophore upon the distance to the metallic nanoparticle can be seen. Importantly, the strongest plasmon induced enhancement of the fluorescence occurs for distances between 10 and 30 nm; for smaller distances non-radiative fluorescence quenching dominates, while for longer distances the fluorophore barely feels the presence of the metallic nanoparticle.

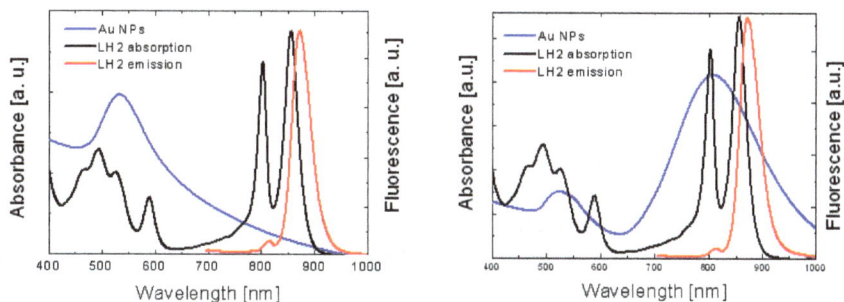

Fig. 2. Comparison between absorption spectra of spherical end elongated gold nanoparticles with the fluorescence and absorption of the LH2 complex. In the first case plasmon excitations should influence mainly the absorption in the visible range, in the second case the effect should be visible for absorption and emission in the infrared.

Another critical parameter that influences the interaction between metallic nanoparticle and fluorophore is the relation of their spectral properties. This is shown schematically in Fig. 2. In the first case scenario the absorption of metallic nanoparticles overlaps significantly with absorption of a biomolecule (in this case this is the LH2 complex from purple bacteria) in the spectral range of about 530-550 nm, while featuring virtually no overlap with the fluorescence. One may expect that for such a combination the major influence of metallic nanoparticles is due to absorption enhancement. In contrast, for a hybrid nanostructures built of components characterized with spectral properties as those displayed in Fig. 2b, there should be absorption enhancement both around 560 nm as well as in the infrared spectral region, around 800 nm. In addition, since there is a spectral overlap between plasmon band and fluorescence emission, the radiative rate should also increase as a result of plasmon excitation.

Optical spectroscopy provides variety of techniques that allows for distinguishing between various processes that determine the net effect of plasmonic excitations in metallic nanoparticles on a fluorophore. Indeed, in an ideal situation, where only absorption rate is affected by the plasmon excitation, there should be no change in the fluorescence decay time, while an additional band should appear in the fluorescence excitation spectrum. In contrast, when only radiative rate increases as a result of plasmon coupling, the fluorescence excitation spectrum for a hybrid nanostructure should be identical to the reference structure, with much shorter fluorescence decay time. Several experimental configurations exhibiting these various aspects of plasmon coupling with pigment – protein complexes are discussed in this contribution.

2. Materials and methods

In this section we introduce the structure and the optical properties of light-harvesting complexes used in our studies as well as present basic characteristics of metallic nanostructures, including their morphology and plasmon characteristics. Next we present experimental techniques employed for investigating the interactions between plasmon excitations and chromophores embedded in the proteins. These include standard absorption and fluorescence/fluorescence excitation spectroscopy, both in solution and in a layered geometry, as well as confocal fluorescence microscopy coupled with time-resolved capability and spectrally-resolved detection. Combination of all these experimental techniques allows for comprehensive description of plasmon-induced effects on the complex biomolecular systems.

2.1 Light-harvesting complexes

Pigment-protein complexes that take part in photosynthesis can be generally divided into two groups: complexes containing reaction centers, which carry out charge separation, and complexes responsible solely for harvesting the sunlight and transferring it to the reaction centers. Large proteins of Photosystem I and Photosystem II fall into the first category, while light harvesting complex 2 (LH2) from purple bacteria and peridinin – chlorophyll – protein (PCP) complex from algae belong to the second group. As the structure and function of all these biomolecules has been described in detail previously, we focus here only on the aspects that are relevant for understanding the influence of plasmon excitations on the optical properties of light-harvesting systems, PCP and LH2.

2.1.1 Peridinin-chlorophyll-protein

Light-harvesting complexes were developed in the course of evolution in order to enhance and broaden the absorption of photosystems for the efficient use of sunlight in photosynthesis. Their major function of these pigment-protein complexes is to harvest the sunlight and transfer the energy to the Photosystems. Peridinin-chlorophyll-protein (PCP) found in Dinoflagellates *Amphidinium carterae* is one of many such complexes. It is a water-soluble protein employed as an antenna external to the membrane. The structure of the PCP complex, shown in Fig. 3, has been determined with 1.3 Å resolution using X-ray crystallography (Hofmann, 1996). The native form of PCP consists of two chlorophyll a (Chl) and eight peridinin (Per) molecules embedded in a protein matrix. All the pigments are arranged in two almost similar clusters and embedded in the hydrophilic protein capsule. The conjugated portion of each Per is close to the chlorophyll tetrapyrrole ring at a van der Waals distance (3.3 to 3.8 Å), the distance between Mg atoms of the two Chl a in one monomer is 17.4 Å and intercluster edge-to-edge distances between Per are in the range of 4-11 Å. The ratio of Per to Chl a of 4:1 indicates that PCP utilizes the carotenoids as its main light-harvesting pigments. It has been shown that the PCP complex can be reconstituted with other Chl derivatives which exhibit different optical properties (Brotosudarmo, 2006). Importantly, the folding of the protein used in the reconstitution procedure takes place over almost identical pathway as in the native system, which results in very similar structures of the reconstituted systems. Since each of these chlorophyll molecules features specific absorption and emission characteristics, it became possible to construct and study the energy transfer dynamics as well as inter-pigment interactions in a well-defined geometry given by the protein (Mackowski, 2007, Polivka, 2005).

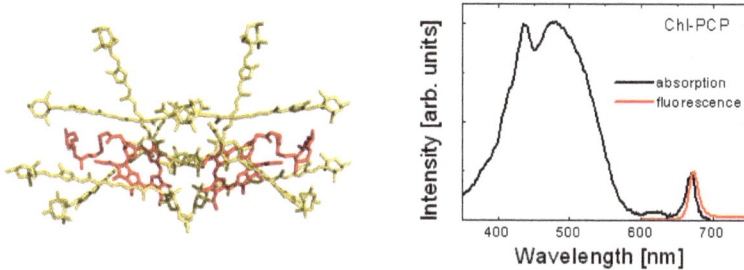

Fig. 3. Pigment structure of the PCP complex reconstituted with Chl *a* together with absorption (black line) and fluorescence (red line) measured in water solution at room temperature.

The absorption spectrum of the Chl-PCP displayed in Fig. 3 has an intense, broad band between 400 to 550 nm that is mainly due to Per absorption, and two Chl – related bands at 440 nm (Soret) and 660 nm (Q_Y). One example of chlorophylls used for reconstituting PCP complexes is [3-acetyl]-chlorophyll a (acChl). Chemically, it differs from Chl only by the C-3 substituent, but the absorption and fluorescence spectra of acChl-PCP, the PCP complex reconstituted with acChl, are red shifted as compared to the PCP complex containing Chl. At the same time the Per absorption in the blue-green spectral range is affected very slightly. For the PCP complexes reconstituted with acChla the absorption of the Qy band of the chlorophyll molecules is shifted by approximately 20 nm to the red. The fluorescence emission of the PCP complex originates from weakly coupled Chl molecules and it appears at 670 nm for Chl-PCP and 690 nm for acChl-PCP. Upon absorption of light, peridinins in PCP transfer their electronic excitation to Chl a. The efficiency of this excitation energy transfer is higher than 90% [20]. Subsequently, Chl a passes the energy on to membrane-bound light-harvesting complexes and the Photosystem II. Clearly, the absorption spectrum of PCP enables the photosynthetic apparatus to harness the sunlight not only in the red spectral range, but it extends it into the blue-green spectral region.

Optical spectroscopy studies of both native and reconstituted PCP complexes have been carried out on the ensemble (Akimoto, 1996; Kleima, 2000; Krueger, 2001) and single-molecule levels (Mackowski, 2007; Wormke, 2007a; Wormke, 2008). Using transient absorption in femtosecond timescale main energy transfer pathways have been described, it has also been demonstrated that the two Chl a molecules interact relatively weakly with characteristic transfer time between them to be of the order to 12 ps (Kleima, 2000). These observations were also corroborated with fluorescence studies of individual PCP complexes: it has been shown that it is possible to distinguish emission originating from each of the two Chl a molecules and using the property of sequential photobleaching of the Chl the energy splitting between the two molecules in the monomer were determined (Wormke, 2007a). Recent work on PCP complexes reconstituted with both Chl a and Chl b provided coherent description of the energy transfer pathways and dynamics in this unique antenna (Mackowski, 2007).

2.1.2 Light-harvesting complex 2

Another example of a pigment-protein complex employed to harness sunlight energy and transfer it efficiently to reaction centers is a light-harvesting complex 2 (LH2) from purple bacteria Rhodopseudomonas palustris. This protein is placed in the thylakoid membrane, where many of LH2 complexes surround relatively widely spaced LH1 complexes, to which they transfer excitation energy. The BChl a molecules in LH1 have a single strong near-infrared absorption band of 875 nm, while LH2 has two strong BChl a absorption bands at 800 nm and 850 nm (Fig. 2). In this way the energy gradient is formed, which facilitates efficient energy transfer from LH2 to LH1 complex, and then further to the reaction center. Both the structure and the optical properties of the LH2 have been a subject of intense studies in recent years (van Oijen, 1999; Hofmann, 2003; Bopp, 1997). It has been shown using atomic force microscopy technique, that LH2 complexes arrange around a LH1 light-harvesting complex, in the middle of which is a reaction center (Scheuring, 2004). The spatial arrangement of Bacteriochlorophylls (BChls) and carotenoids in this complex is displayed in Fig. 4. The X-ray crystallography studies of the LH2 complex have shown that out of the 27 BChl molecules 18 form a strongly coupled ring with average distances between the molecules less than 1 nm (McDermott, 1995). This excitonically coupled ring is responsible for the absorption band at 850 nm. The remaining 9 molecules form a ring of weakly coupled BChls as they are spaced by more than 2 nm. All pigments, BChls and carotenoids, are embedded in a hydrophobic protein (not shown). Single molecule investigations (van Oijen, 1999) proved that the B 850 ring are in fact not fully symmetric, and the exciton levels feature significant splitting.

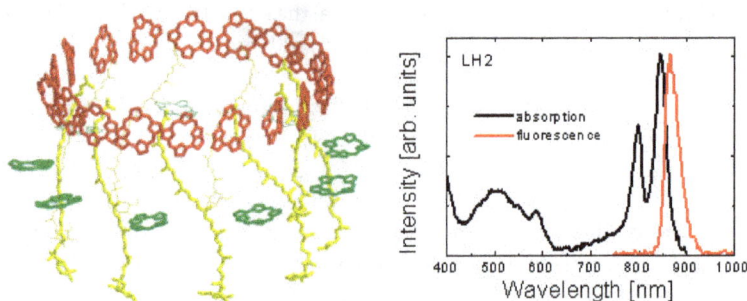

Fig. 4. Pigment structure of the LH2 complex together with absorption (black line) and fluorescence (red line) measured in buffer solution at room temperature.

The absorption spectrum of the LH2 complex is shown in Fig. 4. It consists of two prominent bands at 800 nm and 850 nm which correspond to absorption of the two rings of BChl molecules. The carotenoids are in close contact with both BChl rings, and are mainly responsible for a broad absorption between 390 nm and 550 nm. Importantly, the fluorescence of LH2, which originates exclusively from the strongly coupled ring (named B850) has therefore an excitonic character. The presence of strong absorption bands and fluorescence emission in the infrared spectral range requires – in order to influence the optical properties of the LH2 complex - application of metallic nanoparticles that feature plasmon resonances in the near infrared.

2.2 Metallic nanostructures

Metallic structures with nanometric sizes have been the subject of intense research in recent years due to mainly unique optical properties of these systems that can be used for manipulating light at the nanoscale, designing biosensors and artificial chiral nanostructures, enhancing the optical properties of semiconductor nanocrystals and organic fluorescent dyes, enabling detection of emitters characterized with low fluorescence quantum yields, such as carbon nanotubes or DNA. In addition to efforts aimed at exploiting plasmon effect in metallic nanoparticles, significant research have been carried out to achieve almost perfect control of the morphology of metallic nanoparticles, and thus the plasmon properties thereof. Among many techniques of fabrication of metallic nanostructures are evaporation of metallic film on a corrugated substrate (Chettiar, 2010; Mackowski 2008), nanosphere lithography (Hulteen, 1995), electron beam lithography, electrochemical deposition, direct formation of silver island film (Ray, 2006), and chemical synthesis (Link, 1999). Each of these methods requires particular technical capabilities, frequently the experimental setups for nanostructure fabrication is expensive making them hardly accessible. Chemical synthesis of nanoparticles however, is quite simple, at least on the basic level, and, when mastered, provides a way to obtain highly monodisperse nanoparticles with tailored morphology and surface functionalization. This results in well-defined optical properties such as energies of plasmon resonances, and conjugation capabilities to other nanostructures or surfaces. In this contribution we describe the interactions present in hybrid nanostructures composed of light-harvesting complexes PCP and LH2 and metallic nanostructures fabricated using chemical deposition of silver island film on a glass substrate, electron beam assisted deposition of silver island film on a glass substrate, as well as chemically synthesized gold spherical nanoparticles and nanorods. We show that by careful design of a hybrid nanostructure we can control the impact of plasmon excitations in metallic nanoparticles upon the absorption and emission of the chlorophyll-containing light-harvesting complexes.

2.2.1 Silver island film

One of the simplest to fabricate metallic nanostructures is a corrugated metallic film. The method to obtain such a film with islands characterized with sizes of tens of nanometers has been previously applied to study the impact of plasmon interactions upon the fluorescence of various organic dyes, semiconductor quantum dots, and a few proteins, including the green fluorescent protein (Lakowicz, 2006; Ray, 2006). Silver island films used in our experiment were prepared by reducing an aqueous silver nitrate solution. All chemicals were purchased from Sigma-Aldrich and used as received. First, freshly prepared aqueous NaOH (1.25 M) was added to a silver nitrate solution. The precipitate was re-dissolved by adding NH$_4$OH, and the solution was cooled to ~5°C under stirring. After adding D-glucose, clean microscope cover slips were dipped in the solution, which was then heated up to 30°C. The resulting Ag-covered glass coverslips were examined using absorption spectroscopy and atomic force microscopy (AFM). In order to change the morphology and thus the properties of the silver island film, we fabricated several samples with varied dipping time in the reaction solution: the coverslips were kept in solution for 1 and 3 minutes. In Fig. 5 we show AFM image of the SIF obtained by dipping the coverslip for 1 minute in the reaction solution. The islands are characterized with average sizes of about 40-

50 nm laterally and the surface density is very high. In addition we also include in Fig. 5 the absorption spectrum measured for the SIF structure which features plasmon resonance with a maximum at 450 nm and the linewidth of about 150 nm, and thus matches the absorption of Per in the PCP complexes.

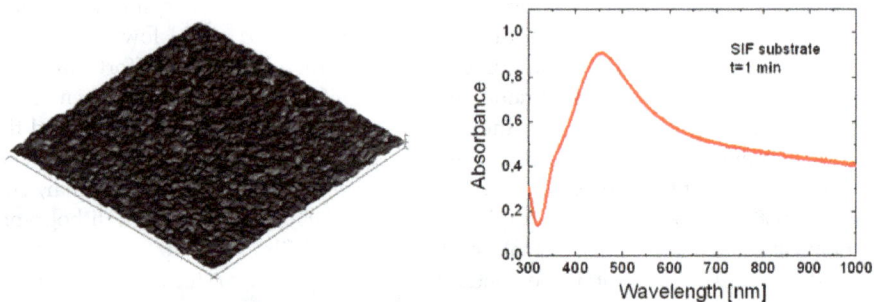

Fig. 5. Atomic force microscopy image of a silver island film fabricated by chemical synthesis together with absorption spectrum of the sample obtained for 1 minute-long dipping time of the glass substrate in the reaction solution. The lateral size of the AFM image is 5 microns.

2.2.2 Semicontinuous metallic layer

The semicontinuous silver film has been fabricated using electron beam-assisted evaporation of silver on glass substrate (Chettiar, 2010). While this nanostructure may look similar to the SIF discussed above, the substrates obtained with the e-bam technique are typically more homogeneous, the sizes and shapes of the silver islands is controlled to higher degree. Proper adjustment of the parameters during the evaporation process leads to corrugated metallic films with designed optical properties. For instance, it has been shown that by changing evaporation time it is possible to obtain morphologies ranging from roughly isolated islands to the strongly coalescing ones. Such differences in morphology resulted in strong shift of plasmon energies towards the red and near infrared spectral ranges, opening thus completely new possibilities for applying these structures for controlling the optical properties of infrared – emitting systems. Yet another important advantage of semicontinuous metallic films fabricated using e-beam assisted evaporation is the capability of uniform coating of such films with dielectric layers with thicknesses ranging from a few nanometers up to tens of nanometers. As plasmon induced effects depend crucially upon the separation between metallic nanoparticles and optically active molecules, such structures render themselves a highly suitable system for investigating processes that occur in plasmonic hybrid nanostructures. In particular, the results included in this contribution have been obtained for a semicontinuous silver film covered with a 25-nm-thick SiO_2 layer evaporated in the same process without exposing the structure to ambient conditions.

2.2.3 Colloidal metallic nanoparticles

Among metallic nanostructures, ones of the most studied are colloidal metallic nanoparticles. It is triggered mainly by enormous variety of metallic nanoparticles that can be synthesized, even with little resources: sizes of the nanoparticles range from a few to a few hundreds of nanometers. They can also be of essentially any shape: spherical, elongated, triangular, cube or star-like. Furthermore, since the nanoparticles are synthesized in the colloidal form, it is possible to functionalize their surface with functional group suitable for specific attachment to surfaces or conjugation with other nanostructures. Lastly, there have been many examples for self assembly of metallic nanoparticles in complex structures with new properties and functions (Link, 1995).

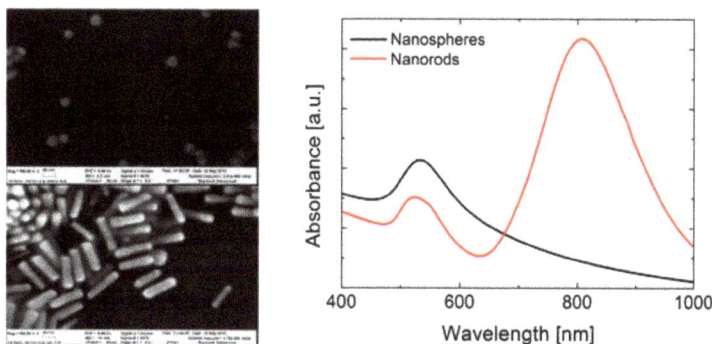

Fig. 6. Scanning electron microscopy images of gold spheres and nanorods. The structural data is accompanied with absorption spectra measured for these two samples at room temperature.

In this contribution we report on synthesis of spherical gold nanoparticles and gold nanorods with the purpose to match absorption bands of light-harvesting complexes. While spherical gold nanoparticles, which feature plasmon resonances around 500-500 nm correspond to carotenoid absorption both in the PCP and LH2 complexes, the nanorods have their resonance also in the infrared, as shown in Fig. 6. By controlling the reaction we tune the energy of the plasmon resonance exactly to 800 nm, thus matching perfectly the B800 absorption of the LH2 complex.

Gold nanoparticles were synthesized using a reduction reaction and dispersed in toluene. The average diameter of the nanocrystals was 5 nm, which results in a plasmon resonance maximum at 530 nm. The synthesis of Au nanorods was based on seed-mediated growth in water solution. All chemicals ($HAuCl_4 \times 3H_2O$ (99.9%), $NaBH_4$ (99%), L-Ascorbic Acid (99+%), hexadecyltrimethylammoniumbromide (CTAB) (99%), and $AgNO_3$ (99+ %)) were purchased from Aldrich and used without further purification. Deionized water (Fluka) was used in all experiments. In order to prepare Au seeds CTAB solution (4.7 ml, 0.1M) was mixed with 25 µl of 0.05 M $HAuCl_4$. To the stirred solution, 0.3 ml of 0.01 M $NaBH_4$ was added, which resulted in the formation of brownish yellow solution. Seeds solution was kept at room temperature until further used. For the synthesis we use Au seeds prepared beforehand. The "seed-mediated" method was developed previously; it is carried out in aqueous solution at atmospheric pressure and near room temperature. Appropriate quantities and molarities of CTAB (150 ml, 0.1 M), $HAuCl_4$ (1.5 ml, 0.05 M), L-Ascorbic acid (1.2 ml, 0.1 M), 0.01 M $AgNO_3$ (1.6 ml, 1.8 ml, 2 ml) and seed (360 µl) water solutions were added one by one in a flask, followed by a gentle mixing. Addition of ascorbic acid, as a mild reduction agent, triggered a mixture color change from dark yellow to colorless. After addition of the seed solution, the mixtures was put into water bath and kept at constant temperature of 28 °C for 2 hours. Obtained products were separated from unreacted substrate and spherical particles by centrifugation at 9.000 rpm for 60 minutes. The supernatant was removed using a pipette and the precipitate was redissolved in pure water.

2.3 Sample preparation

In the research described in this work, we have used several sample architectures, from a very simple ones, where light-harvesting complexes were deposited directly on the surface of the metallic layer being either in the form of silver island film or colloidal gold nanoparticles spin-coated on a glass coverslide, to more advanced, where metallic nanoparticles were separated from the light-harvesting complexes by a thin dielectric layer. For that purpose we used SiO2 deposited using e-beam assisted evaporation, the thickness of the SiO2 layer was changed from 5 nm to 40 nm. The light-harvesting complexes were dispersed in a PVA and then spin-coated on the substrate.

2.4 Experimental techniques

The optical properties of hybrid nanostructures comprising light-harvesting complexes and metallic nanostructures have been studied using absorption and fluorescence spectroscopy in the visible and infrared regions. spectral region. Absorption spectra were obtained using a Cary 50 spectrophotometer. Fluorescence and fluorescence excitation spectra of both structures were measured using the FluoroLog 3 spectrofluorimeter equipped with specially designed mount suitable for holding planar samples. A Xenon lamp source with a double grating monochromator was used for excitation and the signal was detected with a thermoelectrically cooled photomultiplier tube characterized by a dark current of less than 100 cps.

Fluorescence spectra of samples comprising light-harvesting complexes and Au nanoparticles were measured in a standard optical setup with a back-scattering geometry. The laser excitation beam (λ=485 nm, 640 nm, or 405 nm) was focused, using a lens with a focal length of 30 mm, on the sample surface and the excitation power was controlled using notch filters. Typical excitation powers used were in the range of 200 μW. The emission was guided through a 150 μm pinhole and focused on a slit of a 0.5 monochromator (Shamrock 500, Andor) coupled with a charge coupled device detector (iDus 420BV, Andor). Fluorescence decays were studied using time-correlated single photon counting. For excitation, a diode-pumped solid state laser emitting at 405 nm, 640 nm, or 485 nm and generating 30 ps pulses at 80 MHz repetition rate was used. Detection was carried out with an ultrafast avalanche photodiode detector (idQuantique). The experiment was controlled using a time-correlated single photon counting card (SPC 150 Becker & Hickl). Emission spectra as well as fluorescence decays were collected for ten different spots across the sample in order to check for the reproducibility and homogeneity of the sample. Fluorescence of light-harvesting complexes was extracted using appropriate long-pass and band-pass optical filters.

2.4.1 Fluorescence imaging

In order to image fluorescence of light-harvesting complexes coupled to metallic nanoparticles we constructed a confocal fluorescence microscope based on Olympus infinity-corrected microscope objective LMPlan 50x, characterized with a numerical aperture of 0.5 and working distance of 6 mm (Krajnik, 2011). The resulting laser spot size is about 1 µm for the excitation laser of 485 nm. The sample is placed on a XYZ piezoelectric stage (Physik Instrumente) with 1 nm nominal resolution of a single step, which enables us to raster-scan the sample surface in order to collect fluorescence maps. They are formed by combining fluorescence intensity measurements with the motion of the XY translation stage. For excitation of fluorescence, we use one of four diode-pumped solid-state lasers with wavelengths of 405, 485, 532 and 640 nm. Typical optical power of the laser sources is about 5 mW, but in the case of actual measurements it needs to be strongly reduced in order to prevent photobleaching of the molecules. We used the excitation powers in the range of 0.004 to 0.04 mW. Gaussian beams of the lasers are achieved by using a spatial filter. The fluorescence is detected in a back-scattering geometry and focused on a confocal pinhole (150 µm) in order to reduce stray light coming out of the focal plane. The emission of PCP complexes is extracted with HQ 650LP (Chroma) dichroic mirror and HQ 670/10 (Chroma) bandpass filter. In order to extract fluorescence of LH2 complexes we used a longpass filter HQ850LP (Chroma) and a bandpass filter D880/40m (Chroma).

Our experimental configuration, described in detail in (Krajnik, 2011), allows for measuring fluorescence intensity, spectra and lifetimes. The spectrum, dispersed using the Amici prism is measured with a CCD camera (Andor iDus DV 420A-BV). The spectral resolution of the system is about 2 nm. Fluorescence intensity maps are collected with an avalanche photodiode (PerkinElmer SPCM-AQRH-14) with dark count rate of about 80 cps. Fluorescence lifetimes are measured using time-correlated single photon counting module (Becker & Hickl) equipped with fast avalanche photodiode (idQuantique id100-50) triggered by a laser pulse. Time resolution of the TCSPC setup is about 30 ps.

3. Experimental results

In this section we describe experimental results obtained for five architectures of hybrid nanostructures comprising metallic nanoparticles and light-harvesting systems. As for metallic nanostructures we used silver island film, semicontinuous silver film, spherical gold nanoparticles and elongated gold nanoparticles (nanorods). We coupled them with chlorophyll and carotenoid molecules embedded in the PCP complex from *Amphidinium carterae* and in the LH2 complex from *Rhodopseudomonas palustris*. The results of optical spectroscopy and microscopy show that the optical properties of light-harvesting systems are affected by the plasmon excitations in metallic nanoparticles both in th visible and infrared spectral ranges. Depending on the actual design of a nanostructure, either absorption or fluorescence radiative rate enhancement is obtained.

Generally, the effect of plasmon excitations in metallic nanoparticles on the optical properties of nearby emitters is monitored by measuring the fluorescence intensity. When the geometry of a hybrid nanostructure leads to plasmon-induced enhancement, the fluorescence intensity of such a hybrid structure is increased. When, on the other hand, non-radiative energy transfer from the emitter to the metallic nanoparticles plays the dominant role, the emission is efficiently quenched. However, fluorescence spectrum alone gives only limited information about the actual processes responsible for the enhancement of the emission intensity. In order to elucidate the mechanisms in detail, it is important to combine standard fluorescence spectroscopy with fluorescence excitation spectroscopy and time-resolved fluorescence spectroscopy; these two experimental techniques provide a way to separate the plasmon-induced increase of the radiative rate from an induced increase of the absorption.

3.1 Peridinin-chlorophyll-protein on silver island film

Initial experiments on hybrid nanostructures composed of light-harvesting complexes and metallic nanoparticles have been carried out on PCP complexes deposited directly onto the silver island film layer (Mackowski, 2008). In order to change the spectral properties of the metallic film, we fabricated SIF substrates with 1 and 3 second long dipping time in the reaction solution. Next, PCP complexes diluted in PVA were spin coated in ensemble concentration on the SIF layer. Since the thickness of the PVA layer is approximately 100 nm, the structure formed in this way spans over all relevant ranges of plasmon-pigment interaction. For PCP complexes located very close to the SIF the non-radiative energy transfer to the metallic nanostructure should play a dominant role and thus fluorescence quenching is expected. In contrast, when the distance between light-harvesting complexes and the SIF is larger than 40-50 nm, there is virtually no interaction between the two components of the hybrid nanostructure. Yet, the optical properties of all in-between molecules should be affected by the plasmon excitations in metallic nanostructure.

In Fig. 7 we display fluorescence images obtained with our confocal fluorescence microscope for the PCP complexes spin-coated onto two SIF substrates characterized with different time of deposition. Bright areas correspond to the higher fluorescence intensity. In the case of the SIF substrate that was kept in the reaction solution for 1 minute only the image is relatively homogeneous, the variation of fluorescence intensity is moderate. On the other hand, for the second structure, which was kept in solution two minutes longer, the areas of high and low fluorescence intensity are clearly separated from each other. We attribute the areas characterized with high fluorescence intensity to regions where the SIF layer was formed during the reaction, while the low fluorescence intensity suggests that the metallic layer detached from the glass substrate during the reaction.

The structure where both SIF and glass surfaces are present at once provide an easy and straightforward means to compare the fluorescence properties of PCP complexes coupled to plasmon excitation to the uncoupled ones. In Fig. 8 we show fluorescence spectra as well as fluorescence decay curves measured with the laser focused on either one of the two areas. As expected, for the PCP complexes placed on the SIF substrate the intensity of the emission is substantially higher that for the reference structure. The enhancement factor estimated from these two spectra is about fourfold. It correspond well to the average enhancement factor obtained for this structure. Importantly, as demonstrated in previous report (Mackowski, 2008), the maximum emission of the PCP complexes as well as the shape of the fluorescence spectrum remain unchanged upon coupling the light-harvesting complexes to the metallic nanoparticles. Also, since we use a 485 nm laser wavelength for the excitation, the observation of the intact fluorescence emission for both substrates indicates that the efficiency of the energy transfer from carotenoids to Chl molecules is comparable. This indicates that the PCP complexes that interact with metallic nanoparticles preserve their overall functionality. We can also see that the change in the fluorescence intensity is accompanied with sharp reduction of the fluorescence lifetime. In fact, the emission of PCP complexes on the glass substrate features a monoexponential decay while upon coupling to the SIF substrate the fluorescence decay curve is more complex. First a rapid decay takes place, which is probably due to efficient quenching of the PCP complexes that are very close to the metallic layer. After a first nanosecond the decay time of fluorescence gets longer, thus becoming similar to the decay observed for the reference structure.

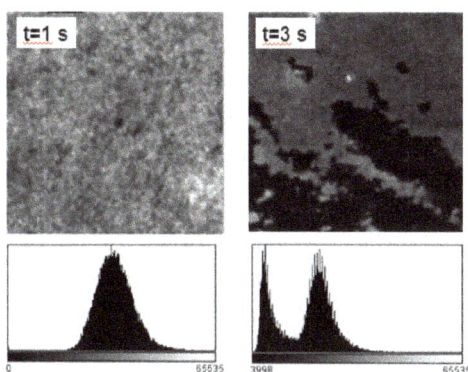

Fig. 7. Images of PCP fluorescence measured for the complexes deposited on the SIF substrate with 1 minute-long dipping time and 3 minute-long dipping time in the reaction solution. The maps were obtained at room temperature for the laser excitation wavelength of 485 nm, the laser power was 40 µW. The size of the images is 100x100 microns.

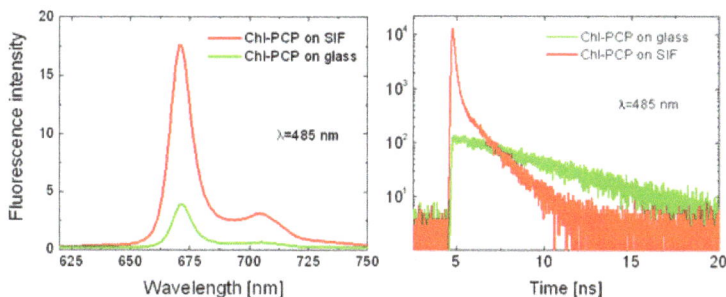

Fig. 8. Comparison between fluorescence spectra and fluorescence decay curves measured for PCP complexes on the glass and SIF substrates. For all measurements the excitation wavelength was 485 nm.

Overall, the results obtained for PCP complexes embedded in PVA matrix on the SIF layer demonstrate that high inhomogeneity of the structure leads to quite complicated behavior. Indeed the enhancement of absorption rate is entangled with enhancement of fluorescence rate, and in addition, signatures of non-radiative energy transfer from the chlorophylls to metallic structure are present. In the case of the hybrid nanostructure studied here, there is no control over the morphology of the SIF itself as well as over the separation between light-harvesting complexes and metallic surface. Therefore, other approaches need to be devised, aimed at better control of sizes or shapes of metallic nanoparticles and the distance between the proteins and metallic structures.

3.2 Peridinin-chlorophyll-protein on semicontinuous silver film

In previous sections we pointed out the important role of the separation between light-harvesting complexes and the metallic nanoparticles. In the case of sample geometry involving inhomogeneous silver island film and PCP complexes spin-coated directly on top of it in a relatively thick PVA layer, we have observed that the increase of the fluorescence emission is a combined product of absorption and fluorescence rate enhancement. In addition, the signatures of non-radiative energy transfer from the PCP complexes to the SIF layer have been observed in the time-resolved spectra.

In order to minimize the influence of the processes that lead to fluorescence quenching and at the same time to achieve uniform distance from the metallic layer to light-harvesting complexes, we have fabricated a hybrid nanostructure based on semicontinuous silver film (Czechowski, 2011). Such a corrugated metallic surface can be made using e-beam assisted evaporation under high-vacuum conditions. Scanning electron microscopy studies of similarly prepared samples have indicated improved uniformity of the islands, that resulted in narrowing of the plasmon resonance measured in the absorption experiment (Chettiar 2010). Furthermore, on top of the silver film we deposited a 25-nm-thick silica layer. The layer serves two purposes: on the one hand it protects the silver surface against oxidation, on the other hand it provides a uniform spacer between metallic nanoparticles and light-harvesting complexes. The final change compared to the structure where PCP complexes were spin-coated in a PVA matrix on top of the SIF, concerned direct deposition of the PCP water solution on the SiO_2 surface of the spacer. In this way we can assume that all the complexes are at approximately identical distances from the silver islands. Here we used PCP complexes reconstituted with acChl a as they offer the largest energy separation between their fluorescence and plasmon resonance of silver islands. In addition, the concentration of PCP complexes is much higher than for samples prepared with spin-coating, which makes it possible to study the plasmon induced effects using standard fluorescence excitation spectroscopy.

The fluorescence excitation spectrum measured for the detection wavelength of 690 nm for acChl-PCP on glass substrate is shown in Fig. 9. It is compared with with the result obtained for acChl-PCP complexes placed on the semicontinuous silver film. The excitation spectrum for the reference structure is similar to previously published (Brotosudarmo, 2008) it features strong absorption due to Per in the spectral range from 400 nm to 550 nm, and corresponds roughly to the absorption spectrum. This suggests that the sample preparation leaves no effect on either the protein or the pigments. In contrast, the maximum of fluorescence excitation spectrum is blue-shifted by ~40 nm for the PCP complexes deposited on the silver island film and separated from the metallic nanostructures by a 25-nm thick SiO_2 layer. The difference between the two cases is seen after subtracting both curves and evaluating the enhancement of the emission. We find that the enhancement curve is a well-defined band with a maximum at 407 nm and linewidth of about 35 nm. We attribute this enhancement to plasmon excitations in the metallic layer that impact the absorption of the PCP complexes.

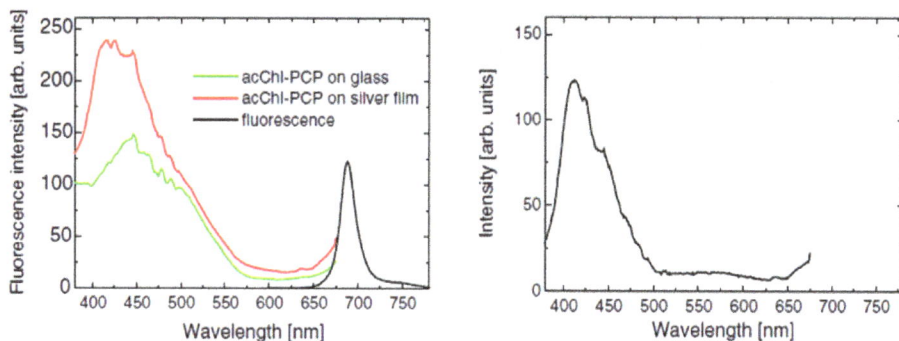

Fig. 9. Comparison between fluorescence excitation spectra measured for acChl-PCP on glass substrate and semicontinuous silver film. The detection energy was 690 nm. An enhancement dependence on the wavelength obtained by subtracting both fluorescence excitation curves is also shown.

It is important to note that the fluorescence excitation spectra measured for the reference structure and for the PCP complexes deposited on the silver film are not in any way adjusted or normalized. Yet, they are very comparable for wavelengths longer than 475 nm, in particular in the absorption range of low energy Per molecules. This suggests that the number of PCP complexes probed in both experiments is almost identical, which makes the estimation of the enhancement factor remarkably straightforward. Also the fluorescence spectrum measured for the hybrid nanostructure is is identical, for all excitation wavelengths, to that of the reference structure, which supports our previous observation that the preparation of the hybrid nanostructure has no measurable effect on the protein or pigment properties.

The fluorescence excitation data point towards increase of the absorption rate of the light-harvesting complex as being the dominant mechanism responsible for the enhancement of the fluorescence intensity. This suggestion is also helped with analysis of the spectral properties of both the PCP complexes and the semicontinuous silver film: they overlap mainly in the blue-green spectral range. In order to verify this we carry out time-resolved fluorescence experiment with the excitation wavelength of 405 nm, which corresponds to the maximum of the enhancement curve displayed in Fig. 9. The result if this experiment in shown in Fig. 10. The decay time of the control sample on glass is equal to 3.2 ns, while for the hybrid nanostructure a shortening of the lifetime to 2.3 ns when plasmons in the silver island film are excited. This less than 30 percent reduction of the lifetime, while measurable, is relatively small compared to previous results on fluorescent dyes (Dulkeith, 2002) and light-harvesting complexes (Mackowski, 2008), where order-of-magnitude changes have been measured. The small change of the fluorescence lifetimes in the case of the acChl-PCP complexes coupled to the semicontinuous silver film supports our conclusion that the enhancement measured in the fluorescence excitation is predominantly due to the enhancement of the excitation rate in the light-harvesting complexes. We also note that the fluorescence decay curve measured for the hybrid nanostructure features a monoexponential behavior, in contrast to the results obtained for PCP complexes deposited on the SIF. Such a uniform characteristics suggests improved homogeneity of the distance between the light-harvesting complexes and the metallic layer, as indeed expected for our preparation procedure. We have also carried out time-resolved experiments with other excitation energies, in particular with 640 nm. This wavelength corresponds to direct excitation of Chl molecules and excites no plasmons. In this case the fluorescence decay shows no dependence upon either glass or metallic substrate.

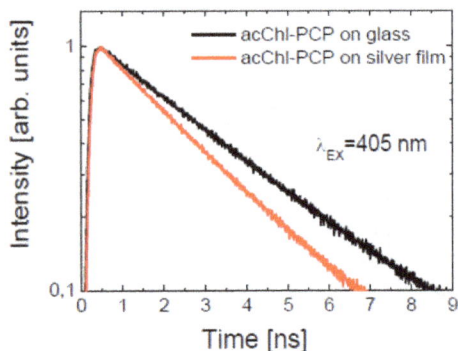

Fig. 10. Comparison of fluorescence decay curves measured for acChl-PCP on glass substrate and semicontinuous silver film. The excitation wavelength was 405 nm.

Finally we comment on another aspect of fluorescence decay time reduction observed for the 405 nm laser excitation. Since this reduction is attributed to the increase of the radiative rate of emission, it implies that there are plasmons excited in the semicontinuous silver film with wavelengths around 690 nm, where acChl-PCP emits. As 405 nm laser excites no such plasmons directly, this observation could be indicative of plasmon propagation in terms of energy relaxation. This hypothesis requires further experimental evidence but when proven correct, it could open another pathway in te field of plasmon engineering.

The results of fluorescence spectroscopy on acChl-PCP complexes deposited on semicontinuous silver film spaced by 25 nm SiO_2 layer confirm that by careful design of plasmonic hybrid nanostructure it is possible to selectively enhance the absorption of the light-harvesting complexes. The next step is to devise and fabricate a hybrid nanostructure, which would allow for systematic studies of plasmon induced effects as a function of the separation layer thickness.

3.3 Peridinin-chlorophyll-protein on spherical gold nanoparticles

The influence of the distance upon the interaction between PCP complexes and metallic nanoparticles requires fabrication of structures with precisely controlled thickness of the SiO_2 spacer. Such structures were fabricated in an analogous way as described previously, with the thickness of SiO_2 layer equal to 4, 12, and 40 nm. In this case however the metallic nanostructure used was a monolayer of uniform gold nanoparticles characterized with plasmon resonance at 530 nm.

While at the distances of 40 nm the influence of plasmon excitations on the optical properties of light-harvesting complexes is expected to be minimal, for thinner spacers the effect should be much stronger. In order to evaluate that we carried out fluorescence imaging experiment on PCP complexes deposited on the three Au nanoparticle samples with varied thickness of the SiO$_2$ spacer. In the first step a fluorescence map was acquired of 100x100 micron sample area. The fluorescence maps were in all cases very uniform, variations of fluorescence intensity were below 15 percent. Next, approximately 50 fluorescence spectra we collected, each off a different spot on the sample surface. Finally, the same procedure was applied for measuring fluorescence decay curves. In this way statistically significant information about fluorescence intensity as well as fluorescence decay time is obtained.

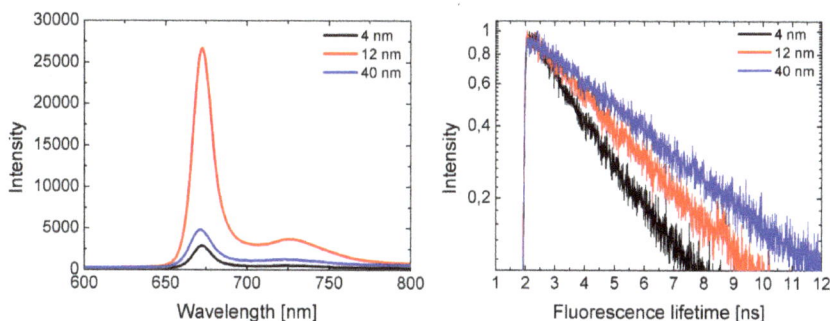

Fig. 11. Typical fluorescence spectra and fluorescence decay curves of Chl-PCP complexes deposited on Au nanoparticle substrates with different thickness of the SiO$_2$ spacer: 40 nm (blue), 12 nm (red), and 4 nm (black). The laser excitation wavelength was 485 nm.

In Fig. 11 we compare representative fluorescence spectra of Chl-PCP deposited on plasmonic substrates with Au spherical nanoparticles. The continuous-wave results are accompanied with time – resolved data. The intensity of fluorescence emission for 12 nm spacer is dramatically (fivefold) enhanced as compared to the reference structure with 40 nm thick SiO$_2$ spacer. For the smallest spacer (4 nm) the fluorescence decreases rapidly due to non-radiative energy transfer from chlorophylls embedded in the PCP complexes to metallic nanoparticles. Importantly, in analogy to all previously described experiments, the fluorescence spectrum of the light-harvesting complexes remains unchanged, indicating that the biomolecules are intact upon interacting with metallic nanoparticles.

Fluorescence decay curves that accompany the spectra provide means for understanding the mechanism of fluorescence enhancement. The decay time measured for the reference structure (40 nm) is equal to 3.3 ns, a typical value for PCP complexes reconstituted with Chl *a* (Mackowski, 2007). As the SiO$_2$ spacer gets thinner, the fluorescence lifetime gets shorter, and for 12 nm thick spacer is equal to 2.5 ns. Further reduction of the fluorescence lifetime is seen for the thinner, 4 nm, spacer. In this case the decay time is approximately 50 percent of the reference value. However, the mechanism of lifetime reduction is in both cases (4 nm and 12 nm) completely different. In the first case the shortening of the fluorescence decay time indicates enhancement of radiative rate of PCP complexes. This effects contributes to the observed increase of the emission intensity seen in the fluorescence spectra. On the other hand, for the 4 nm thick SiO$_2$ spacer, the lifetime reduction is due to excitation quenching. Thee results obtained for PCP complexes coupled to Au nanoparticles demonstrate clear dependence of the fluorescence enhancement upon the distance between chlorophyll-containing proteins and metallic nanoparticles. While most of the effect is due to increase of absorption, there is also significant contribution associated with increase of the radiative rate. This approach can be then used for optimizing the geometry of plasmonic hybrid nanostructure for the most efficient performance.

3.4 Light-harvesting complex 2 on spherical gold nanoparticles

Light-harvesting complex LH2 from the purple bacteria is characterized by relatively weak absorption in the visible spectral range with its main absorption bands appearing in the near infrared, at 800 and 850 nm. By coupling LH2 to spherical gold nanoparticles we attempt to enhance the absorption between 400 and 550 nm. The geometry of the hybrid nanostructure was identical to discussed previously: monolayers of Au nanoparticles were covered with SiO$_2$ dielectric layers with thickness of 4, 12, and 40 nm. During the experiment the fluorescence spectra excited into carotenoid absorption (Wormke, 2007b) were measured at ten different locations across the sample. In this way it was possible to account for any inhomogeneities due to the preparation of the hybrid nanostructures. The fluorescence spectra measured with SiO$_2$ spacers between 4 and 40 nm and are shown in Fig. 12.

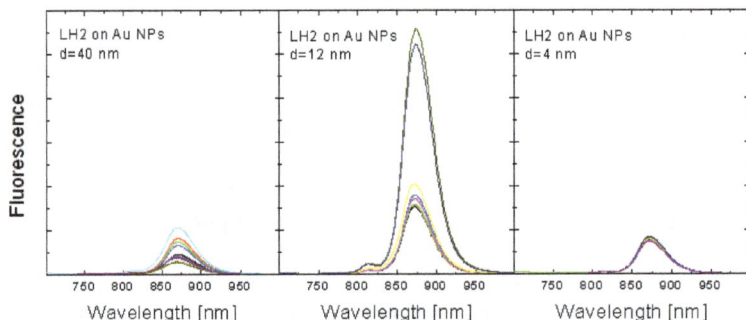

Fig. 12. Fluorescence spectra measured for LH2 complexes deposited on Au spherical nanoparticles on SiO2 spacers with thicknesses as indicated. The spectra were obtained for ten different locations on each sample.

There are several interesting observations worth pointing out. First of all, for the reference sample with 40-nm-thick SiO$_2$ spacer the scattering of the measured intensities can be attributed to local fluctuations in the LH2 concentration due to spin-coating approach. In contrast, for the sample with the 12-nm-thick SiO$_2$ spacer the spread of fluorescence intensities is significantly greater and the observed variation cannot be due to fluctuations of the LH2 concentration. Since plasmon interactions are expected to be significant for such a separation between the metallic nanoparticles and light-harvesting complexes, we attribute the distribution of fluorescence intensity to variation in plasmon coupling between the LH2 complexes and Au nanoparticles. Such variations can be caused for instance by interface roughness of the SiO$_2$ layer, even small variations of the spacer thickness would result in measurable changes of the fluorescence intensity. Finally, for the thinnest SiO$_2$ layer of 4 nm the fluorescence intensities are all very similar. In fact, the measured distribution is even less pronounced than in the case of the reference sample. Such behavior may well be due to the dominant role of the fluorescence quenching caused by metallic nanoparticles, which takes over below a certain thickness of the spacer between the metallic nanoparticles and light-harvesting complexes. In such a case any fluctuations of either LH2 concentration or SiO$_2$ spacer thickness may be of lesser significance.

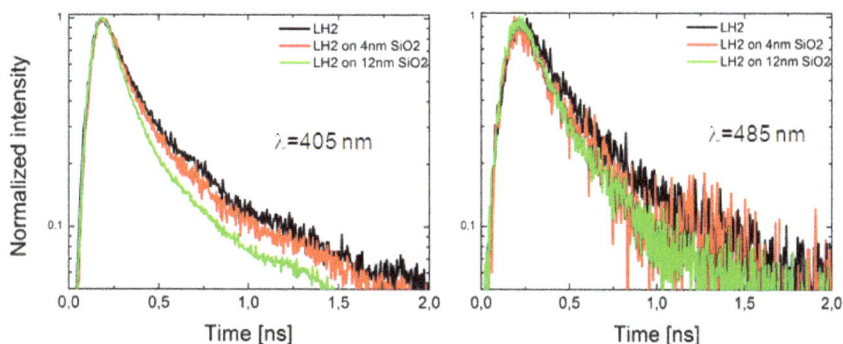

Fig. 13. Fluorescence decay curves measured for LH2 complexes on Au nanoparticles separated by SiO$_2$ spacer. Excitation wavelengths of 405 nm and 485 nm were used.

In order to determine the possible origin of the observed fluorescence enhancement, time-resolved fluorescence was measured on identically prepared samples. The fluorescence decay curves obtained for the structure with 4 and 12 nm thick SiO$_2$ layer is compared in Fig. 13 with the one measured for LH2 complexes deposited directly in glass substrate (Bujak, 2011). Apparently, upon coupling to the plasmons localized in the Au nanoparticles the fluorescence decays show virtually no change. Therefore, we assume that the fluorescence enhancement is predominantly due to an increase in the absorption in the carotenoid region of the LH2. The observation of exclusive increase of the absorption efficiency in the LH2 complexes coupled to Au nanoparticles was rendered by two factors. On the one hand, the difference in energy between plasmon resonance and the fluorescence emission is almost 400 nm, thus the overlap between low-energy tail of the plasmon resonance with the emission spectrum of the LH2 is minimal. This is much larger energy difference than for PCP complexes deposited on the semicontinuous silver film or Au nanoparticles. On the other hand, spherical gold nanoparticles are very uniform in size. This inhibits any possibility of energy relaxation in plasmonic structure, as it was observed for PCP complexes on the highly inhomogeneous SIF substrate.

The results described so far point clearly towards strong dependence of the plasmon induced effects upon the excitation energy. In most cases achieving strong coupling requires direct excitation of plasmons in metallic nanoparticles. In order to illustrate this, the fluorescence lifetimes were measured for LH2 complexes on Au nanoparticles with the excitation energy of 405 nm. In contrast to the 485 nm excitation, this energy excites plasmons very inefficiently while still populating excited states of carotenoids. The results included in Fig. 13 show that the fluorescence lifetime shows no dependence upon the thickness of the SiO$_2$ spacer. However, the actual enhancement factor measured for 405 nm laser is substantially reduced compared to 485 nm laser, which very efficiently excites plasmons in metallic nanoparticles. The comparison is displayed in Fig. 14. For both excitation wavelengths the dependence of the enhancement factor on the distance between light-harvesting complexes and metallic nanoparticles is qualitatively the same. Yet, under the condition of efficient excitation of plasmons the maximum enhancement observed for the spacer with 12 nm thickness is 2.5 times greater.

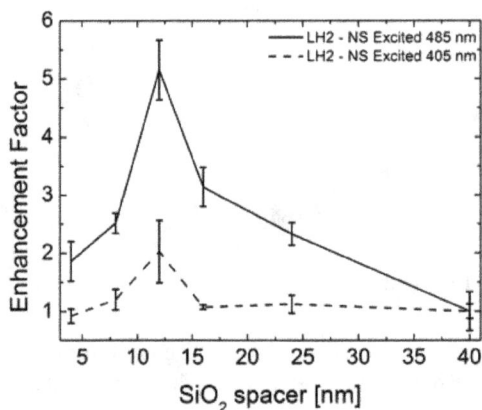

Fig. 14. Comparison of distance dependence of the fluorescence intensity enhancement for PH2 complexes deposited on Au spherical nanoparticles with different spacer thickness. The data was obtained for 485 nm and 405 nm laser excitations.

In conclusion, results of fluorescence spectroscopy carried out on hybrid nanostructures composed of light-harvesting complex LH2 and gold nanoparticles demonstrate the strong impact of plasmon excitations upon the optical properties of the biomolecule. For a spacer with a thickness of 12 nm substantial increase of the fluorescence intensity is observed, which is due to an enhancement of absorption of the carotenoids in this light-harvesting complex. Furthermore, we observe strong dependence of the fluorescence enhancement on the laser wavelength: for efficient excitation of plasmons in metallic nanoparticles ($\lambda=485$ nm) the enhancement is approximately 2.5 times stronger than for the out-of-plasmon-resonance excitation wavelength ($\lambda=405$ nm).

3.5 Light-harvesting complex 2 on gold nanorods

The final example of a hybrid nanostructure composed of light-harvesting complexes and metallic nanoparticles is a system where we combine Au nanorods with LH2 complexes from purple bacteria. From the previous discussion we know that by using Au nanorods we gain a tunability of plasmon resonances that reach near infrared spectral region (Bryant, 2008). In this way then we can affect the spectral properties of 800 and B850 absorption bands of the LH2 complex as well as its fluorescence emission.

In Fig. 15 we show the result of fluorescence imaging experiment carried out on LH2 complexes deposited directly on gold nanorods with plasmon resonances at 550 nm and 800 nm. The maxima of the resonances match ideally with absorption bands of the LH2 complex, attributed to carotenoids and bacteriochlorophylls, respectively. In the experiment we probe the fluorescence enhancement for these two excitation wavelengths, importantly for these two excitations the same sample area was monitored. It can be seen in particular for the maps shown in the upper row of Fig. 15, areas with low fluorescence intensity are clearly correlated. In the case of LH2 complexes on glass substrate fluorescence maps acquired for both excitation wavelengths are very uniform, as shown below the maps with intensity histograms. In both cases the histograms oare of Gaussian shape with maxima at 7500 and 21000 cps for 556 and 808 nm excitation, respectively. The picture changes

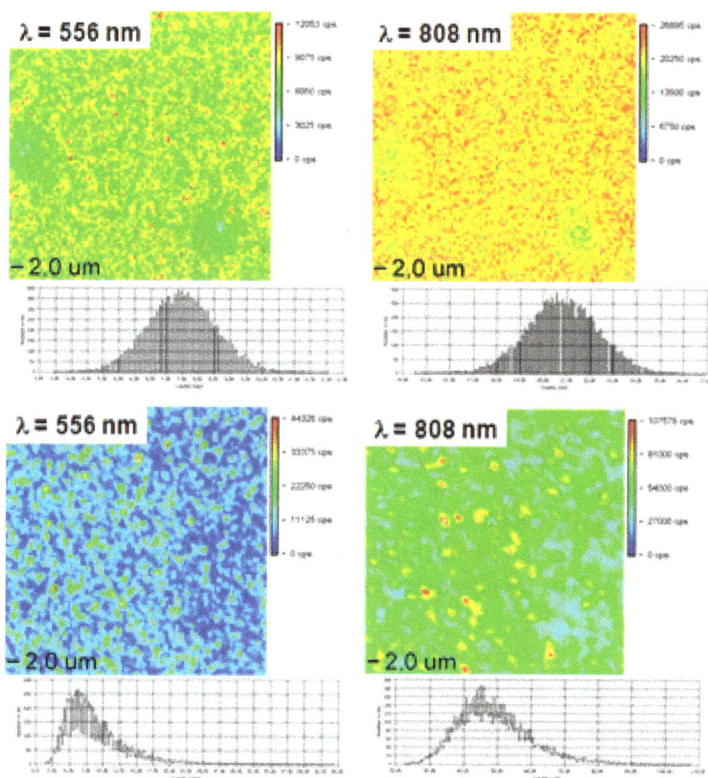

Fig. 15. Fluorescence images of LH2 complexes deposited on glass substrate (upper row) and Au nanorods (lower row). For exciting carotenoid absorption a 556 nm laser was used, whereas for exciting B800 BChl ring – a 808 nm laser was used. The maps for a given structure were obtained from the same sample area. The size of the images is 50 x 50 microns.

qualitatively for the LH2 complexes deposited on gold nanorods. The most pronouncing effect is much larger inhomogeneity of the fluorescence maps. There are regions of a few micron size that feature much stronger emission intensity. We can attribute them to either favorable separation between LH2 and Au nanorods or orientation/geometry of gold nanorods that would lead to formation of hot-spots of strongly localized electromagnetic field.

In addition to highly homogeneous fluorescence images, there are also significant differences of the distribution of fluorescence intensity, in spite of using the same excitation powers for a given laser wavelength. Indeed, the maximum of fluorescence intensity measured for 556 nm appears roughly at the same value as for the reference sample, but the histogram features substantial high-intensity tail of intensities, which is due to plasmon-induced enhancement in the hybrid nanostructure. Conversely, for the excitaiton of 808 nm the we also observe a broad tail towards higher intensities, but in this case the average intensity is also twice the average intensity measured for the reference sample. These preliminary results demonstrate that by using gold nanorods we are able to modulate the optical properties of multi-chromophoric systems such as light-harvesting complexes, which absorb in the infrared spectral range. Further work is required to coherently describe the complexity of plasmon interactions in this system.

4. Summary and conclusions

We have described various geometries of hybrid nanostructures composed of light-harvesting complexes from algae or purple bacteria and metallic nanostructures in the form of silver island films or monolayers of metallic nanoparticles synthesized chemically. The samples were studied with numerous optical spectroscopy and microscopy techniques including fluorescence excitation, time-resolved fluorescence, and fluorescence imaging with high spatial resolution. In all fabricated structures we observe strong effects attributable to plasmon induced effects on the optical properties of the light-harvesting complexes. Depending on the actual geometry we are able to increase fluorescence or absorption rate, in most cases however both effects are entangled. The results demonstrate that plasmon excitations in metallic nanostructures can be efficiently applied for controlling the light-harvesting capability of photosynthetic complexes, possibly paving the road towards novel photovoltaic architectures based – at least in some degree - on natural photosynthesis.

5. Acknowledgment

Research in Poland has been supported by the WELCOME project "Hybrid Nanostructures as a Stepping Stone towards Efficient Artificial Photosynthesis" funded by the Foundation for Polish Science and EUROCORES project "BOLDCATS" funded by the European Science Foundation. I am indebted to my friends and colleauges, with whom I have a great pleasure to collaborate on this project: I thank Wolfgang Heiss (Linz University), Eckhard Hofmann (University of Bochum), Richard J. Cogdell (University of Glasgow), Nicholas A. Kotov (University of Michigan), Hugo Scheer (LMU Munich) and the members of their research groups involved in parts of this research. Last but not least, I also acknowledge members of my research group at the Institute of Physics, Nicolaus Copernicus Unviersity in Torun, in particular Dr. Dawid Piatkowski, Dr. Radek Litvin, Lukasz Bujak, Nikodem Czechowski, Bartosz Krajnik, Maria Olejnik, Kamil Ciszak, and Mikolaj Schmidt for their excellent work and vital contribution.

6. References

Akimoto, S.; Takaichi, S.; Ogata, T.; Nishimura, Y.; Yamazaki, I. & Mimuro, M. (1996), Excitation energy transfer in carotenoid chlorophyll protein complexes probed by femtosecond fluorescence decays, *Chemical Physics Letters*, Vol.260, No.1-2, (September 1996), pp.147-152, ISSN 009-2614

Anger, P; Bharadwaj, P. & Novotny, L. (2006), Enhancement and quenching of single-molecule fluorescence, *Physical Review Letters*, Vol.96, No. 11, (March 2006), pp. 113002, ISSN 0556-2813

Atwater, H & Polman, A.(2010), Plasmonics for improved photovoltaic devices, *Nature Materials*, Vol.9, (February 2010) pp. 205–213, ISSN 1476-1122

Blankenship, R. (2002), *Molecular Mechanisms of Photosynthesis*, Wiley-Blackwell, ISBN 978-0632-04321-7, Oxford, United Kingdom

Bopp, M.; Jia, Y.; Li, L.; Cogdell, R. & Hochstrasser R. (1997), Fluorescence and photobleaching dynamics of single light-harvesting complexes, *Proceedings of the National Academy of Sciences of the United States of America*, Vol.94, No.20, (September 1997), pp.10630-10635, ISSN 0027-8424

Brotosudarmo, T.; Hofmann, E.; Hiller, R.; Wörmke, S.; Mackowski, S.; Zumbusch, A.; Bräuchle, C. & Scheer, H.(2006), Peridinin-chlorophyll-protein reconstituted with chlorophyll mixtures: preparation, bulk and single molecule spectroscopy, *FEBS Letters*, Vol.580, No.22, (October 2006), pp.5257–5262, ISSN 0014-5793

Brotosudarmo, T.; Mackowski, S.; Hofmann, E.; Hiller, R.; Bräuchle, C. & Scheer, H. (2008), Relative Binding Affinities of Chlorophylls in Peridinin - Chlorophyll - Protein Reconstituted with Heterochlorophyllous Mixtures, *Photosynthesis Research, Vol.95*, No.2-3, (February 2008), pp.247-252, ISSN 0166-8595

Bryant, G.; García de Abajo, F. & Aizpurua, J. (2008), Mapping the plasmon resonances of metallic nanoantennas, *Nano letters*, Vol.8, No.2, (January 2008), pp.631-636, ISSN 1530-6984

Bujak, L.; Czechowski, N.; Piatkowski, D.; Litvin, R.; Mackowski, S.; Brotosudarmo, T.; Cogdell, R.; Pichler, S. & Heiß, W. (2011), Absorption Enhancement of LH2 Light-Harvesting Complexes Coupled to Spherical Gold Nanoparticles, *Applied Physics Letters*, Vol.99, No.17, (October 2011), pp. 173701, ISSN 0003-6951.

Carmeli, I; Liberman, I.; Kraverski, L.; Fan, Z.; Govorov, A.; Markovich, G. & Richter, S. (2010), Broad band enhancement of light absorption in photosystem I by metal nanoparticle antennas, *Nano letters*, Vol.10, No.6, (May 2010), pp. 2069-2074, ISSN 1530-6984

Chettiar, U.; Nyga, P.; Thoreson, M.; Kildishev, A.; Drachev, V. & Shalaev, V. (2010), FDTD modeling of realistic semicontinuous metal films, *Applied Physics B: Lasers and Optics*, Vol.100, No.1, (March 2010), pp. 159-168, ISSN 1432-0649

Czechowski, N.; Nyga, P.; Schmidt, M.; Brotosudarmo, T.; Scheer, H.; Piatkowski D. & Mackowski, S. (2011), Absorption Enhancement in Peridinin-Chlorophyll-Protein

Light-Harvesting Complexes Coupled to Semicontinuous Silver Film, *Plasmonics*, Vol., (2011), pp.1-7, ISSN 1557-1955

Dulkeith, E.; Morteani, A.; Niedereichholz, T.; Klar, T.; Feldmann, J.; Levi, S.; van Veggel, F.; Reinhoudt, D.; Möller, M. & Gittins, D. (2002), Fluorescence quenching of dye molecules near gold nanoparticles: radiative and nonradiative effects, *Physical Review Letters*, Vol.89, No.20, (October 2002), pp.203002, ISSN 0556-2813

Govorov, A.; Bryant, G.; Zhang, W.; Skeini, T.; Lee, J.; Kotov, N.; Slocik, J. & Naik, R. (2006), Exciton–plasmon interaction and hybrid excitons in semiconductor–metal nanoparticle assemblies, *Nano letters*, Vol.6, No.5, (April 2006), pp. 984-994, ISSN 1530-6984

Govorov, A. (2008), Enhanced optical properties of a photosynthetic system conjugated with semiconductor nanoparticles: the role of Förster transfer. *Advanced Materials*, Vol.20, No.22, (April 2008), pp.4330-4335, ISSN 1521-4095

Hofmann, C.; Aartsma, T.; Michel, H. & Köhler, J. (2003), Direct observation of tiers in the energy landscape of a chromoprotein: a single-molecule study, *Proceedings of the National Academy of Sciences of the United States of America*, Vol.100, No.26, (December 2003), pp.15534-15538, ISSN 0027-8424

Hofmann, E.; Wrench, P.; Sharples, F.; Hiller, R.; Welte, W. & Diederichs, K. (1996), Structural basis of light harvesting by carotenoids: peridinin-chlorophyll-protein from Amphidinium carterae, *Science*, Vol.272, No.5269, (June 1996), pp.1788–1791, ISSN 0036-8075

Hulteen, J. and van Duyne, R. (1995), Nanosphere Lithography: A Materials General Fabrication Process for Periodic Particle Array Surfaces, *Journal of Vacuum Science and Technology A*, Vol.13, No.3, (1995), pp.1553-1558, ISSN 0022-5355

Kim, I.; Bender, S.; Hranisavljevic, J.; Utschig, L.; Huang, L.; Wiederrecht, G. & Tiede, D. (2011), Metal nanoparticle plasmon-enhanced light-harvesting in a Photosystem I thin film, *Nano letters*, Vol.11, No.8, (August 2011), pp.3091-3098, ISSN 1530-6984

Kleima, F.; Hofmann, E.; Gobets, B.; van Stokkum, I.; van Grondelle, R.; Diederich, K. & van Amerongen, H. (2000), Förster excitation energy transfer in peridinin-chlorophyll a-protein, *Biophysical Journal*, Vol.78, No.1, (January 2000), pp.344–353, ISSN 0006-3495

Krajnik, B.; Schulte, T.; Piatkowski, D.; Czechowski, N.; Hofmann, E. & Mackowski, S. (2011), SIL-based Confocal Fluorescence Microscope for Investigating Individual Nanostructures, *Central European Journal of Physics*, Vol.9, No.2, (2011), 293-299, ISSN 1895-1082

Krueger, B.; Lampoura, S.; van Stokkum, I.; Papagiannakis, E.; Salverda, J.; Gradinaru, C.; Rutkauskas, D.; Hiller, R. & van Grondelle, R. (2001), Energy transfer in the peridinin chlorophyll-a protein of Amphidinium carterae studied by polarized transient absorption and target analysis, *Biophysical Journal*, Vol.80, No.6, (June 2001), pp.2843-2855, ISSN 0006-3495

Lakowicz, J. (2006), Plasmonics in biology and plasmon-controlled fluorescence, *Plasmonics*, Vol.1, No1., (2006), pp. 5-33, ISSN 1557-1955

Link, S. and El-Sayed, M. (1999), Size and Temperature Dependence of the Plasmon Absorption of Colloidal Gold Nanoparticles, *Journal of Physical Chemistry B*, Vol.103, No.21, (May 1999), pp.4212-4217, ISSN 1520-6106

Mackowski, S.; Wörmke, S.; Brotosudarmo, T.; Jung, C.; Hiller, R.; Scheer, H. & Bräuchle, C., *Biophysical Journal*, Vol. 93, No.9, (August 2007), pp. 3249-3258, ISSN 0006-3495

Mackowski, S.; Wörmke, S.; Maier, A.; Brotosudarmo, T.; Harutyunyan, H.; Hartschuh, A.; Govorov, A.; Scheer, H. & Bräuchle, C. (2008), Metal-enhanced fluorescence of chlorophylls in single light - harvesting complexes, *Nano letters*, Vol.8, No.2, (February 2008), pp. 558-564, ISSN 1530-6984

Mackowski, S. (2010), Hybrid Nanostructures for Efficient Light Harvesting, *Journal of Physics: Condensed Matter, Vol.22*, No.19, (April 2010), pp.193102/1-17, ISSN 0953-8984

Maier, S. (2004). *Plasmonics: fundamentals and applications*, Springer Science + Business Media LLC, ISBN 978-0387-37825-1, New York, USA

McDermott, G.; Prince, S.; Freer, A.; Hawthornthwaite-Lawless, A.; Papiz, M.; Cogdell, R. & Isaacs, N. (1995), Crystal structure of an integral membrane light-harvesting complex from photosynthetic bacteria, *Nature*, Vol.374, (April 1994), pp. 517-521, ISSN 0028-0836

Nieder, J.; Bittl, R. & Brecht, M. (2010), Fluorescence studies into the effect of plasmonic interactions on protein function, *Angewandte Chemie International Edition*,Vol.49, No.52, (November 2010), pp. 10217–10220, ISSN 1521-3773

Polivka, T. & Sundström, V. (2004), Ultrafast dynamics of carotenoid excited states-from solution to natural and artificial system, *Chemical Reviews*, Vol.104, No.4, (February 2004), pp.2021–2071, ISSN 0009-2665

Polívka, T.; Pascher, T.; Sundström, V. & Hiller, R. (2005), Tuning energy transfer in the peridinin-chlorophyll complex by reconstitution with different chlorophylls, *Photosynthesis Research*, Vol.86, No.1-2, (2005), pp.217-227, ISSN 0166-8595

Ray, K.; Badugu, R. & Lakowicz, J. (2006), Metal-Enhanced Fluorescence from CdTe Nanocrystals: A Single-Molecule Fluorescence Study, *Journal of American Chemical Society*, Vol.128, No.28, (June 2006), pp.8998-8999, ISSN 0163-3864

Scheuring, S.; Sturgis, J.; Prima, V.; Bernadac, A.; Lévy, D. and Rigaud. J. (2004), Watching the photosynthetic apparatus in native membranes, *Proceedings of the National Academy of Sciences of the United States of America*, Vol.101, No.31, (August 2004), pp.11293-11297, ISSN 0027-8424

van Oijen, A.; Ketelaars, M.; Köhler, J.; Aartsma T. & Schmidt, J. (1999), Unraveling the electronic structure of individual photosynthetic pigment-protein complexes, *Science*, Vol.285, No.5426, (July 1999), pp.400-402, ISSN 0036-8075

Wörmke, S.; Mackowski, S.; Jung, C.; Ehrl, M.; Zumbusch, A.; Brotosudarmo, T.; Scheer, H.; Hofmann, E.; Hiller R. & Bräuchle, C. (2007a), Monitoring Fluorescence of Individual Chromophores in Peridinin-Chlorophyll-Protein Complex Using Single Molecule Spectroscopy, *Biochimica et Biophysica Acta – Bioenergetics*, Vol.1767, No.7, (July 2007), pp.956-964, ISSN 0006-3002

Wörmke, S.; Mackowski, S.; Brotosudarmo, T.; Garcia, T.; Braun, P.; Scheer, H.; Hofmann, E. & Bräuchle, C. (2007b), Detection of single biomolecule fluorescence excited through energy transfer: Application to light-harvesting complexes, *Applied Physics Letters*, Vol.90, No.19, (May 2007), pp.193901, ISSN 0003-6951

Wörmke, S.; Mackowski, S.; Schaller, A.; Brotosudarmo, T.; Johanning, S.; Scheer, H. & Bräuchle C. (2008), Single Molecule Fluorescence of Native and Refolded Peridinin-Chlorophyll-Protein Complexes, *Journal of Fluorescence*, Vol.18, No. 3-4, (2008), pp. 611-617, ISSN 1053-0509

2

Synthesis of Titanate and Titanium Dioxide Nanotube Thin Films and Their Applications to Biomaterials

Mitsunori Yada and Yuko Inoue
Saga University
Japan

1. Introduction

Recently, titanium compounds with one-dimensional nanostructures, such as nanotubes and nanofibers, have recently attracted much attention. Among these 1-D compounds, nanotubes composed of titanium dioxide and titanate are now being studied actively. Titanium dioxide nanotubes can be synthesized using porous anodic alumina membranes (Imai et al., 1999; Yamanaka et al., 2004), organic molecules (Jung et al., 2002), or polycarbonate membranes (Shin et al., 2004) as templates, or methods involving anodization of titanium metals (Macak et al., 2005). Since the interesting reports by Kasuga et al. (Kasuga et al., 1998; Kasuga et al., 1999) and Chen et al. (Chen et al., 2002), titanate and titanium dioxide nanotubes synthesized using the hydrothermal method have found a wide range of potential uses in photocatalysis (Tokudome et al., 2004; Jiang et al., 2008), dye sensitizing solar batteries (Uchida et al., 2002), hydrogen storage (Bavykin et al., 2005), electrochromism (Tokudome et al., 2005), bonelike apatite formation (Kubota et al., 2004), proton conductors (Thorne et al., 2005), electron field emission characteristic (Miyauchi et al., 2006), photoinduced hydrophilicity (Tokudome et al., 2004), etc.

In order to maximize the characteristics of the nanotube and to use them efficiently, preventing their excessive aggregation and arrangement at larger than micrometer or centimeter size are considered important. Especially, it is important to fabricate thin films composed of nanotubes. Kasuga et al. (Kasuga et al., 2003) reported the fabrication of titanate nanotube thin films by coating a titanate nanotube dispersion liquid to a substrate, and then calcinating the substrate. Tokudome et al. (Tokudome et al., 2004) and Ma et al. (Ma et al., 2004) also reported the fabrication of titanate nanotube thin films using a layer-by-layer method. However, neither study had transformed titanate nanotube thin films into titanium dioxide thin films. Kim et al. (Kim et al., 2007) used electrophoretic deposition (EPD) to fabricate 2-μm-thick titanate nanotube thin films, and they transformed the titanate nanotube thin films into titanium dioxide nanotube thin films by calcination. However, these methods involve complicated processes, including (1) synthesis of nanotubes, (2) preparation of a liquid in which the synthesized nanotubes are dispersed, (3) coating of the nanotubes onto a substrate using the prepared liquid, and (4) fixation of the coated nanotubes onto the substrate surface by calcination. Since it is generally difficult to prepare a liquid in which nanotubes are uniformly dispersed and that partial aggregation is inevitable, the homogeneity of thin films thus formed is questionable. Moreover, permanent fixation of the thin films onto the substrates is also doubtful. On the other hand, titanate and titanium dioxide nanotube thin films can also be formed on titanium metal by immersing titanium metal as a raw material into NaOH aqueous solution and then performing hydrothermal treatment (Miyauchi et al., 2006; Tian et al., 2003; Chi et al., 2007; Yada et al., 2007; Guo et al., 2007). The fabrication of titanate nanotube thin films using titanium metal plates was first reported by Tian et al. (Tian et al., 2003). The thin (~10 μm) films were detached from the titanium metal plates by hydrothermal reaction for 20 h. In contrast, thin films obtained by a short (6 h) hydrothermal reaction strongly adhered to the titanium metal plate. Miyauchi et al. (Miyauchi et al. 2006) obtained a titanium dioxide nanotube thin film by hydrothermal treatment on titanium metal, followed by acid treatment and calcination. Although this thin film was fixed onto the substrate, its thickness was only a few hundred nanometers. Therefore, it is clear that titanate and titanium dioxide nanotube thin films tend to detach from the substrates when they become too thick. Chi et al. also reported the fabrication of a sodium titanate nanotube thin film (Chi et al., 2007). However, the thickness of the film was not mentioned in their report, and the sodium titanate nanotubes were not transformed into titanium dioxide nanotubes.

In this chapter, first, we will report the synthesis and organization of sodium titanate nanotube (hereafter referred to as Na-TNT) of size larger than a micrometer, using various titanium metals with controlled shapes of a micrometer size including plate, wire with a diameter of a micrometer, mesh woven from the titanium wire, microspheres, and microtube (Yada et al., 2007). The titanium metal acts as a template for the organization as well as a titanium source. Therefore, the originality of our study is to use titanium metal as a morphology-directing material. In addition, we will report a novel procedure for fixation of Na-TNT thin film on titanium metal (Yada et al., 2007). As a result, the thickness of the sodium titanate nanotube thin film can be adjusted by changing the duration of the hydrothermal reaction and the obtained films are thicker than those reported in previous studies (Miyauchi et al., 2006; Tian et al., 2003). Furthermore, we will also introduce a novel "hydrothermal transcription method" for forming Na-TNT films on various substrates such as Co-Cr alloy and SUS316L (Yada et al., 2008). Transformation of Na-TNT thin films into thin films consisting of anatase nanotube, anatase nanowires, anatase nanoparticles, and rhomboid-shaped anatase nanoparticles are also introduced (Inoue et al. 2010). To obtain an anatase nanotube thin film, it is necessary to slightly modify previously reported methods for synthesizing titanium dioxide nanotube particles.

Next, in this chapter, obtained titanate and titanium dioxide nanotube thin films will be applied to antibacterial biomaterials (Inoue et al., 2010). The nanotube thin film has several advantages: it can be formed on titanium, titanium alloy, Co-Cr alloy, and SUS316L, which are useful for manufacturing surgical instruments and implants such as artificial joints; its thickness can be controlled up to 20 μm or more, in contrast to only 1 μm for the thickness of the previously reported sodium titanate thin film with a porous network structure; and medicines can be incorporated into the nanotube. It is well known that bacterial infection may occur during surgery because of several factors. For example, during hip-replacement arthroplasty, bacterial infections occur in 1% to 2% of operations and usually cause physical and economic burdens for patients, such as re-implantation. As a conventional method for preventing infections, antibiotics are administered even in operation rooms with few pathogens. However, this does not prevent every infection. Therefore, imparting antibacterial properties to implants is currently under investigation. There have been reports of the use of apatite coating containing silver on implants by sputtering (Chen et al., 2006), silver-plated implants (Hardes et al., 2007), and gentamicin–hydroxyapatite coating for cementless joints (Alt et al., 2006), all of which have shown antibacterial properties. However, these methods have drawbacks such as the need for expensive instruments and the use of antibiotics that may cause the emergence of resistant bacteria. Therefore, further research is required. In this study, in order to develop more convenient and inexpensive antibacterial implants, silver ions are studied as an antibacterial component along with titanate nanotube formed on the surface of titanium. Silver is one of the most common antibacterial elements and is considered highly safe with high antibacterial activity. Sodium titanates are composed of a titanate framework with a negative electric charge and Na^+ ions with a positive electric charge. Since they have a cation exchange property, Na^+ ions can be exchanged with several cations (Kim et al., 1997; Chen et al., 2002; Sun et al., 2003; Bavykin et al., 20006). Therefore, it is considered that sodium titanate can be transformed into silver titanate by exchange of Na^+ in sodium titanate with Ag^+, and the in vivo elution of silver ions from the titanates would be promising for application to antibacterial implants. In addition, it is suggested that the titanate nanotube thin film would be able to possess a larger amount of silver and allow the amount of silver to be controlled more widely as compared with the titanate thin film previously reported (Kim et al., 1996). In this chapter, we will describe the synthesis and characterization of titanate nanotube thin films with silver and the behavior of silver ion elution of the thin films in vitro. We will also describe the antibacterial properties against methicillin-resistant Staphylococcus aureus (MRSA) with a biofilm-forming gene, which is a major concern in actual infections, to investigate the possibility of using synthesized thin films as antibacterial implants.

Finally, we will describe the apatite-forming abilities of titanium compound nanotube thin films by comparing the apatite deposition behaviors of a sodium titanate nanotube thin film (Yada et al., 2007), a titanium dioxide nanotube thin film (Inoue et al., 2010), and a silver nanoparticle/silver titanate nanotube nanocomposite thin film (Inoue et al., 2010), in simulated body fluid (SBF) (Yada et al., 2010). In evaluating the in vivo apatite-forming ability or the osteoconductive property of a material, researchers commonly perform experiments in SBF (Kokubo et al., 2006)). Kim et al. (Kim et al., 1996) first reported the formation of a sodium titanate thin film with a porous network structure on a titanium metal plate by alkali and heat treatment and demonstrated the osteoconductive property of the obtained sodium titanate thin film. Since then, researchers have actively performed many studies on the applications of sodium titanate thin films in implants (Kokubo et al., 1996; Kim et al., 1997; Kim et al., 1997; Nishiguchi et al., 1999; Jonášová et al., 2003; Kim et al., 2003; Muramatsu et al., 2003; Wang et al., 2007; Wang et al., 2008). Similar studies have also been performed on calcium titanate thin films (Hanawa et al., 1997; Hamada et al., 2002; Nakagawa et al., 2005; Kon et al., 2007; Ohtsu et al., 2008), titanium dioxide thin films (Ohtsuki et al., 1997; Wang et al., 2001; Wang et al., 2003; Byon et al., 2007), and a nanohydroxyapatite thin film (Xiong et al., 2010), and the excellent biocompatibilities of these films have been reported. Therefore, titanium compound thin films show tremendous promise for use as implant materials.

2. Synthesis and characterization of titanate and titanium dioxide nanotube thin films

2.1 Sodium titanate nanotube thin films formed on various shaped titanium metal templates

2.1.1 Sodium titanate nanotube thin film formed a titanium plate

First, the growth and fixation of Na-TNT on titanium plate were investigated. A titanium plate (20 mm × 20 mm × 2 mm) was immersed in 20 ml of 10 mol/l aqueous NaOH solution in a Teflon container and reactions were carried out at 160°C. After hydrothermal treatment for 20 h, the surface of the plate changed to pale, indicating the formation of a thin film on the titanium plate. In order to wash out NaOH and the particles that adhered to the surface of the thin film, the plate was washed with water after the reaction. The thin film immediately exfoliated as shown in Fig. 1a, and then a surface with a metallic luster similar to that of titanium metal appeared on the surface of the plate. The greater part of the film was posited to consist of nanotubes with an outer diameter of approximately 8 nm (Fig. 1b). Through EDX analysis, the mol fraction of Na/Ti/O for the obtained film was determined to be 1:1.947:4.943. The film was thus assumed to be $Na_2Ti_4O_9 \cdot H_2O$, though some titanate structures, such as $A_2Ti_2O_5 \cdot H_2O$ (Yang et al., 2003), $A_2Ti_3O_7$ (Chen et al., 2002), $H_2Ti_4O_9 \cdot H_2O$ (Nakahira et al., 2004), and lepidocrocite titanates (Ma et al., 2003), have been assigned as nanotube constituents (A=Na and /or H) as summarized by Tasi et al (Tasi et al., 2006). Moreover, by detailed SEM observation of the cross-section of this film, the film thickness was determined to be approximately 20.2 μm, as shown in Fig. 1c. The thickness of the Na-TNT phase was determined to be approximately 19.2 μm, and the thickness of the dense sodium titanate phase without Na-TNT was determined to be approximately 1.0 μm. Although fibrous morphologies were observed on the surface of the film, the back of the film was flat with no visible fibers. Therefore, the film exfoliation was considered to occur at the interface between the titanium metal phase and the sodium titanate phase without Na-TNT. Based on the above results, the formation of the Na-TNT thin film can be explained as follows: (1) titanium dissolves into titanium ions (Ti^{4+}) by oxidizers, H^+ and/or O_2; (2) dissolved Ti^{4+} ions immediately form titanium species (Wu et al., 2006) such as TiO_3^{2-}, $TiO_2(OH)_2^{2-}$, and $Ti_nO_{2n+m}^{2m-}$ and the concentration of titanium species in the reaction solution increases as the dissolution of titanium is accelerated; (2) titanium species are reprecipitated as sodium titanate with an increase in the concentration of titanium species in the reaction solution; (3) since the concentration of the titanium species in the reaction solution is expected to increase with time, the sodium titanate phase without Na-TNT is formed when the concentration of titanium species is low and the Na-TNT phase is formed after the concentration becomes sufficiently high. The Na-TNT-free sodium titanate phase formed at low concentrations of titanium species may be amorphous sodium titanate. Since the concentration of titanium species in the reaction solution is considered to affect the type of sodium titanate cluster and the formation rate of the sodium titanate phase, the concentration of titanium species, together with temperature and other concentrations, are also considered to be factors in determining the type of phase formed.

Moreover, in order to prevent detachment of the thin Na-TNT film, the as-synthesized plate was slowly dried at room temperature after the hydrothermal treatment without washing it with water. Although NaOH crystals were observed on the plate, the thin film still adhered

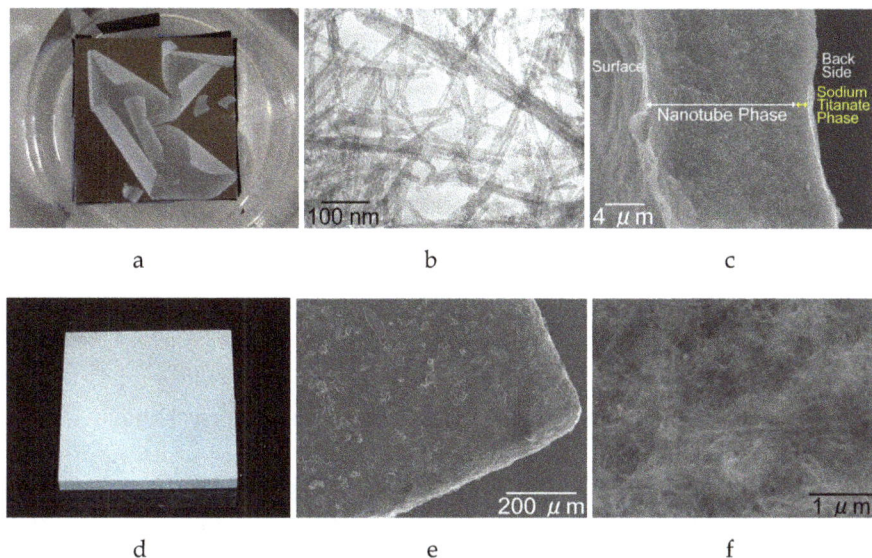

Fig. 1. Photograph (a, d), TEM (b), and SEM (c, e, f) images for the as-grown (a-c) and the 300 °C calcined (d-f) products obtained after the 20 h reaction.

to the plate. When the plate was washed with water after heat treatment at 300°C for 1 h in air, although the NaOH crystals dissolved, the thin Na-TNT film still adhered to the plate firmly and no detachment was observed as shown in Fig. 1d. Na-TNT formation was confirmed by the fibrous morphologies observed on the surface of the thin film in an enlarged SEM image (Figs. 1e, f) and nanotubes observed in a TEM image of the thin film. Moreover, in an XRD pattern of the thin film (Fig. 2), only diffraction peaks characteristic of Na-TNT (Chen et al., 2002) were observed along with peaks assigned to titanium metal. The reason for this stable coating is probably because polycondensation of hydroxyl groups in the interface area between the titanium plate and the Na-TNT-free sodium titanate phase occurred by the heat treatment at 300°C, and Na-TNT being firmly fixed on the plate. The slow drying process is also considered to be important for the fixation of Na-TNT onto the titanium plate, since the thin film detached from the titanium plate by drying at 60°C. The formation and fixation of the Na-TNT thin film were also observed in the reaction after 3 h. Nanotubular structures similar to those of the 20 h product were also observed. The thickness of approximately 5 μm for the film obtained after the 3 h reaction was smaller than that of 20.2 μm for the film obtained after the 20 h reaction. The thickness of the film is thus controllable by the reaction time. On the other hand, when an as-synthesized Na-TNT thin film obtained by hydrothermal reaction in 10 mol/L NaOH solution at 160 °C for 1 h was washed with large amounts of water, the Na-TNT thin film do not detach from the substrates and remains as thin as approximately 1 μm. Therefore, it is clear that sodium titanate nanotube thin films tend to detach from the substrates when they become too thick, but remain firmly fixed on substrates when the obtained samples are dried (without washing with water) and subsequently calcined at 300 °C.

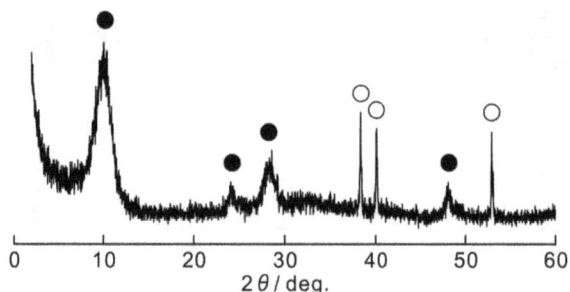

Fig. 2. XRD pattern for the plate obtained after the 20 h reaction. Peak assignment: ○ titanium metal, ● sodium titanate nanotube.

2.1.2 Sodium titanate nanotube thin films formed on titanium wire, titanium mesh, titanium sphere, and titanium microtube

Titanium wire (lengths: 5 cm, 24 cm, and diameters: 53.4 μm, 104.4 μm, 203.7 μm), titanium mesh (woven from the titanium wire with a diameter of 104.4 μm, 20 mm × 20 mm), titanium tube (inner diameter: 800 μm, outer diameter: 1 mm, length: 1 cm), and titanium spheres (diameter: 850–1180 μm, weight: 0.21-0.24 g) were used as metal titanium sources instead of a titanium plate.

After the hydrothermal treatments for 3 h and 20 h, the surfaces of titanium mesh and titanium sphere were completely covered with Na-TNT thin film. Both outer and inner surfaces of the microtube were also covered with uniform nanotubes with an average diameter of 8 nm. Typical digital camera and SEM images are shown in Figs. 3a-c.

On the other hand, the formation of Na-TNT thin film on a titanium wire requires special procedures which are different from those for plate, mesh, sphere, and microtube. First, the synthesis and fixation of Na-TNT were investigated using titanium metal wires of diameters 53.4 and 104.4 μm and length 5 cm as titanium sources. As a result, after the hydrothermal treatment for 3 h, sodium titanate with an irregular morphology was formed on the surface of the titanium wires, and only small amount of nanotubes was observed in the product synthesized using the titanium wire of diameter 104.4 μm. The diameters decreased from 53.4 and 104.4 μm for the original wires to 36.3 and 93.8 μm for the wires after a reaction time of 3 h, respectively. Moreover, after the hydrothermal treatment for 20 h, both wires completely dissolved. In addition, in the experiment using a wire of diameter 53.4 μm and length 24 cm, no nanotubes were observed on the surface of the obtained wire at a reaction time of 3 h and the wire completely dissolved at a reaction time of 20 h. On the other hand, in the experiment using a wire of diameter 104.4 μm and length 24 cm, the amount of Na-TNT formed increased at 3 h reaction time, and the surface of the wire was completely covered with Na-TNT thin film at the 20 h reaction. The reason for the complete dissolution of the original wires is because the dissolution rates of titanium species from the wires were faster than the redeposition rate of sodium titanate on the surface of the wires. On the other hand, the reason for the complete coverage of Na-TNT on the wire without dissolution is that the redeposition rate of sodium titanate nanotubes on the wire's surface became faster than the dissolution rate of titanium species from the wire with an increase in its diameter and length. These differences depending on the diameters of the original wires are

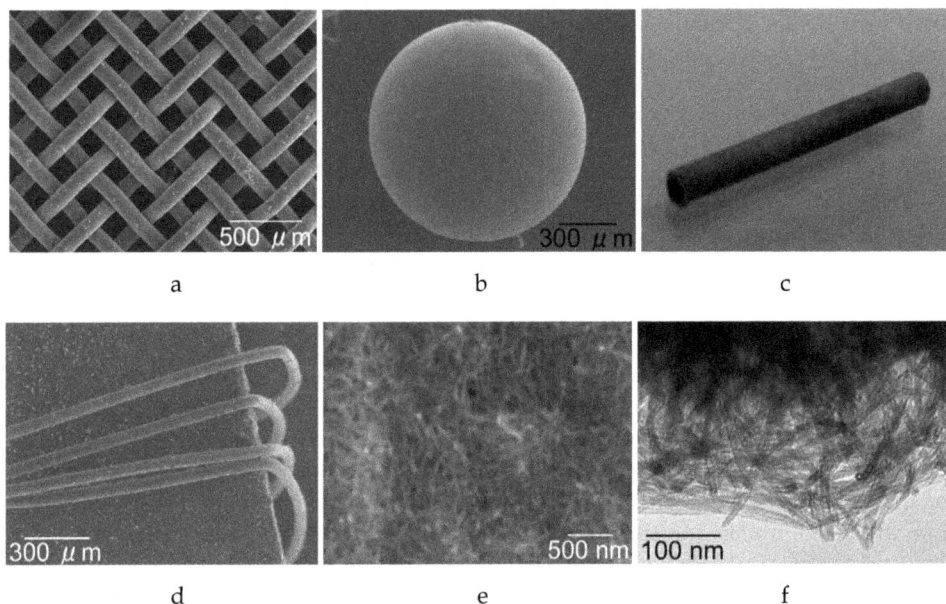

Fig. 3. SEM (a, b, d, e), Photograph (c), and TEM (f) images for the mesh (a), micro-sphere (b), microtube (c), and wire (d-f) obtained after the 3 h reaction.

explained as follows. Surface area and surface texture strongly affect the concentration of titanium species in the reaction solution. The amount of titanium species in the reaction solution increases with an increase in the diameter of the wire, since the surface area of the wire increases with an increase in the diameter. Additionally, the difference in the surface texture of the wires also affects the concentration of titanium species near their surfaces. Detailed SEM observations of the original titanium wires confirmed that the surfaces of the wire of diameter 104.4 µm were porous, but the surface of the wire of diameter 53.4 µm was relatively smooth. The concentration of titanium species would be higher near the wire and lower as the distance from the wire increases. In particular, the concentration of titanium species near the porous surface would be higher than that near the smooth surface. Therefore, in the experiments using the 104.4 µm diameter wire, the amount of titanium species formed per unit of time and the concentration of the titanium species would be large due to their larger diameters and porous surfaces, and consequently the concentration of titanium species would be sufficiently high for the formation of Na-TNT as the dissolution of titanium proceeded. On the other hand, since the surface area of the 53.4 µm diameter wire was predicted to be smaller than the 104.4 µm diameter wire due to its diameter and smooth surface, the concentration of titanium species formed per unit of time would also be small. Therefore, the concentration of titanium species would be too low for the formation of Na-TNT. Consequently, sodium titanate with irregular morphology was formed without Na-TNT at a reaction time of 3 h and the original wire completely dissolved at a reaction time of 20 h. Taking into consideration the above discussion, a similar hydrothermal and fixing treatment was performed using a wire of diameter 53.4 µm and length 24 cm wound onto the above mentioned titanium plate, which could act as a source of titanium species.

Wired morphologies remained for 3 h (Fig. 3d) and 20 h reactions, respectively, and the surfaces of both wires were completely covered with uniform Na-TNT thin films (Figs. 3e, f). It is considered that since the amount of titanium species reprecipitated on the wire, supplied by the dissolution from the titanium plate, was larger than the amount of titanium species dissolved from the wire, the surface of the wire was covered with Na-TNT. These results also indicate that dense concentration of titanium species near the titanium surface is required for the formation of Na-TNT on the titanium wire. Based on the above results, Na-TNT applications can be largely extended by the hydrothermal treatment of a cloth woven with titanium wires and by weaving a cloth with Na-TNT/Ti wires.

2.2 Sodium titanate nanotube thin films formed on Co–Cr alloy and SUS316L plates

We devised a "hydrothermal transcription method" for forming Na-TNT films on various substrates, as shown in Fig. 4. In this method, Na-TNT would be produced by re-depositing or transcribing the titanium species such as TiO_3^{2-}, $TiO_2(OH)_2^{2-}$, and $Ti_nO_{2n+m}^{2m-}$ formed near the surface of the titanium plate by hydrothermal treatment in aqueous NaOH solution on other substrate as Na-TNT, and grown to form dense films on several substrates as well as on the titanium plate. As shown in Fig. 4, under the conditions where a titanium metal plate and a substrate were adjacently placed, the titanium metal plate and substrate were spaced uniformly (about 200 µm) and fixed using titanium wires or SUS316 wires. For the substrate, Co–Cr alloy disk, SUS316L plate, SUS430 plate, tantalum plate, and silicon plate were used. These were immersed in 10 mol/l NaOH aqueous solution and reacted hydrothermally for 20 h at 160 °C. After the reaction, the samples were removed from the container and dried. Then, by washing in water following heat treatment at 300 °C, excessive NaOH adhered on the substrate was removed.

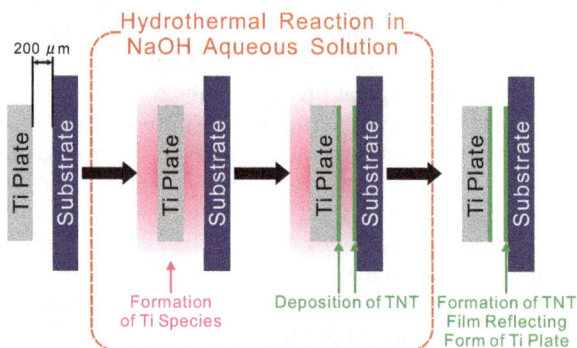

Fig. 4. Schematic representation of a reaction process by the hydrothermal transcription method.

Firstly, the Co–Cr alloy disk was used as a substrate. As shown in Fig. 5a, after the reaction, the formation of a white film whose base is the color of Co–Cr alloy along the square form of counter titanium plate on only the face countered to the titanium plate was observed. This white film strongly adhered to the Co–Cr alloy plate. By SEM images (Figs. 5b, c), the uniform and dense formation of fibrous substances was identified. Also by TEM observation of fibrous substances, nanotubes with an outer diameter of about 8 mm were identified (Fig.

5d). The thickness of this film is about 5 μm. This thickness was less than the 20 μm thickness of the Na-TNT film formed on the titanium plate when reacted singly (Yada et al., 2007). As the XRD pattern of the Co–Cr alloy surface countered to the titanium plate, a diffraction peak characteristic to titanate nanotube near $2\theta = 10°$ as well as the peaks attributed to the Co–Cr alloy of the raw material were observed. From a EDX analysis, it was found that the film contains Na, Ti, O, Co, Cr, and Si, and the molar ratio for the film was Na:Ti:O:Co:Cr:Si=0.322:1:2.401:0.112:0.052:0.045. Sodium titanate nanotube film is thus thought to be formed on the Co–Cr alloy disk. The elements of Co, Cr, and Si would dissolve from the Co–Cr alloy disk and would be incorporated into the titanate framework and/or the interlayer spacing of the titanate. Furthermore, as observed above, the white Na-TNT film reflecting the square form of the titanium plate was observed on the Co–Cr alloy disk countered to the titanium plate (Fig. 5a). Thus, the titanium species capable of forming Na-TNT were present near the surface of titanium, and it can be considered that Na-TNT patterning reflecting the form of the titanium plate was made on the surface of the Co–Cr alloy disk countered to the titanium plate. The above results suggest that by using several forms of the titanium plate as the titanium source, several forms of Na-TNT patterning can be made on heterogeneous substrates.

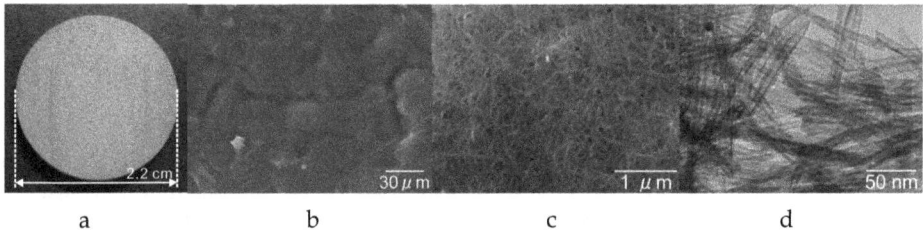

Fig. 5. Digital camera (a), SEM (b, c), and TEM (d) images for the obtained thin film formed on the surface of Co-Cr alloy countered to titanium metal.

When the same experiment was conducted with SUS316L plate instead of the Co–Cr alloy disk, it was found that as in the case of Co–Cr alloy, diluted white Na-TNT thin film was formed on the surface of SUS316L plate countered to the titanium plate. On the other hand, when the same reaction was performed using the SUS430 plate instead of Co–Cr alloy disk, brown and black iron compounds were formed on the SUS430 plate, although white Na-TNT film was formed in part. SUS430 is an industrial grade stainless alloy, whereas SUS316L is a stainless alloy used in implants and has exceptionally high corrosion resistance, and the results reflecting this corrosion resistance were obtained. When the same reaction was performed with the tantalum plate instead of the Co–Cr alloy plate, copious amounts of white products were produced on the tantalum plate, and particles other than nanotubes were observed. In addition, when the same experiment was performed using a silicon plate instead of the Co–Cr alloy disk, the silicon plate was completely dissolved by the hydrothermal reaction. On considering the differences in the responsiveness of substrates, it is thought that the dissolution rate of the titanium plate and substrates and redeposition rate of chemical species that arose from the dissolved titanium plate and substrates should be considered. Particularly, in this experiment system, it is considered that the dissolution rate of substrates has a large effect on the results of the experiments. The dissolution rate of using metals as substrates, as in this study, can be explained by the ionization tendency, i.e., oxidation–reduction potential. It is considered that titanium dissolves into titanium ions (Ti^{4+}) by oxidizers, H^+ and/or O_2, and these dissolved ions immediately form chemical species (Wu et al., 2006) such as TiO_3^{2-}, $TiO_2(OH)_2^{2-}$, and $Ti_nO_{2n+m}^{2m-}$ which are re-deposited as Na-TNT. When a substrate whose ionization tendency is smaller than titanium, especially materials such as SUS316L and Co–Cr alloy, is hydrothermally reacted with titanium simultaneously, the dissolution rate of titanium is higher than that of the substrate. In this reaction, titanium species are immediately produced following the dissolution of titanium and spread and re-deposited on the substrate as Na-TNT film, which predominates the dissolving reaction of the substrates. As a result, the surface of the substrate is covered by Na-TNT film, and the dissolution of the substrate was further minimally suppressed. On the other hand, it is considered that when the substrates with ionization tendencies larger than titanium, i.e., substrates such as silicon and tantalum, and titanium were hydrothermally reacted at the same time, the dissolution reaction of substrates predominate the dissolution reaction of titanium. Na-TNT film was not thus obtained on the substrates.

2.3 Hydrogen titanate and anatase-type titanium dioxide nanotube thin films

H^+ ion-exchange treatment for the sodium titanate nanotube thin film and the subsequent calcination can produce an anatase-type titanium dioxide nanotube thin film.

2.3.1 H⁺ ion-exchange treatment for sodium titanate nanotube thin films

We performed H^+ ion-exchange treatment for Na-TNT thin film obtained after the 3 h reaction using 0.01 mol/l hydrochloric acid solution at 90 and 140 °C for 3 h. The thin films resulting from treatment at these two temperatures remain attached over the entire surface of each sample. EDX analysis reveal that because the molar ratio of Na/Ti decrease from 0.48 before treatment to 0 after treatment at 90 and 140 °C, Na^+ ions between titanate layers are confirmed to be completely exchanged for H^+ ions. We observe nanotubes with an average outer diameter of 8.3 nm and inner diameter of 3.3 nm in the ion-exchange-treated sample at 90 °C using 0.01 mol/l hydrochloric acid solution (Fig. 6a). No change is observed in the porous structure of the thin film before or after treatment. The XRD pattern for the H^+ ion-exchanged sample at 90 °C shows four diffraction peaks (near $2\theta = 9, 24, 29,$ and $48°$) attributed to titanate together with peaks attributed to α-titanium, similar to those for the as-grown sample, as shown in Fig. 7. We therefore believe that the H^+ ion-exchange treatment at 90 °C transforms sodium titanate nanotubes into hydrogen titanate nanotubes while maintaining the crystal structure of titanate, nanotubular morphology, and porous thin-film structure. In contrast, the H^+ ion-exchange treatment at 140 °C replaces the fibrous morphology with rhomboid-shaped particles (average diameter 21 nm) (Fig. 6b) and pores (~ 45 nm diameter) in the interstitial gaps. The XRD pattern of this sample (Fig. 7) shows peaks attributed to anatase. Therefore, a porous thin film consisting of rhomboid-shaped anatase is confirmed to be formed on the titanium metal plate. Change in the crystal structure of sodium titanate nanotubes to anatase nanotubes by acid treatment have been described previously by Tsai et al. (Tsai et al., 2006). They reported that although a nanotube form is maintained by acid treatment at pH 1.6, only irregular-shaped anatase particles are formed by acid treatment at pH 0.38. Zhu et al. (Zhu et al., 2005) reported that hydrogen titanate nanofiber transforms into anatase nanocrystals in dilute (0.05 mol/L) HNO_3 at 80–120 °C. They stated that monodispersed anatase nanocrystals are obtained at 80 °C and aggregates of nanocrystals are obtained at 120 °C. Our results also suggest that the change in the crystal structure change of titanate compounds to anatase is determined not only by pH but also by the temperature of the ion-exchange treatment. We suggest that the high temperature (140 °C) of the ion-exchange treatment is responsible for the change in the crystal structure of hydrogen titanate to the anatase structure, with a high degree of crystallization, and that this change occurs due to polycondensation and dissolution–redeposition reactions.

a   b

Fig. 6. TEM images of the ion-exchange-treated thin films at 90 °C (a) and 140 °C (b).

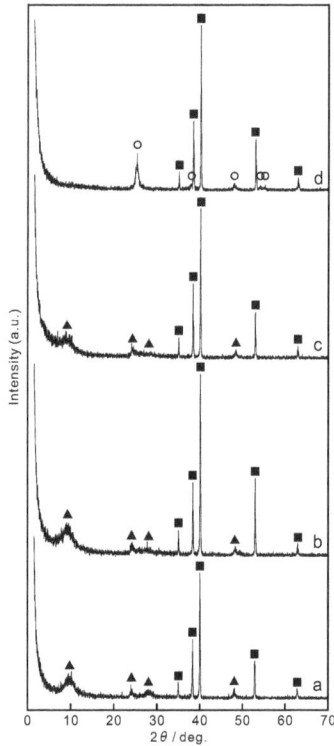

Fig. 7. XRD patterns of the as-grown (a) and the ion-exchange-treated thin films at 40 °C (b), 90 °C (c), and 140 °C (d). Peak assignment: ■, α-titanium; ○, anatase; ▲, titanate.

The temperature required for the complete H[+] ion-exchange reaction would be higher than that previously reported (Kasuga et al., 1998; Kasuga et al., 1999; Tokudome et al., 2004; Uchida et al., 2002; Tokudome et al., 2005; Thorne et al., 2005; Miyauchi et al., 2006; Tokudome et al., 2004; Kasuga et al., 2003), because the H[+] ion-exchange treatments at 40 °C using 0.01, 0.1, and 1 mol/l hydrochloric acid solutions were unsuccessful. H[+] ion-exchange treatment at 40 °C for 3 h using 0.01 and 0.1 mol/l hydrochloric acid solutions resulted in nanotube films respectively. However, substantial amounts of Na[+] ions remained in the samples after treatment. It is considered that elevated temperature assists the diffusion of ions, allowing ion-exchange to occur within the deepest regions of the film. After H[+] ion-exchange treatment at 40 °C using 1.0 mol/l hydrochloric acid solution, the nanotube thin films detached from the titanium metal plate.

2.3.2 Calcination of hydrogen titanate nanotube thin film

The hydrogen titanate nanotube thin films obtained by H[+] ion-exchange treatment at 90 °C using a 0.01 mol/L solution of hydrochloric acid were calcined at 300–900 °C for 3 h in air to transform them into titanium dioxide nanotube thin films. A uniform thin film formed on each sample surface, similar to the sample before calcination.

TEM images (Fig. 8) and XRD patterns (Fig. 9) show that calcination at 300 and 450 °C yields anatase nanotubes. Although the average inner diameter of 3.3 nm for nanotubes synthesized by calcination at 450 °C is similar to that of the as-grown sodium titanate nanotubes, the average outer diameter of the nanotubes decreased from 8.3 nm for the as-grown thin film to 8.1 nm for the anatase nanotubes synthesized by calcination at 450 °C. This slight decrease in the average outer diameter may be due to a phase transition from titanate into anatase. An cross-section image of the thin film calcined at 450 °C is similar to that of the as-grown sodium titanate nanotube thin film. Although a dense phase is observed at the bottom of the thin film (i.e., at the interface between the nanotube phase and the titanium metal), the porous structure composed of fibrous particles is observed in the film itself. Calcination at 600 °C yields anatase nanofibers approximately 11 nm thick, but not nanotubes (Figs. 8c and 9d). The porous structure consisting of fibrous particles are maintained until calcination at 600 °C. Calcination at 750 °C changes the thin film into a porous thin film consisting of particles (with 50-nm average diameter) and interstitial pores (with 79-nm average size) as shown in Fig. 8d. We attribute these changes in morphology to a progressive sintering reaction caused by the high calcination temperature. Furthermore,

a   b   c   d

Fig. 8. TEM images of the 90 °C ion-exchange-treated thin films calcined at 300 °C (a), 450 °C (b), 600 °C (c), and 750 °C (d).

calcination at 900 °C yields a dense rutile thin film because of the densification and phase transition caused by the sintering of anatase nanoparticles (Fig. 9f).

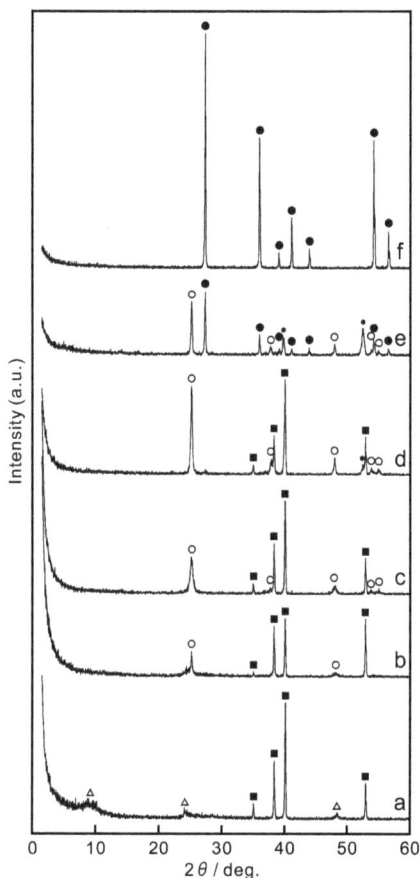

Fig. 9. XRD patterns of the 90 °C ion-exchange-treated (a) and the 90 °C ion-exchange-treated thin films calcined at 300 °C (b), 450 °C (c), 600 °C (d), 750 °C (e), and 900 °C (f). Peak assignment: ■, α-titanium; ○, anatase; ●, rutile; △, hydrogen titanate; *, distorted titanium.

2.4 Silver nanoparticle / silver titanate nanotube nanocomposite thin film

Na-TNT thin film obtained after the 3 h reaction with dimensions of 20 mm × 20 mm × 2 mm (hereafter referred to as Na-TNT-TF) was immersed in 12 mL of 0.05 M silver acetate solution at 40 °C for 3 h, then repeatedly washed with distilled water and dried in a cool dark place, to exchange the Na+ in the sodium titanate with Ag+. Hereafter, the sample obtained by the silver ion-exchange treatment of Na-TNT-TF was called Ag-TNT-TF.

The EDX spectra of the samples before and after the silver ion-exchange treatment were then compared. Since the peaks attributed to Na, observed in the samples before the silver ion-exchange treatment (Na-TNT-TF), disappeared in the samples after the silver ion-exchange treatment (Ag-TNT-TF), and the peaks attributed to Ag appeared after the silver ion-exchange treatment, Na$^+$ in the sodium titanate seemed to be exchanged with Ag$^+$ during the silver ion-exchange treatment. However, for the composition calculated from these spectra, the molar ratio of Ag/Ti was 0.67 for Ag-TNT-TF, while the molar ratio of Na/Ti was 0.50 for Na-TNT-TF. This confirmed presence of Ag in an excess compared with the exchangeable cations in the sample. SEM observation at the micrometer scale did not show changes in the morphologies before and after the silver ion-exchange treatment. However, the TEM observations of Ag-TNT-TF (Fig. 10) show particles with sizes ranging from several nanometers to a few dozen nanometers, which were not observable before the ion-exchange treatment. These are considered to be silver nanoparticles, since the color of Ag-TNT-TF was slightly yellow, indicating the formation of silver nanoparticles. The silver nanoparticles were deposited on titanates by the photoreduction of silver ions that were adsorbed on the titanate surface. The excess silver determined through the exchangeable mass of ions observed via EDX analysis is thus attributed to these silver nanoparticles. In other word, in the silver ion-exchange treatment, Ag$^+$ ion was not only incorporated into the titanate by ion exchange with Na$^+$ ion, but also deposited on the outer surface of titanate as silver nanoparticles.

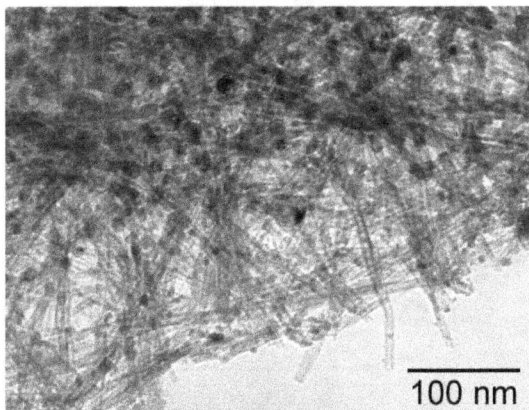

Fig. 10. TEM image of silver nanoparticle/silver titanate nanotube nanocomposite thin film.

Furthermore, TF-XRD patterns of Na-TNT-TF and Ag-TNT-TF shown in Fig. 11 also indicate the transformation of sodium titanate thin film into silver titanate thin film. When silver ion-exchange treatment was performed for Na-TNT-TF, a diffraction peak expressing the interlayer distance of 10 Å that was observed in Na-TNT-TF, disappeared in Ag-TNT-TF. The disappearance of the diffraction peak expressing the interlayer distance of 10 Å is considered to be due to the insertion of Ag$^+$ ions into an interlayer of titanate and disappearance of the layered structure of titanate. A further reason is a strong and peculiar interaction between the inserted Ag$^+$ ions and the titanate layer, which would cause a structural change of the layered structure into a three-dimensional structure. This structural change can also be confirmed, as the diffraction peaks in Na-TNT-TF due to the crystal structure of titanate, observed at $2\theta = 24.2°$ and $28.3°$, disappeared in Ag-TNT-TF, concomitant with the appearance of a new diffraction peak at $2\theta = 29.3°$ in Ag-TNT-TF. These results indicate the formation of silver titanate nanotube.

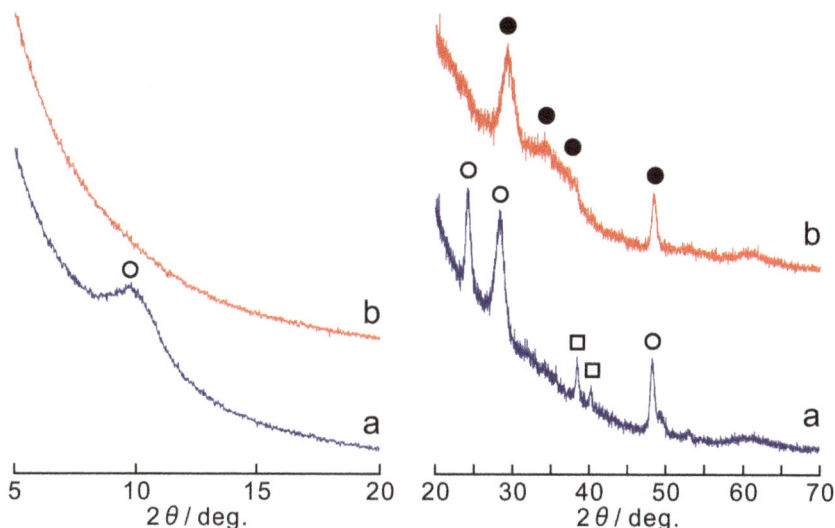

Fig. 11. TF-XRD patterns of sodium titanate nanotube thin film (a) and silver nanoparticle/ silver titanate nanotube nanocomposite thin film (b). Peak assignment: □, titanium; ○, sodium titanate; ●, silver titanate.

3. Antibacterial activities of titanate nanotube thin films

3.1 Elution properties of silver ions from the silver nanoparticle / silver titanate nanotube nanocomposite thin film

The elution properties of silver ions from the samples were examined in various solutions to determine the behavior of silver in MRSA environment or in the body. Ag-TNT-TF with dimensions of 20 mm × 20 mm × 2 mm was immersed in 15 mL physiological saline, phosphate buffered saline (+) (PBS(+)), phosphate buffered saline (−) (PBS(−)), and fetal bovine serum solution, maintained at 37 °C for 24 h. Then, the eluates were collected, centrifuged, and filtrated through a 0.22-μm filter. After filtration, Ag concentration in the eluates was determined by inductively coupled plasma mass spectrometry (ICP-MS). In physiological saline, PBS(+), and PBS(−), almost the same average concentration of eluted silver was measured − 300, 320, and 440 ppb, respectively. The eluted silver ions may originate from metallic silver and silver titanate. Since the solubility of metallic silver is known to be very small, the large portion of eluted silver was eluted by exchanging silver ions in titanates with Na^+, K^+, and H^+ in the solutions. On the other hand, in fetal bovine serum, the average eluted silver concentration was measured in large amounts − 82000 ppb for Ag-TNT-TF. This was because a large quantity of a compound composed of silver and a protein was formed together with AgCl, since the protein that exists in fetal bovine serum has very high affinity with Ag^+ through the –SH group or –NH group in the protein, and the amount of exchangeable cations in fetal bovine serum was larger than that in physiological saline and PBS. Moreover, when the silver elution test in fetal bovine serum was performed for a silver metal plate under similar conditions as that for Ag-TNT-TF, silver of 7900 ppb was eluted. This amount was also significantly smaller than that for of Ag-TNT-TF. In the silver elution test of Ag-TNT-TF, although silver is eluted from silver nanoparticles deposited on the surface of titanate, its elution amount is thus considered small. Consequently, in the XRD and TF-XRD patterns of the sample obtained by the silver elution test of Ag-TNT-TF, diffraction peaks appeared near $2\theta = 10°$, $24°$ and $28°$, which were not observed in Ag-TNT-TF. These diffraction peaks of the sample obtained by the silver elution test of Ag-TNT-TF appeared at the same locations as the diffraction peaks observed in Na-TNT-TF. This indicates that the crystal structure of the sample obtained by the silver elution test of Ag-TNT-TF is similar to that of Na-TNT-TF. The likely reasons are as follows: (1) the layered structure of titanate of Na-TNT-TF transformed into a three-dimensional structure because Ag^+ ions were inserted into the interlayer of titanate by the silver ion-exchange treatment to form silver titanate, (2) Ag^+ ions were eluted from the silver titanate and Na^+ ions were reinserted into the titanate during the silver elution test; and (3) the three-dimensional structure of titanate returned to the original condition as in Na-TNT-TF. Thus, it is clearly demonstrated that insertion (intercalation) and elimination (deintercalation) of Ag^+ ions occurs during the silver ion-exchange treatment and the silver elution test, respectively. Therefore, this experiment indicated that Ag^+ ions in silver titanate greatly contributed to the elution of Ag^+. Diffraction peaks attributable to AgCl were also observed in the sample obtained by the silver elution test of Ag-TNT-TF, because AgCl particles were formed by the reaction between eluted Ag^+ and Cl^- in fetal bovine serum. Since fetal bovine serum solution is considered as the system most similar to MRSA environment, silver elution tests in fetal bovine serum solution were repeated (Fig. 12). In Ag-TNT-TF, the elution concentration slowly decreased from 94000 ppb for the first test to 11000 ppb for the tenth, indicating that a large amount of eluted silver was measured in the tenth test for Ag-TNT-TF. The 2-step elution curve was obtained from Ag-TNT-TF. A rapid elution of a large amount of silver at the initial stage of the repeated elution test (between the first and third time) was considered to be mainly due to Ag^+ ion elution from the silver titanate based on the ion-exchange reaction. Since the elution reaction (ion-exchange reaction) of Ag^+ from silver titanate is rapid, the elution of Ag^+ is considered to be almost completed at the initial stage of the repeated elution test. These discussions are also supported by the above described TF-XRD data, indicating that crystal structure of the sample obtained by the silver elution test of Ag-TNT-TF is similar to that of Na-TNT-TF. Therefore, a slow elution of a small amount of silver after the forth repetition of the elution test was considered to be mainly because of silver elution from the silver nanoparticles. This two-step elution curve is difficult to explain if it is considered that only silver nanoparticles are formed in the thin film, but it is reasonably explained if two types of Ag (silver nanoparticles and silver titanate) exhibiting different elution behaviors are present in Ag-TNT-TF. The silver titanates loading silver nanoparticles would be promising as a novel antibacterial material, because they have two silver sources. The silver-ion elution property of silver titanate would be different from that of the silver nanoparticles, i.e., the elution speed of silver ions from silver titanate would be greater than that from silver nanoparticles. Therefore, silver titanate would be effective for short-term bacterial killing, and silver nanoparticles would be effective for long-term antibacterial action. Since we have already found that a thicker (i.e., 20 μm thick) titanate nanotube film can be formed after a longer reaction time or 20 h in NaOH solution, it would be possible to prolong the elution period of silver ions and to increase the amount of eluted silver ions or the duration.

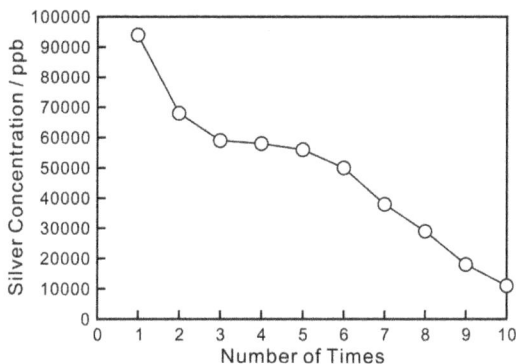

Fig. 12. Repeated silver ion elution test of silver nanoparticle/silver titanate nanotube nanocomposite thin film.

3.2 Antibacterial property of silver nanoparticle / silver titanate nanotube nanocomposite thin film

The modified Japanese Industrial Standard test (JIS Z 2801) was performed as an antibacterial test as follows. To approximate an infection environment within an actual organism, an inactivated bovine serum (0.4 mL) was used as a solvent of bacterial suspension to create a eutrophic condition, and the antibacterial test was conducted using MRSA with a biofilm-forming gene. A bacterial suspension (0.4 mL) was dropped on a 50 mm × 50 mm × 2 mm test piece, covered with a 40 mm × 40 mm polyethylene film (Elmex Corp.), and cultured at 37 °C for 24 h. The test piece was washed, and the viable MRSA count was determined. The antibacterial test was performed thrice for each of the samples of Na-TNT-TF and Ag-TNT-TF, to obtain averaged values of viable MRSA counts. The antibacterial activity value (R) for the sample was calculated as follows.

$$R = \{\log(B/A) - \log(C/A)\} = \log(B/C)$$

Here, A, B, and C are the average viable MRSA counts just after inoculation, after 24 h for a blank and after 24 h for a sample, respectively. The viable MRSA count just after the inoculation was 2.1×10^5 CFU/sample, and for a blank, the average viable MRSA count after 24 h increased to 5.9×10^8 CFU/sample. In Na-TNT-TF, the average viable MRSA count after 24 h was slightly less than that of the blank: 1.1×10^7 CFU/sample. On the other hand, in Ag-TNT-TF, the average viable MRSA count after 24 h was markedly small: 3.3×10^2 CFU/sample. R increased from 1.7 for Na-TNT-TF to 6.3 for Ag-TNT-TF through the silver ion-exchange treatment. These results indicate that the silver ion-exchanged titanate thin films display high antibacterial activity against MRSA. It was also revealed that although the crystal structure of titanate itself does not have a large antibacterial effect, higher antibacterial activity arises in the silver in the titanate. The conversion of sodium titanates into antibacterial materials through the silver ion-exchange treatment can apply to other nanostructured sodium titanates. For example, by the same silver ion-exchange treatment, porous sodium titanate film calcined at 600 °C reported by Kim et al. (Kim et al., 1997) can also be converted into silver nanoparticle / silver titanate nanocomposite thin film with high antibacterial activity for MRSA of $R=6.7$.

4. Apatite-forming ability of titanate and titanium dioxide nanotube thin films

The three thin films (Na-TNT-TF, TiO$_2$-NT-FT (anatase type titanium dioxide nanotube thin film formed by the H$^+$ ion-exchange treatment and calcination at 450 °C of Na-TNT-TF), Ag-TNT-TF) formed on a titanium metal were immersed in simulated body fluid (SBF) and monitored the development of apatite formation on their surfaces. In accordance with ISO 23317, "Implants for surgery-*In vitro* evaluation for apatite-forming ability of implant materials," we evaluated the apatite-forming ability on the surface of the coating in SBF. A plate was placed in 96.0 mL SBF at 36.5°C. After 2, 4, and 14 days, the plate was removed and gently rinsed with water. The surface of the plate was dried in air.

For Na-TNT-TF, after immersing the film for 4 days, the SEM images showed only nanotubes and no substances with foliaceous morphology peculiar to apatite; the XRD patterns remained unchanged. However, when the SBF immersion was extended to 14 days, the SEM images showed the surface of the film to be completely covered with a dome-shaped form consisting of foliaceous particles peculiar to apatite (Figs. 13a, b); the XRD pattern showed diffraction peaks attributable to apatite. Thus, apatite is confirmed to be formed on the sodium titanate thin films. In addition, after immersing the hydrogen titanate nanotube thin film formed by exchanging Na$^+$ ions between the layers of the layered sodium titanate for H$^+$ ions in SBF for 4 days, no apatite was evident. This lack of apatite indicates that ions (Na$^+$ and H$^+$) between the titanate layers do not particularly contribute to the acceleration of the apatite formation. In contrast, for TiO$_2$-NT-FT formed by the 450°C calcination of the hydrogen titanate nanotube thin film and Ag-TNT-TF, after immersing the

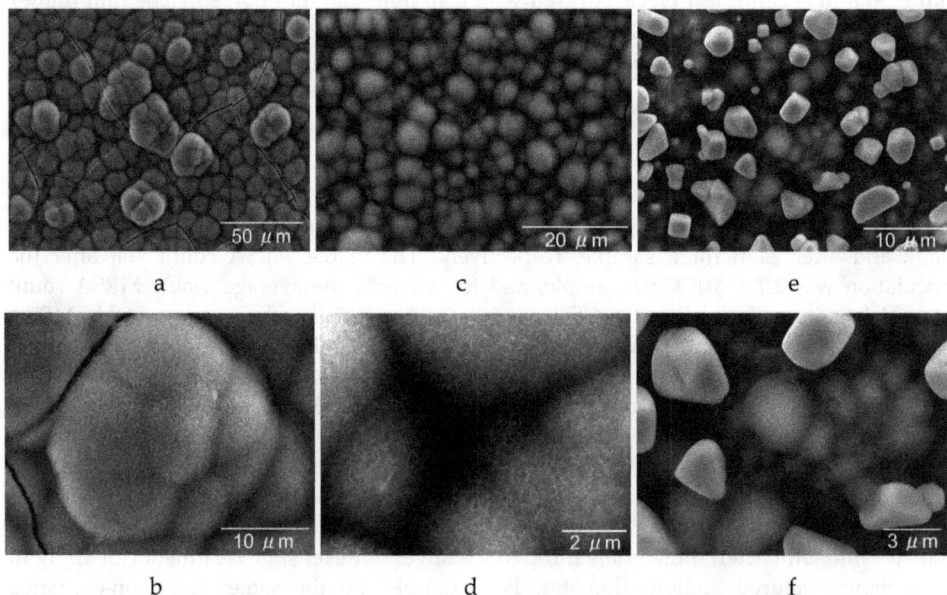

Fig. 13. Low magnification (a, c, e) and high magnification (b, d, f) SEM images of Na-TNT-TF (a, b), TiO$_2$-NT-FT (c, d), and Ag-TNT-TF (e, f) after immersions in SBF.

films for 4 days, the SEM images showed a stretch of the dome-shaped form consisting of foliaceous particles peculiar to apatite (Figs. 13c, d). XRD patterns of the two films showed diffraction peaks attributable to apatite, respectively. These results indicate that the surfaces of these two thin films are completely covered with apatite and that the apatite-forming ability of the two films is greater than that of Na-TNT-TF having layered structure. In contrast, for TiO_2-NT-FT after immersing the film in SBF for 2 days, the surface is thinly covered with apatite. However immersing Ag-TNT-TF for 2 days, the surface is almost covered with a dome-shaped form consisting of foliaceous particles peculiar to the apatite (Figs. 13e, f). The apatite-forming ability of Ag-TNT-TF is, thus, slightly higher than that of TiO_2-NT-FT.

We then investigated the newly observed high apatite-forming ability of silver nanoparticle/silver titanate nanocomposite thin film (Ag-TNT-TF). After immersing the film in SBF for 4 days, we observed bulky particles of a few micrometers in diameter together with apatite. XRD pattern shows diffraction peaks attributable to silver chloride. EDX element mapping analysis shows the bulky particles to be composed of Ag and Cl (Fig. 14) and, therefore, to be silver chloride particles. Therefore, Ag^+ ions are eluted from silver titanate mainly by ion-exchange reaction with cations, resulting in deposition of silver chloride particles. After immersing the film in SBF for 1 day, SEM images reveal only bulky AgCl particles on the film surface, however immersing the film for 2 days, the surface is almost covered with a dome-shaped form consisting of foliaceous particles peculiar to the apatite (Figs. 13e, f), as mentioned above. We clarified whether silver nanoparticle or silver titanate contributes more to the apatite formation by investigating the apatite-forming ability of a silver metal plate. After immersing the plate in SBF for 4 days, no apatite formation was evident; thus, the silver titanate nanotubes are responsible for the high apatite-forming ability. Researchers have reported that the effects of crystal structure (Uchida et al., 2003) and surface hydroxyl groups such as Ti-OH (Kasuga et al., 2002) influence the apatite formation on the titanium compounds immersed in SBF. Kokubo et al. reported that the apatite-forming ability is improved by the crystal structure transformation of sodium titanate into titanium dioxide (Fujibayashi et al., 2001; Uchida et al., 2002; Takemoto et al., 2006). Although the detailed crystal structure of silver titanate is not yet known, we speculate that the surface atomic arrangement and surface functional groups of silver titanate might be suitable for rapid apatite formation. We further investigated the high apatite forming ability by considering –OH groups that influence apatite formation using the FT-IR measurements. As shown in Fig. 15, Na-TNT-TF and TiO_2-NT-FT exhibited similar absorption spectra in a wide range of 3000−3700 cm^{-1}. These absorption spectra are considered to be mainly due to water molecules adsorbed on the inner and outer surfaces of nanotubes and partially due to –OH groups on the surface. Unlike Na-TNT-TF and TiO_2-NT-FT, strong absorption was observed at 3000−3400 cm^{-1} in addition to 3400−3700 cm^{-1} in Ag-TNT-TF. This absorption at 3000−3400 cm^{-1} is considered to indicate the existence of surface –OH groups due to silver titanate. A surface atomic arrangement peculiar to silver titanate would arise and a large number of –OH groups would be generated on the nanotube surfaces, which would stimulate apatite formation.

Oh et al. (Oh et al., 2005) and Tsuchiya et al. (Tsuchiya et al., 2006) reported anatase-type titanium dioxide nanotube thin films synthesized by anodization and heat treatment of the titanium metal. We compared the apatite-forming ability of these nanotube thin films with

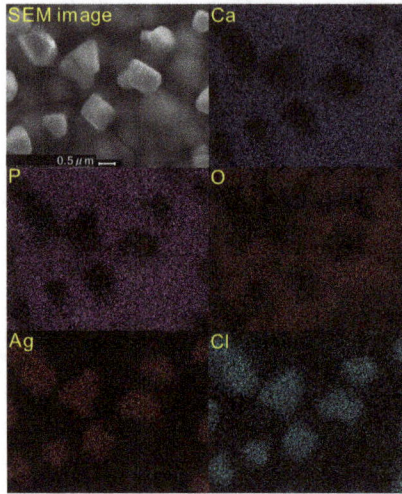

Fig. 14. Elemental mapping performed on Ag-TNT-TF after immersion in SBF for 4 days using EDX analysis.

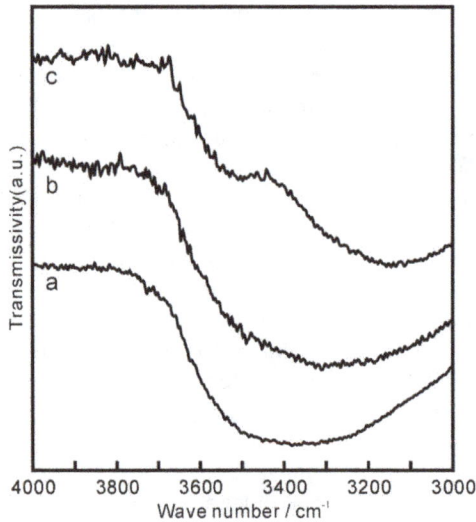

Fig. 15. FT-IR spectra of Na-TNT-TF (a), TiO$_2$-NT-FT (b), and Ag-TNT-TF (c).

that of TiO$_2$-NT-FT obtained in this study by immersing the film in SBF for 2 days. SEM images show not only a small amount of the dome-shaped form consisting of foliaceous particles peculiar to apatite, but also several slightly swelled and whitish areas. EDX element mapping on this thin film revealed that titanium dioxide nanotubes exist in the blackish areas and apatite exists in the whitish areas. At this point, after 2 days of immersion, the apatite phase has grown slightly but not yet achieved its dome-shaped form. Therefore, the apatite-forming ability of TiO$_2$-NT-FT is slightly superior though still similar to that of the thin film with 2-μm long nanotube synthesized by Tsuchiya et al. (Tsuchiya et al., 2006) and clearly superior to that of the nanotube thin film synthesized by Oh et al. (Oh et al., 2005) and the thin film with 500-nm long nanotube synthesized by Tsuchiya et al. (Tsuchiya et al., 2006). While comparing the ratio of the void parts to the anatase-type titanium dioxide part on the surface using SEM images, we found the proportion consisting anatase-type titanium dioxide to be larger in the surface of our synthesized thin film as compared to that in the surfaces of the thin films reported by Oh et al. and Tsuchiya et al. Hence, the apatite-forming ability of our film is also correspondingly higher.

5. Conclusion

In this study, novel procedure of synthesis and fixation of Na-TNT onto titanium metals with various morphologies such as plate, wire, mesh, tube, and sphere was reported. Especially, since the Na-TNT/Ti composite wires have softness and flexibility peculiar to metal titanium because of the existence of titanium metal in its core part, this wire can be fabricated into various shapes of cloth, fiber, etc. with centimeter or meter size by using conventional spinning techniques. The Na-TNT thin films can be transformed into anatase-type titanium dioxide nanotube thin films. Another advantage of the proposed procedure is that the thickness of the thin films produced is greater than that of the thin films reported by other researchers. Therefore, Na-TNT's applications mentioned in the introduction would be remarkably expanded. In addition, the novel growth of the Na-TNT film on substrates such as Co–Cr alloy and SUS316L and simple patterning of the Na-TNT phase by the hydrothermal transcription method were also reported. As these substrates including titanium metal, Co-Cr alloy, and SUS316L have superior mechanical properties and corrosion resistance, they are frequently used as implants such as artificial joints. Generally, the coating of films to implants with complex shapes requires thin films with uniform and controlled thickness, and a high fixing strength to the implants. Because of the direct growth of the nanotubes from the substrate, our proposed method is very simple and the fixing strength to the substrate is expected to be higher. Therefore, the method proposed in the present study has excellent potential for these biomedical applications.

Next, through a silver ion-exchange treatment, Na+ ions in sodium titanate nanotube were exchanged with Ag+ ions in silver acetate solution, along with the loading of silver nanoparticles on the titanate surfaces, and the layered structure of titanate transformed into a new three-dimensional crystal structure. Results of silver ion elution tests of the obtained thin films in fetal bovine serum solution indicate that the release period and the number of silver ions released from the silver titanate thin films can be controlled. The silver ion-exchanged titanate thin films showed high antibacterial activity against MRSA. It was also revealed that although the crystal structure of titanate itself has no large antibacterial effect, higher antibacterial activity mainly arises from the silver ions held in the titanate. The samples coated with apatite containing silver and silver plate have already been reported as possessing antibacterial properties through metallic silver with low solubility. In contrast, in this study, the antibacterial properties were mainly caused by the elution of silver ions from a titanate with an ion-exchange property. Since the thin film obtained by this study has a higher silver-ion elution speed, greater and more rapid antibacterial effects than in metallic silver can be expected. Since we have also revealed that the morphology, thickness, and crystal structure of the titanate phase can be controlled, we think that this can also promise control of the antibacterial properties, i.e., the duration and amount of antibacterial activity, for the future. The obtained results should aid the development of more convenient and inexpensive antibacterial implants.

Furthermore, the present study compares the apatite-forming ability of a sodium titanate nanotube thin film, an anatase-type titanium dioxide nanotube thin film, and a silver nanoparticle/silver titanate nanotube nanocomposite thin film. Of these, the apatite-forming ability of the silver titanate nanotube was higher than that of the titanium dioxide nanotubes or the sodium titanate nanotubes, in that order. This superior apatite-forming ability of the silver nanoparticle/silver titanate nanotube nanocomposite thin film is a novel phenomenon and is presumably due to the surface atomic arrangement of silver titanate, the large amount of Ti-OH formed on the nanotube surface, or both. In conclusion, the silver nanoparticle/silver titanate nanotube nanocomposite thin film, which have the antibacterial property and the ability to form bone-like apatite, i.e., the osteoconductive property, may have bright prospects for future use in implant materials such as artificial joints.

6. Acknowledgment

This research was partially supported by KAKENHI (16685021, 19750172) and Saga University Dean's Grant 2010 For Promising Young Researchers. Figures are reproduced with permission from American Chemical Society, Elsevier, and John Wiley & Sons.

7. References

Alt, V.; Bitschnau, A.; Österling, J.; Sewing, A.; Meyer, C.; Kraus, R.; Meissner, S. A.; Wenisch, S.; Domann, E. & Schnettler, R. (2006). The effects of combined gentamicin–hydroxyapatite coating for cementless joint prostheses on the reduction of infection rates in a rabbit infection prophylaxis model. *Biomaterials*, Vol. 27, pp. 4627-4634.

Bavykin, D. V.; Lapkin, A. A.; Plucinski, P. K.; Friedrich, J. M. & Walsh, F. C. (2005). Reversible storage of molecular hydrogen by sorption into multilayered TiO_2 nanotubes. *The Journal of Physical Chemistry B*, Vol. 109, pp. 19422-19427.

Bavykin, D. V.; Friedrich, J. M. & Walsh, F. C. (2006). Protonated Titanates and TiO_2 Nanostructured Materials: Synthesis, Properties, and Applications. *Advanced Materials*, Vol. 18, pp. 2807-2824.

Byon, E.; Jeong, Y.; Takeuchi, A.; Kamitakahara, M. & Ohtsuki, C. (2007). Apatite-forming ability of micro-arc plasma oxidized layer of titanium in simulated body fluids. *Surface and Coat Technology*, Vol. 201. pp. 5651-5654.

Chen, Q.; Zhou, W.; Du, G. & Peng, L.-M. (2002). Trititanate nanotubes made via a single alkali treatment. *Advanced Materials*, Vol. 14, pp. 1208-1211.

Chen, W.; Liu, Y.; Courtney, H. S.; Bettenga, M.; Agrawal, C. M.; Bumgardner, J. D. & Ong, J. L. (2006). In vitro anti-bacterial and biological properties of magnetron co-sputtered silver-containing hydroxyapatite coating. *Biomaterials*, Vol. 27, pp. 5512-5517.

Chi, B.; Victorio, E. S. & Jin, T. (2007). Synthesis of TiO_2-Based Nanotube on Ti Substrate by Hydrothermal Treatment. *Journal of Nanoscience and Nanotechnology*, Vol. 7, pp. 668-672.

Fujibayahsi, S.; Nakamura, T.; Nishiguchi, S.; Tamura, J.; Uchida, M.; Kim, H. M. & Kokubo, T. (2001). Bioactive titanium : effect of sodium removal on the bone-bonding ability of bioactive titanium prepared by alkali and heat treatment. *Journal of Biomedical Materials Research*, Vol. 56, pp. 562-570.

Guo, Y.; Lee, N.-H.; Oh, H.-G.; Yoon, C.-R.; Park, K.-S.; Lee, H.-G.; Lee, K.-S. & Kim, S.-J. (2007). Structure-tunable synthesis of titanate nanotube thin films via a simple hydrothermal process. *Nanotechnology*, Vol. 18, pp. 295608-295616.

Hanawa, T.; Kon, M.; Ukai, H.; Murakami, K.; Miyamoto, Y. & Asaoka, K. (1997). Surface Modification of Titanium in Calcium-Ion-Containing Solutions. *Journal of Biomedical Materials Research*, Vol. 34, pp. 273-278.

Hamada, K.; Kon, M.; Hanawa, T.; Yokoyama, K.; Miyamoto, Y. & Asaoka, K. (2002). Hydrothermal modification of titanium surface in calcium solutions. *Biomaterials*, Vol. 23, pp. 2265-2272.

Hardes, J.; Ahrens, H.; Gebert, C.; Streitbuerger, A.; Buerger, H.; Erren, M.; Gunsel, A.; Wedemeyer, C.; Saxler, G & Winkelmann, W. (2007). Lack of toxicological side-effects in silver-coated megaprostheses in humans. *Biomaterials*, Vol. 28, pp. 2869-2875.

Imai, H.; Takei, Y.; Shimizu, K.; Matsuda, M. & Hirashima, H. (1999). Direct preparation of anatase TiO_2 nanotubes in porous alumina membranes. *Journal of Materials Chemistry*, Vol. 9, pp. 2971-2972.

Inoue, Y.; Noda, I.; Torikai, T.; Watari, T.; Hotokebuchi, T. & Yada, M. (2010). TiO_2 nanotube, nanowire, and rhomboid-shaped particle thin films fixed on a titanium metal plate. *Journal of Solid State Chemistry*, Vol. 183, pp. 57-64.

Inoue, Y.; Uota, M.; Torikai, T.; Watari, T.; Noda, I.; Hotokebuchi, T. & Yada, M. (2010). Antibacterial properties of nanostructured silver titanate thin films formed on a titanium plate, *Journal of Biomedical Materials Research Part A*, Vol. 92A, pp. 1171-1180.

Jiang, Z.; Yang, F.; Luo, N.; Chu, B. T. T.; Sun, D.; Shi, H.; Xiao, T. & Edwards, P. P. (2008). Solvothermal synthesis of N-doped TiO_2 nanotubes for visible-light-responsive photocatalysis. *Chemical Communication*, pp. 6372-6374.

Jung, J. H.; Kobayashi, H.; van Bommel, K. J. C.; Shinkai, S. & Shimizu, T. (2002). Creation of Novel Helical Ribbon and Double-Layered Nanotube TiO_2 Structures Using an Organogel Template. *Chemistry of Materials*, Vol. 14, pp. 1445-1447.

Jonášová, L.; Müller, F. A.; Helebrant, A.; Strnad, J. & Greil, P. (2003). Biomimetic apatite formation on chemically treated titanium. *Biomaterials*, Vol. 25, pp. 1187-1194.

Kasuga, T.; Hiramatsu, M.; Hoson, A.; Sekino, T. & Niihara, K. (1998). Formation of titanium oxide nanotube. *Langmuir*, Vol. 14, pp. 3160-3163.

Kasuga, T.; Hiramatsu, M.; Hoson, A.; Sekino, T. & Niihara, K. (1999). Titania nanotubes prepared by chemical processing. *Advanced Materials*, Vol. 11, pp. 1307-1311.

Kasuga, T.; Kondo, H. & Nogami, M. (2002). Apatite formation on TiO_2 in simulated body fluid. *Journal of Crystal Growth*, Vol. 235, pp. 235-240.

Kasuga, T. (2003). *Jpn. Kokai Tokkyo Koho*, P2003-220127A.

Kim, G.-S.; Ansari, S. G.; Seo, H.-K.; Kim, Y.-S. & Shin, H.-S. (2007). Effect of annealing temperature on structural and bonded states of titanate nanotube films. *Journal of Applied Physics*, Vol. 101, pp. 024314-024320.

Kim, H. M.; Miyaji, F.; Kokubo, T. & Nakamura, T. (1996). Preparation of bioactive Ti and its alloys via simple chemical surface treatment. *Journal of Biomedical Materials Research*, Vol. 32, pp. 409-417.

Kim, H. M.; Miyaji, F.; Kokubo, T. & Nakamura, T. (1997). Effect of heat treatment on apatite-forming ability of Ti metal induced by alkali treatment. *Journal of Materials Science: Materials in Medicine*, Vol. 8, pp. 341-347.

Kim, H. M.; Miyaji, F.; Kokubo, T.; Nishiguchi, S. & Nakamura, T. (1997). Graded surface structure of bioactive titanium prepared by chemical treatment. *Journal of Biomedical Materials Research*, Vol. 45, pp. 100-107.

Kim, H. M.; Himeno, T.; Kawashita, M.; Lee, J. H. & Kokubo, T. (2003). Surface Potential Change in Bioactive Titanium Metal during the Process of Apatite Formation in Simulated Body Fluid. *Journal of Biomedical Materials Research Part A*, Vol. 67A, pp. 1305-1309.

Kokubo, T.; Miyaji, F.; Kim, H. M. & Nakamura, T. (1996). Spontaneous Formation of Bonelike Apatite Layer on Chemically Treated Titanium Metals. *Journal of the American Ceramic Society*, Vol. 79, pp. 1127-1129.

Kokubo, T. & Takadama, H. (2006). How useful is SBF inpredicting in vivo bone bioactivity ? *Biomaterials*, Vol. 27, pp. 2907-2915.

Kon, M.; Sultana, R.; Fujihara, E.; Asaoka, K. & Ichikawa, T. (2007). Hydroxyapaite Morphology Control by Hydrothermal Treatment. *Key Engineering Materials*, Vol. 330-332, pp. 737-740.

Kubota, S.; Johkura, K.; Asanuma, K.; Okouchi, Y.; Ogiwara, N.; Sasaki, K. & Kasuga, T. (2004). Titanium oxide nanotubes for bone regeneration. *Journal of Materials Science: Materials in Medicine*, Vol. 15, pp. 1031-1035.

Ma, R.; Bando, Y. & Sasaki, T. (2003). Nanotubes of lepidocrocite titanates. *Chemical Physics Letters*, Vol. 380, pp.577-582.

Ma, R.; Sasaki, T. & Bando, Y. (2004). Layer-by-Layer Assembled Multilayer Films of Titanate Nanotubes, Ag- or Au-Loaded Nanotubes, and Nanotubes/Nanosheets with Polycations. *Journal of the American Chemical Society*, Vol. 126, pp. 10382-10388.

Macak, J. M.; Tsuchiya, H. & Schmuki, P. (2005). High-Aspect-Ratio TiO_2 Nanotubes by Anodization of Titania. *Angewandte Chemie International Edition*, Vol. 44, pp. 2100-2102.

Miyauchi, M.; Tokudome, H. Toda, Y.; Kamiya, T. & Hosono, H. (2006). Electron Field Emission from TiO_2 Nanotube Arrays Synthesized by Hydrothermal Reaction. *Applied Physics Letters*, Vol. 89, pp.043114-043116.

Muramatsu, K.; Uchida, M.; Kim, H. M.; Fujisawa, A. & Kokubo, T. (2003). Thromboresistance of alkali- and heat-treated titanium metal formed with apatite. *Journal of Biomedical Materials Research Part A*, Vol. 65A, pp. 409-416.

Nakagawa, M.; Zhang, L.; Udou, K.; Matsuya, S. & Ishikawa, K. (2005). Effects of hydrothermal treatment with $CaCl_2$ solution on surface property and cell response of titanium implants. *Journal of Materials Science: Materials in Medicine*, Vol. 16, pp. 985-991.

Nakahira, A.; Kato, W.; Tamai, M.; Isshiki, T.; Nishio, K. & Aritani, H. (2004). Synthesis of nanotube from a layered $H_2Ti_4O_9H_2O$ in a hydrothermal treatment using various titania sources. *Journal of Materials Science*, Vol. 39, pp. 4239-4245.

Nishiguchi, S.; Nakamura, T.; Kobayashi, M.; Kim, H. M.; Miyaji, F. & Kokubo, T. (1999). The effect of heat treatment on bone-bonding ability of alkali-treated titanium. *Biomaterials*, Vol. 20, pp. 491-500.

Oh, S. H.; Finõnes, R. R.; Daraio, C.; Chen, L. H. & Jin, S. (2005). Growth of nano-scale hydroxyapatite using chemically treated titanium oxide nanotubes. *Biomaterials*, Vol. 26, pp. 4938-4943.

Ohtsu, N.; Abe, C.; Ashino, T.; Semboshi, S. & Wagatsuma, K. (2008). Calcium-hydroxide slurry processing for bioactive calcium-titanate coating on titanium. *Surface and Coat Technology*, Vol. 202, pp. 5110-5115.

Ohtsuki, C.; Iida, H.; Hayakawa, S. & Osaka, A. (1997). Bioactivity of titanium treated with hydrogen peroxide solutions containing metal chlorides. *Journal of Biomedical Materials Research*, Vol. 35, pp. 39-47.

Shin, H.; Jeong, D.-K.; Lee, K.; Sung, M. M. & Kim, J. (2004). Formation of TiO_2 and ZrO_2 Nanotubes Using Atomic Layer Deposition with Ultra-Precise Wall Thickness Control, *Advanced Materials*, Vol. 16, pp. 1197-1200.

Sun, X. & Li, Y. (2003). Synthesis and Characterization of Ion-Exchangeable Titanate Nanotubes. *Chemistry - A European Journal*, Vol. 9, pp. 2229-2238.

Tsai, C.-C. & Teng, H. (2006). Structural features of nanotubes synthesized from NaOH treatment on TiO_2 with different post-treatments. *Chemistry of Materials*, Vol. 18, pp. 367-373.

Takemoto, M.; Fujibayashi, S.; Neo, M.; Suzuki, J.; Matsushita, T.; Kokubo, T. & Nakamura, T. (2006). Effect of sodium removal by dilute HCl treatment. *Biomaterials*, Vol. 27, pp. 2682-2691.

Thorne, A.; Kruth, A.; Tunstall, D.; Irvine, J. T. S. & Zhou, W. (2005). Formation, Structure, and Stability of Titanate Nanotubes and their Proton Conductivity. *The Journal of Physical Chemistry B*, Vol. 109, pp. 5439-5444.

Tian, Z. R.; Voigt, J. A.; Liu, J.; Mckenzie, B. & Xu, H. (2003). Large oriented arrays and continuous films of TiO_2-based nanotubes. *Journal of the American Chemical Society*, Vol. 125, pp. 12384-12385.

Tokudome, H. & Miyauchi, M. (2004). N-doped TiO_2 Nanotube with Visible Light Activity. *Chemistry Letter*, Vol. 33, pp. 1108-1109.

Tokudome, H. & Miyauchi, M. (2004). Titanate nanotube thin films via alternate layer deposition. *Chemical Communication*, pp. 958-959.

Tokudome, H. & Miyauchi, M. (2005). Electrochromism of titanate-based nanotubes. *Angewandte Chemie International Edition*, Vol. 44, pp. 1974-1977.

Tsuchiya, H.; Macak, J. M.; Müller, L.; Kunze, J.; Müller, F.; Greil, P.; Virtanen, S. & Schmuki, P. (2006). Hydroxyapatite growth on anodic TiO_2 nanotubes. *Journal of Biomedical Materials Research Part A*, Vol. 77A, pp. 534-541.

Uchida, S.; Chiba, R.; Tomiha, M.; Masaki, N. & Shirai, M. (2002). Application of Titania Nanotubes to a Dye-Sensitized Solar Cell. *Electrochemistry*, Vol. 70, pp. 418-420.

Uchida, M.; Kim, H. M.; Kokubo, T.; Fujibayashi, S. & Nakamura, T. (2002). Effect of water treatment on the apatite-forming ability of NaOH-treated titanium metal. *Journal of Biomedical Materials Research*, Vol. 63, pp. 522-530.

Uchida, M.; Kim, H. M.; Kokubo, T.; Fujibayashi, S. & Nakamura, T. (2003). Structural dependence of apatite formation on titania gels in a simulated body fluid. *Journal of Biomedical Materials Research Part A*, Vol. 64A, pp. 164-70.

Wang, F.; Liao, Y.; Wang, M.; Gong, P.; Li, X.; Tang, H.; Man, Y.; Yuan, Q.; Wei, N.; Tan, Z. & Ban, Y. (2007). Evaluation of Sodium Titanate Coating on Titanium by Sol-Gel Method In Vitro. *Key Engineering Materials*, Vol. 330-332, pp. 777-780.

Wang, X. J.; Li, Y. C.; Lin, J. G.; Yamada, Y.; Hodgson, P. D. & Wen, C. E. (2008). In vitro bioactivity evaluation of titanium and niobium metals with different surface morphologies. *Acta Biomaterialia*, Vol. 4, pp. 1530-1535.

Wang, X. X.; Hayakawa, S.; Tsuru, K. & Osaka, A. (2001). A comparative study of in vitro apatite deposition on heat-, H_2O_2-, and NaOH- treated titanium surfaces. *Journal of Biomedical Materials Research*, Vol. 54, pp. 172-178.

Wang, X. X.; Yan, W.; Hayakawa, S.; Tsuru, K. & Osaka, A. (2003). Apatite deposition on thermally and anodically oxidized titanium surfaces in a simulated body fluid. *Biomaterials*, Vol. 24, pp. 4631-4637.

Wu, D.; Liu, J.; Zhao, X.; Li, A.; Chen, Y. & Ming, N. (2006). Sequence of events for the formation of titanate nanotubes, nanofibers, nanowires, and nanobelts. *Chemistry of Materials*, Vol. 18, pp. 547-553.

Xiong, J.; Li, Y.; Hodgson, P. D. & Wen, V. (2010). Nanohydroxyapatite coating on a titanium–niobium alloy by a hydrothermal process. *Acta Biomaterialia*, Vol. 6, pp. 1584-1590.

Yada, M.; Inoue, Y.; Uota, M.; Torikai, T.; Watari, T.; Noda, I. & Hotokebuchi, T. (2007). Plate, Wire, Mesh, Microsphere, and Microtube Composed of Sodium Titanate Nanotubes on a Titanium Metal Template. *Langmuir*, Vol. 23, pp. 2815-2823.

Yada, M.; Inoue, Y.; Uota, M.; Torikai, T.; Watari, T.; Noda, I. & Hotokebuchi, T. (2008). Formation of Sodium Titanate Nanotube Films by Hydrothermal Transcription. *Chemistry of Materials*, Vol. 20, pp. 364-366.

Yada, M.; Inoue, Y.; Gyoutoku, A.; Noda, I.; Torikai, T.; Watari, T. & Hotokebuchi, T. (2010). Apatite-forming ability of titanium compound nanotube thin films formed on a titanium metal plate in a simulated body fluid. *Colloids and Surfaces B: Biointerfaces*, Vol. 80, pp. 116-124.

Yamanaka, S.; Hamaguchi, T.; Muta, H.; Kurosaki, K. & Uno, M. (2004). Fabrication of Oxide Nanohole Arrays by a Liquid Phase Deposition Method. *Journal of Alloys and Compounds*, Vol. 373, pp. 312-315.

Yang, J.; Jin, Z.; Wang, X.; Li, W.; Zhang, S.; Guo, X. & Zhang, Z. (2003). Study on composition, structure and formation process of nanotube $Na_2Ti_2O_4(OH)_2$. *Journal of the Chemical Society Dalton Transactions*, Vol. 3, pp. 3898-3901.

Zhu, H. Y.; Lan, Y.; Gao, X. P.; Ringer, S., P.; Zheng, Z. F.; Song, D. Y. & Zhao, J. C. (2005). Phase Transition between Nanostructures of Titanate and Titanium Dioxides via Simple Wet-Chemical Reactions. *Journal of the American Ceramic Society*, Vol. 127, pp. 6730-6736.

3

Utilization of Nanoparticles Produced by Aqueous-Solution Methods – Formation of Acid Sites on CeO$_2$-TiO$_2$ Composite and 1-D TiO$_2$ for Dye-Sensitized Solar Cells

Motonari Adachi et al.[*]

Fuji Chemical Co., Ltd., 1-35-1 Deyashikinishi-Machi, Hirakata,
Japan

1. Introduction

Nanoparticles with well-defined nanostructures with unique physical properties are assembled into optoelectronic (Colvin et al. 1994), and nano electronic (Fuhrer et al. 2000) devices and other functional materials (Morris et al. 1999). Highly crystallized nanoparticles can be produced by aqueous-solution methods which provide low cost and ease of fabrication.

In this chapter two utilizations of nanoparticles are presented. First one is formation of acid sites on CeO$_2$-TiO$_2$ composite. Cerium dioxide has an unusual ability to shift easily between the reduced and oxidized states (Ce^{3+} $\rightleftarrows$ Ce^{4+}). This ability coupled with a high oxygen transport capacity gives a unique property of catalysis. Based on the remarkable properties of cerium dioxide, catalytic activity of nanoscale composite of CeO$_2$-TiO$_2$ was studied with variation in composition and formation temperature, which brought change in the number of Lewis acid site together with morphological changes.

The second one is 1-D TiO$_2$ for dye-sensitized solar cells (DSSCs). We succeeded in the preparation of titania nanorods (Jiu et al. 2006), network structure of titania nanowires (Adachi et al. 2004) and one-dimensional titania nanochains. All cells composed of these highly crystallized 1-dimensional titania nanoscale materials (1DTNM) show high power

---

[*] Keizo Nakagawa[2], Yusuke Murata[3], Masahiro Kishida[4], Masahiko Hiro[5], Kenzo Susa[6],
Jun Adachi[7], Jinting Jiu[8] and Fumio Uchida[1]
[1]*Fuji Chemical Co., Ltd., 1-35-1 Deyashikinishi-machi, Hirakata, Japan*
[2]*Department of Advanced Materials, Institute of Technology and Science, The University of Tokushima,*
Minami-josanjima, Tokushima, Japan
[3]*Toyo Tanso Co., Ltd., 5-7-12 Takeshima, Nishiyodogawa-ku, Osaka, Japan*
[4]*Graduate School of Engineering, Kyushu University,744 Motooka, Nishi-ku, Fukuoka, Japan*
[5]*Hitachi Chemical Co., Ltd., 2-1-1 Nishishinjuku, Shinjuku, Tokyo, Japan*
[6]*Trial Corporation., 2-195 Asahi, Kitamoto, Japan*
[7]*National Instituite of Biomedical Innovation, 7-6-8 Asagi Saito, Ibaraki, Japan*
[8]*The Institute of Scientific and Industrial Research (ISIR), Osaka University, 8-1 Mihogaoka, Ibaraki, Japan*

conversion efficiency about 9 %. We also present necessity of 1DTNM for attainment high efficiency theoretically based on the consideration of electron transport processes in the titania electrode and then present that it is indispensable to use highly crystallized 1DTNM for attainment of higher efficient DSSCs based on the analysis of experimental results obtained by electrochemical impedance spectroscopy (EIS) and $I-V$ measurements.

2. CeO$_2$-TiO$_2$ composite as a catalyst

Ceria-based materials are major compounds of the rare earth family, and these have been extensively studied and found application as ultraviolet absorbers (Masui et al., 1997, 2000), solid electrolytes (Inaba & Tagawa, 1996), so-called three-way catalysts for automotive exhaust catalysts (Bekyarova et al., 1998), and soot oxidation catalysts (Pisarello et al., 2002; Aneggi et al., 2006). Nanocrystalline ceria materials have received much attention owing to their physical and chemical properties, which are markedly different from those of the bulk materials. Of particularly interest, the electronic conductivity of CeO$_2$ can be enhanced four orders of magnitude when its microstructure is changed from the micro- to nanocrystalline region (Chiang et al., 1996). Various aqueous solution-based methods for synthesizing crystallized CeO$_2$ nanoparticles (Masui et al., 2002a; Hirano et al., 2000; Li et al., 2001; Wu et al., 2002; Zhou et al., 2003; Bumajdad et al., 2004) and 1D, 2D and 3D CeO$_2$ nanostructures with different morphologies (Vantomme et al., 2005, Zhou et al., 2005, Kuiry et al., 2005, Ho et al., 2005; Han et al., 2005; Sun et al., 2006; Zhong et al., 2007) have been investigated. Some of the properties of these materials, such as the dispersibility of the particles (Masui et al., 2002a) and their catalytic properties (Masui et al., 1997; Sun et al., 2006; Zhong et al., 2007) have also been studied.

The features of CeO$_2$ in these applications are mainly due to the unique combination of its elevated oxygen transport capacity, coupled with its ability to shift easily between the reduced and oxidized states (Ce^{3+}↔Ce^{4+}). To increase the temperature stability and ability of ceria to store and release oxygen, other transition and non transition metal ions (such as Al^{3+}, Si^{4+}, Ti^{4+} and Zr^{4+}) are normally introduced into the ceria cubic structure (Reddy et al., 2003, 2005; Rynkowski et al., 2000; Masui et al., 2002b). The redox and catalytic properties of CeO$_2$ are strongly influenced when it is combined with other transition metals. In addition, when the particle size is decreased below 100 nm, the materials become nanophasic, where the density of defects increases, such that up to half (50%) of the atoms are situated in the cores of the defects, promoting fast catalyst activation and reaction kinetics (Reddy et al., 2005). Thus, a study of the synthesis and reaction characteristics of nano-sized ceria-based mixed oxides is very important for utilizing the oxygen transport capacity and redox properties. One of the main disadvantages of ceria-based nanoparticles prepared in aqueous solution, however, is the resultant hard agglomeration of the fine particles, which has posed a major challenge to the realization of the full potential of nanocrystalline CeO$_2$ powders.

In this section, first we present the preparation of cubic CeO$_2$ nanoparticles using an alkoxide-primary amine surfactant in an aqueous solution and the existence of a clear potential to make 1D, 2D or 3D CeO$_2$ materials by assembling cubic-shape CeO$_2$ nanoparticle building blocks. Amine surfactant works as a colloidal stabilizer through the adsorption on the CeO$_2$ nanoparticles. Second, the preparation of CeO$_2$-TiO$_2$ nanocomposite nanostructures is presented. The morphologies and redox reactivities of CeO$_2$-TiO$_2$

composite nanostructures are influenced by changing the mole ratio of cerium/titanium alkoxides and by changing the calcination temperature.

2.1 Preparation of CeO₂ nanoparticles and CeO₂-TiO₂ composite nanostructure

The preparation method of CeO_2 nanoparticles and CeO_2-TiO_2 composite are based on the aqueous solution system including metal alkoxides and amine surfactant molecules. The experimental procedure has been described in detail in our previous papers (Murata & Adachi, 2004; Nakagawa et al., 2007). The typical synthesis was as follows: first, laurylamine hydrochloride (LAHC) was dissolved in distilled water. Cerium tri-isopropoxide (CTIP) or cerium *n*-butoxide (CeBu) was used as a cerium source. Tetraisopropyl orthotitanate (TIPT) was used as a titanium source. In the synthesis of CeO_2 nanoparticles, CTIP or CeBu was mixed with acetylacetone (ACA) in a beaker and immediately added to an aqueous LAHC solution at pH 4.6. In the case of the synthesis of CeO_2 – TiO_2 composite nanostructures, the mole ratio of CeBu to TIPT (CeBu/TIPT) was changed to 100/0, 75/25, 25/75 and 0/100. Each mixed alkoxide solution was mixed with acetylacetone. In all cases, the mole ratio of metal alkoxides to ACA and metal alkoxides to LAHC were 1 and 4, respectively. After stirring at room temperature for 1 h, the reaction temperature was then changed to 353 K. When the two solutions were mixed, precipitation occurred immediately. After 1 week, the precipitates were separated by centrifugation. After washing with 2-propanol and successive centrifugation, the obtained products were dried through a combination of freeze-drying and vacuum drying, and calcined in air at different temperatures.

2.2 Cubic CeO₂ nanoparticles and their assembled structures

The formation yield of CeO_2 particles for the surfactant assisted-process was 100% approximately. The structure of CeO_2 nanoparticles was studied by TEM image of CeO_2 sample in a dried state. During the formation process, we observed systematic changes in color of the precipitated particles. After mixing of the solution of CTIP or CeBu modified with ACA with the aqueous solution of LAHC, the brown transparent original solution immediately became dark brown. The white colloidal suspension was formed after stirring for 1 h at room temperature. A brown and clear supernatant was formed after the precipitation. Further color change of the precipitate was observed. First, the color of the CeO_2 particles changed from white to dark blue in about 1 day at 353 K. Subsequently, the color of the precipitate gradually turned into pale purple for 1 week, but the color change was slower than the first change. Moreover the wet centrifuged precipitate appeared dark blue, and the freeze-dried powders were gray. But final CeO_2 particles calcined at 673 K was light yellow. These changes in color were observed in the cases that the particles were synthesized in LAHC surfactant aqueous solution at pH 4.2. On the other hand, there was no color change of the precipitate without LAHC surfactant. These color changes are related to the valence state of the Ce; most likely purple corresponds to Ce^{3+} and yellow corresponds to Ce^{4+}. Therefore, it is clear that the Ce^{3+} oxide is stabilized by existence of LAHC in aqueous solution.

We succeeded in the preparation of CeO_2 nanoparticles with cubic structures and 1D, 2D or 3D CeO_2 nanostructures by assembling the cubic-shape CeO_2 nanoparticle building blocks (Murata & Adachi, 2004, Nakagawa et al., 2007) as shown in Fig. 1, 2 and 3. It is evident from this figure that the particle shape was square, and the particle size was calculated to be 2.7-3.8 nm. Furthermore, it seems that the particles were aligned. TEM image of Figure 1a clearly showed the mono-dispersed CeO_2 nanoparticles. The inset picture shows the SAED pattern and Debye-Scherrer rings of the nanoparticles, which can be indexed as those of cerium oxide with the cubic fluorite structure. The HRTEM images and FFT pattern as shown in Figure 1b show that the CeO_2 cubic nanoparticles had a single crystalline structure and high crystallinity; these lattice images were observed for many particles. The main lattice spacing of the crystalline structure was calculated to be 3.11 Å according to FFT analysis. This lattice spacing corresponds to the (111) planes of CeO_2 with a cubic phase, which coincides with the SAED analysis.

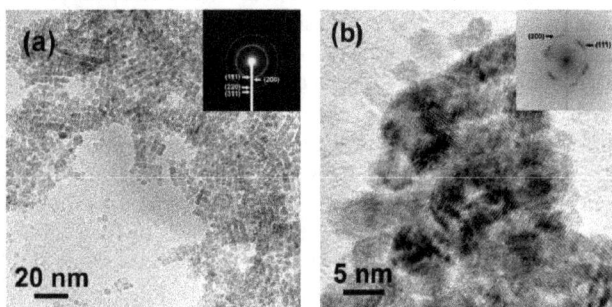

Fig. 1. (a) Low-magnification TEM images of the freeze-dried CeO_2 nanoparticles prepared at 353 K for 1 week. Inset: SAED pattern. (b) High-Resolution TEM images of the aggregated CeO_2. The lattice images were observed. Inset: FFT pattern obtained from HRTEM.

1D rod-like CeO_2 structures are obtained after calcination at 673 K. Rod-like CeO_2 with diameters of 30 nm and lengths of 180 nm are observed although the majority of CeO_2 samples were assembled into aggregates as shown in Fig.2. The HRTEM image show that the principal axis of the crystal growth of CeO_2 was aligned along the rod axis.

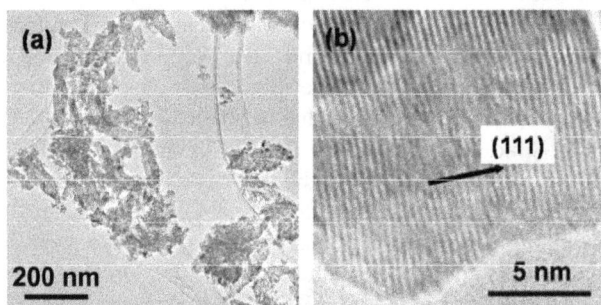

Fig. 2. (a) TEM images of CeO_2 calcined at 673 K for 4 h. (b) HRTEM image of rod-like CeO_2 with a clearly lattice image of (111) planes (d = 3.11 Å).

An ordered structure (2D or 3D superlattice-like structure) are also obtained from the freeze-dried CeO_2 nanoparticles. Figure 3 shows an array of cubic nanocrystals with a mean inter-particle (center-to-center) distance of 2.9 nm, as determined from direct imaging and the FFT pattern. We believe this assembly with an ordered structure is formed to minimize the total surface energy, which is attained by the association of the cubic CeO$_2$ with a face-to-face structure.

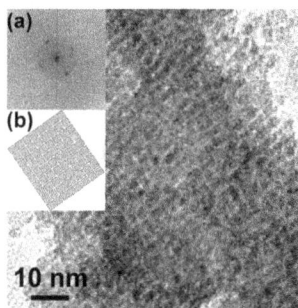

Fig. 3. TEM image of CeO$_2$ nanocrystals self-assembled into a superlattice-like arrangement with dimensions of the order on the nano-scale, Inset a: the FFT pattern confirms the orientational order of the superlattice-like structure, Inset b: the model structure demonstrates the superlattice-like assembled CeO$_2$ nanoparticles.

The thermodynamics of hydrolysis and condensation depend on the strength of the entering nucleophile and electrophilicity of the metal, and on the partial charge. Transition metals are very electropositive, and the hydrolysis and condensation kinetics of the transition metal alkoxides are affected by the positive partial charge δ^+ (Livage et al., 1988). Positive partial charge δ^+ for metals in various alkoxides have been reported; for example, cerium alkoxide: 0.75, titanium alkoxides: 0.63, and silicon alkoxides: 0.32. Since a large positive partial charge corresponds to a rapid reaction rate, the precursor for the complex formed from cerium alkoxide was not generated gradually, but the nano-sized particles were formed by the rapid hydrolysis and condensation reactions. In our systems using LAHC and CTIP (or CeBu) modified with ACA, the resulting suspensions of CeO$_2$ nanoparticles were exceptionally mono-dispersive without aggregation, demonstrating the high power of LAHC as a colloidal stabilizer through the adsorption of LAHC on the surface of the CeO$_2$ nanoparticles, in accordance with the results of Sugimoto et al. who reported the effect of primary amines as shape controllers for the synthesis of TiO$_2$ (Sugimoto et al., 2003). Since the shape of the CeO$_2$ particles is nearly cubic even if the cubic shape has somewhat rounded edges and corners, the LAHC would control the morphology of the CeO$_2$ particles.

For hydrous oxides in aqueous solution systems, the charge-determing ions are H$^+$ and OH$^-$, which establish the charge on the particles by protonating or deprotonating the MOH bonds on the surface of the particles.

$$M\text{-}OH + H^+ \rightarrow M\text{-}OH_2^+ \tag{1}$$

$$M\text{-}OH + OH^- \rightarrow M\text{-}O^- + H_2O \tag{2}$$

The ease of protonation and deprotonation on the surface of the oxide depends on the metal atom. The pH at which the particle is neutrally charged is called the point of zero charge (PZC). At pH > PZC, Eq. 2 predominates, and the particle is negatively charged, whereas at pH < PZC, Eq.1 makes the particle positive. Value of the PZC for CeO$_2$ particles is 8.1 (De Faria and Trasatti, 1994). The magnitude of the surface potential depends on the departure of the pH from the PZC, and that potential attracts oppositely charged ions that present in the solution. Therefore, at pH 4.2, the hydrolyzed and condensed CeO_2 particle is positively charged. LAHC molecules also have a positively charged amine group under acidic condition. Hence, there seems to be no driving force for adsorption by electrostatic attraction. However, chloride ion (Cl^-) mediates the interaction between the laurylamine surfactant and charged CeO_2 by weak H-bonding forces, and CeO_2 particles are covered by surfactant molecules, resulting in the formation of cube crystals. Since the adsorption of LAHC takes place to a specific crystal face, anisotropic structures such as cubes would be formed.

2.3 Morphology of CeO₂-TiO₂ composite

A few studies on CeO_2 - TiO_2 composite nanoparticles (Reddy et al., 2003, 2005; Rynkowski et al., 2000; Masui et al., 2002b) have been reported. Reddy et al. obtained CeO_2 - TiO_2 composites comprised of relatively larger nanocrystals of CeO_2 and TiO_2 (anatase), and some overlapped regions (Reddy et al., 2005). Rynkowski et al. studied the redox properties of CeO_2 - TiO_2 composites (Rynkowski et al., 2000), and stated the existence of the CeO_2 - TiO_2 composite. Masui et al. also synthesized CeO_2 - TiO_2 composite nanoparticles, and reported the deactivation of the thermal and photocatalytic properties of this species by the formation of the CeO_2 - TiO_2 composite (Masui et al., 2002b).

We also studied the preparation of CeO_2-TiO_2 composite nanostructures by changing the mole ratio of cerium/titanium alkoxides and found the effective redox reactivities of CeO_2-TiO_2 composite nanostructures (Nakagawa et al., 2007). During the synthesis of CeO_2-TiO_2 composite, the reaction behavior of each solution was observed. When the surfactant solution and metal alkoxide solutions were mixed, precipitation occurred immediately. When the mole ratio of CeBu/TIPT = 75/25 and 25/75, dark brown-gels and dark purple-precipitates formed, while purple-precipitates with a transparent liquid layer were observed at the mole ratios of CeBu/TIPT = 100/0, that was the same behavior using CTIP. The morphology and crystalline structure of the CeO_2-TiO_2 composite nanostructures varied according to the change in the mole ratio of CeBu to TIPT. When CeBu/TIPT was 75/25, the nano-network structure with a diameter of 3–9 nm was observed and the SAED pattern indicated a cubic fluorite structure (Figure 4). Whereas, when CeBu/TIPT was 25/75, aggregate structures of rod-like morphology with an average diameter of 20 nm and length of 80 nm were observed, and the SAED pattern showed several spots corresponding to the lattice plane of the anatase phase of TiO_2 (Figure 5). In the case of the synthesis with only TIPT, a TiO_2 nano-network structure of connecting nanowires with diameter of 5–15 nm formed by an oriented attachment mechanism (Adachi et al., 2004, Nakagawa et al., 2005).

Figure 6 shows the variation in XRD patterns of the CeO_2-TiO_2 composite calcined at 673 K for 4 h (Nakagawa et al., 2007). The peaks at CeBu/TIPT = 100/0 are sharp and can be indexed to a CeO_2 cubic fluorite structure. When CeBu/TIPT was 75/25, the XRD peaks were indexed to a CeO_2 cubic fluorite structure, although the peaks became very broad. The reason for the broad peak is due to the formation of composite materials. In the HRTEM image shown in Figure 4b, the lattice image of the (111) plane of the cubic fluorite structure could be observed. These observations indicate that the crystalline structure of the nano-network at CeBu/TIPT = 75/25 consists of a CeO_2 cubic fluorite structure, which is different

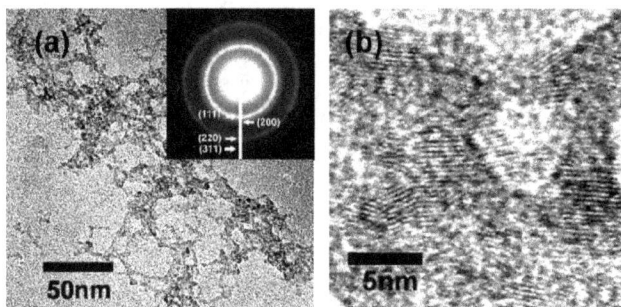

Fig. 4. (a) TEM and (b) HRTEM image of CeO$_2$ - TiO$_2$ composite nanostructures
(CeBu/TIPT = 75/25) after reaction at 353 K for 1 week, inset: SAED patterns.

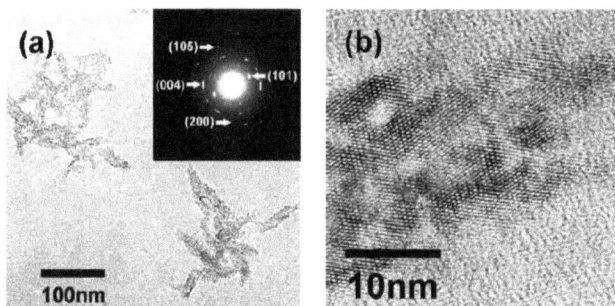

Fig. 5. (a) TEM and (b) HRTEM image of CeO$_2$ - TiO$_2$ composite nanostructures
(CeBu/TIPT = 25/75) after reaction at 353 K for 1 week, inset: SAED patterns.

from that of pure CeO$_2$. No TiO$_2$ anatase peaks were observed. Therefore, the formed
materials under CeBu/TIPT = 75/25 constitute the composite materials of CeO$_2$ and TiO$_2$,
i.e., the formed materials are not a simple mixture of pure CeO$_2$ and TiO$_2$. As a characteristic
of our reaction system, the initial solution, including the two metal alkoxides is uniformly
well mixed on a molecular scale, easily leading to the formation of composite materials.
Since the positive partial charge δ^+ of cerium alkoxide is larger than that of titanium
alkoxide, as mentioned above, it is inferred that the reaction rate of CeBu is faster than TIPT.
Moreover, the content of cerium is much higher than titanium. From these facts, the
crystalline structure of the composite materials is inferred as a CeO$_2$ cubic fluorite structure,
which is different from that of pure CeO$_2$. The different crystalline structure creates a new
morphology, i.e., a nano-network structure, which also leads to the formation of Lewis acid
sites, as described later. The XRD patterns at CeBu/TIPT = 25/75 show mainly broad peaks
of the TiO$_2$ anatase phase and also show a broad peak of CeO$_2$ around $2\theta = 30°$. The broad
peaks indicate the formation of composite materials, which lead to a nanorod structure.
Since the content of titanium is much higher than cerium, the main crystalline structure
corresponds to the TiO$_2$ anatase phase, but a small amount of CeO$_2$ crystalline structure is
also included, because the reaction rate of CeBu is faster than TIPT.

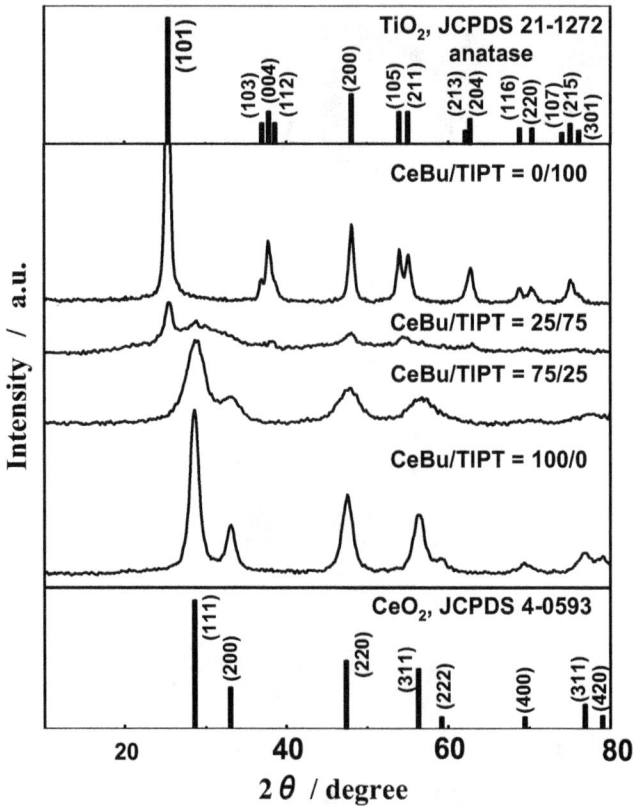

Fig. 6. XRD patterns of the CeO_2–TiO_2 composite nanostructures at the mole ratio of CeBu/TIPT = 100/0, 75/25, 25/75 and 0/100 after calcination at 673K for 4 h.

2.4 Surface properties of CeO_2-TiO_2 composites

The reaction activity of CeO_2-TiO_2 composite nanostructures was investigated through the formation rate of I_3^-, formed due to the oxidation of I^- to I_2 in excess KI aqueous solution (Nakagawa et al., 2007). Nanostructured CeO_2-TiO_2 (10 mg) was suspended by magnetic stirring in 10 ml of 0.2 M KI aqueous solution without light irradiation. After initiation of the reaction, 0.3 ml of the reaction solution was taken, and the concentration diluted to one tenth. The concentration of I_3^- was measured using a Shimadzu UV-2450 spectrometer from the absorbance at 288 nm. Figure 7 shows the I_3^- formation results of CeO_2-TiO_2 composite nanostructures after calcination at 673 K. It was found that CeO_2 nanoparticles and CeO_2-TiO_2 composite nanostructures have the ability to oxidize I^- to I_2 although the TiO_2 nanostructure shows little activity. The activity of the CeO_2-TiO_2 composite nanostructure reaches a maximum at CeBu/TIPT = 75/25 at 623 K.

It is known that cerium oxide shows a high oxidation ability and oxygen storage capacity, and the appearance of these functions is attributed to the following two reasons. One is the redox couple Ce^{3+}/Ce^{4+}, which shows the ability of cerium oxide to shift between CeO_2 and

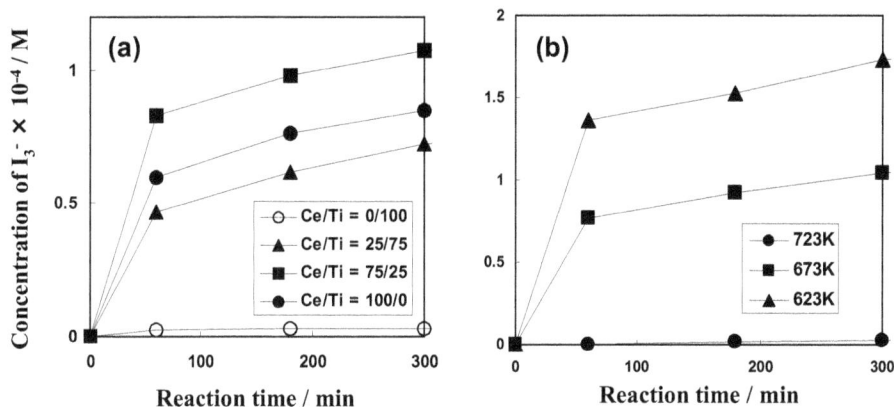

Fig. 7. The variation of concentration of I$_3$- with reaction time, (a) the effect of the mole ratio of CeBu/TIPT calcined at 673 K, (b) the effect of calcination temperature at the mole ratio of CeBu/TIPT = 75/25.

Ce$_2$O$_3$ under oxidizing and reducing conditions, respectively. Another is its structure: the stable structure of cerium oxide at room temperature under atmospheric pressure is the cubic fluorite structure in which oxygen ions do not have a close-packed structure. Owing to this structure, cerium oxide can easily form many oxygen vacancies while maintaining the basic crystal structure (Reddy et al., 2003). Cerium has a family of related mixed-valency binary oxides, which are anion-deficient and fluorite-related Ce$_2$O$_{2n-2m}$ between Ce$_2$O$_3$ and CeO$_2$ at lower temperatures (Kang & Eyring, 1997). It is considered that many vacant oxygen sites exist in cerium oxide; the cerium cation (Ce^{n+}) acts as the Lewis acid site and robs the electron of I-. Additionally, the number of Lewis acid sites could be altered by changing the composition of the CeO$_2$-TiO$_2$ composite, because mixed oxides, e.g., SiO$_2$-TiO$_2$ composites, have been frequently reported to exhibit higher catalytic activity than the pure metal oxide (Méndez-Román & Cardona-Martínez, 1998; Hu et al., 2003). As pointed out above, the uniformly mixed solution of the metal alkoxides led to homogeneously mixed composite oxides on the atomic scale in our preparation method.

We confirmed the formation and number of Lewis acid sites from the pyridine adsorption on the surface of the CeO$_2$-TiO$_2$ composite nanostructure (Nakagawa et al., 2007) as shown in Figure 8. In the results of IR spectra, two peaks at 1620 and 1350 cm^{-1} were assigned to the antisymmetric and symmetric stretching vibrations of the carboxyl group, respectively. A peak at 1595 cm^{-1} and two peaks at 1480 and 1440 cm^{-1} were observed, and these peaks were assignable to hydrogen-bonded pyridine and pyridine bonded to a Lewis site, respectively (Zaki et al., 1989, 2001). It was found that Lewis acid sites evidently exist in the CeO$_2$-TiO$_2$ composite nanostructures and these results show a good correlation between the reaction activity (Figure 7a) and the peak area as determined from the Lewis acid sites (Figure 8).

2.5 Conclusions of 2nd section

1. The preparation method of cubic CeO$_2$ nanoparticles using an alkoxide-primary amine surfactant in an aqueous solution was presented. In additoion, a clear potential to make

1D, 2D or 3D CeO_2 materials by assembling cubic-shape CeO_2 nanoparticle building blocks was also revealed.

2. CeO_2-TiO_2 composite nanostructures could be prepared by changing the mole ratio of cerium/titanium alkoxides. The morphology and crystalline structure of the CeO_2-TiO_2 composite nanostructures were influenced with the mole ratio of the metal alkoxides. These composite nanostructures showed effective reaction activity to oxidize I- to I_2 because of the formation of the Lewis acid sites.

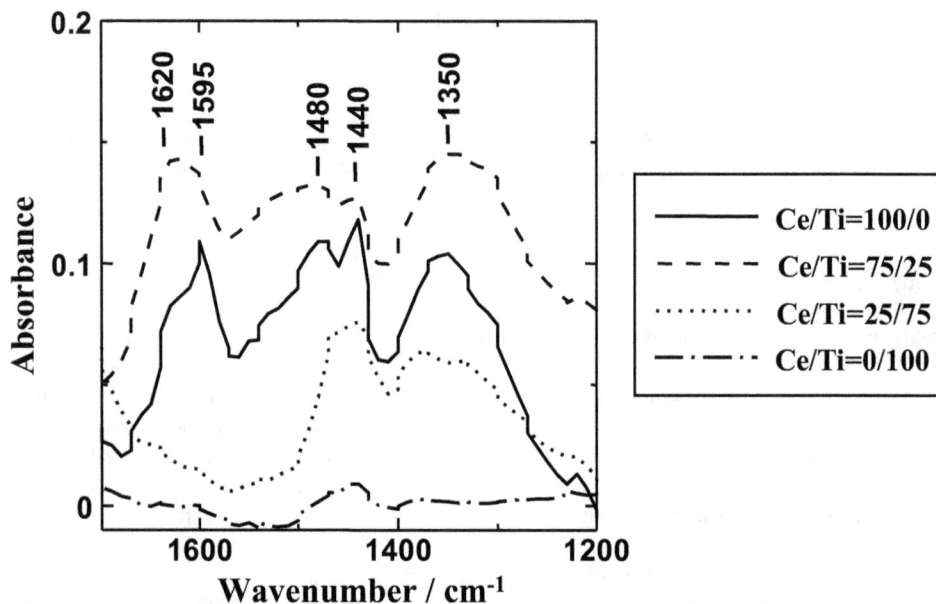

Fig. 8. Pyridine adsorption results (at room temperature) of the CeO_2-TiO_2 composite nanostructure at the different mole ratio of CeBu/TIPT calcined at 673 K.

3. 1-D TiO$_2$ for dye-sensitized solar cells

Dye-sensitized solar cells (DSSCs) have attracted much attention as they offer the possibility of extremely inexpensive and efficient solar energy conversion, because light from the sun is the ideal source of energy, and the supply of energy is gigantic, i.e., 3×10^{24} J/year or about 10^4 times more than what mankind consumes currently. In 1991, O'Regan and Grätzel (O'Regan & Grätzel 1991) published a remarkable report, and the Grätzel group attained 10 % efficiency in 1993 (Nazeeruddin et al. 1993). The system already reached conversion efficiency 11.5 % (Chen, C-Y. et al. 2009), and recently even 12.3 % was reported by Grätzel in Hybrid Organic Photovoltaics Conference in Valencia Spain. These conversion efficiencies exceed the level to supply electricity at the rate of home use, i.e., 10 %. Nevertheless, the energy conversion efficiency of the cells for commercial devices has not yet reached the level, which provides lower cost than that of conventional methods of electricity generation using fossil fuel. Therefore, attainment of higher efficient cells is still one of the most important challenges for the dye-sensitized solar cells.

Utilization of Nanoparticles Produced by Aqueous-Solution Methods – Formation of Acid Sites on CeO$_2$-TiO$_2$
Composite and 1-D TiO$_2$ for Dye-Sensitized Solar Cells

65

Titania dioxide is the most promising material for the electrode of DSSCs. Many investigators have improved the anodic electrode over 10 years (Kim et al. 2009, Ito et al. 2008, Grinis et al. 2008, Hamann et al. 2008, Chen, D. et al. 2009, Miyashita et al. 2008, Wang, M. et al. 2009, Youngblood et al. 2009). One-dimensional titania nanoscale materials (1DTNM) have been investigated for attainment of highly efficient solar cells (Colodrero et al. 2009, Kar et al. 2009, Shankar et al. 2009, Wang, D. et al. 2009, Kang, T-S. et al. 2009, Kuang, D. et al. 2008, Shankar et al. 2008, Adachi et al. 2004, Jiu et al. 2006). In this section we present the clear reason for necessity of 1DTNM for attainment of higher efficient dye-sensitized solar cells through theoretical consideration and based on the experimental evidences verifying the consideration.

First we present necessity of 1DTNM theoretically based on the consideration of the electron transport processes obtained from electrochemical impedance spectroscopy (EIS), together with I−V measurement of the same cell. We present then experimentally that it is indispensable to use highly crystallized 1DTNM for attainment of higher efficient DSSCs based on the analysis of experimental results obtained by EIS and I−V measurements. Also we present that all electrodes composed of our three kinds of 1DTNM showed high light-to-electricity conversion efficiency around 9%.

3.1 Experimental procedure

In order to elucidate the relationship between the composition of titania thin film electrode and performance of the electrode, we made three kinds of electrodes, i.e., an electrode made of P-25 only, an electrode made of P-25 with polyethylene glycol (PEG) and an electrode made of network structure of titania nanowires (TNW) mixed with P-25 with PEG (TNW 28%) first. Since the electrode containing TNW was the best one, we made electrodes made of various amount of TNW mixed with the mixture of P-25 and PEG. The percentage of TNW to (TNW + P-25) in Ti atom content was varied from 0 % to 100 %. Furthermore, we made DSSCs with electrodes composed of all our three kinds of 1DTNM.

3.1.1 Synthesis of highly crystallized TiO$_2$ nanoscale materials

The procedure of TiO$_2$ single crystalline nanowires with network structure has been reported in our previous paper (Adachi et al. 2004). The synthesis procedure of highly crystallized titania nanorods has been described in our previous paper (Jiu et al. 2006). The procedure of titania nanochains is almost the same as that of titania nanorods, except usage of HCl instead of ethylenediamine (EDA) to adjust pH values to 1.3 to 5. Titania nanochains can be synthesized using P123 (triblock copolymer of (poly(ethylene oxide)$_{20}$-poly(propylene oxide) $_{70}$-poly(ethylene oxide)$_{20}$) instead of F127 (triblock copolymer of (poly(ethylene oxide)$_{106}$-poly(propylene oxide) $_{70}$-poly(ethylene oxide)$_{106}$)).

3.1.2 Preparation of titania electrods and dye-sensitized solar cells

We synthesized highly crystallized titania nanoparticles (TNP) with diameter of 3-5 nm other than 1DTNM mentioned above (Jiu et al. 2004, Jiu et al. 2007). Titania electrodes with thin film were made by applying titania samples on an electric conducting glass plate. Fluorine doped tin oxide (FTO) was used as an electric conducting oxide. Dilute solution of

TNP with diameter 3-5 nm was applied on the surface of FTO as a blocking layer. The three kinds of electrodes made of P-25 only, P-25 with PEG and titania nanowire network (TNW) mixed with P-25 with PEG were prepared by coating each gel solution containing these titania materials on the FTO glass by doctor blade method. The gel solution of P-25 only was made by dissolving P-25 powder into water. The aqueous gel solution of P-25 with PEG was made after the procedure reported by Grätzel's group (Nazeeruddin et al. 1993). The gel solution of TNW mixed with P-25 with PEG was made by mixing the gel solution of P-25 with PEG with the reaction products TNW after centrifugation and washing by 2-propanol.

The higher efficient cells constituted with 1DTNM were fabricated as follows. First, the gel solution of TNP with diameter of 3-5 nm was coated three times by doctor blade method on a FTO glass, making 3 layers of TNP. In the case of cells made of TNW, the gel solution of TNW mixed with P-25 with PEG was coated by 8-10 times. The ratio of TNW to P-25 in Ti atom content was around 0.3. In the case of cells made of titania nanorods, the reaction products after centrifugation was mixed with the two gel solutions of P-25 with PEG and the solution of TNP. The mixed gel solution was coated 7-10 times. In the case of titania nanochains, the procedure was the same as the case of titania nanorods.

After each coating, the sample was calcined at 773 K for 10 min. The last calcination was made at 773 K for 30 min. Dye was introduced to the titania thin films by soaking the film $1-3$ days in 3×10^{-4} M solution of ruthenium dye in the mixed solvent of tert-butanol and acetnitryl. Cis-di(thiocyanate) bis(2,2'-bipyridyl-4,4'-di-carboxylate)-ruthenium(II) bis-tetra-butyl-ammonium (N719) (Solaronix SA) produced by Grätzel's group (Nazeeruddin et al. 1993) was used as the dye.

The DSSCs were comprised of a titania thin film electrode on a conducting glass plate, and a platinum electrode made by sputtering on the conducting glass and electrolyte between the titania thin film and the platinum. The composition of the used electrolyte was 0.1 M Guanidium thiocyanate, 0.6 M 1-butyl-3-methylimidazolium iodide, 0.03 M I_2, and 0.5 M TBP (4-tert-butyl pyridine) in the mixed solvent of acetonitrile + n-valeronitrile (volume 85 : 15).

3.1.3 Characterization of titania materials and solar cells

Characterization of the produced materials was made by X-ray diffraction (XRD) (Rigaku Goniometer PMG-A2, CN2155D2), transmission electron microscopy (TEM) (JEOL 200 CX and JEM-2100F), fast Fourier transform (FFT), selected-area electron diffraction (SAED), scanning electron microscopy (SEM) (JEOL JSM 7500FA) and isotherm of nitrogen adsorption (BEL SORP 18 PLUS). The photo-current-voltage characteristics were measured using an AM 1.5 solar simulator (YSS-E40, Yamashita Denso) and in which the light intensity is 100 mW/cm^2 calibrated with a secondary reference solar cell standardized by JET (Japan Electrical Safety & Environmental Technology Laboratories). Electron transport processes were measured by electrochemical impedance spectroscopy (EIS) (Solartron 1255B). The cell size was 0.25 cm^2.

3.2 Necessity of highly crystallized titania nanoscale materials

First, let us consider the reason why highly crystallized one-dimensional titania materials are needed. Fig. 9a shows a typical Nyquist plot obtained by EIS. Total direct current (dc) resistance is given by the length from 0 to the point at ω=0 on the real axis as shown by Fig. 9a. This fact is confirmed later by reproduction of I−V curve using measured total dc resistances at various bias voltages as shown in Fig. 10. Total dc resistance is also obtained from the slope of the tangent line at the point of Voc. (Fig. 9b) When the total dc resistance becomes small, the slope becomes steep, and the fill factor becomes larger, resulting in a high light-to-electricity conversion efficiency. Thus, the total dc resistance should be small.

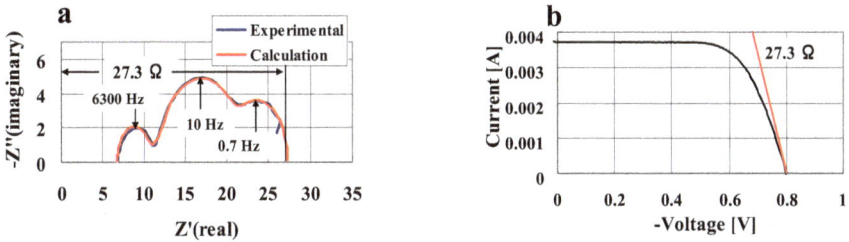

Fig. 9. (a) Typical Nyquist plot obtained by EIS, (b) I−V curve for the same cell

However, the largest arc of around 10 Hz in Fig. 9a represents the resistance of recombination reactions between electrons in the titania electrode and I$_3^-$ ions in the electrolyte. Small total dc resistance means small resistance for recombination reactions, indicating rapid reaction rate of recombination. Thus, small total dc resistance seems an obstacle for attainment of highly efficient solar cells. But, whether electrons in the titania electrode are properly collected by the transparent conducting glass electrode or react with I$_3^-$ ions in the electrolyte by recombination reactions is determined by the ratio of the resistance for the transport rate to the conducting glass electrode against the resistance for the recombination reactions. When the resistance for the transport rate to the conducting glass electrode is much smaller than that of the recombination reactions, almost all electrons are properly collected by the conducting glass electrode. This means that the transport rate of electrons in the titania electrode should be very rapid, indicating that we need nice titania materials with high electron transport rate, i.e., highly crystallized one-dimensional nanoscale TiO$_2$ materials are needed.

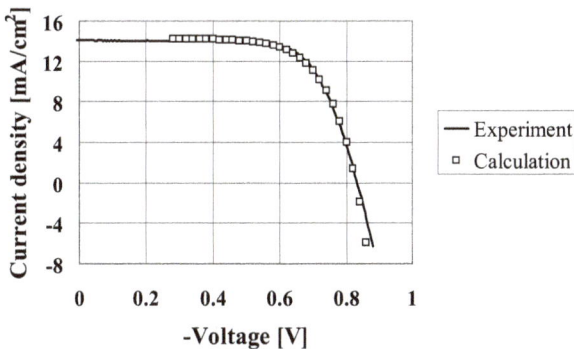

Fig. 10. Reproduction of I−V curve by total dc resistances at various bias voltages.

Solid line in Fig. 10 shows experimentally obtained I−V curve under illumination. The square keys show calculated curve based on the observed total dc resistances at various bias voltages by EIS and the following relationship between current density and voltage,

$$di = \frac{dV}{R_t} \tag{3}$$

where R_t stands for total dc resistance. The calculated curve reproduces experimentally obtained I−V curve very well, confirming that the total dc resistances can be determined accurately from Nyquist plot of EIS analysis.

3.3 Comparison of three kinds of electrodes (P-25 only, P-25 with PEG and TNW mixed with mixture of P-25 and PEG)

Fig. 11 shows I−V curves of the cells made of three kinds of electrodes, i.e., (a) P-25 only, (b) P-25 with PEG and (c) titania nanowire network (TNW) mixed with mixture of P-25 and PEG. The cell made of P-25 only showed the lowest power conversion efficiency (PCE) 4.02 %. PCE of 6.86 % was obtained for the cell made of P-25 with PEG. The highest PCE 8.64 % was obtained for the cell made of TNW mixed with mixture of P-25 and PEG, in which the percentage of titanium atoms of TNW was 28 % for the total titanium atoms, i.e., TNW + P-25. Table 1 shows the current density J_{sc}, open circuit voltage V_{oc}, fill factor FF and power conversion efficiency η of the three kinds of cells.

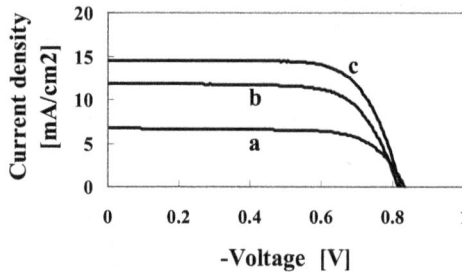

Fig. 11. I−V curves of the cells made of three kinds of electrodes, i.e., (a) P-25 only, (b) P-25 with PEG and (c) titania nanowire network (TNW) mixed with mixture of P-25 and PEG.

	J_{sc} [mA/cm2]	V_{oc} [V]	FF	η [%]
P-25	6.73	0.84	0.72	4.02
P-25 + PEG	11.38	0.83	0.73	6.86
TNW + (P-25 + PEG)	14.56	0.82	0.72	8.64

Table 1. Current density J_{sc}, open circuit voltage V_{oc}, fill factor FF and power conversion efficiency η of the three kinds of cells.

The results of incident photon to current efficiency (IPCE) for the three kinds of cells are shown in Fig. 12. IPCE of the cell made of P-25 only was lowest because of the small amount of dye adsorption. The cell made of TNW with P-25 with PEG showed highest IPCE because

Utilization of Nanoparticles Produced by Aqueous-Solution Methods – Formation of Acid Sites on CeO$_2$-TiO$_2$
Composite and 1-D TiO$_2$ for Dye-Sensitized Solar Cells

69

of the largest amount of dye adsorption. Also IPCE in the range of 600 nm to 700 nm shows shoulder like increase because of the strong scattering of TNW.

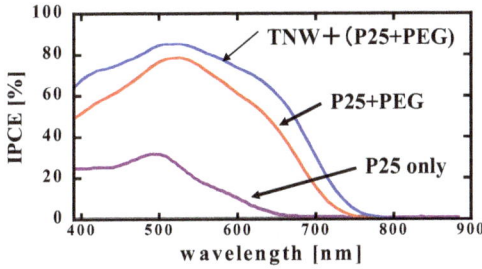

Fig. 12. Results of IPCE for three kinds of cells.

Fig. 13 shows Nyquist plots of the three kinds of cells under open circuit conditions. The total resistance of each cell was obtained as 49 Ω for P-25 only cell, 32 Ω for P-25+PEG cell and 27 Ω for TNW+(P-25+PEG) cell, respectively. Since total resistance corresponds to the slope of the tangent line at Voc, the slope of the tangent line in $I-V$ curves in Fig. 11 became steeper with decreasing of the total resistance of the cell. The plotted squares in Fig. 13 represent experimental results and the solid curves show the calculated spectra from equations (4) to (7) using parameters shown in Table 2 for each cell (Adachi et al. 2006).

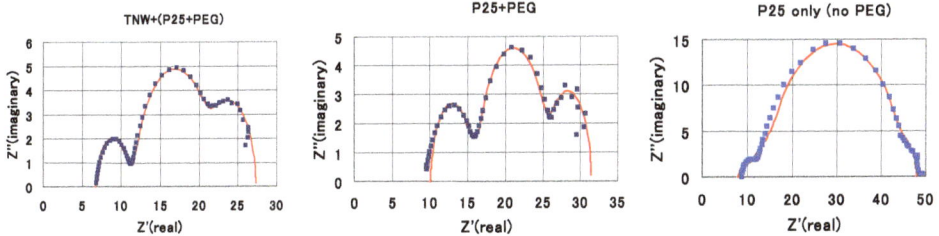

Fig. 13. Nyquist plot of three kinds of cells under open circuit conditions.

Impedance equations for electron transport processes are given as follows (Adachi et al. 2006). For the impedance concerning with titania electrode, equation (4) was derived:

$$Z = R_w \left(\frac{1}{\left(\dfrac{\omega_k}{\omega_d} \right)\left(1 + \dfrac{i\omega}{\omega_k} \right)} \right)^{1/2} \coth\left[\left(\frac{\omega_k}{\omega_d} \right)\left(1 + \frac{i\omega}{\omega_k} \right) \right]^{1/2} \tag{4}$$

where,

$$\omega_d = \frac{D_{eff}}{L^2}, \ \omega_k = k_{eff}, \ R_w = \frac{k_B T}{q^2 A n_s} \frac{L}{D_{eff}} = Con \frac{L}{D_{eff}}, \ R_k = Con \frac{1}{L k_{eff}} \tag{5}$$

For the impedance concerning with platinum electrode, equation (6) was assumed:

$$Z_{Pt} = \frac{1}{1 + i\omega r_{pt} C_{pt}}$$

(6)

where, r_{pt} and C_{pt} represent the resistance at the Pt surface and the capacitance at the Pt surface, respectively. For the impedance concerning with tri-iodide diffusion, finite Warburg impedance equation, i.e., equation (7) was assumed:

$$Z_N = R_{I3-} \frac{1}{\sqrt{\left(D_{I3-} / \delta^2\right)}} \tanh \sqrt{\frac{i\omega}{\left(D_{I3-} / \delta^2\right)}}$$

(7)

The calculated solid curves in Fig. 13 agree quite well with the plotted experimental data. The characteristics shown in Table 2 are following three points which show strong tendency for the highly efficient cells. 1) The resistance for the electron transport from the titania electrode to the conducting glass electrode Rw becomes smaller with increasing conversion efficiency. 2) The ratio of the resistance for the recombination reactions against the resistance for the transport rate to the conducting glass electrode (Rk/Rw) becomes large, and the rate constant of recombination reactions k_{eff} becomes smaller with increasing conversion efficiency. 3) The values of Con, which represents constant inversely proportional to the

	P25 only	P25+PEG	TNW+(P25+PEG)
Con=	2.65	0.28	0.163
$D_{eff}=$	0.0006	0.00014	0.00008
$L =$	0.0042	0.0025	0.002
$k_{eff}=$	23	13.8	10
$R_{I3}=$	5.1	7	7.5
$D_{I3-}=$	0.000015	0.000003	0.000005
$\delta =$	0.005	0.005	0.005
$R_{Pt}=$	1.5	4.7	3.65
$C_{Pt}=$	0.00005	0.00007	0.00005
$R_{sub}=$	7.9	10	6.7
$Rk/Rw=$	1.48	1.62	2
$Rw=$	14.8	5	4.08
$Rk=$	27.4	8.11	8.15
$n=$	2.42×10^{17}	2.29×10^{18}	3.94×10^{18}

Where, Con=$k_B T/qAn$ [Ωcms^{-1}], where k_B [JK^{-1}] represents Boltzmann constant, T [K] is absolute temperature, q [C] is elementary charge, A [cm^2] is area of the cell and n [cm^{-3}] is electron density. Con: constant inversely proportional to the electron density, D_{eff} [cm^{2}s^{-1}]: diffusion coefficient of electron, L [cm]: film thickness of TiO$_2$ electrode, k_{eff} [s^{-1}]: reaction rat constant of recombination reactions, R_{I3-} [Ω]: diffusion resistance of I$_3^-$, D_{I3-} [cm^{2}s^{-1}]: diffusion coefficient of I$_3^-$, δ [cm]: thickness of the electrolyte phase, R_{Pt} [Ω]: resistance of Pt electrode, C_{Pt} [F]: capacity of Pt electrode, R_{sub} [Ω]: resistance of substrate, Rw[Ω]: resistance for electron transport in the TiO$_2$ electrode, Rk [Ω]: resistance for recombination reaction.

Table 2. Determined parameters concerning with electron transport by impedance spectroscopy for three kinds of cells

	J_{sc} [mA/cm^2]	V_{oc} [V]	FF	η [%]	thickness [μm]
0wt%	10.93	0.85	0.74	6.85	10
0wt%	11.82	0.82	0.71	6.87	26
0wt%	10.92	0.80	0.71	6.20	35
5wt%	11.98	0.85	0.73	7.48	19
5wt%	10.96	0.81	0.73	6.50	32
10wt%	12.84	0.84	0.73	7.87	18
10wt%	12.68	0.84	0.72	7.64	20
28wt%	14.88	0.82	0.71	8.66	24
28wt%	13.09	0.83	0.74	8.04	27
50wt%	13.39	0.85	0.71	8.02	14
50wt%	15.18	0.80	0.70	8.51	22
100wt%	11.94	0.84	0.73	7.28	5
100wt%	9.93	0.82	0.71	5.75	9
100wt%	10.16	0.84	0.70	5.97	11

Table 3. Performance of dye sensitized solar cells with various TNW content.

electron density, becomes smaller with increasing conversion efficiency, i.e., Con value increases in the order of P-25 only > (P-25+PEG) > TNW+(P-25+PEG). Therefore the electron density n increases with increasing conversion efficiency. So, the characteristics of highly efficient cells are high electron density, small resistance for the electron transport to the conducting glass electrode, and large ratio of the resistances Rk/Rw with small rate constant for recombination reactions.

3.4 Effects of content of TNW on the properties of dye-sensitized solar cells

Since the cells containing TNW gave high conversion efficiencies, we examined the effects of content of TNW in the electrode composed of TNW and P-25 upon the conversion efficiency with variation in TNW from 0 % to 100 %. Content of TNW was defined as percentage of titanium atoms of TNW in the total titanium atoms included in the titania electrode. Table 3 shows performance of DSSCs with various TNW content, i.e., J_{sc}, V_{oc}, FF and η, together with film thickness. Effect of TNW content on PCE is shown in Fig. 14.

PCE of the cells including TNW are higher than those cells without TNW, indicating that TNW is useful to attain high efficiency except 100% TNW case. When the film thickness of 100% TNW cells increased larger than 5 μm, peel off of the films with cracks was observed by SEM images as shown in Fig. 15, resulting that less than 6% of PCE were observed as shown in Table 3 and Fig. 14. Thus, mixing of TNW with P-25 nanoparticles is important to make robust films.

Since the amount of adsorbed dyes is another important factor to affect PCE, the amounts of adsorbed dyes for the cells with various TNW contents are shown against film thickness in Fig. 16. The amount of adsorbed dye in the cells containing TNW from 0 % to 50 % locates in the same straight line regardless of the difference in TNW contents, except 100 % TNW which shows higher adsorbed amounts.

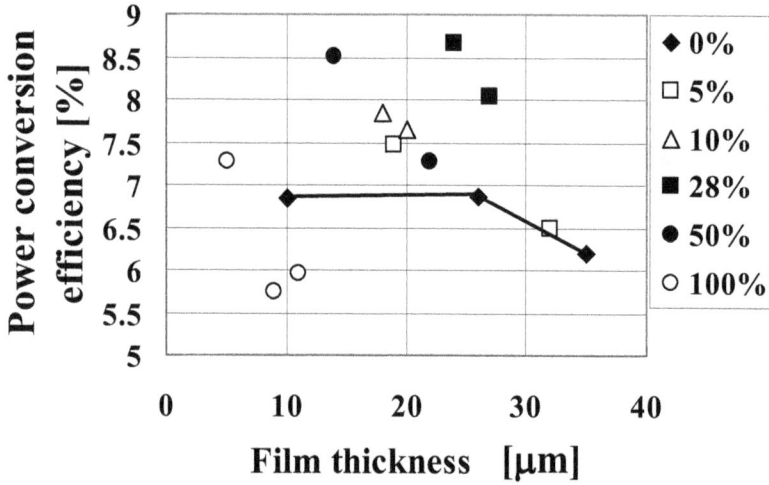

Fig. 14. Effect of content of TNW on power conversion efficiency

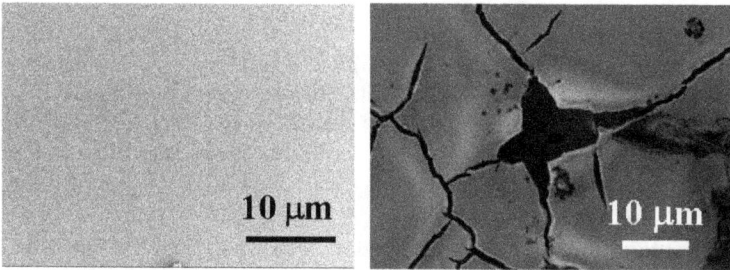

Fig. 15. Top views of 100% TNW films. Left: 5 μm, right: 11μm thickness.

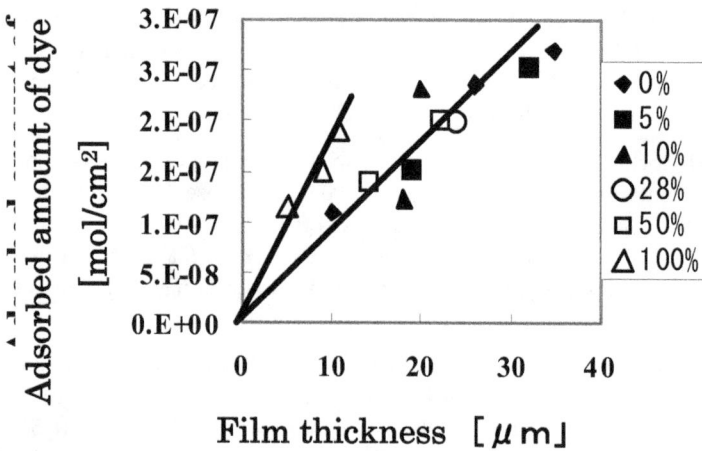

Fig. 16. Relationship between film thickness and the amount of dye.

TNW [%]	0	0	0	5	5	5	10	10	10	10
Con=	0.309	0.397	0.4	0.316	0.257	0.33	0.22	0.224	0.277	0.23
D_{eff} =	0.000038	0.000093	0.00024	0.0000747	0.000095	0.00025	0.00009	0.000085	0.00022	0.00015
L =	0.001	0.00264	0.0035	0.00191	0.00191	0.0032	0.0018	0.0018	0.002	0.002
k_{eff} =	13.8	7.67	13.8	7.67	7.67	13.8	7.67	7.67	16.3	13.8
$R_{D.}$ =	7.7	7	7	7.6	8	6.5	9.1	8.1	5.1	6.4
$D_{D.}$ =	0.00001	0.000004	0.00000295	0.000007	0.0000068	0.00000328	0.0000027	0.0000057	0.0000045	0.0000045
δ =	0.005	0.005	0.005	0.005	0.005	0.005	0.005	0.005	0.005	0.005
R_{pt} =	3.5	6.7	4.8	15	5.8	3.5	4.25	3.4	7	4.6
C_{pt} =	0.00006	0.000046	0.000055	0.0000313	0.0000675	0.00005	0.00005	0.00006	0.00004	0.000065
R_{sub} =	7.96	8.9	9.8	12	11.79	10.4	8.69	8.92	8.5	9.8
Rk/Rw=	2.75	1.74	1.42	2.67	3.4	1.77	3.62	3.42	3.37	2.72
Rw=	8.13	11.3	5.83	8.08	5.17	4.22	4.4	4.74	2.52	3.07
Rk=	22.4	19.6	8.28	21.6	17.5	7.47	15.9	16.2	8.5	8.33
n =	2.17×10^{18}	1.69×10^{18}	1.67×10^{18}	2.11×10^{18}	2.6×10^{18}	2.03×10^{18}	3.04×10^{18}	2.99×10^{18}	2.42×10^{18}	2.91×10^{18}

TNW [%]	28	28	50	50	50	50	100	100	100
Con=	0.165	0.225	0.206	0.217	0.251	0.15	0.161	0.154	0.185
D_{eff} =	0.0001	0.00012	0.00014	0.00018	0.00015	0.000104	0.000072	0.000023	0.00015
L =	0.002	0.002	0.00139	0.00139	0.0022	0.0022	0.000925	0.000505	0.0009
k_{eff} =	10	12.5	18.65	19.9	7.67	5	8	10.32	18
$R_{D.}$ =	5.6	5.8	6.9	6.9	9.5	11.5	14.5	13.4	11
$D_{D.}$ =	0.000007	0.0000055	0.000005	0.0000053	0.000003	0.0000035	0.000011	0.0000096	3.52E-06
δ =	0.005	0.005	0.005	0.005	0.005	0.005	0.005	0.005	0.005
R_{pt} =	2	2.8	3	9.85	5	3.88	4	5.8	5.7
C_{pt} =	0.0001	0.00006	0.000038	0.00006	0.0000577	0.000045	0.000052	0.00005	0.0000433
R_{sub} =	5.6	5.8	9.12	10.1	8.21	8.07	8	10.7	11
Rk/Rw=	2.5	2.4	3.89	4.68	4.04	4.3	10.5	8.74	10.3
Rw=	3.3	3.75	2.04	1.68	3.68	3.17	2.07	3.38	1.11
Rk=	8.25	9	7.94	7.84	14.9	13.6	21.8	29.5	11.4
n =	3.89×10^{18}	2.85×10^{18}	3.25×10^{18}	3.08×10^{18}	2.67×10^{18}	4.46×10^{18}	4.16×10^{18}	4.35×10^{18}	3.62×10^{18}

Table 4. Parameters determined by EIS analysis for the cells with various TNW contents.

This higher adsorption of 100 % TNW is attributed to the smaller diameter of TNW of 3-7 nm, which is much smaller than the diameter of P-25 of 23 nm. The specific surface area of P-25 and the mixture of 28 % TNW with P-25 after calcinations at 773 K for 30 min. were 45 m²/g and 48 m²/g, respectively. These values of specific surface area are much smaller than that of 100 % TNW which is 78 m²/g after calcinations. This difference in specific surface area between 28 % TNW with P-25 and pure 100 % TNW corresponds well to the difference in adsorbed dye amount between from 0 % to 50 % TNW with P-25 and pure 100 % TNW. These findings suggest some interesting structural change in the surface of the mixture of TNW and P-25. However, the reason why the cells containing different TNW content from 0% to 50 % locates in the same straight line in Fig. 16 is not well understood at present.

Resistance for electron transport from titania electrode to the transparent conducting glass electrode Rw are plotted against TNW content in Fig. 17a. Rw values decrease steeply up to 10 % and become gradual decrease after 20 % of TNW content. This decrease indicates clearly that electron transport in the titania electrode is improved by mixing TNW with P-25 nanoparticles.

The ratios of Rk representing the resistance for the recombination reactions between electrons in the titania electrode and I_3^- in the electrolyte to Rw are plotted against TNW content in Fig. 17b. The ratio of Rk/Rw increases with increase in TNW content. This shows that TNW restrains the recombination reactions between electrons and I_3^- and contributes to collect electrons properly to the transparent conducting glass electrode. The findings shown in Fig. 17 a, b bring the high electron density in the titania electrode as shown in Fig. 17c.

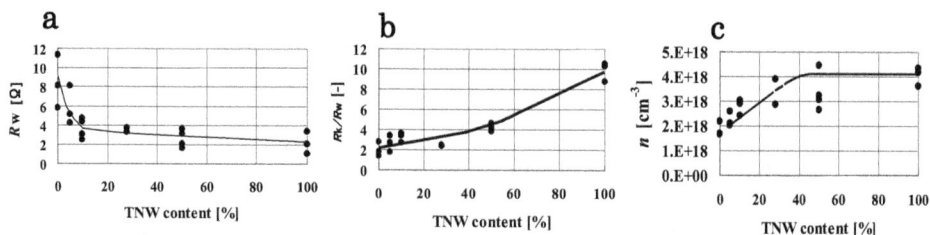

Fig. 17. a) Relationship between Rw and TNW content, b) relationship between Rk/Rw and TNW cointent and c) relationship between electron density and TNW content.

Thus, the conclusion deduced from the experiments of three kinds of cells, i.e., small resistance for the electron transport to the conducting glass electrode, large value of resistance ratio Rk/Rw, and high electron density in the titania electrode as the characteristics of highly efficient cells, was confirmed again by the experiments of variation in TNW content.

3.5 Some examples of our highly crystallized one-dimensional TiO₂ nanoscale materials for fabricating highly efficient dye-sensitized solar cells

We succeeded in the preparation of titania nanorods (TNR) (Jiu et al. 2006), network structure of titania nanowires (Adachi et al. 2004) and one-dimensional titania nanochains (see Fig. 18), which have been newly synthesized. We applied these materials for DSSCs.

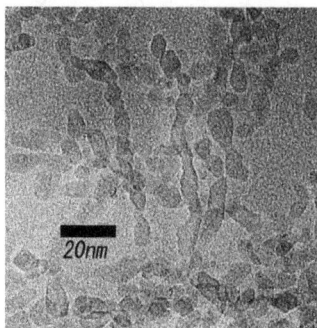

Fig. 18. TEM image of titania nanochains.

We present highly crystallized one-dimensional titania nanoscale materials are effective to attain high light-to-electricity conversion yield. As shown in our previous paper (Adachi et al. 2004), network structure of single crystal-like titania nanowires can be synthesized successfully by the oriented attachment mechanism. We attained 9.33 % conversion

Utilization of Nanoparticles Produced by Aqueous-Solution Methods – Formation of Acid Sites on CeO$_2$-TiO$_2$
Composite and 1-D TiO$_2$ for Dye-Sensitized Solar Cells

75

efficiency with complex titania electrode made of titania nanowires and P-25. Recently, we attained the same conversion efficiency 9.33 % using different electrolyte, i.e., 0.6M 1-butyl 3-methyl imidazolium iodide, 0.1M guanidium thiocyanate, 0.05M I$_2$, 0.5M tert-butylpyridine in a mixture of acetonitrile and valeronitrile (85:15) for a complex titania electrode made of titania nanowires, titania nanoparticles (3-5 nm in diameter) and P-25. (Fig. 19)

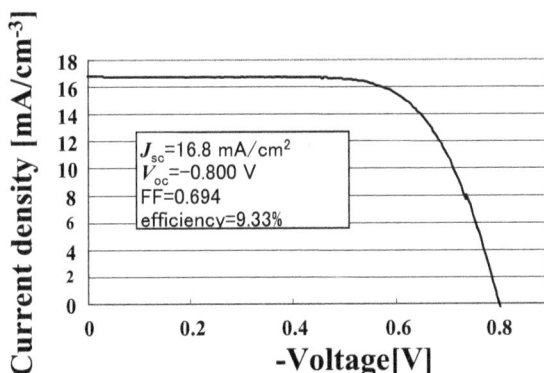

Fig. 19. I−V curve obtained for a cell with a complex electrode composed of network structure of single-crystal-like titania nanowires, titania nanoparticles and P-25

In our previous paper (Adachi et al. 2004), we used an electrolyte composed of 0.1 M of LiI, 0.6 M of1,2-dimethyl-3-n-propylimidazolium iodide, 0.05 M of I$_2$, 1 M of 4-tert-butylpyridine in methoxyacetonitrile and got 9.33 % conversion efficiency with short circuit current density Jsc=19.2 mA/cm^2, open circuit voltage Voc=0.72 V and fill factor 0.675. In the recent results, Voc value 0.8 V is larger than that of previous one 0.72 V, because guanidium thiocyanate decreased redox potential of I$^-$/I$_3^-$ in the electrolyte. Unfortunately, we got lower short circuit current density Jsc=16.8 mA/cm^2 than that of our previous one, and the same efficiency was obtained.

Highly crystallized titania nanorods (TNR) have been synthesized by hydrothermal process using blockcopolymer (F127) and surfactant cetyltrimethylammonium bromide (CTAB) as a mixed template (Jiu et al. 2006). TNR with 100-300 nm in length and 20-30 nm in diameter was obtained. A high-resolution TEM (HRTEM) image of single TNR shows that titanium atoms align perfectly in titania anatase crystalline structure with no lattice defect, and the surface of TNR is facetted with the TiO$_2$ anatase {101} faces (Yoshida et al. 2008). The fringes are {101} planes of anatase TiO$_2$ with a lattice spacing of about 0.351 nm, which agrees with the value recorded in JCPDS card. The highly crystallized titania nanorods prepared successfully were used to fabricate a titania electrode of DSSCs. The complex electrodes were made by the repetitive coating-calcining process: 3 layers of titania nanoparticles (3-5 nm in diameter) were first coated on FTO conducting glass, followed by 8 layers of mixed gel composed of titania nanorods and titania nanoparticles. A high light-to-electricity conversion efficiency of 8.93 % was achieved (Yoshida et al. 2008).

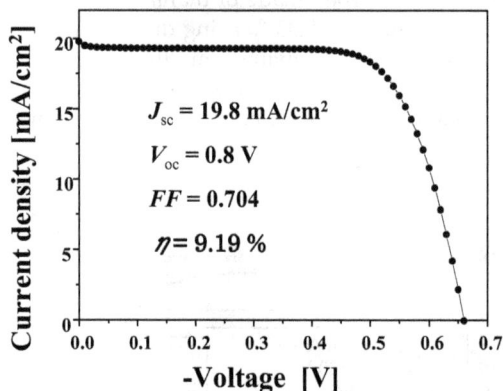

Fig. 20. I−V curve obtained for the cell composed of one-dimensional chains of titania nanoparticles mixed with fine titania nanoparticles (3 - 5 nm in diameter).

We have newly synthesized titania nanochains as shown in Fig. 18. Highly crystallized titania nanoparticles with diameter of around 10 nm combine with each other and make chains. The obtained white solid product was mixed with spherical titania nanoparticles (3-5 nm in diameter) synthesized using F127 reported in our previous paper (Jiu et al. 2004, Jiu et al. 2007) to fabricate titania film electrodes. The I−V curve of the cell is shown in Fig. 20. The obtained light-to-electricity conversion yield of the cell was 9.2%.

All three kinds of one-dimensional titania nanoscale materials mentioned above show high light-to-electricity conversion yield around 9%, suggesting strongly that highly crystallized one-dimensional titania materials are essentially important for attainment of high efficient dye-sensitized solar cells.

3.6 Conclusions of 3rd section

1. Many researchers familiar with EIS measurement know that highly efficient dye-sensitized solar cells show small total resistance of the cell, i.e., small Nyquist spectrum. They also know that largest arc of Nyquist plot represents the resistance for recombination reactions Rk. This apparent conflict is solved clearly by theoretical consideration through recognition that the large value of the ratio Rk/Rw is essentially important for the highly efficient cells, and the absolute value of Rk is not important, i.e., very small Rw is indispensable for the highly efficient cells.

2. The experimental results of I−V and EIS measurements of the three kinds of cells made of P-25 only, P-25+PEG, and TNW+P-25+PEG and also cells made of various content of TNW with P-25+PEG clearly showed the following three points as characteristics of highly crystallized 1-dimensional titania nanoscale material TNW. 1) Resistance of electron transport in the titania electrode Rw is small. 2) The ratio of resistance Rk/Rw is large. 3) Electron density n in the titania electrode is high.

3. All cells composed of three kinds of highly crystallized 1-dimensional titania nanoscale materials, i. e., network structure of titania nanowires, titania nanorods, and titania nanochains, show high power conversion efficiency about 9 %.

4. References

Adachi, M. Murata, Y. Takao, J. Jiu, J. Sakamoto, M. & Wang, F. (2004). Highly Efficient Dye-Sensitized Solar Cells with Titania Thin Film Electrode Composed of Network Structure of Single-Crystal-Like TiO₂ Nanowires Made by "Oriented Attachment" Mechanism. *J. Am. Chem. Soc.*, 126, pp. 14943-14949

Adachi, M. Sakamoto, M. Jiu, J. Ogata, Y. Isoda, S. (2006). Determination of Parameters of Electron Transport in Dye-Sensitized Solar Cells Using Electrochemical Impedance Spectroscopy: J. Phys. Chem. B 110, pp. 13872-13880

Aneggi, E.; de Leitenburg, C. Dolcetti, G & Trovarelli, A. (2006) Promotional effect of rare earths and transition metals in the combustion of diesel soot over CeO₂ and CeO₂–ZrO₂. *Catalysis Today.* 114, pp. 40-47.

Bekyarova, E.; Fornasiero, P.; Kaspar, J. & Graziani, M. (1998) CO oxidation on Pd/CeO₂–ZrO₂ catalysts. *Catalysis Today.* 45, pp. 179-183.

Bumajdad, A.; Zaki, M. I.; Eastoe, J. & Pasupulety, L. (2004) Microemulsion-Based Synthesis of CeO₂ Powders with High Surface Area and High-Temperature Stabilities. *Langmuir.* 20, pp. 11223-11233.

Chen, C-Y., Wang, M., Li, J-Y., Pootrakulchote, N., Alibabael, L., Ngoc-le, C-H., Decoppetr, J-D., Tsai, J-H., Grätzel, C., Wu, C-G., Zakeeruddin, M., & Grätzel, M. (2009) Highly efficient light-harvesting ruthenium sensitizer for thin-film dye-sensitized solar cells. *ACS NANO*, 3, pp. 3103-3109

Chen, D., Huang, F. Cheng, Y-B. & Caruso, R. A. (2009) Mesoporous anatase TiO₂ beads with high surface areas and controllable pore sizes: A superior candidate for high-performance dye-sensitized solar cells. *Adv. Mater.*, 21, pp. 2206-2210

Chiang, Y. M.; Lavik, E. B., Kosacki, I.; Tuller, H. L. & Ying, J. Y. (1996) Defect and transport properties of nanocrystalline CeO₂₋ₓ. *Appl. Phys. Lett.* 69, pp. 185-187.

Colodrero, S., Mihi, A., Häggman, L. Ocana, M. Boschloo, G. Hagfeldt, A., & Miguez, H. (2009) Porous one-dimensional photonic crystals improve the power-conversion efficiency of dye-sensitized solar cells. *Adv. Mater.*, 21, pp. 764-770

Colvin, V. L., Schlamp, M. C., & Alivisatos, A. P. (1994) Light-emitting diodes made from cadmium selenide nanocrystals and a semiconducting polymer Nature, 370, pp. 354-357.

De Faria, L. A. & Trasatti, S. (1994) The Point of Zero Charge of CeO₂. J. Colloid Interface Sci., 167, pp. 352-357.

Fuhrer, M. S., Nygard, J., Shih, L., Forero, M., Yoon, Y. -G., Mazzoni, M. S. C., Choi, H. J., Ihm, J., Louie, S. G., Zettl, A., McEuen, P. L. (2000) Crossed nanotube junctions Science, 288, 494-497.

Grinis, L., Dor, S., Ofir, A., & Zaban, A. (2008) Electrophoretic deposition and compression of titania nanoparticle films for dye-sensitized solar cells. *J. Photochem. Photobio. A: Chemistry*, 198, pp. 52-59

Hamann, T. W., Farha, O. K. & Hupp, J. T. (2008) Atomic layer deposition of TiO₂ on aerogel templates: New photoanodes for dye-sensitized solar cells. *J. Phys. Chem. C*, 112, pp. 19756-19764

Han, W-Q.; Wu, L. & Zhu, Y. (2005) Formation and Oxidation State of CeO₂₋ₓ Nanotubes. *J. Am. Chem. Soc.* 127, pp. 12814-12815.

Hirano, M.; Fukuda, Y.; Iwata, H.; Hotta, Y. & Inagaki, M. (2000) Preparation and Spherical Agglomeration of Crystalline Cerium(IV) Oxide Nanoparticles by Thermal Hydrolysis. *J. Am. Ceram. Soc.* 83, pp. 1287-1289.

Ho, C.; Yu, J. C.; Kwong, T.; Mak, A. C. & Lai, S. (2005) Morphology-Controllable Synthesis of Mesoporous CeO2 Nano- and Microstructures. *Chem. Mater.* 17, pp. 4514-4522.

Hu, S.; Willey, R. J. & Notari, B. (2003) An investigation on the catalytic properties of titania–silica materials. *J. Catal.*, 220, pp. 240-248.

Inaba, H. & Tagawa, H. (1996) Ceria-based solid electrolytes. *Solid State Ionics.*, 83, pp. 1-16.

Ito, S., Murakami, T. N., Comte, P., Liska, P., Grätzel, C., Nazeerudin, M. K. & Grätzel, M. (2008) Fabrication of thin film dye sensitized solar cells with solar to electric power conversion efficiency over 10%. *Thin Solid Film*, 516, pp. 4613-4619

Jiu, J. Isoda, S. Adachi, M. & Wang, F. (2007). Preparation of TiO2 nanocrystalline with 3-5 nm and application for dye-sensitized solar cell. *J. Photochem. Photobio. A: Cemistry*, 189, pp. 314-321

Jiu, J. Isoda, S. Wang, F. & Adachi, M. (2006). Dye-Sensitized Solar Cells Based on a Single-Crystalline TiO2 Nanorod Film. *J. Phys. Chem. B*, 110, pp. 2087-2092

Jiu, J. Wang, F., Sakamoto, M., Takao, J. & Adachi, M. (2004). Preparation of nanocrystal TiO_2 with mixed template and application for dye-sensitized solar cell. *J. Electrochem. Soc.* 151, pp. A1653-A1658

Kang, T-S., Smith, A. P., Taylor, B. E. & Durstock, M. F. (2009) Fabrication of highly-ordered TiO_2 nanotube arrays and their use in dye-sensitized solar cells. *Nano Letter*, 9, pp. 601-606

Kang, Z. C. & Eyring, L. (1997) A compositional and structural rationalization of the higher oxides of Ce, Pr, and Tb. *J. Alloys and Comp.* 249, pp. 206-212.

Kar, A., Smith, Y.R., & Subramanian, V. (2009) Improved photocatalytic degradation of textile dye using titanium dioxide nanotubes formed over titanium wires. *Environmental Sci. and Technol.*, 43, pp. 3260-3265

Kim, Y. J., Lee, M. H., Kim, H. J., Lim, G., Choi, Y. S., Park, N-G., Kim, K., & Lee, W. I. (2009) Formation of highly efficient dye-sensitized solar cells by hierarchical pore generation with nanoporous TiO2spheres. *Adv. Mater.*, 21, pp. 3668-3673

Kuang, D., Brillet, J., Chen, P., Takata, M. Uchida, S. Miura, H., Sumioka, K., Zakeeruddin, S. M. & Grätzel, M. (2008) Application of highly ordered TiO_2 nanotube arrays in flexible dye-sensitized solar cells. *ACS Nano*, 2, pp. 1113-1116

Kuiry, S.; Patil, S.; Deshpande, S. & Seal, S. (2005) Spontaneous Self-Assembly of Cerium Oxide Nanoparticles to Nanorods through Supraaggregate Formation. *J. Phys. Chem. B*, 109, pp. 6936-6939.

Li, J. –G.; Ikegami, T.; Lee, J. –H. & Mori, T. (2001) Characterization and sintering of nanocrystalline CeO_2 powders synthesized by a mimic alkoxide method. *Acta. Mater.* 49, pp. 419-426.

Livage, J.; Henry, M. & Sanchez, C. (1988) Sol-gel chemistry of transition metal oxides. *Prog. Solid St. Chem.*, 18, pp. 259-341.

Masui, T.; Fujiwara, K.; Machida, K.; Adachi, G.; Sakata, T. & Mori, H. (1997) Characterization of Cerium(IV) Oxide Ultrafine Particles Prepared Using Reversed Micelles. *Chem. Mater.*, 9 pp. 2197-2204.

Masui, T.; Yamamoto, M.; Sakata, T.; Mori, H. & Adachi, G. (2000) Synthesis of BN-coated CeO_2 fine powder as a new UV blocking material. *J. Mater. Chem.*, 10, pp. 353-357.

Masui, T.; Hirai, H.; Imanaka, N. & Adachi, G. (2002) Synthesis of cerium oxide nanoparticles by hydrothermal crystallization with citric acid. *J. Mater. Sci. Lett.* 21, pp. 489-491.

Masui, T.; Fukuhara, K.; Imanaka, N.; Sakata, T.; Mori, H. & Adachi, G. (2002) Effects of Titanium Oxide on the Optical Properties of Cerium Oxide. *Chem. Lett.* 31, pp. 474-745.

Méndez-Román, R. & Cardona-Martínez, N. (1998) Relationship between the formation of surface species and catalyst deactivation during the gas-phase photocatalytic oxidation of toluene. *Catal. Today.*, 40, pp. 353-365.

Miyashita, M., Sunahara, K., Nishikawa, T. Uemura, Y., Koumura, N., Hara, K., Mori, A., Abe, T. Suzuki, E. & Mori, S. (2008) Interfacial electron-transfer kinetics in metal-free organic dye-sensitized solar cells: Combined effects of molecular structure of dyes and electrolytes. *J. Am. Chem. Soc.*, 130, pp. 17874-17881

Morris, C. A., Anderson, M. L., Stroud, R. M., Merzbacher, C. I., & Rolison, D. R., (1999) Silica sol as a nanoglue: Flexible synthesis of composite aerogels *Science*, 284, 622-624.

Murata, Y. & Adachi, M. (2004) Formation of highly dispersed cerium oxide with cubic structure prepared alkoxide-surfactant system. *J. Mater. Sci.*, 39, 7397-7399.

Nakagawa, K.; Wang, F.; Murata, Y. & Adachi, M. (2005) Effect of Acetylacetone on Morphology and Crystalline Structure of Nanostructured TiO₂ in Titanium Alkoxide Aqueous Solution System. *Chem. Lett.* 34, pp. 736-737.

Nakagawa, K.; Murata, Y.; Kishida, M.; Adachi, M.; Hiro, M. & Susa, K. (2007) Formation and reaction activity of CeO₂ nanoparticles of cubic structure and various shaped CeO₂–TiO₂ composite nanostructures. *Mater. Chem. Phys.*, 104, pp. 30-39.

Nazeeruddin, M. K., Kay, A., Rodicio, I., Humphry, B. R., Mueller, E., Liska, P., Vlachopoulous, N., & Grätzel, M. (1993) Conversion of light to electricity by cis-X₂bis(2,2′-bipyridyl-4,4′-dicarboxylate)ruthenium(II) charge-transfer sensitizers (X = Cl⁻, Br⁻, I⁻, CN⁻, and SCN⁻) on nanocrystalline TiO₂ electrodes. *J. Am. Chem. Soc.*, 115, PP. 6382-6390

O'Regan, B & Grätzel, M. (1991) A low-cost, high-efficiency solar cell based on dye-sensitized colloidal TiO₂ films. *Nature*, 353, pp. 737-740

Pisarello, M. L.; Milt, V.; Peralta, M. A.; Querini, C.A. & Miró. E. E. (2002) Simultaneous removal of soot and nitrogen oxides from diesel engine exhausts. Catalysis Today. 75, pp. 456-470.

Reddy, B. M.; Khan, A.; Yamada, Y.; Kobayashi, T.; Loridant, S. & Volta, J-C. (2003) Structural Characterization of CeO₂–TiO₂ and V₂O₅/CeO₂–TiO₂ Catalysts by Raman and XPS Techniques. *J. Phys. Chem. B*, 107, pp. 5162-5167.

Reddy, B. M.; Khan, A.; Lakshmanan, P.; Aouine, M.; Loridant, S. & Volta, J-C. (2005) Structural Characterization of Nanosized CeO₂–SiO₂, CeO₂–TiO₂, and CeO₂–ZrO₂ Catalysts by XRD, Raman, and HREM Techniques. *J. Phys. Chem. B*, 109, pp. 3355-3363.

Rynkowski, J.; Farbotko, J.; Touroude, R. & Hilaire, L. (2000) Redox behaviour of ceria–titania mixed oxides. *Appl. Catal. A*, 203, pp.335-348.

Shankar, K., Basham, J. I., Allan, N. K., Varghese, O. K., Mor, G. K., Feng, X., Paulose, M. J., Seabold, A., Choi, K-S., & Grimes, C. A. (2009) Recent advances In the use of TiO₂ nanotube and nanowire arrays for oxidative photoelectrochemistry. *J. Phys. Chem. C*, 113, pp. 6327-6359

Shankar, K., Bandara, J., Paulose, M., Wietasch, H., Varghese, O. K., Mor, G. K., LaTempa, T. J., Thelakkat, M., & Grimes, C. A. (2008) Vertically aligned single crystal TiO^2 nanowire arrays grown directly on transparent conducting oxide coated glass: Synthesis details and applications. *Nano Letter*, 8, pp. 1654-1659

Sugimoto, T.; Zhou, X. & Muramatsu, A. (2003) Synthesis of uniform anatase TiO_2 nanoparticles by gel–sol method: 4. Shape control. *J. Colloid Interface Sci.*, 259, pp. 53-61.

Sun, C.; Sun, Jie.; Xiao, G.; Zhang, H.; Qiu, X.; Li, H. & Chen, L. (2006) Mesoscale Organization of Nearly Monodisperse Flowerlike Ceria Microspheres. J. Phys. Chem. B, 110, pp. 13445-13452.

Vantomme, A.; Yuan, Z-Y.; Du, G. & Su, B-L. (2005) Surfactant-Assisted Large-Scale Preparation of Crystalline CeO_2 Nanorods. *Langmuir.* 21, pp. 1132-1135.

Wang, D., Liu, Y., Yu, B., Zhou, F., & Liu, W. (2009) TiO_2 nanotubes with tunable morphology, diameter, and length: Synthesis and photo-electrical/catalytic performance. *Chem. Mater.*, 21, pp. 1198-1206

Wang, M., Chen, P., Humphry-Baker, R., Zakeeruddin, S. M., & Grätzel, M. (2009) The influence of charge transport and recombination on the performance of dye-sensitized solar cells. *ChemPhysChem*, 10, pp. 290-299

Wu, N-C.; Shi, E-W.; Zheng, Y-Q. & Li, W-J. (2002) Effect of pH of Medium on Hydrothermal Synthesis of Nanocrystalline Cerium(IV) Oxide Powders. *J. Am. Ceram. Soc.* 85, pp. 2462-2468.

Yoshida, K. Jiu, J. Nagamatsu, D. Nemoto, T. Kurata, H. Adachi, M. & Isoda, S. (2008). Structure of TiO2 Nanorods Formed with Doiuble Surfactants, Molecular Crystals and Liquid Crystals. 491, pp. 14-20

Youngblood, J. W., Lee, S-H. A., Kobayashi, Y., Hernandez-Pagan, E. A., Hoertz, P. G., Moor, T. A. Moor, A. L., Gust, D., & Mallouk, T. E. (2009) Photoassisted overall water splitting in a visible light-absorbing dye-sensitized photoelectrochemical cell. *J. Am. Chem. Soc.*, 131, pp. 926-927

Zaki, M. I.; M Hussein, G. A.; Mansour, S. A. A. & El-Ammawy, H. A. (1989) Adsorption and surface reactions of pyridine on pure and doped ceria catalysts as studied by infrared spectroscopy. *J. Mol. Catal.* 51, pp. 209-220.

Zaki, M. I.; Hasan, M. A. & Pasupulety, L. (2001) Surface Reactions of Acetone on Al_2O_3, TiO_2, ZrO_2, and CeO_2: IR Spectroscopic Assessment of Impacts of the Surface Acid–Base Properties. *Langmuir.*, 17, pp. 768-774.

Zhou, X. -D.; Huebner, W. & Anderson, H. U. (2003) Processing of Nanometer-Scale CeO_2 Particles. *Chem. Mater.* 15, pp. 378-382.

Zhou, K.; Wang, X.; Peng, Q. & Li, Y. (2005) Enhanced catalytic activity of ceria nanorods from well-defined reactive crystal planes. *J. Catal.* 229, pp. 206-212.

Zhong, L-S.; Hu, J-S.; Cao, A-M.; Liu, Q.; Song, W-G.; & Wan, L-J. (2007) 3D Flowerlike Ceria Micro/Nanocomposite Structure and Its Application for Water Treatment and CO Removal. *Chem. Mater.* 19, pp. 1648-1655.

4

Hollow Nano Silica: Synthesis, Characterization and Applications

N. Venkatathri

Department of Chemistry, National Institute of Technology, Andhra Pradesh,
India

1. Introduction

Since the discovery of mesoporous silica molecular sieves by Beck et al. (Beck et al., 1992; Kresge et al., 1992), mesoporous materials have opened many new possibilities for application in the fields of catalysis (Tanev et al., 1994), separation, and nanoscience (Wu & Bein, 1994; Agger et al., 1998; Li et al., 2003; Yu et al., 2005]. In recent years, fabrication of silica materials with designed structure (e.g. thin films, monoliths, hexagonal prisms, toroids, discoids, spirals, dodecahedron and hollow sphere shapes) is an important research in modern materials chemistry. Among them the fabrication of monodispersed hollow spheres with control size and shape is fastest developing area (Schacht et al., 1996; Bruinsma et al., 1997; Fowler et al., 2001). It is generally accepted that hollow sphere with mesopores will exhibit more advantages in mass diffusion and transportation as compared with conventional hollow spheres with solid shell. They can serve as a small container for application in catalysis and control release studies (Mathlowitz et al., 1997; Huang & Remsen, 1999). The methods currently used to fabricate a wide range of stable hollow spheres include nozzle reactor processes, emulsion/phase separation, sol-gel processing, and sacrificial core techniques. The fabrication of hollow spheres has been greatly impacted by the layer-by-layer (LbL) self-assembly technique (Decher, 1997). This method allows the construction of composite multilayer assemblies based on the electrostatic attraction between nanoparticles and oppositely charged polyions. By varying the synthetic methodology and reactants, it is highly probable to achieve the materials with interesting morphology and properties.

The presence of pores of uniform size lined with silanol groups confers these mesoporous materials as a potential candidate for hosting a variety of guest chemical species, such as organic molecules, semiconductor clusters, and polymers (Moller & Bein, 1998). For example, MCM-41 was reported as a drug delivery system (Vallet-Regi et al., 2001). Ibuprofen has been shown to readily adsorb from an *n*-hexane solution into the porous matrix of MCM-41, and to slowly release into a solution simulating physiological fluid. Furthermore, it has been found that in this host/guest system there is a strong interaction between the silanol groups and the carboxylic acid of the ibuprofen molecule. Having proven the feasibility of this system for drug retention and delivery, further effort should be made in gaining control of the amount of drug delivered, and its release rate. It can be thought that this delivery rate could be modulated by modifying the interaction between the confined molecule and the mesoporous matrix with different morphology. Here, one of the advantages of nanocuboids compared to conventional mesoporous materials is reflected in their much higher storage capacity. Ibuprofen with the molecules size of 1.0 x 0.6 nm was used to examine the storage capacity.

Nanomaterials are the talk of today's Materials researchers. Mesoporous hollow silica spheres were recently invented. It is important due to the drug storage property. Synthesis of mesoporous silica nano hollow cuboids is the very recent advancement (Venkatathri et al., 2008) in this category. In the present invention, the physicochemical property of mesoporous silica's, Nanocrystalline MCM-41 and Nanohollow cuboids were compared. It is found that Nanohollow cuboids store much more drug molecules say Ibuprofen.

2. Experimental

Silica Nanohollow cuboids are synthesized as follows. 3.57 ml of triethanolamine (TEtA, 98%, Aldrich, USA) was added to a solution containing 74 ml of ethanol (99%, Aldrich, USA) and 10 ml of deionized water. 6 ml of tetraethoxyorthosilicates (TEOS, 98%, Aldrich, USA) was added to the above prepared mixture at 298 K with vigorous stirring. The reaction mixture was stirred for another 1 h. A solution containing 5 ml of TEOS and 2 ml of octadecyltrimethoxy silane (C18TMS, 90 %, Aldrich, USA) was added to the above solution (11.4 SiO_2: 6 TEtA: 1 C18TMS: 149 H_2O: 297.5 EtOH) and further reacted for 24 h. The resulting octadecyl group incorporated silica nanocomposite was retrieved by centrifugation. The sample was washed several times with distilled water, dried and calcined at 823 K for 8 h in air to obtain hollow cuboids silica material.

Nanocrystalline Silica MCM-41 is synthesized as follows. Cetyltrimethylammonium bromide was dissolved in 120 g of deionized water to yield a 0.055 mol l^{-1} solution, and 9.5 g of aqueous ammonia (25 wt%, 0.14 mol) was added to the solution. While stirring, 10 g of tetraethoxy silane (0.05 mol) was added slowly to the surfactant solution over a period of 15 min resulting in a gel with the following molar composition: 1 TEOS: 0.152 cetyltrimethylammonium bromide; 2.8 NH_3: 141.2 H_2O. The mixture was stirred for one hour then the white precipitate was filtered and washed with 100 ml of deionized water. After drying at 363 K for 12 h, the sample was heated to 823 K (rate 1 K min^{-1}) in air and kept at this temperature for 5 h to remove the template.

X-ray diffractograms (XRD) were recorded on Rigaku Multiplex diffractometer using Cu Kα radiation and a proportional counter as detector. A divergence slit of 1/328 on the primary optics and an anti-scatter slit of 1/168 on the secondary optics were employed to measure data in the low angle region. The particle size and shape were analyzed by a Scanning electron microscope (SEM), Topcon, SM-300. Transmission electron micrographs (TEM) of the samples were scanned on a on a JEOL JSM-2000 EX electron microscope operated at 200 kV. The samples for TEM were dispersed in isopropyl alcohol, deposited on a Cu-grid and dried. Thermogravimetry (TG) analysis of the crystalline phase was performed on an automatic derivatograph (Setaram TG 92). The specific surface area (BET) of the samples was determined using a Micromeritics ASAP 2010 volumetric adsorption analyzer. Before N_2 adsorption samples was evacuated in vacuum at 573 K. The data points of p/p_0 in the range of about 0.05–0.3 were used in the calculations. The

Fourier transform Infrared (FT-IR) spectra in the framework region were recorded in the diffuse reflectance mode (Nicolet 60SXB) using 1:300 ratio of sample with KBr, pellet. Ultraviolet – visible (UV-Vis.) spectroscopic analysis were carried out using Shimadzu, UV-2450 spectrometer.

Ibuprofen (IBU) drug (Ranbaxy Chem. LTD., 99%) was dissolved in hexane solution at a concentration of 30 mg/ml. 1.0 g nanocuboids or MCM-41 was added into 50 ml IBU hexane solution at room temperature. Sealing the vials to prevent the evaporation of hexane, then the mixture was stirred for 24 h. The nanocuboids or MCM-41 adsorbed with IBU was separated from this solution by centrifugation and dried under vacuum at 60 °C. Filtrates (1.0 ml) was extracted from the vial and diluted to 10 ml, and then was analyzed by UV/vis spectroscopy at a wavelength of 235-320 nm.

3. Results and discussion

The X-ray diffraction pattern calcined MCM-41 and hollow cuboids are given in Fig. 1a,b. The pattern shows their identity. The pattern from as-synthesized sample did not change much on calcinations. Both the as-synthesized and calcined patterns of hollow cuboid shows three Bragg diffraction peaks, which can be assigned to the (1 0 0), (1 1 0) and (2 0 0) reflections of a hexagonal symmetry structure (*P6mm*) typical for MCM-41. *d* spacing and unit cell parameter (a_0) calculated from the XRD data are 12.6 nm and 14.54 nm respectively (Grun et al., 1999).

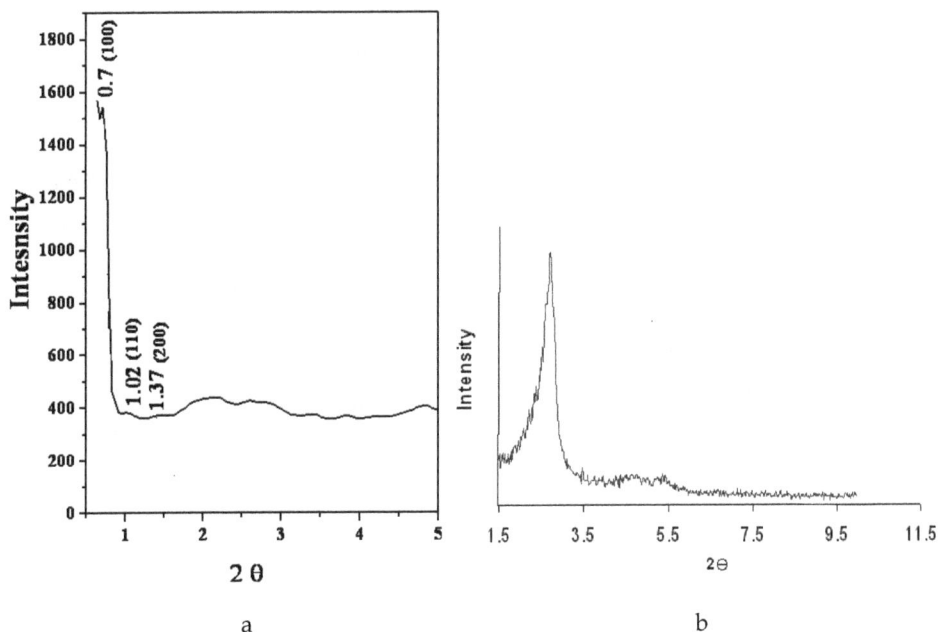

a   b

Fig. 1. X-ray diffraction pattern of calcined mesoporous silicas a) Nanohollow cuboids and b) Nanocrystalline MCM-41

Fig.2a,b shows the scanning electron micrograph of MCM-41 and hollow cuboids. MCM-41 particle size is 200 – 500 nm with spherical shape. Hollow cuboids are aggregate of cuboids with 500 nm particle size.

a   b

Fig. 2. Scanning electron micrograph of calcined mesoporous silicas
a) NanocrystallineMCM-41 and b) Nanohollow cuboids

Transmission electron micrograph of MCM-41 and hollow cuboids are given in Fig. 3a,b. MCM-41 shows hexagonal array of channels characteristic of Mesoporous structure. By Fast Fourier Transform (FFT) of the TEM images, we estimate a unit cell dimension of 3.3 nm. TEM of cuboids shows core and shell structure. It can be seen from the images that the average inner diameter of the cuboids are nearly 100 nm, with outer shell thickness 50 nm. The particle sizes are uniform similar to SEM results. This distinguished pore channel arrangement with most of them running through the shell, are favorable for the access of guest molecules.

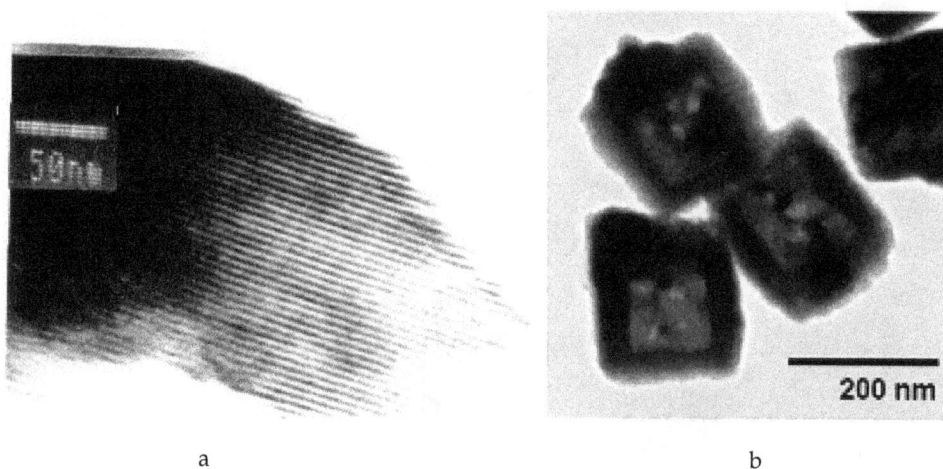

a   b

Fig. 3. Transmission electron micrograph of calcined mesoporous silicas
a) NanocrystallineMCM-41 and b) Nanohollow cuboids

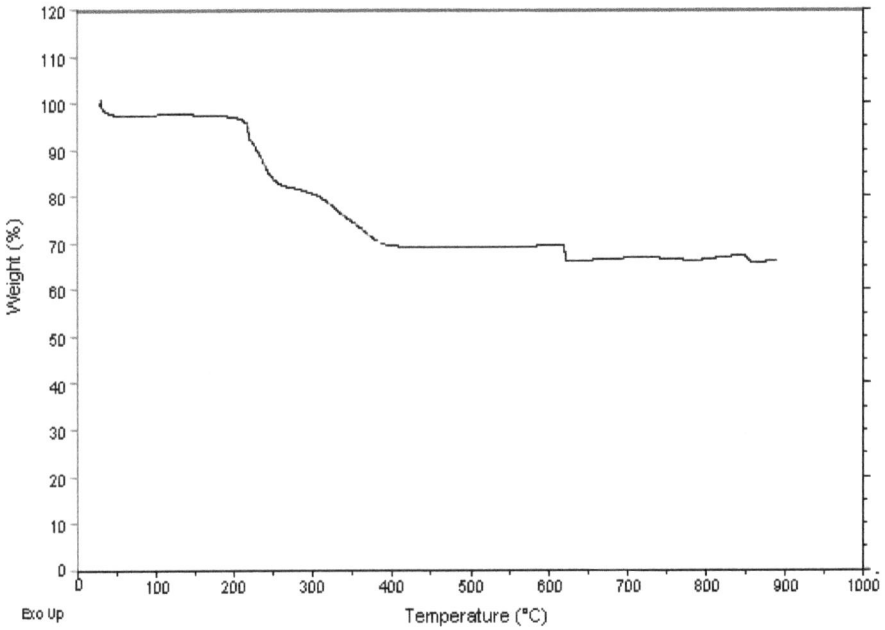

a

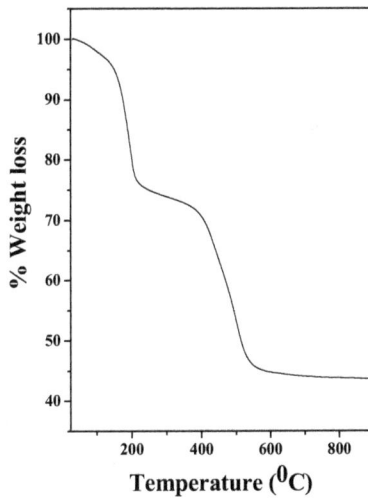

b

Fig. 4. Thermogravimetric profile of as-synthesized mesoporous silicas,
a) Nanocrystalline MCM-41 and b) Nanohollow cuboids.

The Thermogravimetry of MCM-41 and hollow cuboids were given in Fig. 3a,b. MCM-41 shows the 30 % loss at 25 - 625 $^{\circ}$C due to the loss of template. The initial endothermic loss is due to loss of physisorbed water. Later the exothermic loss is due to oxidative decomposition of template. According to the curve the cuboids began to lose its weight at the beginning of heating, likely because of desorption of the physisorbed water and ethanol. It eliminates almost 25 % of its weight in the temperature range 25-200°C and losses almost 30 % weight in the temperature range of 200-500 $^{\circ}$C. The later weight loss is due to the oxidative decomposition of the template.

Typical nitrogen sorption isotherms for MCM-41 and hollow cuboids are shown in Fig. 5a,b. In case of MCM-41, the nitrogen isotherms indicate a linear increase of the amount of adsorbed nitrogen at low pressures (P/Po = 0.35). The resulting isotherm can be classified as a type IV isotherm with a type H2 hysteresis, according to the IUPAC nomenclature (Fujiwara et al., 2004; Brunauer et al., 1940; de Boer, 1958; IUPAC, 1957). The steep increase in nitrogen uptake at relative pressures in the range between P/Po = 0.40 and 0.60 is reflected in a narrow pore size distribution. Thus, the variation of the catalyst in the solution during the growth process enables one to adjust and to control pore structural parameters such as the specific surface area (900 m^2/g), the specific pore volume (1.29 cm^3/g), and the average pore diameter (239 A$^{\circ}$) and medium pore width (302 A$^{\circ}$). The nitrogen adsorption/desorption isotherms of nanocuboid is of type IV nature (Fig. 5b) and exhibited a H1 hysteresis loop, which is typical of mesoporous solids (Wu et al., 2002). Furthermore, the adsorption branch of the isotherm showed a sharp inflection at a relative pressure value of about 0.68. This is characteristic of capillary condensation within uniform pores. The position of the inflection point indicates mesoporous structure, and the sharpness of these steps indicates the uniformity of the mesoporous size distribution. Correspondingly, the pore size distribution of the calcined sample shows a narrow pore distribution with a mean value of 1.90 nm. The sample with a specific surface area of 792 m^2/g and pore volume of 0.51 cm^3/g was obtained using the Brunauer–Emmett–Teller (BET) and Barrett– Joyner–Halenda (BJH) methods, respectively.

The Fourier transform Infrared spectra of as-synthesized MCM-41 and hollow cuboids are shown in Fig. 6a,b. Peaks around 1700 and 3430 cm^{-1} corresponding to the carboxyl and hydroxyl groups (Li et al., 2002) respectively. The adsorption peak belonging to the Si-O stretching vibration of Si-OH bond appears at 960 cm^{-1}(Shan et al., 2004). The weak peaks at 2855 and 2920 cm^{-1} belong to the stretching vibrations of C-H bonds, which show a few organic groups are adsorbed on the spheres. The peaks for carboxyl, hydroxyl and C-H vibrations are weak in MCM-41, shows the lesser organics, resulting of organic template. The strong peaks near 1100, 802 and 467 cm^{-1} agree to the Si-O-Si bond which implies the condensation of silicon source (Agger et al., 1998).

Fig. 7 shows the UV ray absorbance spectra of 30 mg/ml ibuprofen hexane solutions (Zhu et al., 2005) before (a) and after (b) the interaction with nanocuboid and (c) MCM-41. The drug put in contact with nanocuboid and MCM-41 does not show any sign of degradation, since the positions of the absorbance maxima remain unchanged after the interaction and no new bands appear. The Ultraviolet ray absorbance intensity of filtrate decreases after Ibuprofen solution interaction with nanocuboids and MCM-41. This shows the remaining Ibuprofen is adsorbed over the molecular sieves. It was calculated that 561.8 mg and 270.5 mg ibuprofen

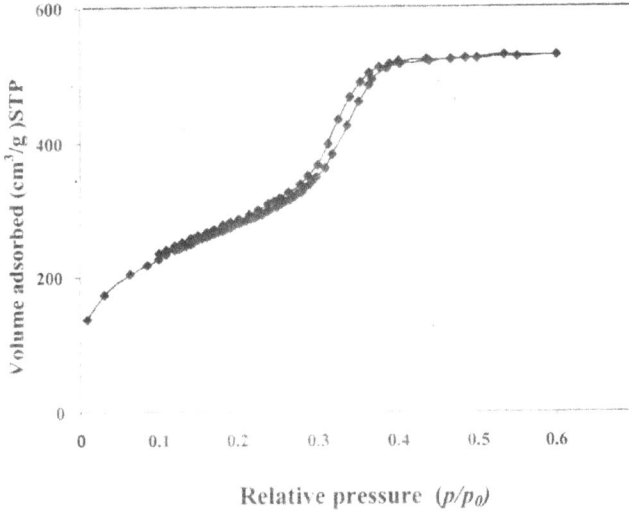

a

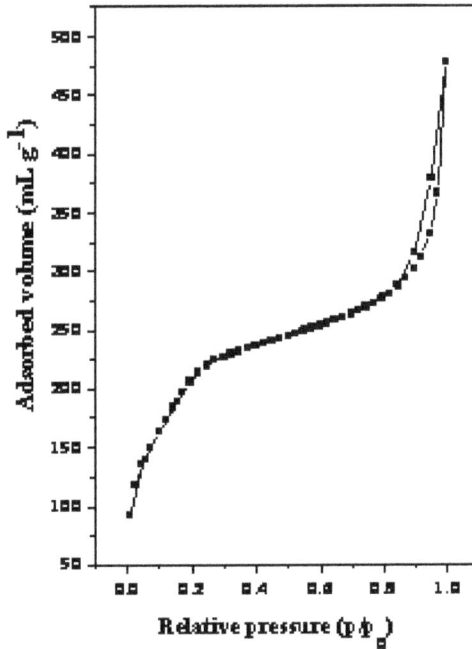

b

Fig. 5. Nitrogen adsorption/desorption isotherms of calcined mesoporous silicas, a) Nanocrystalline MCM-41 and b) Nanohollow cuboids.

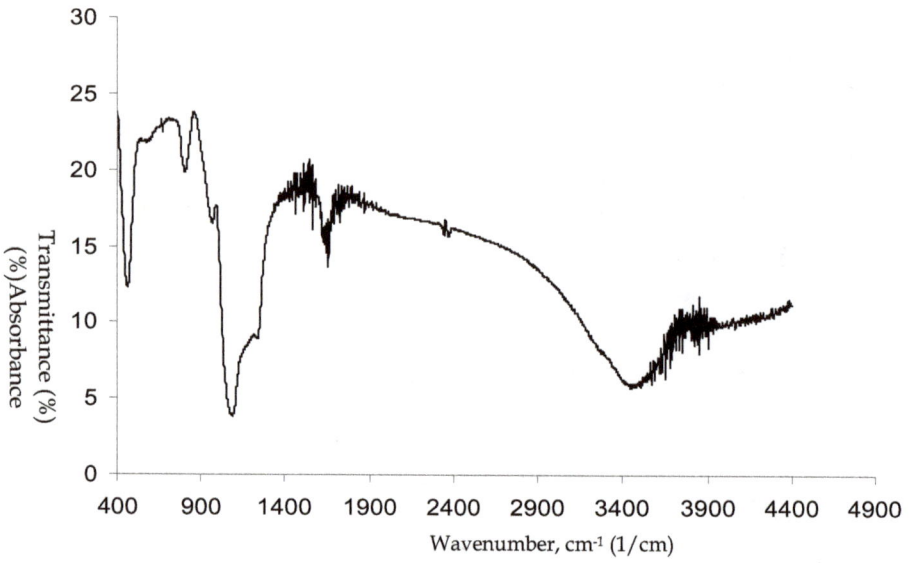

a

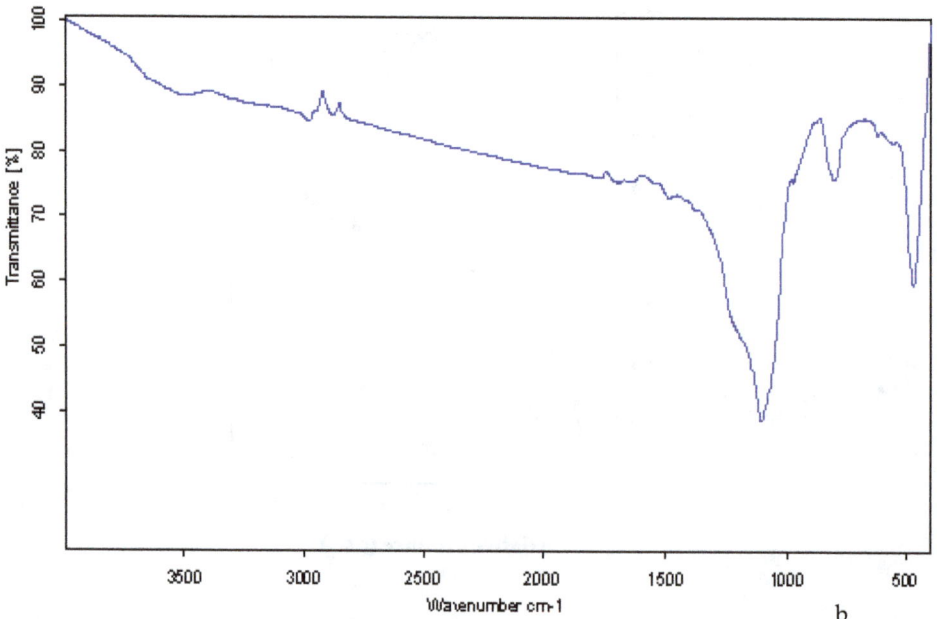

b

Fig. 6. Fourier transform Infrared spectroscopic analysis of as-synthesized mesoporous silicas, a) Nanohollow cuboids and b) a) Nanocrystalline MCM-41.

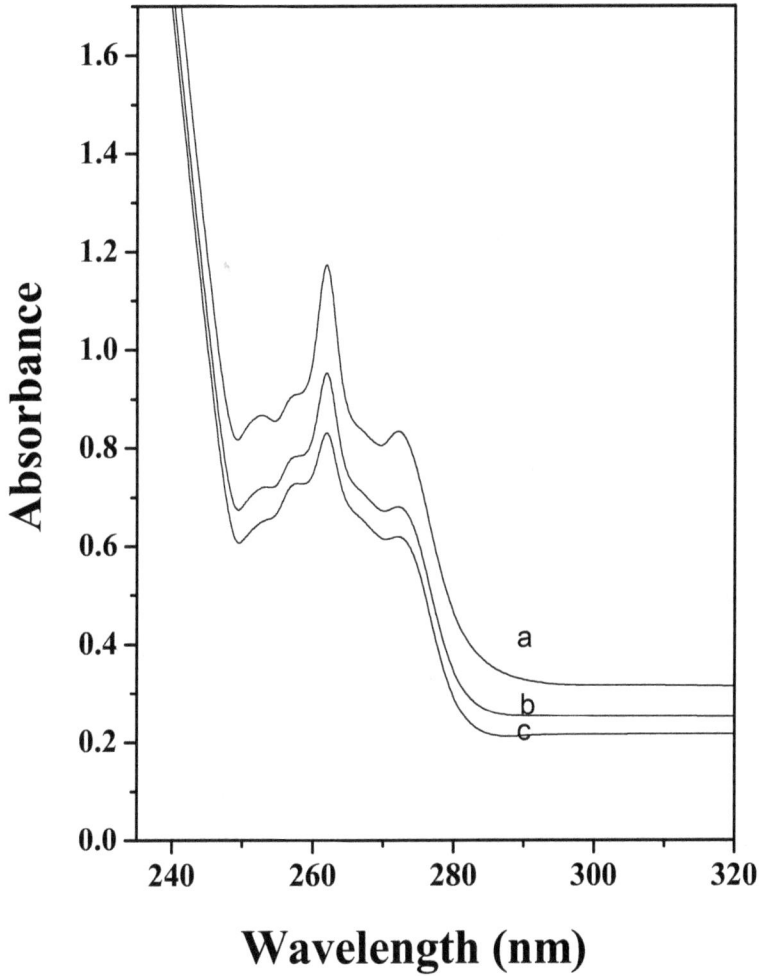

Fig. 7. The Ultraviolet – visible absorbance spectra of 30 mg/ml ibuprofen hexane solutions before (a) and after (b) the interaction with calcined mesoporous silica Nanohollow cuboids and (c) mesoporous Silica Nanocrystalline MCM-41.

molecules can be stored in per gram nanocuboid and MCM-41, respectively from Ultraviolet ray absorbance according to Beer–Lambert Law (Jeffery et al., 1997). The surface area and pore volume of MCM-41 and nanocuboid are very close to each other, but much more ibuprofen molecules can be stored into nanocuboid than into MCM-41. This illustrates that the hollow cores could hold more than half drug molecules of total storage amount.

Tetraethylorthosilicate (TEOS) was hydrolyzed in the presence of basic triethanolamine. However the hydrolysis rate of TEOS using triethanolamine is very slow as compared to hydrolysis with NH_3. For example, using the molar ratio described above, TEOS can be hydrolyzed in 2h using NH_3 whereas triethanolamine took 24 h to hydrolyze the TEOS. In the present synthetic recipe, triethanolamine not only act as a catalyst for the hydrolysis but also it acts as a reactant. The hydrolyzed silica monomers react with triethanolamine to give respective oxide. Such silicate-triethanolamine adduct are held together with hydrogen bonding. The triethanolamine sandwiched silica layer condensed and form nanocuboids. MCM-41 is reported to crystallize by self assembly of surfactant/template (Grun et al., 1999) in similar to nanocuboids.

4. Conclusion

A novel procedure was invented to synthesize mesoporous Silica Nano hollow cuboids with uniform size and morphology. It is characterized by various physicochemical techniques. The results are compared with Nanocrystalline silica MCM-41. Transmission electron micrographs shows, 150 nm hollow diameter and 50 nm shell thickness in hollow cuboids. Further, the mesoporous silica Nanohollow cuboids were found to store much more guest molecules than conventional mesoporous silica Nanocrystalline MCM-41.

5. Acknowledgement

The author thanks Director, National Institute of Technology, Warangal, India for constant encouragement throughout the course of work.

6. References

Agger J.R., Anderson M.W., & Pemble M.E., 1998, Growth of Quantum – confined Indium Phosphide inside MCM-41. *J. Phys. Chem. B, vol.* 102, 1998, pp. 3345-3353.

Beck J.S., Vartuli J.C., . Roth W.J., Leonawicz M.E., & Kresge C.T., 1992, A new family of mesoporous molecular sieves prepared with liquid crystal templates. *J. Am. Chem. Soc.,* vol. 114, 1992, pp. 10834-10843.

Bruinsma P.J., Kim A.Y., & Liu J., 1997, Mesoporous silica synthesized by solvent Evaporation: spun Fibers and spray-Dried Hollow spheres. *Chem. Mater.,* vol. 9, 1997, pp. 2507-2512.

Brunauer S., Deming L.S., Deming W.S., & Teller E., 1940, On a Theory of the vander waals Adsorption of Gases. *J. Am. Chem. Soc.,* vol. 62, 1940, pp. 1723-1732.

de Boer J.H., 1958, *The Structure and Properties of Porous Materials,* Butterworths, London, 1958.

Decher G., 1997, Fuzzy Nano assemblies : Toward Layered polymeric Multicomposites. *Science,* vol. 277, 1997, pp. 1232-1237,.

Fowler C.E., Khushalani D., & Mann S., 2001, Interfacial synthesis of hollow microsphere of meso structured silica. *Chem. Comm.*, 2001, pp. 2028-2029.

Fujiwara M., Shiokawa K., Tanaka Y., & Nakahara Y., 2004, Preparation and formation Mechanism of Silica Microcapsules (Hollow sphere) by water/oil/water Interfacial Reaction. *Chem. Mater.*, vol. 16, 2004, pp. 5420-5426.

Grun M., Unger K.K., Matsumoto A., & Tsutsumi K., 1999, Novel pathways for the preparation of mesoporous MCM-41 materials : Control of porosity and morphology. *Micropor. Mesopor. Mat.*, vol. 27, 1999, pp. 207-216.

Huang H., & Remsen E.E., 1999, Nanocages Derived from shell cross-linked Micelle Templates. *J. Am. Chem. Soc.*, vol. 121, 1999, pp. 3805-3806.

IUPAC, Reporting Physisorption Data for Gas/Solid Systems, 1957, *Pure Appl. Chem.*, vol. 87, 1957, pp. 603-608.

Jeffery G.H., Bassett J., Mendham J., & Denney R.C., 1997, *Vogel's Textbook of Quantitative Chemical Analysis*, Addition Wesley Longman Limited, Edinburgh Gate, Harlow, Essex CM20 2JE, England, 1997, pp. 649-655.

Kresge C.T., Leonowicz M.E., Roth W.J., Vartuli J.C., & Beck J.S., 1992, Ordered mesoporous molecular sieves synthesized by a liquid – crystal template mechanism. *Nature*, vol. 359, 1992, pp. 710-712.

Li Y., Shi J., Hua Z., Chen H., Ruan M., & Yan D., 2003, Hollow spheres of Mesoporous Aluminosilicate with a three- dimensional pore Network and Extraordinarily High Hydrothermal stability. *Nano Lett.*, vol. 3, 2003, pp. 609-612.

Li Y., Xu C., Wei B., Zhang X., Zheng M., Wu D., & Ajayan P.M., 2002, Self – organized Ribbons of Aligned carbon Nanotubes. *Chem. Mater.*, vol. 14, 2002, pp. 483-485.

Mathlowitz E., Jacob J.S., Jong Y.S., Carino G.P., Chickering D.E., Chaturvedl P., Santos C.A., Vijayaraghavan K., Montgomery S., Bassett M., & Morrell C., 1997, Biologically erodable microspheres as potential oral drug delivery systems. *Nature*, vol. 386, 1997, pp. 410-414.

Moller K., & Bein T., 1998, Inclusion chemistry in Periodic Mesoporous Hosts, *Chem. Mater.*, vol. 10, 1998, pp. 2950-2963.

Schacht S., Huo Q., Voigt-Martin I.G., Stucky G.D., & Schuth F., 1996, Oil-Water Interface Templating of Mesoporous Macroscale Structures. *Science*, vol. 273, 1996, pp. 768-771.

Shan Y., Gao L., & Zheng S., 2004, A facile approach to load CdSe nanocrystallites into mesoporous SBA-15. *Mater. Chem. Phys.*, vol. 88, 2004, pp. 192-196.

Tanev P.T., Chibwe M., & Pinnavaia P.J., 1994, Titanium – containing mesoporous molecular sieves for catalytic oxidation of aromatic compounds. *Nature*, vol. 368, 1994, pp. 321-323,.

Vallet-Regi M., Ra´mila A., del Real R.P., & Pe´rez-Pariente J., 2001, A New Property of MCM-41 : Drug Delivery System. *Chem. Mater.*, vol. 13, 2001, pp. 308-311.

Venkatathri N., Srivastava R., Yun D.S., & Yoo J.W., 2008, Synthesis of a novel class of mesoporous hollow silica from organic templates. *Micropor. Mesopor. Mat.*, vol. 112, 2008, pp. 147-152.

Wu C.G., & Bein T., 1994, Polyaniline wires in oxidant-containing Mesoporous channel hosts. *Chem. Mater.*, vol. 6, 1994, pp. 1109-1112.

Wu P., Tatsumi T., Komatsu T., & Yashima T., 2002, Postsynthesis, Characterization and Catalytic Properties in Alkene Epoxidation of Hydrothermally stable Mesoporous Ti-SBA-15. *Chem. Mater.*, vol. 14, 2002, pp. 1657-1664.

Yu K., Guo Y., Ding X., Zhao J., & Wang Z., 2005, Synthesis of silica nanocubes by sol-gel method. *Mat. Lett.*, vol. 59, 2005, pp. 4013-4015.

Zhu Y., Shi J., Chen H., Shen W., & Dong X., 2005, A facile method to synthesize novel hollow mesoporous silica spheres and advanced storage property. Micropor. Mesopor. Mat., vol. 84, 2005, pp. 218-222.

5

Self-Organization of Silver-Core Bimetallic Nanoparticles and Their Application for Catalytic Reaction

Kazutaka Hirakawa
Faculty of Engineering, Shizuoka University
Japan

1. Introduction

Metal nanoparticles have received much attentions as a building block of advanced materials for nanoscience and nanotechnology (Bönnemann & Richards, 2001). Their optical, (Fukumi et al., 1994; Lu et al., 1999; Link et al., 1999; Shipway et al., 2000), magnetic (Sun et al., 1999; Teranishi & Miyake, 1999), and catalytic (Kiely et al., 1998; Pileni, 1998; Bradley, 1994; Harriman, 1990; Lee et al., 1995; Toshima et al., 1995; Bonilla et al., 2000; Siepen et al., 2000) properties have been reported with great interests. The character of metal nanoparticle can be altered by the addition of other metals. Bimetallic nanoparticles, composed of two different metallic elements, have been reported to show outstanding characters different from the corresponding monometallic nanoparticles (Harriman, 1990; Yonezawa & Toshima, 1993; Toshima & Hirakawa, 1997, 1999; Toshima & Wang, 1994; Lee et al., 1995). For example, catalytic activities of gold (Au)-core structured bimetallic nanoparticles, gold/platinum (Au/Pt) (Harriman, 1990; Yonezawa & Toshima, 1993; Toshima & Hirakawa, 1999), gold/palladium (Au/Pd) (Toshima & Hirakawa, 1999; Lee et al., 1995), and gold/rhodium (Au/Rh) (Toshima & Hirakawa, 1999), for hydrogenation and/or water reduction are higher than platinum (Pt), palladium (Pd), and rhodium (Rh) monometallic nanoparticles, respectively. Surprisingly, in some cases, a physical mixture of monometallic nanoparticles such as Pt and ruthenium (Ru) nanoparticles in solution shows higher catalytic activity than the corresponding monometallic nanoparticles under a certain condition (Toshima et al., 1995; Toshima & Hirakawa, 1997). This suggests that an interaction between two kinds of monometallic nanoparticles can produce novel nanoparticles. Further, it has been reported that physical mixture of silver (Ag) and other metal nanoparticles, such as Pt, Rh, and Pd, spontaneously forms the bimetallic nanoparticles with Ag-core structure in aqueous solution. This reaction can be used to construct the core-shell structured novel bimetallic nanoparticles. The formed nanoparticles demonstrate superior character for certain catalytic reactions.

In this chapter, the simple method of the preparation of core-shell structured bimetallic nanoparticles by the physical mixing and the application of the formed novel metal nanoparticles for catalytic reaction are described. The topics of the catalytic reaction presented in this chapter are the visible light induced hydrogen generation (Toshima &

Hirakawa, 2003), the removal of reactive oxygen species (Hirakawa & Sano, 2009), and its application to the chemoprevention of ultraviolet induced biomolecules damage (Hirakwa et al., 2008, 2009).

2. Spontaneous formation of silver-core bimetallic nanoparticles

Much attention has been paid to bimetallic nanoparticles, especially those having a core/shell structure (Toshima et al., 2007). From the view point of Au catalysts, bimetallic nanoparticles have received much attention recently. On the other hand, a physical mixture of monometallic nanoparticles such as Pt and Ru nanoparticles in solution shows higher catalytic activity than the corresponding monometallic nanoparticles under a certain condition (Toshima et al., 1995; Toshima & Hirakawa, 1997). Further, it has been reported that physical mixture of Ag and other metal nanoparticles, such as Pt, Rh, and Pd, spontaneously forms the bimetallic nanoparticles with Ag-core structure in aqueous solution (Figure 1). In this section, the spontaneous formation of the Ag-core bimetallic nanoparticles is reviewed.

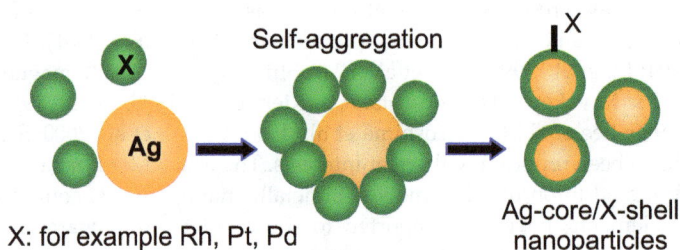

Fig. 1. Schematic diagram of the spontaneous formation of Ag-core bimetallic nanoparticles

2.1 Siver-core/rhodium-shell bimetallic nanoparticles

The interaction between Ag and Rh monometallic nanoparticles in solution by physical mixing was reported. The main reason for using Ag and Rh nanoparticles is the reported prominent characteristics of Rh nanoparticles as a catalyst (Toshima & Hirakawa, 1999), and the expected electronic effect of Ag similar to Au upon enhancement of the catalytic activity of Rh. Furthermore, Ag is inexpensive metal compared with Au. The colloidal dispersions of Ag and Rh monometallic nanoparticles protected by poly(N-vinyl-2-pyrrolidone) (PVP), a water soluble polymer, were prepared by an alcohol reduction method (Hirai et al., 1979). Average diameters of Ag and Rh monometallic nanoparticles were 7.5 nm and 2.2 nm, respectively.

2.1.1 Surface plasmon absorption of siver-core bimetallic nanoparticles

Colloidal sol of Ag nanoparticles shows characteristic plasmon absorption aeound 400 nm (Henglein, 1979). The plasmon absorption band of Ag nanoparticles decreased by addition of Rh nanoparticles and was almost completely extinguished within 30 min after mixing (Figure 2). The parts of plasmon absorption in larger wavelength region were preferentially extinguished within 10 min, suggesting that influences of Rh nanopartilces on Ag nanoparticles depend on the size of the Ag nanoparticles. When relatively smaller molar quantity of Rh to Ag was added, the plasmon absorption was not completely extinguished. More than 40 atom-mol% of Rh against to Ag was required to extinguish the plasmon absorption band completely.

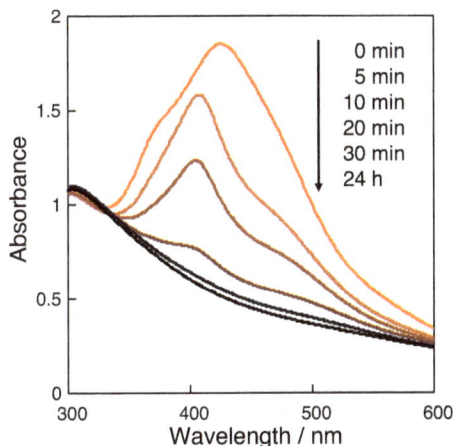

Fig. 2. UV-Vis spectral change of the physical mixtures of dispersions of Ag and Rh nanoparticles. The aqueous solutions of Ag (1 atom-mmol L^{-1}, 50 mL) and Rh (1 atom-mmol L^{-1}, 50 mL) nanoparticles were mixed.

2.1.2 Transmission electron microgram of the siver-core bimetallic nanoparticles

Figure 3 shows transmission electron microscopy (TEM) photographs of the physical mixtures of Ag and Rh monometallic nanoparticles. The samples for TEM measurement were prepared by drying the aqueous dispersions of the physical mixtures of Ag and Rh nanoparticles under vacuum in 0, 10, and 30 min, and 24 h, respectively, after mixing. Relatively large particles are attributed to Ag nanoparticles, and rather small ones are Rh nanoparticles. The TEM photographs showed that Rh particles gathered around Ag particle to surround within several minutes, comparable period of the extinction of plasmon absorption. Interestingly, these aggregated particles changed into homogeneous small particles (average diameter =2.7 nm) after 24 h. Preliminary study has shown that the

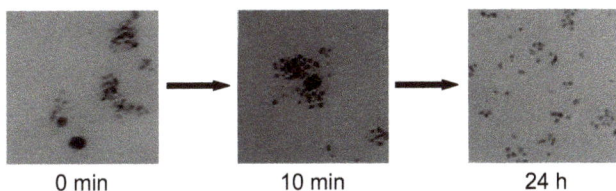

Fig. 3. TEM photographs of the physical mixtures of Ag and Rh nanoparticles. The aqueous solutions of Ag and Rh nanoparticles (1/1, atom-mol/atom-mol) were mixed, and dried after indicated periods.

increase of Rh/Ag molar ratio reduces the average diameter and the size distribution of the nanoparticles. The elemental analysis using characteristic X-ray in high-resolution TEM measurement has shown that the particles produced from their physical mixtures in 24 h are composed of Ag and Rh.

2.1.3 X-ray diffraction of the of siver-core bimetallic nanoparticles

Figure 4 shows X-ray diffraction (XRD) patterns of poly(N-vinyl-2-pyrrolidone)-protected Ag and Rh monometallic nanoparticles, and their physical mixture. The sample of the physical mixture of Ag and Rh nanoparticles was prepared by drying the mixtures of their aqueous solutions under vacuum for 24 h after mixing. The XRD pattern of the mixtures of Ag and Rh nanoparticles was similar to that of Rh nanoparticle, suggesting that the surface of the particle produced by mixing Ag and Rh nanoparticles is composed of Rh. Similarly, Au-core/Pt-shell and Au-core/Pd-shell structured nanoparticles have shown the XRD pattern quite similar to that of their surface metals (Yonezawa & Toshima, 1995). These findings suggest that the aggregation of Rh particles around the Ag particle is involved in the extinction of the plasmon absorption.

2.1.4 Mechanism of the formation of the siver-core/rhodium shell bimetallic nanoparticles

Henglein *et al.* reported that lead (Pb) atoms transfer from Pb colloidal particle onto the surface of Ag colloidal particle in physical mixing of Ag and Pb colloidal sols (Henglein et al., 1992). If the extinction of the plasmon absorption is due to coating of the surface of Ag particle by Rh atoms transferred from Rh nanoparticle, at least 28 mol% of Rh to Ag is required assuming that a Ag particle (average diameter = 7.5 nm) is uniformly coated by Rh atoms in a one-atom layer. In the present experiments about 40 atom-mol% of Rh to Ag was

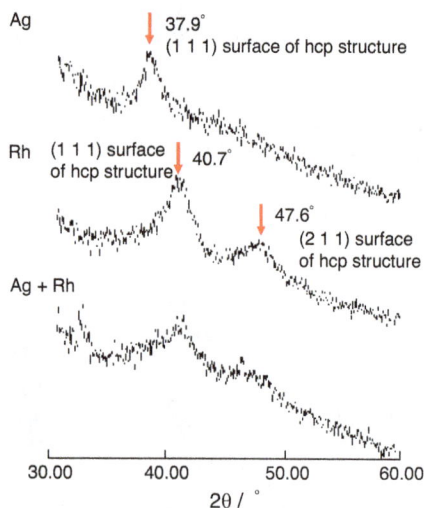

Fig. 4. XRD patterns of Ag and Rh monometallic nanoparticles, and their physical mixture (1/1, atom-mol/atom-mol)

required to completely extinguish the plasmon absorption, which is reasonably supporting the above assumption. These observations suggest that the physical mixture of Ag and Rh nanoparticles spontaneously generates Ag/Rh bimetallic nanoparticles with an Ag-core/Rh-shell structure. The disappearance of the XRD peak of Ag nanoparticles suggests that the core of this bimetallic nanoparticles is not complete Ag, but possibly has a partial Ag/Rh alloy structure. The driving force of the formation of this Ag/Rh bimetallic nanoparticles may be due to the larger binding energy between Ag and Rh atoms than between Rh atoms (Peiner & Kopitzki, 1998). Reduction of diameter of the nanoparticle increases not only its surface energy but also number of the binding sites between Ag and Rh atoms, which stabilizes the total energy. Therefore, the shrinking of Ag/Rh bimetallic nanoparticles might be explained by the balance between the binding energy and the surface energy. The size and the rate of formation of the bimetallic nanoparticles can be controlled by the kind and concentration of protective agents. The self-assembling formation of bimetallic nanoparticle using Ag nanoparticle is applicable to construction of novel nanoparticles.

2.2 Silver-core/noble metal-shell bimatallic nanoparticles

The above mentioned procedure can be used to prepare the Ag-core/noble metal shell nanoparticles, other than Ag-core/Rh-shell nanoparticles. The physical mixing of Ag and other metal nanoparticles, such as Au, Pt, Rh, and Pd particles, produces Ag-core bimetallic particles. The interaction rate between Ag and other metal nanoparticles was determined by the extinction of the surface plasmon absorption of Ag nanoparticle. The initial step of this reaction was investigated by isothermal titration calorimetry (Toshima et al., 2005). This study revealed that the strength of the interaction between Ag and other metals increases in the order of Rh/Ag > Pd/Ag > Pt/Ag.

The formed Ag-core/Pt-shell nanoparticle catalyzed the decomposition of hydrogen peroxide (described later). On the other hand, Au and Au/Ag nanoparticles showed an activity of photocatalytic decomposition of methylene blue (Hirakawa, 2007), although their activities were significantly smaller than that of well-known titanium dioxide photocatalyst (Fujishima et al., 2000, 2008). The physical mixing method is simple and useful to prepare novel bimetallic nanoparticles. These nanoparticles may be used as catalyst and photocatalyst.

2.3 Application to the preparation of trimetallic nanoparticles

This method can be applied to the preparation of trimetallic nanoparticles (Toshima et al., 2007, 2011). It has been reported that the synthesis of trimetallic nanoparticles having a Au-core structure by a combination of the preparation of bimetallic nanoparticles by co-reduction with the formation of core/shell-structured bimetallic nanoparticles by self-organization in physical mixture (Figure 5). The formation of trimetallic nanoparticles has been suggested by UV–Vis spectral change, TEM image change, FT-IR spectra of adsorbed carbon monoxide, XPS spectra and calorimetric studies. The catalytic activity of trimetallic nanoparticles in the molar ratio of Au/Pd/Rh = 1/4/20 was higher than the corresponding monometallic and bimetallic nanoparticles for hydrogenation of methyl acrylate. This high catalytic activity can be understood by sequential electronic charge transfer from surface Rh atoms to interlayered Pt atoms and then to core Au atoms (Toshima et al., 2011).

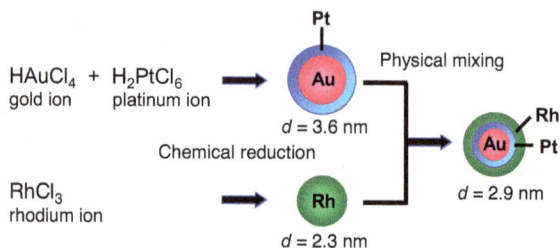

Fig. 5. Schematic diagram of the formation of the trimetallic Au/Pt/Rh nanoparticles

3. Visible-light-induced hydrogen generation by metal nanoparticle catalytic system

Metal nanoparticles are very important materials for nanoscience and nanotechnology (Fukumi et al., 1994; Lu et al., 1999; Link et al., 1999; Sun et al., 1999; Teranishi & Miyake, 1999; Akinaga, 2002). A particularly large number of reports have been published on their applications to catalysts (Kiely et al., 1998; Pileni, 1998; Bradley, 1994; Widegren & Finke 2003; Willner et al., 1987; Toshima et al., 1995; Yonezawa & Toshima, 1993). As the catalyst in the homogeneous system, the colloidal dispersions of metal nanoparticles have the advantage that they are soluble or homogeneous in an aqueous solution and transparent to visible light (Kiely et al., 1998; Pileni, 1998; Bradley, 1994; Widegren & Finke 2003; Willner et al., 1987). Thus, colloidal metal nanoparticles are useful for photocatalytic reaction systems. For example, colloidal metal nanoparticles catalyze the water reduction in the visible-light-induced electron transfer system composed of ethylenediaminetetraacetic acid disodium salt (EDTA), tris(bipyridine)ruthenium(II) dichloride ([Ru(bpy)$_3$]$^{2+}$), and 1,1'-dimethyl-4,4'-bipyridium dichloride (methyl viologen, MV^{2+}) (Yonezawa & Toshima, 1993) (Figure 6).

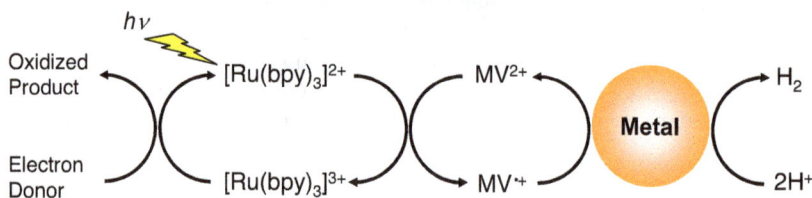

Fig. 6. Schematic diagram of the visible-light induced hydrogen generation using the electron transfer system and metal nanoparticle catalyst

3.1 Catalytic activity of gold-core/platinum-shell bimetallic nanoparticles

The bimetallization of metal nanoparticle can improve the catalytic activity of surface metal. Especially, core-shell structured nanoparticles are important. Several study demonstrated the Au-core/Pt shell metal nanoparticles show higher catalytic activity for the visible-light-induced hydrogen generation than Pt monometallic nanoparticles. The following study is an example of the hydrogen generation using Au/Pt nanoparticle catalyst (Yonezawa & Toshima, 1993). In this study, the Au/Pt bimetallic systems stabilized by polymer and micelle were obtained by alcohol- and photo-reduction of the corresponding metal ions in the presence of water-soluble polymers and non-ionic surfactant-micelles, respectively. The UV-Vis spectra and the transmission electron micrographs suggest that the polymer-protected Au/Pt bimetallic systems are composed of bimetallic alloy clusters, but the micelle-protected ones are mostly composed of the mixtures of the monometallic Au and Pt particles. The *in-situ* UV-Vis spectra during the reductions can elucidate the formation processes of the bimetallic dispersions which are different from each other depending on the protective reagent. The Au/Pt bimetallic systems can be used as the catalyst for visible light-induced hydrogen generation. The bimetallic system stabilized by the polymer at a molar ratio of Au/Pt = 2/3 is the most active catalyst.

3.2 Application of siver-core/rhodium-shell bimetallic nanoparticles

It has been reported that the catalytic activity of the Ag/Rh bimetallic nanoparticles for visible-light-induced hydrogen generation (Toshima & Hirakawa, 1999) in an aqueous solution composed of ethylenediaminetetraacetic acid, tris(bipyridine)ruthenium(II), methyl viologen, and metal nanoparticle catalyst. The activity is clearly higher than the corresponding monometallic nanoparticles and alloy-structured Ag/Rh nanoparticles, suggesting that the Ag-core shows an electronic effect on the surface Rh as in the case of the Au-core (Yonezawa & Toshima, 1993) and enhances the catalytic activity of the surface Rh. The highest catalytic activity was observed at 1:9 ratio of Ag and Rh atoms (Figure 7). Similar results reported on the other catalytic reactions.

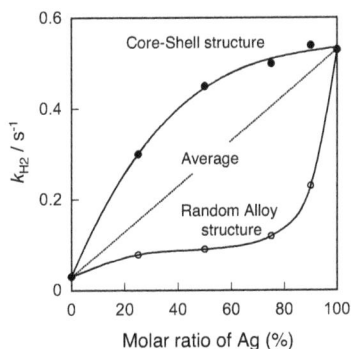

Fig. 7. Hydrogen generation rate coefficient (k_{H2}) depending on the molar ratio of Ag of Ag/Rh bimetallic nanoparticles. The k_{H2} indicates the number of generated H_2 molecules on a surface metal atom per one second. The average is the calculated activity of the simple mixture of Ag and Rh monometallic nanoparticles.

3.3 Carbon dioxide reduction by visible-light-induced electron transfer system using metal nanoparticle

A photochemical reduction of CO_2 can be applied to a novel energy storage process for the utilization of solar energy in the future. The above mentioned catalytic system can be applied to CO_2 reduction. The strategy is the catalytic reduction of CO_2 using electrons gathered by an electron transfer system (Willner et al., 1987, Toshima et al., 1995). It has been reported that nanoparticles catalyzes the reduction of CO_2 and the generation of methane (Toshima et al., 1995) (Figure 8).

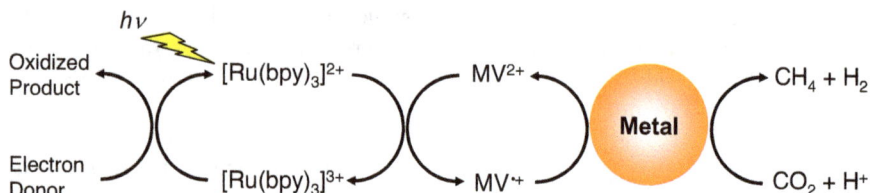

Fig. 8. Schematic diagram of the visible-light induced CO_2 reduction using the electron transfer system and metal nanoparticle catalyst

The possible reaction scheme of the CO_2 reduction is as follows:

$$CO_2 + 8H^+ + 8e^- \rightarrow CH_4 + 2H_2O \tag{1}.$$

This eight-electron reduction of CO_2 is advantageous process compared with other possible CO_2 reduction process from the thermodynamic point of view. Although it is not a study using the silver–core bimetallic nanoprticles, this topic is closely related to the applications of bimetallic nanoparticles to catalytic reaction. Thus, the topic of the CO_2 reduction using metal nanoparticle catalyst is presented here.

Typical reactions were performed by the similar manner to the hydrogen generation. A 20-cm^3 Pyrex Schlenk tube was charged with a 10 cm^3 aqueous solution, containing EDTA (a sacrificial electron donor), [Ru(bpy)$_3$]$^{2+}$ (a photosensitizer), MV^{2+} (an electron mediator), NaHCO$_3$ (a pH adjuster and a CO_2 source), and colloidal dispersion of metal nanoparticles. The mixtures were degassed by freeze-thaw cycles and the tubes were then filled with 1 atm of CO_2. The photo-irradiation was carried out for 3 or 4 h with a 500 W super-high-pressure mercury lamp through a UV cut filter (> 390 nm) in a water bath maintained at 30 °C. About 100 μmol of methane was detected in this system (Toshima et al., 1995). However, it has not been confirmed that methane was actually the reduction product of CO_2.

3.3.1 Strategy for the demonstration of the methane generation from carbon dioxide

In a heterogeneous system, photoreduction of CO_2 was confirmed by experiments using an isotope (Ishitani et al. 1993). To our knowledge, however, the isotopic method has not been applied to the confirmation of the photoreduction of CO_2 in a homogeneous system using the colloidal dispersion of metal nanoparticles. To confirm the above mentioned methane generation, the following study was carried out. In this study, photoreduction of CO_2 was carried out in a similar system to one reported previously (Toshima et al., 1995), and the generation of methane from CO_2 was confirmed by isotopic experiments. As the catalysts, novel metal nanoparticles, i.e., liposome-protected Pt nanoparticles, were prepared and used in the present system. Colloidal dispersions of Pt and Ru nanoparticles were prepared by photoreduction without using ethanol (Yamaji et al. 1995). Preparation of nanoparticles without ethanol is required, because the coexisting ethanol is decomposed during the photochemical reaction, leading to the formation methane. This methane formation cannot be distinguished from the actual methane generation from CO_2. Protective agents used for the metal nanoparticles were poly(N-vinyl-2-pyrrolidone), C$_{12}$EO, and liposome. The products in the gas phase were analyzed with a gas chromatograph. The characterization of gaseous products was carried out with a gas chromatograph mass-spectrometer.

3.3.2 Methane generation from carbon dioxide reduction

The formation of methane was then clearly detected by gas chromatography (about 19 nmol in the case of the liposome protected Pt nanoparticles system). In order to confirm the methane generation from CO_2, isotope experiments were carried out using $NaH^{13}CO_3$ as a CO_2 source and analyzed by a gas chromatograph mass-spectrometer. Since $NaHCO_3$ is equilibrated with CO_2 in solution and easily treated, it was a good source of CO_2 in the present experiments. In this experiment, $^{13}CH_4$ was clearly detected, though the produced methane was not pure $^{13}CH_4$ and it did contain $^{12}CH_4$. In the same experiment, the mole ratio of $^{13}CO_2$ to $^{12}CO_2$ in the gas phase was about 57:43, which is nearly the same as the isotopic ratio of the generated methane. EDTA works as an electron donor in the system and is known to decompose into CO_2. Therefore, $^{12}CH_4$ generation possibly occurs through the reduction of $^{12}CO_2$ generated from EDTA. The effect of EDTA on methane generation was examined in the Pt-liposome system. Methane was detected on visible-light irradiation of the system involving EDTA without CO_2 or $NaHCO_3$ but could not be detected in the absence of EDTA. These results suggest that the detected $^{12}CH_4$ is generated by the reduction of $^{12}CO_2$ originated from EDTA.

3.3.3 Liposome-protected metal nanoparticle catalyst

Liposome was better than other protective-colloid of Pt nanoparticles for methane generation. This is probably explained by assuming that liposome can form a larger and stronger hydrophobic region to concentrate CO_2 around a Pt nanoparticle than $C_{12}EO$ micelle and poly(N-vinyl-2-pyrrolidone). In addition, Ru-$C_{12}EO$ showed higher catalytic activity than Pt-$C_{12}EO$. Thus, Ru-liposome was considered to be an active catalyst for methane generation in the system tested here. The synthesis of Ru-liposome was tried in a way similar to that of Pt-liposome, but the suspension of the Ru-liposome was not active as a catalyst. The resulting Ru-liposome was not as homogeneous, probably because the Ru ion is not miscible with liposome in water.

3.3.4 Summary of the carbon dioxide photo-reduction by metal nanoparticle catalyst

The Pt and Ru nanoparticle catalysts, which were prepared by a photoreduction method of metal salt in water without ethanol, successfully generated methane from CO_2. The methane generation suggests that the eight-electron reduction of CO_2 easily proceeds on metal nanoparticles possibly due to a thermodynamic advantage. This is different from an electrochemical CO_2 reduction using Pt electrodes, on which CO_2 is reduced to CO with adsorbed hydrogen atoms. In the present system using metal nanoparticles, the competition reaction, i.e., the kinetically favorably hydrogen generation, inhibits the methane generation. An increase of CO_2 concentration, the electron supply rate, or both may enhance CO_2 reduction.

4. Catalytic decomposition of hydrogen peroxide by metal nanoparticle

The modification of biomacromolecules upon exposure to reactive oxygen species, including hydrogen peroxide (H_2O_2), dioxide(1-) (superoxide $O_2^{\cdot-}$), hydroxyl radical (HO$^\cdot$), and singlet oxygen ($^1O_2$), is the likely initial event involved in the induction of the mutagenic and lethal effects of various oxidative stress agents (Kawanishi et al. 2001; Cadet et al., 2003;

Drechsel & Patel, 2008). Therefore, the activity of reactive oxygen species generation by various chemical compounds is closely related to their toxicity, carcinogenicity, or both. For example, hydroquinone, a metabolite of carcinogenic benzene, causes DNA damage via H_2O_2 generation (Hirakawa et al., 2002). Many studies have addressed the role of antioxidants, such as vitamins (Slaga, 1995; Sohmiya et al., 2004) and catechins (Weyant et al., 2001), in protection against cancers and cardiovascular diseases. These antioxidants can scavenge reactive oxygen species and protect against cancer occurrence. On the other hand, every antioxidant is in fact, a redox agent, protecting against reactive oxygen species in some circumstances and promoting free radical or secondary reactive oxygen species generation in others. Indeed, an excess of these antioxidants elevates the incidence of cancer (Nitta et al. 1991; Omenn et al., 1996). Solovieva et al. reported that antioxidants, ascorbic acid (Solovieva et al., 2007) and dithiothreitol (Solovieva et al., 2008), exhibit cytotoxicity via H_2O_2 generation. Relevantly, it has been reported that vitamins A (Murata & Kawanishi, 2000) and E (Yamashita et al., 1998) and catechins (Oikawa et al., 2003) induce DNA oxidation through H_2O_2 generation during their oxidation. H_2O_2 is a long-lived reactive oxygen species which plays an important role in biomacromolecular damage induced by various chemical compounds (Kawanishi et al., 2001; Hirakawa et al., 2002).

4.1 Metal catalyzes decomposition of hydrogen peroxide

Various studies have demonstrated the catalytic decomposition of H_2O_2 by noble metals such as Pt (Keating et al., 1965; McKee, 1969; Bianchi et al., 1962), Pd (Keating et al., 1965; McKee, 1969; Bianchi et al., 1962; Eley & Macmahon, 1972) Ag (Baumgartner et al., 1963; Goszner et al., 1972; Goszner & Bischof, 1974), and Au (Eley & Macmahon, 1972; Goszner & Bischof, 1974). These metals themselves are hardly oxidized by reactive oxygen species, however, it is difficult to use metal powder or foils as anti-oxidative drugs. Recently, Kajita et al. reported that Pt nanoparticles catalyze the decomposition of reactive oxygen species (Kajita et al., 2007). These nanoparticles can be dispersed in water and used as homogenous solutions. Because this removal mechanism is catalytic decomposition, no oxidized product is formed through this reaction. Platinum metal is used as a food additive and is not considered to be a toxic material. This result led us to the idea that inorganic materials, in particular noble metals, rather than organic antioxidants, can be used as novel chemopreventive agents against reactive oxygen species-mediated biomolecules damage. In this section, the examination of the removal of H_2O_2 generated from a chemical compound, hydroquione, using water-soluble polymer-protected Pt and Ag/Pt nanoparticles are reviewed.

4.2 Catalytic activity of monometallic nanoparticles

4.2.1 Preparation of metal nanoparticles for reactive oxygen scavenger

Colloidal dispersions of poly(N-vinyl-2-pyrrolidone)-protected Pt, Pd, Rh, and Au nanoparticles were prepared using an alcohol reduction method (Hirai et al., 1979). 50 mL of water/ethanol (1/1, v/v) solution containing 1 mM metal salts and 40 mM poly(N-vinyl-2-pyrrolidone) (monomer unit) was refluxed for 2 h, resulting in the formation of typical colored sols of metal nanoparticles. The solvent was removed by vacuum evaporation, and the nanoparticles were dispersed into water to prepare 1 mM/atom (atomic concentration) metal colloidal sols. An aqueous solution of poly(N-vinyl-2-pyrrolidone)-protected Ag nanoparticles (Shiraishi & Toshima, 1999) was prepared from reduction of 1 mM AgNO$_3$ with NaBH$_4$ in the presence of 40 mM poly(N-vinyl-2-pyrrolidone). The obtained Ag colloidal dispersion was purified with an ultra-filter.

These poly(N-vinyl-2-pyrrolidone)-protected metal nanoparticles formed water-soluble sols. The average diameters (d) and standard deviations (σ) of monometallic nanoparticles determined by TEM measurement were as follows: Pt (d = 2.2 nm, σ= 1.0 nm), Pd (d = 2.0 nm, σ= 0.9 nm), Rh (d = 2.2 nm, σ= 1.0 nm), Ag (d = 10.0 nm, σ=1.9 nm), and Au (d = 10.2 nm, σ= 2.0 nm).

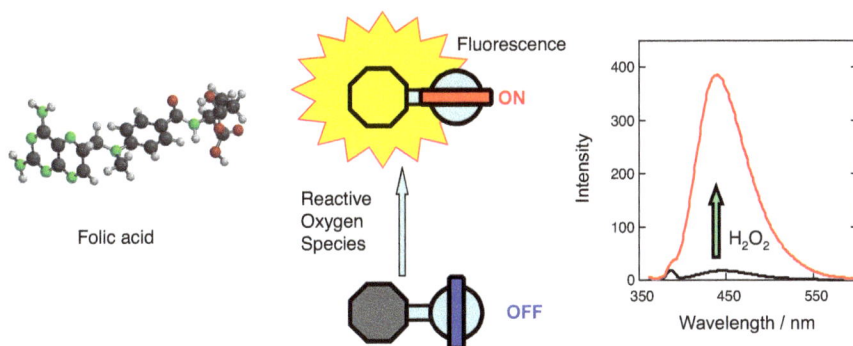

Fig. 9. Fluorometry of reactive oxygen species (hydrogen peroxide) using folic acid

4.2.2 Method of the detection of hydrogen peroxide

The generated H$_2$O$_2$ was measured by a previously reported method using folic acid (Hirakawa, 2006). This assay is based on the fluorescence enhancement of less-fluorescent folic acid via oxidative decomposition by H$_2$O$_2$ and copper(II) ion into strong-fluorescent 2-amino-4-oxo-3H-pterine-6-carboxylic acid (Figure 9). The concentration of H$_2$O$_2$ ([H$_2$O$_2$]) can be determined using a calibration curve. A reaction mixture containing folic acid, copper(II) chloride, and the H$_2$O$_2$ sample (or H$_2$O$_2$ generator [4]) with or without the metal nanoparticle in a sodium phosphate buffer (pH 7.6) was incubated in a microtube for 30 min. After incubation at 37 °C, the fluorescence intensity of the reaction mixture at 450 nm was measured using a fluorescence spectrophotometer with 350-nm excitation.

4.2.3 Platinum nanoparticles effectively scavenge hydrogen peroxide

Platinum nanoparticles effectively scavenged H$_2$O$_2$ in a dose-dependent manner and showed the highest activity among the metal nanoparticles used in this study (Figure 10). A sample solution of 5 µM/atom Pt nanoparticles, among which 1 µg Pt metal is included, exhibits comparable activity for H$_2$O$_2$ decomposition to that of 10 units of catalase. One unit of catalase can remove 1.0 µmol H$_2$O$_2$ per min in water (pH 7.0, 25 °C). Poly(N-vinyl-2-pyrrolidone) itself did not scavenge H$_2$O$_2$. This experiment confirmed that poly(N-vinyl-2-pyrrolidone)-protected Pt nanoparticles can remove H$_2$O$_2$. The mechanism of H$_2$O$_2$ removal by Pt nanoparticles can be explained by catalytic decomposition into water and molecular oxygen as follows:

$$H_2O_2 \rightarrow H_2O + 1/2\,O_2 \tag{2}.$$

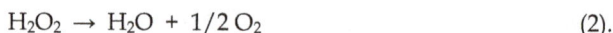

The generation of O_2 gas through the H_2O_2 decomposition was confirmed with a gas-burette as following procedure. The 10 mL of aqueous solution containing 0.1 M H_2O_2 was treated by 10 μg Pt nanoparticles and generated O_2 gas was measured with a gas-burette. The volume of detected gas coincided with that of the theoretically calculated value of O_2 generation from the decomposition of H_2O_2 in the sample solution.

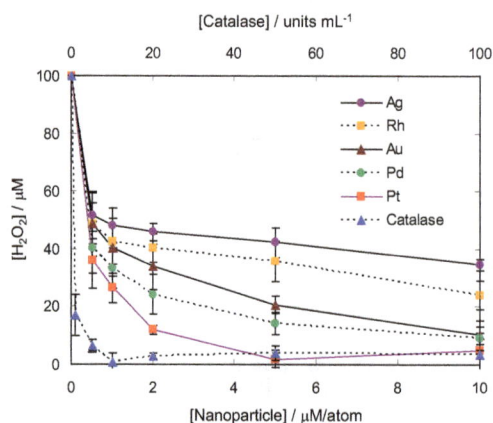

Fig. 10. Removal of H_2O_2 by metal nanoparticles and catalase. The 1 mL of sample solution containing 100 μM H_2O_2, 10 μM folic acid, 20 μM copper(II) chloride, and indicated concentration of metal nanoparticles or catalase was incubated for 30 min. The concentration of H_2O_2 was estimated from the fluorescence measurement.

4.3 Application of siver-core/platinum-shell bimetallic nanoparticles to catalytic decomposition of hydrogen peroxide generated by chemical compound

4.3.1 Preparation of silver-core bimetallic nanoparticles for hydrogen peroxide scavenger

The catalytic activity of Pt and its modified particles with Ag (Ag/Pt) on the decomposition of H_2O_2 generated from chemical compounds was evaluated, since Pt showed the highest activity. The Ag/Pt nanoparticles were prepared from the following procedure. The absorption spectrum of the sol of Pt nanoparticles is a flat curve (Figure 11), indicating the formation of homogenous particles. Ag nanoparticles exhibited a typical yellow color due to surface plasmon absorption around 400 nm. It has been reported that a physical mixture of Ag and Pt nanoparticles spontaneously forms bimetallic nanoparticles, possibly Ag-core/Pt-shell structured particles (Toshima et al., 2005). The time-course of the absorption spectra of this physical mixture showed the extinction of Ag surface plasmon absorption, and the absorption was completely extinguished within 24 h (Figure 11), suggesting that the surface of the formed bimetallic nanoparticles is composed of Pt atoms. Typical TEM images showed the formation of relatively small particles of Pt and large particles of Ag (Figure 12). TEM photographs showed that the large Ag particles disappeared through interaction with Pt particles, resulting in the formation of bimetallic particles smaller than the parent Ag particles (Figure 12). A similar result has been observed in the case of Ag/Rh bimetallic nanopaticles (Toshima & Hirakawa, 2003). These findings suggest the formation of self-organized Ag/Pt bimetallic nanoparticles. These metal nanoparticles are stable in water for several months. The Ag/Pt (Ag-atom/Pt-atom, 1/1) bimetallic nanoparticles were prepared using a self-organization method to mix Pt and Ag monometallic nanoparticles according to previous reports (Toshima & Hirakawa, 2003; Toshima et al., 2002, 2005; Matsushita et al., 2007).

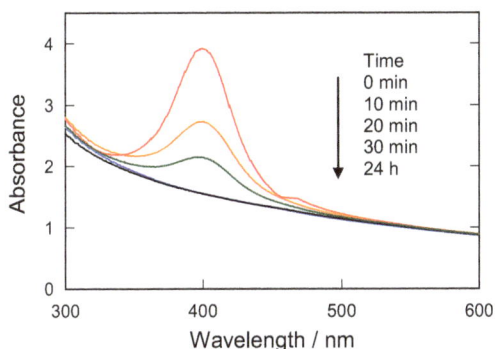

Fig. 11. Absorption spectral change of the physical mixture of dispersions of Ag and Pt nanoparticles. The aqueous solutions of Ag (1 mM/atom, 10 mL) and Pt (1 mM/atom, 10 mL) nanoparticles were mixed and measured at 0, 10, 20, and 30 min, and 24 h after mixing.

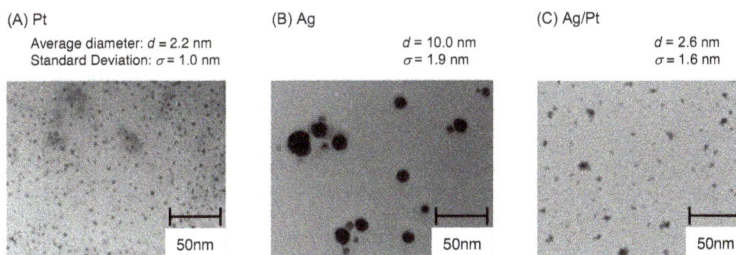

Fig. 12. TEM photographs of metal nanoparticles. The sample of Ag/Pt nanoparticles was prepared by drying the mixtures of the aqueous solutions of Pt and Ag monometallic nanoparticles 24 hour after mixing.

4.3.2 Hydrogen peroxide formation from hydroquinone and its removal by metal nanoparticles

Hydroquinone, which is a metabolite of carcinogenic benzene, was used as H_2O_2 source. This compound can generate H_2O_2 through autooxidation (Figure 13) (Hirakwa et al., 2002). Under these experimental conditions, hydroquinone generated H_2O_2 in a dose-dependent manner (Figure 14). Twenty units/mL catalase effectively removed H_2O_2 generated from this system, and 10 μM/atom (2 μg/mL) Pt nanoparticles exhibited a comparable activity to that of this catalase. Silver nanoparticles showed apparently weaker activity for H_2O_2 removal than Pt nanoparticles. The bimetallization of Pt with Ag apparently suppressed the catalytic activity per unit atom.

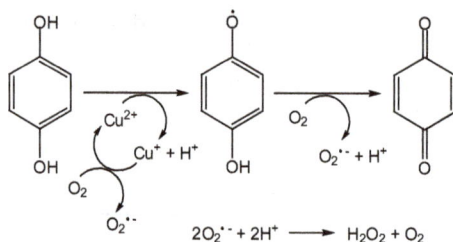

Fig. 13. Schematic diagram of hydrogen peroxide formation by the autooxidation of hydroquinone

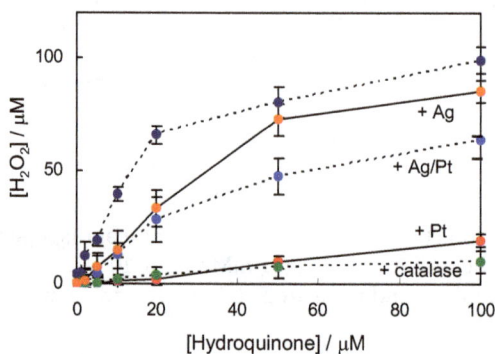

Fig. 14. H_2O_2 generation through autooxidation of hydroquinone in the absence or presence of metal nanoparticles and catalase. The 1 mL of sample solution containing 10 μM folic acid, 20 μM copper(II) chloride, and indicated concentration of hydroquinone with or without 10 μM/atom metal nanoparticles or 20 units/mL catalase was incubated for 30 min. The concentration of generated H_2O_2 was estimated from the fluorescence measurement.

4.3.3 Activity of silver-core/platinum-shell nanoparticles on hydrogen peroxide decomposition

Figure 15 shows the removal activity of H_2O_2 generated from a high concentration of hydroquinone (50 μM) by metal nanoparticles. These metal nanoparticles and catalase scavenged H_2O_2 in a dose-dependent manner. The activity of the 10 μM/atom (2 μg/mL) Pt nanoparticles was comparable to that of 20 units/mL catalase, and Pt completely scavenged H_2O_2 over 20 μM/atom (4 μg/mL). The activity per atom of the Ag/Pt bimetallic nanoparticles was almost the same as that of the Ag monometallic nanoparticles.

To investigate the effect of Pt nanoparticles on H_2O_2 generation through the autooxidation of hydroquinone, NADH consumption during this autooxidation was measured. The consumption of NADH during the autooxidation of hydroquinone was measured by a previously reported method (Oikawa et al., 2003). A sample solution containing 100 μM NADH, 50 μM hydroquinone, and 20 μM copper(II) chloride was incubated at 37 °C in the absence or presence of 20 μM/atom Pt nanoparticles. The concentration of NADH was determined by the measurement of absorbance of NADH at 340 nm using a microplate absorbance reader. The oxidized form of hydroquinone can be reduced into the parent hydroquinone by NADH (Hirakwa et al., 2002). The concentration of NADH was gradually decreased through the redox of hydroquinone and Pt nanoparticles hardly inhibited NADH consumption (data not shown). This result indicated that Pt nanoparticles do not inhibit the H_2O_2 generation itself, because H_2O_2 is produced through the autooxidation of hydroquinone.

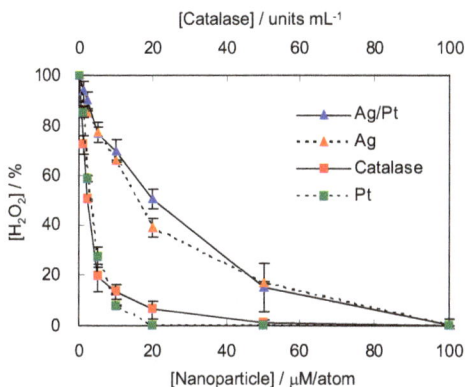

Fig. 15. Removal of H_2O_2 generated through the autooxidation of hydroquinone by metal nanoparticles and catalase. The 1 mL of sample solution containing 10 µM folic acid, 20 µM copper(II) chloride, 50 µM hydroquinone, and indicated concentration of metal nanoparticles or catalase was incubated for 30 min. The concentration of H_2O_2 was estimated from the fluorescence measurement.

4.4 Summary and possible mechanism of hydrogen peroxide decomposition by metal nanoparticles

Poly(N-vinyl-2-pyrrolidone)-protected metal nanoparticles, in particular Pt nanoparticles, exhibited a removal effect on H_2O_2 generated through autooxidation of hydroquinone (Figure 16). The removal of H_2O_2 by these metal nanoparticles can be explained by a catalytic reaction similar to that by catalase, which decomposes H_2O_2 into H_2O and O_2. The formation of H_2O_2 during autooxidation of hydroquinone is through $O_2^{\bullet-}$, which is generated from a reduction of O_2 by hydroquinone (Hirakawa et al., 2002). Because the lifetime of $O_2^{\bullet-}$, which dismutates into H_2O_2 through reaction with H^+, is short ($\sim$ 0.1 ms), the scavenging of $O_2^{\bullet-}$ by a metal nanoparticle can be negligible. The H_2O_2 removal activity per metal atom of these metal nanoparticles occurred in the following order: Pt > Ag $\approx$ Ag/Pt. The activities of H_2O_2 decomposition per metal atom consisting of these metal nanoparticles (µM-H_2O_2/µM-nanometal) have been estimated, and the resulting values are 4.2, 12.2, and 3.8 for Ag, Pt, and Ag/Pt, respectively. Further, the activity on the surface area of the Ag/Pt nanoparticles (17 µM-H_2O_2/cm²-nanometal) was also smaller than that of Pt (49 µM-H_2O_2/cm²-nanometal). These findings showed that the Pt nanoparticles have the highest catalytic activity for H_2O_2 decomposition in the metal nanoparticles used in this experiment and the activity of Pt nanoparticles is suppressed by modification with Ag.

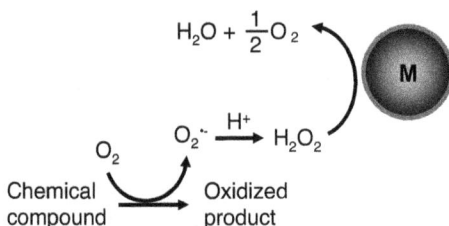

Fig. 16. Hydrogen peroxide generation from an autooxidation of chemical compound and its catalytic decomposition by metal nanoparticle

H_2O_2 is a long-lived reactive oxygen species and plays an important role in DNA damage (Kawanishi et al., 2001, Hirakawa et al., 2002). Indeed, various chemical compounds, including carcinogens, generate H_2O_2 during redox reaction (Kawanishi et al., 2001, Hirakawa et al., 2002). Molecular oxygen is easily reduced by various compounds, leading to the formation of $O_2^{\bullet-}$. Formed $O_2^{\bullet-}$ is rapidly dismutated into H_2O_2. Although H_2O_2 itself is not a strong reactive species, it can generate highly reactive $HO^{\bullet}$ through a Fenton reaction or a Haber-Weiss reaction. Furthermore, H_2O_2 can penetrate a cytoplasm membrane and be incorporated into the cell nucleus. Therefore, H_2O_2 is considered to be one of the most important reactive species or a precursor participating in carcinogenesis. The removal of H_2O_2 is an effective method for cancer chemoprevention. Furthermore, protective agents against H_2O_2 are important to treat *acatalasemia,* a genetic deficiency of erythrocyte catalase inherited as an autosomal recessive trait. Antioxidants, such as vitamins A and E, are effective protective agents. However, the oxidized products of antioxidants or these molecules themselves promote the formation of secondary H_2O_2 (Yamashita et al., 1998; Murata & Kawanishi, 2000). Indeed, an excess of these antioxidants elevates the incidence of cancer (Nitta et al., 1991; Omenn et al., 1996). A catalyst consisting of an inorganic stable material is not oxidized and does not generate secondary reactive oxygen species. Water-soluble nanoparticles of noble metal may become novel protective agents against reactive oxygen species.

In summary, Pt, Ag, and Ag/Pt nanoparticles effectively scavenge H_2O_2 generated from autooxidation of a highly concentrated hydroquinone. Platinum nanoparticles exhibited the highest catalytic activity among these nanoparticles. Pt is a very stable metal against various chemical compounds and permitted as a food additives. The noble metal nanoparticles may be used as novel chemopreventive agents for cancer or other non-malignant conditions induced by chemical compounds through H_2O_2 generation.

5. Application of metal nanoparticles to prevention of ultraviolet radiation induced biomolecules damage

Exposure to solar ultraviolet radiation is undoubtedly linked to skin carcinogenesis and phototoxic effect. Photosensitized reaction by ultraviolet radiation, especially ultraviolet-A (UVA) radiation (320~400 nm), is considered to cause toxic effect through oxidative biomolecules damage including DNA damage (Hiraku et al., 2007). Photosensitized formation of reactive oxygen species, such as hydrogen peroxide, superoxide, hydroxyl radicals, and singlet oxygen, is involved in UVA-induced biomolecules damage. As mentioned above, the application of metal nanoparticles to scavenge reactive oxygen species through catalytic decomposition.

5.1 Traditional methods of chemoprevention to biomolecules damage by ultraviolet radiation and its problem

Many studies have addressed the role of antioxidants, such as vitamins and catechins, in protection against cancers and cardiovascular diseases. These antioxidants can scavenge reactive oxygen species and protect against cancer occurrence. On the other hand, every antioxidant is, in fact, a redox agent, protecting against reactive oxygen species in some circumstances and promoting free radical or secondary reactive oxygen species generation in others. Indeed, an excess of these antioxidants elevates the incidence of cancer. It has been reported that antioxidants, ascorbic acid and dithiothreitol, exhibit cytotoxicity via H_2O_2 generation, and their toxic effects are significantly enhanced by vitamin B_{12}. H_2O_2 is a long-lived reactive oxygen species which plays an important role in biomacromolecules damage induced by various chemical compounds.

5.2 Preventive action of metal nanoparticles on ultraviolet-sensitized oxidation of molecules

As mentioned above, metal nanoparticles catalyze the decomposition of reactive oxygen species. Because this removal mechanism is catalytic decomposition, no oxidized product is formed through this reaction. Platinum metal is used as a food additive and is not considered to be a toxic material. This result led us to the idea that inorganic materials, in particular noble metals, rather than organic antioxidants, can be used as novel chemopreventive agents against UVA-induced biomolecules damage.

Recently, it has been reported that the removal of reactive oxygen species generated from a photocatalytic reaction of titanium dioxide (TiO_2) particles using water-soluble polymer-protected Pt, Rh, and Pt/Ag bimetallic nanoparticles. Silver, a relatively inexpensive noble metal, is also used as a food additive, and bimetallization with Ag may improve the catalytic activity of other metal nanoparticles.

5.2.1 Preparation of metal nanoparticles for ultraviolet protection

The colloidal dispersions of poly(N-vinyl-2-pyrrolidone)-protected Pt and Rh nanoparticles were prepared from an alcohol reduction. The size (particle diameter) of these nanoparticles is about 2 nm. The aqueous solution of poly(N-vinyl-2-pyrrolidone)-protected Ag nanoparticle was prepared from a reduction of silver nitrate by sodium borohydride in the presence of poly(N-vinyl-2-pyrrolidone). The Ag-core/Pt-shell (Ag-atom/Pt-atom, 1/1) bimetallic nanoparticle was prepared using a physical method to mix Pt and Ag monometallic nanoparticles according to the previous reports (Toshima et al., 2005).

5.2.2 Evaluation model for the biomolecules damage by ultraviolet radiation

TiO_2 (anatase) and methylene blue were used as a model of the UVA-induced reaction. The sample solution containing methylene blue and TiO_2 dispersion in sodium phosphate buffer (pH 7.6) with or without metal nanoparticle was irradiated with a UVA lamp (365 nm, 1 mW cm^{-2}). The decomposition of methylene blue was evaluated by absorption measurement at 659 nm. TiO$_2$ is a well-known photocatalyst (Fujishima et al., 2000, 2008). When exposing to UVA light, the reduction-oxidation activity of TiO$_2$ has a significant biological impact, as is exemplified by its bactericidal activity. Photo-irradiated TiO$_2$ effectively decomposed methylene blue (Figure 17). Various reactive oxygen species contribute to the photocatalytic reaction of TiO$_2$. Especially, hydrogen peroxide is long-lived reactive oxygen species and plays an important role in oxidative biomolecules damage. Molecular oxygen is reduced by photoexcited materials, leading to the formation of superoxide. Formed superoxide is rapidly dismutated into hydrogen peroxide. Although hydrogen peroxide itself is not a strong reactive species, it can generate highly reactive hydroxyl radicals through a Fenton reaction or a Haber-Weiss reaction. Furthermore, hydrogen peroxide can penetrate a cytoplasm membrane and be incorporated into the cell nucleous. Therefore, hydrogen peroxide is considered to be one of the most important reactive oxygen species participating in UVA carcinogenesis and phototoxicity. Since other reactive oxygen species, such as directly produced hydroxyl radicals (Hirakawa et al., 2004) and singlet oxygen (Hirakawa & Hirano, 2006), rapidly quenched in aqueous solution, hydrogen peroxide should be key reactive species in this experiment. The TiO$_2$ and methylene blue could be used as a simple model of UVA-induced oxidation.

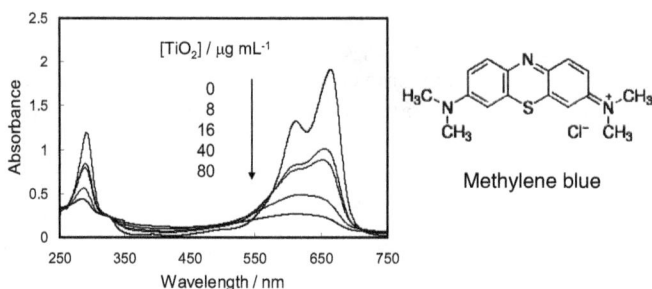

Fig. 17. UV-Vis absorption spectra of methylene blue photocatalyzed by TiO$_2$. The sample solution containing 10 μM methylene blue and indicated concentration of TiO$_2$ in 10 mM sodium phosphate buffer (pH 7.6) was irradiated (Ex = 365 nm, 1 mW cm^{-2}) for 30 min.

5.2.3 Preventive action of metal nanoparticles on ultraviolet radiation induced biomolecules damage

Poly(N-vinyl-2-pyrrolidone)-protected metal nanoparticles, in particular, the Pt nanoparticle, inhibited the methylene blue decomposition photocatalyzed by TiO$_2$ (Figure 18). Poly(N-vinyl-2-pyrrolidone) itself did not inhibit the methylene blue decomposition. This inhibitory effect can be explained by the catalytic decomposition of H$_2$O$_2$ generated through TiO$_2$ photocatalysis. These nanoparticles decomposed H$_2$O$_2$ into H$_2$O and O$_2$ similar to catalase. In the case of H$_2$O$_2$ decomposition, the Pt nanoparticle showed the highest catalytic activity per unit atom. The activity of a 1 μg Pt nanoparticle was comparable to that of 5 units of catalase. One unit of catalase can remove 1.0 μmol H$_2$O$_2$ per 1 min in water (pH 7.0, 25°C). Unexpectedly, the bimetallization with Ag did not show improvement effect and rather decreased the inhibitory effect of Pt nanoparticle on the decomposition of methylene blue.

The UV-Vis absorption spectra of these metal nanoparticles were hardly changed by the photocatalytic reaction, suggesting that the noble metal nanoparticles are stable for reactive oxygen species and UVA irradiation. Organic antioxidant undergoes oxidation in the removal process of reactive oxygen species, leading to the formation of various oxidized products and may produce secondary reactive oxygen species. In the case of noble metal catalyst, these effects can be negligible.

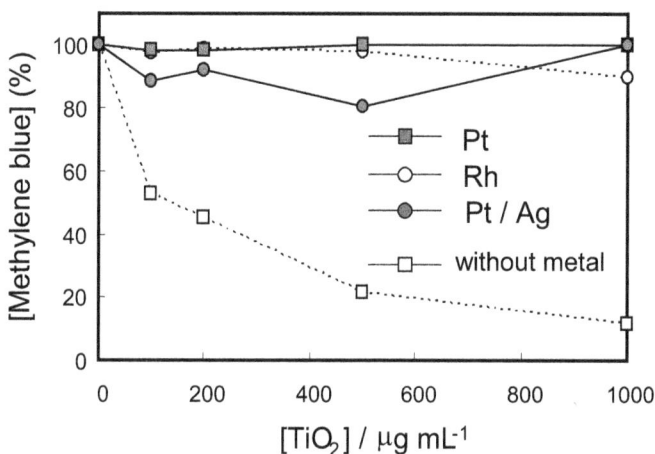

Fig. 18. Inhibitory effect of metal nanoparticles on methylene blue decomposition photocatalyzed by TiO_2. The sample solution containing 20 µg mL^{-1} metal nanoparticle, TiO_2, and 10 µM methylene blue in 10 mM sodium phosphate buffer (pH 7.6) was irradiated (Ex = 365 nm, 1 mW cm^{-2}) for 30 min.

5.3 Summary of the ultraviolet protection by metal nanoparticles

Pt, Rh, and Pt/Ag nanoparticles effectively inhibited the methylene blue decomposition photocatalyzed by TiO_2. TiO_2 photocatalytic system was used as a UVA-induced reactive oxygen species generation. The most important reactive oxygen species in this photocatalytic reaction is H_2O_2, because of its long lifetime in aqueous solution. This inhibitory effect of metal nanoparticle can be explained by the removal of H_2O_2. Unexpectedly, the activity of Pt nanoparticle was not improved by the bimetallization with Ag. Platinum is a very stable metal against various chemical compounds and is used as food additive. A poly(N-vinyl-2-pyrrolidone)-protected Pt nanoparticle may be used as a novel preventive agent for UVA-induced biomolecules damage through reactive oxygen species generation.

6. Conclusion

Physical mixture of Ag and other metal nanoparticles, such as Pt, Rh, and Pd, spontaneously forms the bimetallic nanoparticles with Ag-core structure in aqueous solution. These monometallic nanoparticles can be easily prepared from an alcohol reduction of the corresponding metal ions in the presence of water-soluble polymer such as poly(*N*-vinyl-2-pyrrolidone), a protective colloid. Aqueous sol of Ag nanoparticles exhibits the surface plasmon absorption around 400 nm. The surface plasmon absorption was diminished through interaction with other metal nanoparticle in the physical mixture of these nanoparticles. This phenomenon was explained by that the Ag nanoparticle was coated by other metal. The transmission electron micrograph and X-ray diffraction measurement confirmed the formation of the Ag-core bimetallic nanoparticles. This reaction can be used to construct the core-shell structured novel bimetallic nanoparticles. The formed nanoparticles act superior character for certain catalytic reactions. The catalytic activity of the silver/rhodium bimetallic nanoparticles for visible-light-induced hydrogen generation in an aqueous solution was examined. This system composed of an electron source, a photosensitizer, an electron relay, and metal nanoparticle catalyst. The activity is clearly higher than the corresponding monometallic nanoparticles, suggesting that the silver-core enhances the catalytic activity of the surface rhodium. On the other hand, the catalytic activity of the decomposition of hydrogen peroxide was decreased by this bimetallization. Platinum nanoparticle effectively catalyzes hydrogen peroxide decomposition. The Ag-core/platinum shell bimetallic nanoparticle, which was prepared by the physical mixing of Ag and Pt nanoparticles, demonstrated lower activity of the decomposition of hydrogen peroxide than the monometallic Pt nanoparticle. Metal nanoparticles can be applied to various catalytic reactions. The bimetallic and trimetallic nanoparticles demonstrate superior activity in the certain reaction. The self-assembly formation of Ag-cored nanoparticle may be convenient method to prepare novel metal nanoparticle catalyst.

7. Acknowledgments

The author wish to thank Professor Naoki Toshima (Tokyo University of Science, Yamaguchi) for his helpful discussion and Professor Kenji Murakami (Research Institute of Electronics, Shizuoka University) for his helpful advice on TEM measurement. These works were supported by a Grant-in-Aid for Scientific Research from the Ministry of Education, Culture, Sports, Science and Technology (MEXT) of the Japanese Government.

8. References

Akinaga, H. (2002). Metal-nanocluster Equipped GaAs Surfaces Designed for High-sensitive Magnetic Field Sensors. *Surface Science*, Vol.514, No.1-3, (August 2002), pp.145-150, ISSN 0039-6028

Baumgartner, H. J.; Hood, G. C.; Monger, J. M.; Roberts, R. M. & Sanborn, C. E. (1963). Decomposition of Concentrated Hydrogen Peroxide on Silver I. Low Temperature Reaction and Kinetics. *Journal of Catalysis*, Vol.2, No.5, (October 1963), pp.405-414, ISSN 0021-9517

Bianchi, G.; Mazza, F. & Mussini, T. (1962). Catalytic Decomposition of Acid Hydrogen Peroxide Solutions on Platinum, Iridium, Palladium and Gold Surfaces. *Electrochimica Acta*, Vol.7, No.4, (July-August 1962), pp.457-473, ISSN 0013-4686

Bonilla, R. J.; James, B. R. & Jessop, P. G. (2000). Colloid-catalyzed Arene Hydrogenation in Aqueous/Supercritical Fluid Biphasic Media. *Chemical Communications*, No.11, (May 2000), pp.941-942, ISSN 1359-7345

Bönnemann, H. & Richards, R. M. (2001). Nanoscopic Metal Particles-Synthetic Methods and Potential Applications. *Europian Journal of Inorganic Chemistry*, Vol.2001, No.10, (October 2001), pp.2455-2480, ISSN 1099-0682

Bradley, J. S. (1994). In: *Clusters and Colloids: From Theory and Applications*, G. Schmid (Ed), 456-544, VCH, ISBN 3527290435, Weinheim

Cadet, J.; Douki, T.; Gasparutto, D. And Ravanat, J. L. (2003). Oxidative Damage to DNA: Formation, Measurement and Biochemical Features. *Mutation Research*, Vol.531, No.1-2, (October 2003), pp.5-23, ISSN 0027-5107

Drechsel, D. A. & Patel, M. (2008). Mechanisms of Environmental Neurotoxicant-induced Oxidative Stress. *Free Radical Biology and Medicine*, Vol.44, No.11, (June 2008), pp.1873-1886, ISSN 0891-5849

Eley, D. D. & Macmahon D. M., (1972). The Decomposition of Hydrogen Peroxide Catalyzed by Palladium-gold Alloy Wires. *Journal of Colloid and Interface Science*, Vol.38, No.2, (February 1972), pp.502-510. ISSN: 0021-9797

Fujishima, A.; Rao, T. N. & Tryk, D. A. (2000). Titanium Dioxide Photocatalysis. *Journal of Photochemistry and Photobiology C: Photochemistry Reviews*, Vol.1, No.1, (June 2000), pp.1-21, ISSN 1389-5567

Fujishima, A.; Zhang, X. & Tryk, D. A. TiO_2. (2008). Photocatalysis and Related Surface Phenomena. *Surface Science Reports*, Vol.63, No.12, (December 2008), pp.515-582, ISSN 0167-5729

Fukumi, K.; Chayahara, A,; Kadono, K.; Sakaguchi, T.; Horino, Y.; Miya, M.; Fujii, K,.; Hayakawa, J. & Satou, M. (1994). Gold Nanoparticles Ion Implanted in Glass with Enhanced Nonlinear Optical Properties, *Journal of Applied Physics*; Vol.75, No.6, (March 1994), pp.3075-3080, ISSN 0021-8979

Goszner, K.; Körner, D. & Hite, R. (1972). On the Catalytic Activity of Silver: I. Activity, Poisoning, and Regeneration during the Decomposition of Hydrogen Peroxide. *Journal of Catalysis*, Vol.25, No.2, (May 1973), pp.245-253, ISSN 0021-9517

Goszner, K. & Bischof, H. (1974). The Decomposition of Hydrogen Peroxide on Silver-gold Alloys. *Journal of Catalysis*, Vol.32, No.2, (February 1974), pp.175-1820, ISSN 0021-9517

Harriman, A. (1990). Bimetallic Pt–Au Colloids as Catalysts for Photochemical Dehydrogenation. *Journal of the Chemical Society, Chemical Communications*, No.1, (January 1990), pp.24-26, ISSN 0022-4936

Henglein, A. (1979). Catalysis of the Reduction of Thallium(1+) and of Dichloromethane by Colloidal Silver in Aqueous Solution. *The Journal of Physical Chemistry*, Vol.83, No.22, (November 1979), pp.2858-2862, ISSN 0022-3654

Henglein, A.; Holzwarth, A. & Malvaney, P. (1992). Fermi Level Equilibration between Colloidal lead and Silver Particles in Aqueous Solution. *The Journal of Physical Chemistry*, Vol.96, No.22, (October 1992), pp.8700-8702, ISSN 0022-3654

Hirai, H.; Nakao, Y. & Toshima, N. (1979). Preparation of Colloidal Transition Metals in Polymers by Reduction with Alcohols or Ethers. *Journal of Macromolecular Science: Part A–Chemistry*, Vol.13, No.6, (1979), pp.727-750, ISSN 0022-233X

Yamaji, Y.; Hirakawa, K.; Yonezawa, T. & Toshima, N. (1995). Visible-light-induced Reduction of Carbon Dioxide Using Platinum Colloidal Catalyst. *Proceedings of Annual Meeting on Photochemistry 1995*, pp.B1080, Fukuoka, Japan, October 6-9, 1995

Hirakawa, K.; Oikawa, S.; Hiraku, Y.; Hirosawa, I. & Kawanishi, S. (2002). Catechol and Hydroquinone Have Different Redox Properties Responsible for Their Differential DNA-damaging Ability. *Chemical Research in Toxicology*, Vol.15, No.1, (January 2002), pp.76-82, ISSN 0893-228X

Hirakawa, K. & Toshima, N. (2003). Ag/Rh Bimetallic Nanoparticles Formed by Self-assembly from Ag and Rh Monometallic Nanoparticles in Solution. *Chemistry Letters*, Vol.32, No.1, (January 2003), pp.78-79, ISSN 0366-7022

Hirakawa, K.; Mori, M.; Yoshida, M.; Oikawa, S. & Kawanishi, S. (2004). Photo-irradiated Titanium Dioxide Catalyzes Site Specific DNA Damage via Generation of Hydrogen Peroxide, *Free Radical Research*, Vol.38, No.5, (May 2004), pp.439-447, ISSN 1071-5762

Hirakawa, K. & Hirano, T. (2006). Singlet Oxygen Generation Photocatalyzed by TiO_2 Particles and Its Contribution to Biomolecule Damage. *Chemistry Letters*, Vo.35, No.8, (August 2006), pp.832-833, ISSN 0366-7022

Hirakawa, K. (2006). Fluorometry of Hydrogen Peroxide Using Oxidative Decomposition of Folic Acid. *Analytical and Bioanalytical Chemistry*, Vol.386, No.2, (September 2006), pp.244-248, ISSN 1618-2642

Hirakawa, K. (2007). Preparation of Novel Metal Nanoparticles and Their Catalytic and Photocatalytic Properties. *Reports of Researches Assisted by the Asahi Glass Foundation (2007)*, No.53, (October 2007) pp.1-10, ISSN 0919-9179

Hirakawa, K.; Shiota, K. & Sano, S. (2008). Preventive Action of Metal Nanoparticles on UVA-sensitized Oxidation through Hydrogen Peroxide Formation. *Photomedicine and Photobiology*, Vol.30, (July 2008), pp.27-28, ISSN 0912-232X

Hirakawa, K. & Sano, S. (2009). Platinum Nanoparticle Catalyst Scavenges Hydrogen Peroxide Generated from Hydroquinone. *Bulletin of the Chemical Society of Japan*, Vol.82, No.10, (October 2009), pp.1299-1303, ISSN 0009-2673

Hirakawa, K. (2009). Evaluation and Chemoprevention of Phototoxic Effect by the Novel Materials. *Photomedicine and Photobiology*, Vo.31, (July 2009), pp.33-34, ISSN 0912-232X

Hiraku, Y.; Ito, K.; Hirakawa, K. & Kawanishi, S. (2007). Photosensitized DNA Damage and Its Protection via a Novel Mechanism. *Photochemistry and Photobiology*, Vo.83, No.1, (January 2007), pp.205-212, ISSN 1751-1097

Ishitani, O.; Inoue, C.; Suzuki, Y. & Ibusuki, T. (1993). Photocatalytic Reduction of Carbon Dioxide to Methane and Acetic Acid by an Aqueous Suspension of Metal-Deposited TiO_2. *Journal of Photochemistry and Photobiology A: Chemistry*, Vol.72, No.3, (June 1993), pp.269-271, ISSN 1010-6030

Kajita, M.; Hikosaka, K.; Iitsuka, M.; Kanayama, A.; Toshima, N. & Miyamoto, Y. (2007). Platinum Nanoparticle Is a Useful Scavenger of Superoxide Anion and Hydrogen Peroxide. *Free Radical Research*, Vol.41, No.6, (January 2007), pp.615-626, ISSN 1071-5762

Kawanishi, S.; Hiraku, Y. & Oikawa, S. (2001). Mechanism of Guanine-specific DNA Damage by Oxidative Stress and Its Role in Carcinogenesis and Aging. *Mutation Research*, Vol.488, No.1, (March 2001), pp.65-76, ISSN 0027-5107

Keating, K. B.; Rozner, A. G. & Youngblood, J. L. (1965). The Effect of Deformation on Catalytic Activity of Platinum in the Decomposition of Hydrogen Peroxide. *Journal of Catalysis*, Vol.4, No.5, (October 1965), pp.608-619, ISSN 0021-9517

Kiely, C. J.; Fink, J.; Brust, M.; Bethell, D. & Schiffrin, D. J. (1998). Spontaneous Ordering of Bimodal Ensembles of Nanoscopic Gold Clusters. *Nature*, Vol.396, pp.444-446, (December 1998), ISSN 0028-0836

Lee, A. F.; Baddeley, C. J.; Hardacre, C.; Ormerod, R. M.; Lambert, R. M.; Schmid, G. & West, H. (1995). Structural and Catalytic Properties of Novel Au/Pd Bimetallic Colloid Particles: EXAFS, XRD, and Acetylene Coupling. *The Journal of Physical Chemistry*, Vol.99, No.16, (Aprile 1995), pp.6096-6102, ISSN 0022-3654

Link, S.; Wang, Z. L. & El-Sayed, M. A. (1999). Alloy Formation of Gold–Silver Nanoparticles and the Dependence of the Plasmon Absorption on Their Composition. *The Journal of Physical Chemistry B*, Vol.103, No.18, (Aprile 1999), pp.3529-3533, ISSN 1520-6106

Lu, P.; Dong, J. & Toshima, N. (1999). Surface-Enhanced Raman Scattering of a Cu/Pd Alloy Colloid Protected by Poly(N-vinyl-2-pyrrolidone). *Langmuir*, Vol.15, No.23, (September 1999), pp.7980-7992, ISSN 0743-7463

Matsushita, T.; Shiraishi, Y.; Horiuchi, S. & Toshima, N. (2007). Synthesis and Catalysis of Polymer-Protected Pd/Ag/Rh Trimetallic Nanoparticles with a Core–Shell Structure. *Bulletin of the Chemical Society of Japan*, Vol.80, No.6, (June 2007), 1217-1225, ISSN 0009-2673

McKee, D. W. (1969). Catalytic Decomposition of Hydrogen Peroxide by Metals and Alloys of the Platinum Group. *Journal of Catalysis*, Vol.14, No.4, (August 1969), pp.355-364, ISSN 0021-9517

Murata, M. & Kawanishi, S. (2000). Oxidative DNA Damage by Vitamin A and Its Derivative via Superoxide Generation. *The Journal of Biological Chemistry*, Vol.275, No.3, (January 2000), pp.2003-2008, ISSN 0021-9258

Nitta, Y.; Kamiya, K.; Tanimoto, M.; Sadamoto, S.; Niwa, O. & Yokoro, K. (1991). Induction of Transplantable Tumors by Repeated Subcutaneous Injections of Natural and Synthetic Vitamin E in Mice and Rats. *Japanese Journal of Cancer Research*, Vol.82, No.5, (May 1991), pp.511-517, ISSN 0910-5050

Oikawa, S.; Furukawa, A.; Asada, H.; Hirakawa, K. & Kawanishi, S. (2003). Catechins Induce Oxidative Damage to Cellular and Isolated DNA through the Generation of Reactive Oxygen Species. *Free Radical Research*, Vol. 37, No.8, pp.881-890, (August 2003), ISSN 1071-5762

Omenn, G. S.; Goodman, G. E.; Thornquist, M. D.; Balmes, J.; Cullen, M. R.; Glass, A.; Keogh, J. P.; Meyskens Jr, F. L.; Valanis, B.; Williams Jr, J. H.; Barnhart, S.;

Cherniack, M. G.; Brodkin, C. A. & Hammar, S. (1996). Risk Factors for Lung Cancer and for Intervention Effects in CARET, the Beta-Carotene and Retinol Efficacy Trial. *Journal of the National Cancer Institute*, Vol.88, No.21, (Novenber 1996), pp.1550-1559, ISSN 0027-8874

Peiner, E. & Kopitzki, K. (1988). Metastable Phases Formed by Ion Beam Mixing of Binary Metal Systems with Positive Heats of Hormation. *Nuclear Instruments and Methods in Physics Research Section B: Beam Interactions with Materials and Atoms*, Vol.l34, No.2, (August 1988), pp.173-180, ISSN 0168-583X

Pileni, M. P. (1998). Preparation, Characterization and Application, In: Nanoparticles and Nanostructured Films:, J. H. Fendler (Ed.), 71-100, Wiely-VCH, ISBN 3527294430, Weinheim

Shipway, A. N.; Katz, E. & Willner, I. (2000). Nanoparticle Arrays on Surfaces for Electronic, Optical, and Sensor Applications. *ChemPhysChem*, Vol.1, No.1, (August 2000), pp.18-52, ISSN 1439-7641

Shiraishi, Y. & Toshima, N. (1999). Colloidal Silver Catalysts for Oxidation of Ethylene. *Journal of Molecular Catalysis A: Chemical*, Vol.141, No.1-3, (May 1999), pp.187-192, ISSN 1381-1169

Siepen, K.; Bönnemann, H.; Brijoux, W.; Rothe, J. & Hormes, J. (2000). EXAFS/XANES, Chemisorption and IR Investigations of Colloidal Pt/Rh Bimetallic Catalysts. *Applied Organometallic Chemistry*, Vol.14, No.10, (October 2000), pp.549-556, ISSN 1099-0739

T. J. Slaga, (1995). Inhibition of Skin Tumor Initiation, Promotion, and Progression by Antioxidants and Related Compounds. *Critical Reviews in Food Science and Nutrition*, Vol.35, No.1-2, (January 1995), pp.51-57, ISSN 1040-8398

Sohmiya, M.; Tanaka, M.; Okamoto, K.; Fujisawa, A. & Yamamoto, Y. (2004). Synergistic Inhibition of Lipid Peroxidation by Vitamin E and a Dopamine agonist, Cabergoline. *Neurological Research*, Vol.26, No.4, (June 2004), pp.418-421, ISSN 1743-1328

Solovieva, M. E.; Soloviev, V. V. & Akatov, V. S. (2007). Vitamin B12b Increases the Cytotoxicity of Short-time Exposure to Ascorbic acid, Inducing Oxidative Burst and Iron-Dependent DNA Damage. *European Journal of Pharmacology*, Vol.566, No.1-3, (July 2007), pp.206-214, ISSN 0014-2999

Solovieva, M. E.; Soloviev, V. V.; Kudryavtsev, A. A.; Trizna, Y. A. & Akatov, V. S. (2008). Vitamin B12b Enhances the Cytotoxicity of Dithiothreitol. *Free Radical Biology and Medicine*, Vol.44, No.10, (May 2008), pp.1846-1856, ISSN 0891-5849

Sun, S.; Murray, C. B.; Weller, D.; Folks, L. & Moser, A. (1999). Monodisperse FePt Nanoparticles and Ferromagnetic FePt Nanocrystal Superlattices. *Science*, Vol.287, No. 5460, (March 1999), pp.1989-1992, ISSN 0036-8075

Teranishi ,T. & Miyake, M. (1999). Novel Synthesis of Monodispersed Pd/Ni Nanoparticles. *Chemistry of Materials*, Vol.11, No.12, (November 1999), pp.3414-3416, ISSN 0897-4756

Toshima, N. & Wang, Y. (1994). Polymer-protected Cu/Pd Bimetallic Clusters. *Advanced Materials*, Vol.6, No.3, (March 1994), pp.245-247, ISSN 1521-4095

Toshima, N.; Yamaji, Y.; Teranishi, T. & Yonezawa, T. (1995). Photosensitized Reduction of Carbon Dioxide in Solution Using Noble Metal Clusters for Electron Transfer. *Zeitschrift für Naturforschung A*, Vol.50a, pp.283-291, ISSN 0932-0784

Toshima, N. & Hirakawa, K. (1997). Polymer-protected Pt/Ru Bimetallic Cluster Catalysts for Visible-light-induced Hydrogen Generation from Water and Electron Transfer Dynamics. *Applied Surface Science*, Vol.121/122, (November 1997), pp.534-537, ISSN 0169-4332

Toshima, N. & Hirakawa, K. (1999). Polymer-Protected Bimetallic Nanocluster Catalysts Having Core/Shell Structure for Accelerated Electron Transfer in Visible-Light-Induced Hydrogen Generation. *Polymer Journal*, Vol.31, No.11, (November 1999), pp.1127-1132, ISSN 0032-3896

Toshima, N.; Shiraishi, Y.; Matsushita, T.; Mukai, H. & Hirakawa, K. (2002). Self-organization of Metal Nanoparticles and Its Application to Synthesis of Pd/Ag/Rh Trimetallic Nanoparticle Catalysts with Triple Core/Shell Structures. *International Journal of Nanoscience*, Vol.1, No.5-6, (December 2002), pp.397-401, ISSN 0219-581X

Toshima, N.; Kanemaru, M.; Shiraishi, Y. & Koga, Y. (2005). Spontaneous Formation of Core/Shell Bimetallic Nanoparticles: A Calorimetric Study. *The Journal of Physical Chemistry B*, Vol.109, No.24, (August 2005), pp.16326-16331, ISSN 1520-6106

Toshima, N.; Ito, R.; Matsushita, T. & Shiraishi, Y. (2007), Trimetallic Nanoparticles Having a Au-core Structure. *Catalysis Today*, Vol.122, No.3-4, (Aprile 2007), pp.239-244, ISSN 0920-5861

Toshima, N. & Zhang, H. (2011). Preparation of Novel Au/Pt/Ag Trimetallic Nanoparticles and Their High Catalytic Activity for Aerobic Glucose Oxidation. *Applied Catalysis A: General*, Vol.400, No.1-2, (June 2011), pp.9-13, ISSN 0926-860X

Weyant, M. J.; Carothers, A. M.; Dannenberg, A. J. & Bertagnolli, M. M. (2001). (+)-Catechin Inhibits Intestinal Tumor Formation and Suppresses Focal Adhesion Kinase Activation in the Min/+ Mouse. *Cancer Research*, Vol.61, No.1, (January 2001), pp.118-125, ISSN 0008-5472

Widegren, J. A. & Finke, R. G. (2003). A Review of Soluble Transition-metal Nanoclusters as Arene Hydrogenation Catalysts. *Journal of Molecular Catalysis A: Chemical*, Vol.191, No.2, (January 2003), pp.187-207, ISSN 1381-1169

Willner, I.; Maiden, R.; Mandler, D.; Dürr, H.; Dörr, G. & Zengerle, K. (1987). Photosensitized Reduction of Carbon Dioxide to Methane and Hydrogen Evolution in the Presence of Ruthenium and Osmium Colloids: Strategies to Design Selectivity of Products Distribution. *Journal of the American Chemical Society*, Vol.109, No.20, (September 1987), pp.6080-6086, ISSN 0002-7863

Yamashita, N.; Murata, M.; Inoue, S.; Burkitt, M. J.; Milne, L. & Kawanishi, S. (1998). Alpha-tocopherol Induces Oxidative Damage to DNA in the Presence of Copper(II) Ions. *Chemical Research in Toxicology*, Vol.11, No.8, (August 1998), pp.855-862, ISSN 0893-228X

Yonezawa, T. & Toshima, N. (1993). Polymer- and Micelle-protected Gold/Platinum Bimetallic Systems. Preparation, Application to Catalysis for Visible-light-induced Hydrogen Evolution, and Analysis of Formation Process with Optical

Methods. *Journal of Molecular Catalysis*, Vol.83, No.1-2, (July 1993), pp.167-181, ISSN 1381-1169

Yonezawa, T. & Toshima, N. (1995). Mechanistic Consideration of Formation of Polymer-protected Nanoscopic Bimetallic Clusters. *Journal of the Chemical Society, Faraday Transactions*, Vol.91, No.22, (November 1995), pp.4111-4119, ISSN 0956-5000

6

Experimental and Theoretical Study of Low-Dimensional Iron Oxide Nanostructures

Jeffrey Yue, Xuchuan Jiang*, Yusuf Valentino Kaneti and Aibing Yu
*School of Materials Science and Engineering, University of New South Wales, Sydney,
Australia*

1. Introduction

Iron oxide has many phases, including 16 pure phases (e.g., FeO, Fe_3O_4), 5 polymorphs of FeOOH (e.g., α-FeOOH, β-FeOOH) and 4 kinds of Fe_2O_3 (e.g., α-Fe_2O_3, γ-Fe_2O_3). Because of their unique properties (optical, electronic, magnetic), they have found many applications in the areas of catalysts, magnetic recording, sorbents, pigments, flocculants, coatings, gas sensors, lubrications, and biomedical applications (e.g., magnetic resonance imaging, drug delivery and therapy).

Many efforts have been made in the synthesis (co-precipitation, hydrothermal, micro-emulsion, and sol-gel method), structural characterization, and functional exploration, as well as fundamental understandings of iron oxide nanostructures. Despite some success, several challenges still exist regarding the synthesis, strcture, properties, and fundamental understanding of the iron oxides. the grand challenge is how to efficiently synthesize iron oxides with controlled morphology, size and functionality, and how to fundamentally understand the formation and growth mechanisms, structure, and interaction forces. Therefore, the development of simple but effective experimental and theoretical strategies to overcome the challenges is still imperative.

To fundamentally understand the nanoscale system, theoretical methods should exist. Computational modeling is one of the most important enabling techniques in nanotechnology and material research. It can increase the pace of discovery across the entire scientific scope, and reduce the cost in the development and commercialization of technologies and materials. Various computational approaches have been developed and used to predict the materials properties (e.g., electronic, magnetic, optical) at different length and time scales. For example, at an atomic scale, density functional theory (DFT) is widely used for binding energy calculation, while at a microscopic scale, molecular dynamics (MD) are able to provide insights into atomic/molecular systems.

This Chapter will give a brief overview of the experimental and theoretical methods conducted on iron oxide nanostructures, particularly for low-dimensional iron oxide nanoparticles. This includes: (i) several representative methods for iron oxides nanomaterials in Sections 2 and 3; (ii) surface modified iron oxide nanostructures by

---

* Corresponding Author

surfactants, polymers, silica or metals in Section 4; (iii) functional properties of such nanomaterials in gas sensing, catalysis, and biotechnology in Section 5. Moreover, in Section 6, the discussion will be extended to the theoretical modeling and simulation methods that can predict the formation and performance of nanomaterials, such as MD and DFT methods.

2. Iron oxide materials

Iron is the fourth most abundant element in the Earth's crust, and iron oxides are commonly found in nature and have become the most plentiful transition metal oxides (Morrissey and Guerinot, 2009; Ilani et al., 1999). The complicated phases and features of iron oxides have been listed in **Table 1**.

Some crystalline phases of iron oxides are not very stable and can convert into others. Much work has been conducted to convert akaganéite to hematite and/or magnetite phases to pursuit good performance in catalysis and gas sensing applications. Magnetite nanorods can be produced by the conversion of iron oxyhydroxides into a thermally stable structure of hematite by heating above 400 °C in air, or magnetite in a mixture of H_2 and Ar gas (Bomati-Miguel et al., 2008). Recently, our group has simplified the phase conversion procedures among the iron oxides. The iron oxyhydroxides can directly convert into magnetite by using hydrazine as a reducing agent, and the morphology was maintained. Using this method, magnetite nanorods could be directly synthesized from akaganéite rather than using hematite as an intermediate (Yue et al., 2010).

By using hydrazine, iron(III) ions can be reduced to iron(II). The change in coordination number to the iron atom will therefore transfer from Fe–OH to Fe–O following dehydration. The structure change caused by loss of H_2O will create pores or holes within the nanorod framework. Continuous reaction with hydrazine can form larger defects in the 1-D nanostructure, leading to the collapse of the framework. At the same time, the FeO_6 units will reconstruct into other crystals, and the broken fractions could fuse with neighboring particles to form larger ones. However, this does not happen to hematite because of its thermally stable structure under the considered conditions. The nature of such a conversion needs further investigation. Nevertheless, the proposed approach could be used for a controlled conversion of akaganéite to magnetite nanostructures (**Fig. 1**) without high-temperature treatment. These porous magnetic structures would find more applications in electronic and magnetic areas (Yue et al., 2011).

3. Synthesis methods

A variety of methods have been reported to synthesize iron oxide nanoparticles, including solid-state, liquid-phase, and gas-phase syntheses, as listed in **Table 2**. Among the synthesis approaches, liquid-phase synthesis is the most popular. The iron salts are highly soluble in water and different additives can be used in conjunction to modify the structure of the nanoparticles. Moreover, the liquid-phase synthesis is convenient for understanding ageing, recrystallization, and evolution into other shapes and sizes. It is also available for controlling experimental conditions in liquid (e.g., concentration, salt precursor, pH, temperature, surface modifiers). A few representative synthesis methods are briefly introduced in this Section, such as co-precipitation, hydrothermal and microemulsion.

Iron oxides	Crystallographic system	Crystallographic structural features	References
Goethite (α-FeOOH)	Orthorhombica = 0.9956 nm, b = 0.30215 nm and c = 0.4608 nm	3D-structure built up with $FeO_3(OH)$ octahedra spreading along the (010) direction, with each octahedron linked to eight neighbouring octahedral by four edges and three vertices. Oxygen atoms are in tetrahedral surroundings, either OFe_3H or OFe_3H (bond).	Cornell and Schwertmann, 1991; Cudennec and Lecerf, 2005
Akaganéite (β-FeOOH)	Monoclinic a = 1.056 nm, b = 0.3031 nm and c = 1.0483 nm	Double chains of edge linked Fe(O, OH) octahedral that share corners to form a framework containing large tunnels with square cross sections.	Cornell and Schuwertmann, 1991; Garcia et al., 2009
Lepidocrocite (γ-FeOOH)	Orthorhombic a = 0.3071 nm, b = 1.2520 nm and c = 0.3873 nm	Arrays of close cubic -packed anions (O^{2-}/OH^-) stacked along the [150] direction with Fe^{3+} ions occupying the octahedral interstices.	Cornell and Schuwertmann, 1991
Feroxyhyte (δ'-FeOOH)	Hexagonal a = 0.293 nm and c = 0.456 nm	Disordered hexagonally close-packed array of anions with Fe^{3+} ions distributed over half the octahedral sites in an orderly manner.	Cornell and Schuwertmann, 1991
Hematite (α-Fe$_2$O$_3$)	Hexagonal a= 0.5035 nm and c = 1.375 nm	Stacking of sheets of octahedrally (six-fold) coordinated Fe^{3+} ions. Between two close-packed layers of oxygen ions. Each oxygen ion is bonded to only two Fe ions.	Mohapatra and Anand,2010
Maghemite (γ-Fe$_2$O$_3$)	Cubic a= 0.83474 nm	Each cell of maghemite contains 32 O^{2-} ions, 21(1/3) Fe^{3+} ions and 2 (1/3) O vacancies. Eight cations occupy tetrahedral sites and the remaining cations are distributed over the octahedral sites. The vacancies are confined to octahedral sites.	Cornell and Schuwertmann, 1991; Weckler and Lutz, 1998
Magnetite (Fe$_3$O$_4$)	Cubic a = 0.8396 nm	Inverse spinel structure with a face-centered cubic cell based on 32 O^{2-} ions, regularly close-cubic packed along [111], with Fe^{2+} ions and half of the Fe^{3+} occupying the octahedral sites and the other half of Fe^{3+} ions, occupying the tetrahedral sites.	Cornell and Schuwertmann, 1991 Mohapatra and Anand,2010

Table 1. Complicated phases and polymorphs of iron oxides in nature

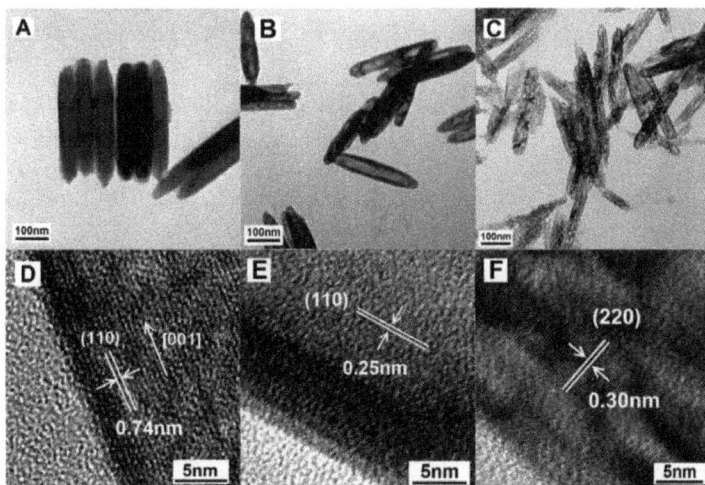

Fig. 1. TEM images showing the conversion of nanorods: (A) β-FeOOH nanorods, (B) β-FeOOH nanorods calcined at 300 °C, (C) β-FeOOH nanorods reduced with N_2H_4 at 80 °C; and (D-F) the corresponding HRTEM images with labeled lattice spacing and crystal planes. Reprinted with permission from (Yue et al., 2011).

3.1 Co-precipitation

One simple and efficient way is to use co-precipitation technique in solution. By this approach, iron(II) and/or iron(III) salts are first dissolved in aqueous solution, and then one alkaline media (e.g., NaOH, Na_2CO_3) solution is added to form precipitate. The prepared particles can be tuned to be uniform in size, shape as well as pure in its composition. Various crystalline phases of iron oxides can be produced using this method, which is controlled by experimental parameters such as types of iron salts (e.g., chloride, sulphate and nitrate), alkaline media, concentration, temperature, and pH (Iida et al., 2007).

Moreover, the phase of iron oxide(s) formed through the co-precipitate approach is often reported as goethite or hematite if iron(III) salt is used. However, the initially precipitated material is usually found as ferrihydrite, which is a thermodynamically unstable phase. The precipitate can further convert into other phases (e.g., hematite, magnetite) depending on the pH, ionic medium, and temperature. For example, Varada et al. (2002) prepared monodispersed acicular goethite particles by precipitating Fe(III) using sodium carbonate. If sodium hydroxide was used, the axial ratio of particles will increase from 60 to 230 nm. It was proposed that different bases have different ability to maintain the solution at a constant pH, where other pH levels would produce polydispersed and hematite particles. The mechanism of the growth of spherical hematite nanoparticles has been explored by Liu et al.(2007). The variation in the final pH of the solution plays a key role in the formation of hematite at different sizes. They found that the particles with diameter of 60-80 nm were obtained at pH 7, while reduced to 30-40 nm in diameter at pH 9.

Synthesis media	Synthesis methods	Common products	Particle shape/size	Features	References
Solid state	Mechanical milling	δ-Fe_2O_3, Fe_3O_4	Spheres D= 2.9-3.6 nm	Mechanical energy to smash.	Lu et al. (2007)
Liquid state	Co-precipitation	α-Fe_2O_3, δ-Fe_2O_3, Fe_3O_4	Nanospheres (d = 30 - 80 nm)	Ageing of ferric and ferrous salts in a basic medium.	Liu et al. (2007)
	Hydrothermal	α-Fe_2O_3, δ-Fe_2O_3, Fe_3O_4 (α-FeOOH β-FeOOH	Nanorods (l: 400-600 nm, w =20-30 nm) Nanodiscs (d = 50 nm, thickness = 6.5 nm)	Low temperature, reaction, commonly conducted in autoclaves, and high efficiency.	Li et al. (2006); Yue et al. (2010, 2011); Jiang et al. (2009)
	Thermal decomposition	α-Fe_2O_3, δ-Fe_2O_3, Fe_3O_4	Nanospheres (d = 16 nm)	High-temperature decomposition of iron organic precursors.	Sun et al. (2004)
	Sol-gel	α-Fe_2O_3, δ-Fe_2O_3, Fe_3O_4	Nanoparticles (24-52 nm)	Dissolve, condensation, and calcinations of alkoxides.	Dong and Zhu (2004)
	Microemulsion	α-Fe_2O_3, δ-Fe_2O_3, Fe_3O_4	Nanoparticles (3-5 nm)	Reaction in two immiscible phases (water and oil).	Vidal-Vidal et al. (2006)
	Sonochemical	Fe_2O_3, Fe_3O_4	Nanorods 10-80 nm in diameter	Ultrasound to promote chemical reaction.	Vijayakumar et al. (2000, 2001)
	Electrochemical	γ-Fe_2O_3	Nanoparticles (d= 3-20 nm)	Electrons act as reactant with no pollution.	Zhang et al (2007); Pascal et al (1999)
Gas state	Spray pyrolysis	δ-Fe_2O_3, Fe_3O_4	Hollow nanospheres, (d =300 nm)	Spraying, aerosol evaporation, condensation, drying, and thermolysis.	González-Carreño et al. (1993)
	Laser pyrolysis or deposition	α-Fe_2O_3, δ-Fe_2O_3, ε-Fe_2O_3 Fe_3O_4	Nanowires (30 nm ×1-5 μm) Nanobelts (100 nm × 7 μm)	Heating of a gaseous mixture of iron precursor.	Morber et al. (2006)

Table 2. Several typical synthesis methods for iron oxides

3.2 Hydrothermal and thermal decomposition methods

Hydrothermal technique is defined as any heterogeneous reaction in the presence of aqueous solvents or mineralizers under a high pressure and a temperature (6-10 atm, 100-200 °C). A hydrothermal reaction requires the iron(III) salt (e.g., iron chloride, nitrate, or sulphate), which can be dissolved in solution followed by reaction with water. This is different from the thermal decomposition reaction that generally takes place for those iron organic precursors ($Fe(CO)_5$, $Fe(acac)_3$, and $Fe(cup)_3$)in an organic solvent at high temperatures (Hyeon et al., 2001; Li et al., 2004; Rockenberger et al., 1999). Both hydrothermal and thermal decomposition methods are commonly used for the synthesis of iron oxide nanoparticles.

The hydrothermal method is often performed in an autoclave, where the reaction system can exceed the boiling point of liquid(s) at normal atmospheric pressure (Jia et al., 2005). The temperature can alter the system in such a way that disrupts the thermodynamics of a material, which is governed by enthalpy (ΔH) and entropy (ΔS), and hence Gibbs free energy (ΔG). The essential role of a fluid under high temperatures is that it changes the vapor pressure of the fluid. This is also beneficial for diverse choices of solvents (polar and non-polar). The morphology and crystalline phase of iron oxides produced through this approach can vary by simply tuning reaction temperature, concentration, and additive(s) (Almeida et al., 2009; Jiang et al., 2010).

The synthesis of iron oxide nanoparticles via a hydrothermal approach can be conducted with or without the use of surfactant(s). Hematite nanoparticles have been prepared by Sahu et al. (1997) under conditions of pH (3-10) and 180 °C in autoclaves. In this study, the average particle size of hematite nanoparticles was found to decrease with an increase of pH. In our recent work (Jiang et al. 2010), we reported a facile hydrothermal route for the synthesis of monodispersed hematite nanodiscs with diameters of ~ 50 nm and thickness of ~6.5 nm in the absence of any surfactants in water at around 90 °C (**Fig. 2**). The nanodiscs exhibited interesting paramagnetic property at a low temperature (20 K), but ferromagnetic at room temperature (~300 K). In addition, the hematite nanodiscs also showed low-temperature catalytic activity in CO oxidation to CO_2.

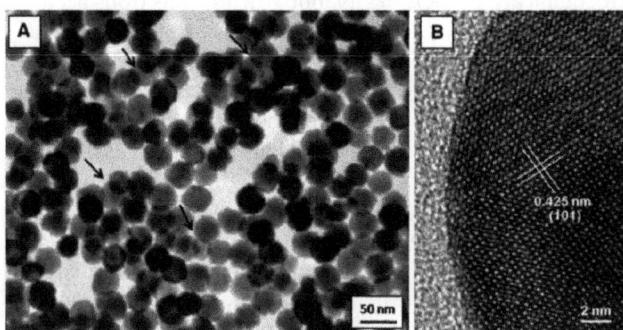

Fig. 2. A) TEM image of α-Fe_2O_3 nanodiscs with overlapping as pointed by arrows; B) HRTEM image showing the lattice fringe of {110} plates with spacing between two adjacent planes of 0.411 nm. Reprinted with permission from (Jiang et al. 2010).

3.3 Microemulsion

Microemulsion method (surfactant-stabilized water/oil (W/O) microemulsion) has been widely used to prepare shape- and size-controlled iron oxide nanoparticles. Generally, a microemulsion is transparent, isotropic and thermodynamically stable dispersion of two immiscible phases (e.g., water and oil). When a surfactant is present in W/O system, the surfactant molecules may form a monolayer at the interface of oil and water, with the hydrophobic tails of the surfactant molecules dissolved in the oil phase and the hydrophilic head groups in the aqueous phase (Wu et al., 2008). In a binary system such as water/surfactant or oil/surfactant, a variety of self-assembled structures can be formed, ranging from spherical and cylindrical micelles to lamellar phases or bi-continuous microemulsions depending on the molar ratio of water, oil and surfactant(s). This will be useful for the generation of nanoparticles with different shapes and sizes.

For example, magnetite nanoparticles ~4 nm in diameter have been prepared by the controlled hydrolysis of ammonium hydroxide with $FeCl_2$ and $FeCl_3$ aqueous solution within the reverse micelles nanocavities generated by sodium bis(2-ethylhexyl) sulfosuccinate (AOT) as a surfactant and heptane as a continuous oil phase (López-Quintela and Rivas, 1993). Lee and co-workers (2005) have successfully synthesized uniform and highly crystalline magnetite nanoparticles in microemulsion nanoreactors. The particle size of the prepared magnetite nanoparticles could be adjusted from 2-10 nm by varying the relative concentrations of iron salt, surfactant, and solvent. Li et al (2009) demonstrated the effect of volumetric ratios of aqueous $FeCl_3$ solution to 1,2-propanediamine on the formation of magnetic particles, as shown in **Fig. 3**. Chin and Yaacob (2007) reported the synthesis of

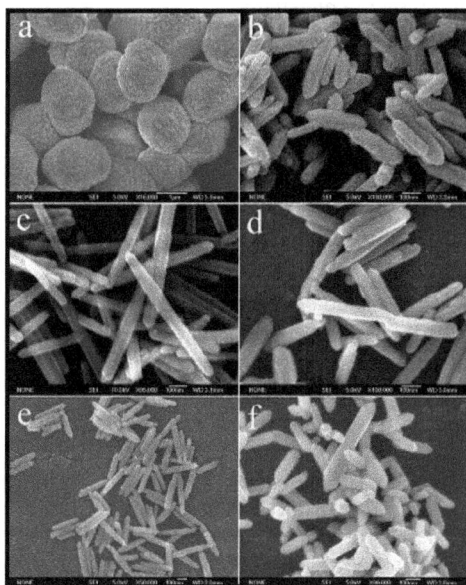

Fig. 3. SEM images of the products obtained at different volume ratios of aqueous $FeCl_3$ solution to 1,2 propanediamine: (a) without 1,2 propanediamine, (b) 3:1, (c) 1:1, (d) 1:2, (e) 1:4, and (f) 1:5. Reprinted with permission from (Li et al., 2009).

magnetic iron oxide nanoparticles with an average particle size of <10 nm by mixing two microemulsion systems, one containing Fe^{2+} ions and the other containing OH- ions. The study reveals that the nanoparticles prepared by the microemulsion technique were smaller in size and higher in saturation magnetization than those nanoparticles prepared by Massart's procedure (Massart et al., 1981).

Despite some success, this microemulsion approach has some drawbacks, such as the difficulty in scale-up production, the adverse effect of residual surfactants on the properties of the nanoparticles, and the aggregation of the produced nanoparticles. Repeated wash processes and further stabilization treatment are usually required for such a reaction approach (Wu et al., 2008).

4. Surface modifications

The surface modifications of nanoparticles have attracted much more attention, which can improve the surface-related properties like hydrophobic or hydrophilic. This can be achieved by using surfactants, polymers, and inorganic materials (silica).

4.1 Surfactants

Surface modification with surfactant(s) is widely used for altering surface properties such as hydrophobic or hydrophilic. The use of surfactant molecules, such as oleic acid, oleylamine, or thiols (Wang et al., 2005), can easily functionalize iron oxide nanoparticles to be hydrophobic surfaces. These molecules can covalently bond to the iron atoms or clusters against particle degradation (Soler et al., 2007).

Many researches focus on the synthesis of water-soluble iron oxide nanoparticles with biocompatibility and biodegradability for biological applications. For example, one is to directly introduce the biocompatible organic molecules, e.g., amino acid (Sousa et al., 2001), vitamin (Mornet et al., 2004), and citric acid (Morais et al., 2003). Despite some advantages, the instability of small organic molecules in alkaline or acidic environment may result in agglomeration of the functionalized iron oxide nanoparticles.

Another alternative technique is to transform the oil-soluble type into water-soluble one via a ligand exchange reaction (Chen et al., 2008). The ligand exchange involves the addition of an excess of ligand(s) to nanoparticle suspension, which has stronger interaction with the nanoparticles than the original ones. Sun et al. (2003) converted the synthesized hydrophobic maghemite nanoparticles into hydrophilic ones by mixing with bipolar surfactants such as tetramethylammonium 11-aminoudecanoate. Lattuada and Hatton (2006) reported that the oleic groups initially present on the surface of magnetite nanoparticles were replaced by various capping agents containing reactive hydroxyl moieties. They also tuned the particle size in the range of 6-11 nm by varying the heating rate.

4.2 Polymers

Polymer-functionalized iron oxide nanoparticles have gained much more attention due to the benefits offered by polymeric coating, which may increase repulsive forces to balance the magnetic and van der Waals attractive forces acting on the nanoparticles (Wu et al.

2008). It has shown that through careful choice of the passivating and activating polymers and/or reaction conditions, polymer-stabilized iron oxide nanoparticles with tailored and desired properties can be synthesized.

The iron oxide particles by ionic properties can be modified with functional polymer groups with –COOH, –NH$_2$ (Chibowski et al., 2009; Kandori et al., 2005; Li et al., 2004). The polymer coated particles can be synthesized by the *ex situ* method, i.e. dispersion of the nanoparticles in a polymeric solution, or *in situ* method, i.e. monomer polymerization in the presence of the synthesized nanoparticles (Mammeri et al., 2005; Guo et al., 2007).

Polymeric coating materials can be classified into two main classes: natural (e.g., dextran, starch, gelatin, chitosan) and synthetic (e.g., polyethylene glycol, PEG; polymethylmethacrylate, PMMA; polyacrylic acid, PAA). However, the saturation magnetization value of iron oxide nanoparticles will decrease after polymer-fictionalization.

Dextran is often utilized as a coating polymer because of its stability and biocompatibility (Laurent et al. 2008). Molday and Mackenzie (Molday and Mackenzie, 1982) have reported the formation of Fe$_3$O$_4$ in the presence of dextran with molecular weight (MW) of 40,000. In the synthesis of dextran-coated ultra-small superparamagnetic iron oxides (USPIO), the reduction of the terminal glucose of dextran was found to be significant for controlling particle size, stability, and magnetic properties. For low molecular weight dextrans (MW, <10,000), it is difficult to obtain nanoparticles with a small size of <20 nm.

Polyvinyl alcohol (PVA) is a hydrophilic and biocompatible polymer that can be used for particle surface modification to prevent particle agglomeration (Laurent et al. 2008). Lee et al. (1996) have modified the surface of magnetite nanoparticles with PVA by precipitation of iron salts at a high pH (13.8) to form stable magnetite colloidal dispersions, and particle size is around 4 nm. The investigators noted that the crystallinity of the magnetite nanoparticles decreased with PVA concentration increasing, although morphology and particle sizes remained. When PVA is introduced, it reacts with the surface through hydrogen bonding between polar functional groups of the polymer and hydroxylated and/or protonated surface of the iron oxide. In addition to the polymer-surface interactions, PVA is known for its hydrogen bonding interaction, resulting in hydrogel structure embedding the nanoparticles. When the PVA concentration is over the critical saturation value, agglomeration may occur for PVA-coated particles via bridging interactions.

4.3 Polymerized amorphous silica

Polymerized tetraethoxysilane (TEOS) network is often used as a surface coating material for iron oxide nanoparticles as this coating can prevent aggregation in solution, improve the chemical stability, and provide better protection against toxicity (Laurent et al. 2008). Additionally, polymerized silica-coated iron oxide nanoparticles exhibited good biocompatibility and solubility in water. Silica coating can stabilize the magnetite nanoparticles in two different ways: one is by shielding the magnetic dipole interaction with the silica shell, and another one is by enhancing the coulomb repulsion of the magnetic nanoparticles. Such a silica coating increases the size of the particles and decreases the saturation magnetization value.

A commonly used method to coat iron oxide nanoparticles with silica is the well-known Stöber method, in which silica is formed *in situ* via hydrolysis and condensation of a sol-gel precursor such as TEOS. For example, Im et al.(2005) have reported the synthesis of silica colloids loaded with superparamagnetic iron oxide nanoparticles, which revealed that the final size of silica colloids depended upon the concentration of iron oxide nanoparticles because the size of silica was closely related to the number of seeds (emulsion drops). The lower concentration the iron oxide nanoparticles in alcohol, the larger size the obtained colloids.

Another one is aerosol-pyrolysis method, in which silica-coated magnetic nanoparticles were prepared by pyrolysis of a mixed precursor of silicon alkoxides and metal compound in a flame environment (Deng et al. 2005). Tartaj *et al.*(2001) synthesized silica-coated γ-Fe_2O_3 hollow spheres with size of 150 ± 100 nm by aerosol pyrolysis of methanol solution containing iron ammonium citrate and silicon ethoxide.

4.4 Metals

Noble metals (e.g., Au, Ag, Pt, and Pd), possessing unique electronic and catalytic properties, can be utilized to improve the physicochemical properties of magnetic nanoparticles and applications in biomedicine. The coating of iron oxide nanoparticles with noble metals can be helpful to improve stability from aggregation, however, decrease the saturation magnetization value in some cases (Wu et al. 2008).

Several procedures have been employed to synthesize such core-shell nanostructures. For example, Mikhaylova *et al.* (2004) have prepared gold-coated superparamagnetic iron oxide nanoparticles (SPION) using a reverse micelle method. In their study, the reverse micelles were formed from surfactant, cetyltrimethylammonium bromide (CTAB), octane (the oil phase), butanol (the co-surfactant), and an aqueous mixture of $FeCl_3$, $FeCl_2$ and $HAuCl_4$ solutions. They found that the Au-coated SPION retained the superparamagnetic properties for a longer period than those of starch-coated and multi-arm polyethylene glycol (MPEG)-coated ones. Wang et al.(2005) obtained gold coated iron oxide nanoparticles, in which the pre-synthesized Fe_3O_4 nanoparticles were used as seeds during the reduction of gold precursor, $Au(OOCCH_3)_3$. The average size of Fe_3O_4 nanoparticles increases from 5.2 ± 0.5 nm to 6.7 ± 0.7 nm after coating with gold (**Fig. 4**). Fe_3O_4/Au and Fe_3O_4/Au/Ag core/shell nanoparticles with tuneable plasmonic and magnetic properties have been developed by controlling the coating thickness and materials (Xu et al. 2007).

A facile and one-pot synthesis approach has been developed by Zhang et al. (Zhang et al., 2010) for generating metal (Au, Pt, Ag and Au-Pt)/Fe_2O_3 nanocomposites assisted by lysine. Lysine, containing functional groups -NH_2 and –COOH, acts as both a linking molecule to the Fe_2O_3 matrix and a capping agent to stabilize the noble metal nanoparticles for a good dispersion. Jiang et al. (Jiang and Yu, 2009) have demonstrated a facile synthetic method for the preparation of Pd/α-Fe_2O_3 nanocomposites by adding citric acid into a mixture of iron oxide nanoparticles and palladium precursor, $Pd(CH_3CN)_2Cl_2$) under a reflux heating at 90°C for 2 hours. The synthesized Pd/α-Fe_2O_3 nanocomposites inherited the rod-like morphology of the α-Fe_2O_3 nanoparticles and they exhibited superior catalytic activity in CO oxidation compared with pure α-Fe_2O_3 nanoparticles. UV-vis measurement of the nanocomposites revealed the presence of two plasma bands centered at around 383 and 552 nm, which can be assigned to the synergistic effect of both Pd and α-Fe_2O_3 nanoparticles.

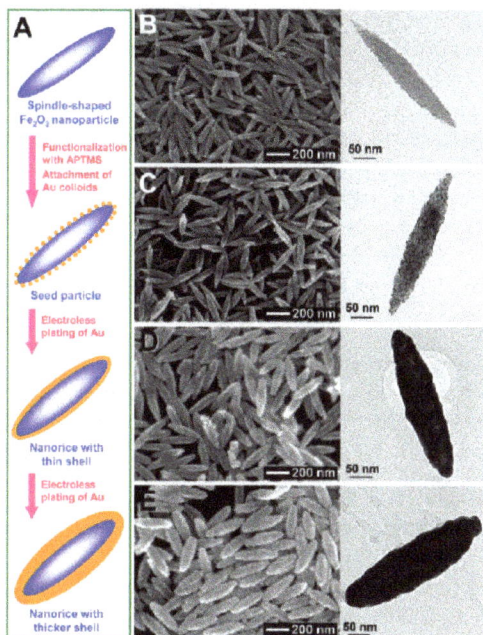

Fig. 4. (A) Schematics of the hematite-gold core-shell nanorice particles. SEM (left) and TEM (right) of (B) hematite core, (C) seed particles, (D) nanorice with thin shells, and (E) nanorice with thick shells. Reprinted with permission from (Wang et al., 2006).

4.5 Carbon

Carbon has been widely studied since its poly-morphologies as active carbon, graphite, graphene, carbon nanotubes, and fullerene bucky ball structures. They have exhibited extraordinary tensile strengths and electrical conductivity due to their covalent $sp2$ hybridized network structure. The combination of semi-conductive iron oxides and carbon may therefore enhance the electrical properties of the nanocomposite material. The method to coat carbon on the surface of iron oxide is often performed by the decomposition of a carbon source (i.e., hydrocarbons, polymer or glucose) at high temperatures under oxygen-free environments (Tristão et al., 2010; Tristão et al., 2009; Zhang et al., 2008, 2010).

Carbon coated iron oxide particles have attracted much more attention. Zhang et al. (2008) demonstrated that carbon coated magnetite nanorods can be synthesized through a series of procedures. In this process, hematite nanorods were firstly synthesized by a hydrothermal method as previously mentioned. Secondly, glucose was coated onto the hematite nanorods by pyrolysis under hydrothermal conditions. Finally, the product was heated at 600 °C under N_2 to carbonize glucose and reduce hematite into magnetite simultaneously. Boguslavsky et al.(2008) reported a similar procedure, in which polydivinylbenzene (PDVB) was used as the carbon source. The PDVB coating was formed by emulsion polymerization of DVB in the presence of γ-Fe_2O_3, followed by annealing of the powder in a quartz tube at 1050 °C under flowing Ar gas for 2 hours. The decomposition of the polymer in this case reduced γ-Fe_2O_3 to metallic Fe, which finally forms carbon coated iron (Fe/C) nanoparticles.

In addition, Wang et al. (2006) have reported the synthesis of Fe_3O_4/C nanocomposites by heating the aqueous solution of glucose and oleic acid-stabilized Fe_3O_4 nanoparticles at 170 °C for 3 hours. The results revealed that without prior surface hydrophobic modification, the magnetite nanoparticles could not be encapsulated by the carbon nanospheres, but instead only bare carbon nanospheres with the size of ~200 nm and Fe_3O_4 nanoparticles were obtained. The variation of glucose concentration (0.3-0.6 M) and the reaction temperature (160-180 °C) were found to have no significant effect on the morphology of the product, however, both reaction time and the amount of oleic acid-stabilized Fe_3O_4 nanoparticles showed significant effects. The increase in the concentration of oleic-acid stabilized Fe_3O_4 nanoparticles from 2.5 to 6 g/L was found to generate a product that has more embedded Fe_3O_4 nanoparticles increasing from 41 to 63%).

Although carbon-coated iron oxide nanoparticles may offer some advantages, such particles are often obtained as agglomerated clusters due to the lack of effective synthetic control, and lack of proper understanding on the formation mechanism. The synthesis of dispersible carbon-coated nanoparticles in isolated forms still remains a challenge in this field.

Moreover, the surface modification of iron oxide allows the attachment of biomolecules such as proteins and drugs (Mohapatra et al. 2007; Sun et al. 2007). The design of the surface modifications may be determined by factors such as ion energy and ion flux of depositing species, interface volume, crystalline size, coating thickness, surface and interfacial energy (Kim et al. 2003; Pinho et al. 2010).

5. Functionalities of iron oxide nanostructures

5.1 Magnetic property

The magnetic property has been extensively studied since it was discovered and explained through electronic structures of atoms. The magnetic dipole moments generated by the spin and orbital angular momenta of electrons in the Fe atom may vary between each phase of the iron oxide material. In general, magnetic behavior of a material depends on the electron spin vector or the total magnetic dipole moment. One important aspect in iron oxide nanoparticles is the unique form of magnetism called superparamagnetism. At temperature of above the blocking temperature, the magnetization behavior is identical to that of atomic paramagnets. This phenomenon will occur if particles reach below a certain size (10-20 nm), when the particle consists of a single magnetic domain, even though the material is ferro- or ferri-magnetic in bulk form (Ye et al., 2007), as shown in **Fig. 5**. Particles with this type of magnetism show high field irreversibility, high saturation field, extra anisotropy contributions, and shifted loops (Pedro et al., 2003).

For noble gold and silver nanoparticles with unique surface plasmon resonance (SPR) properties, they are often used to modify the iron oxide surfaces for generating coupled or multiple functionalities. At the nanoscale, the metallic electron cloud oscillates on the particle surface and absorbs electromagnetic radiation at a particular energy. The surface geometry of the iron oxide particles such as spheres, cubes, triangles, or rods, can therefore influence the absorption of radiation from the ultra-violet up to the near infrared spectrum (350-1200 nm). Other factors that affect the absorption are the solvent and surface functionalization. They are important contributors that can tune the exact frequency and intensity of the plasmon resonance band, which attracts them to the surface enhanced resonance spectroscopy (SERS) for sensing devices (Zhai et al., 2009). This effect is also of importance for bimetallic core/shell nanoparticles. As the ratio of gold to iron oxide increases, the gold character increases and the iron oxide becomes buried beneath and suppresses the dielectric effect. The increasing thickness of the shell structure will therefore cause blue-shifting in the surface plasmon resonance (Lyon et al. 2004).

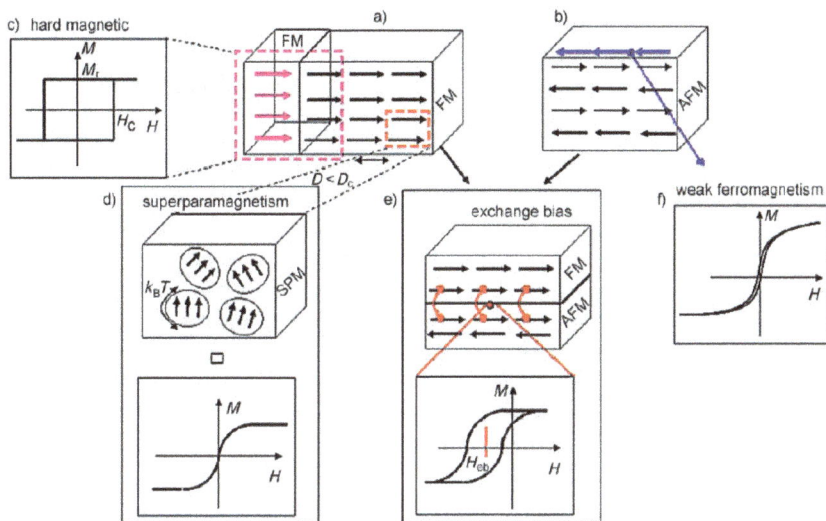

Fig. 5. Diagram of different spin arrangements in magnetic nanoparticles:
a) Ferromagnetism (FM), b) Antiferromagnetism (AFM), D = diameter, D_c = critical diameter, c) a combination of two different ferromagnetic phases in permanent magnets, which are materials with high remanence magnetization (M_r) and high coercively (H_c), d) Superparamagnetism (SPM), e) the interaction at the interface between a ferromagnet and an antiferromagnet producing an exchange bias effect, and f) pure anti-ferromagnetic nanoparticles with superparamagnetic relaxation arising from uncompensated surface spins. Reprinted with permission from (Lu et al., 2007).

5.2 Biomedical applications

Many investigations have been reported the application of nanoparticles for biomedicine, such as magnetic nanoparticles for improving the quality of magnetic resonance imaging (MRI), hyperthermic treatment for malignant cells, site-specific drug delivery, cell labeling, and manipulating cell membranes (Babič et al. 2008; Catherine and Adam, 2003). These magnetic particles can also be used for diagnosis, imaging, and drug delivery.

Iron oxide nanocomposites or particle coated with biocompatible polymer(s) have shown some advantages, e.g., reducing aggregation, maintain magnetic stability, slowdown degrading process under physiological conditions, and lower toxicity (Mahmoudi et al. 2009). So far, they have shown promise for monitoring living cells by both MR and fluorescence imaging, as well as for drug delivery (Liong et al., 2008).

As mentioned previously, iron oxide nanoparticles exhibit paramagnetic or superparamagnetic properties in a limited size range. Particles larger than 50 nm show superparamagnetic iron oxides (SPIO), whereas particles smaller than 50 nm show ultrasmall superparamagnetic property (USP). The smaller ones have the ability to enhance signal detection and increase resolution in the MRI (Foy et al., 2010; Tong et al., 2010). Therefore, the SPIO particles can be used for imaging tumors in the liver and spleen, while superparamagnetic particles for contrast agents for lymphography and angiography. However, the superparamagnetic particles do not retain their magnetism when the external magnetic field is removed, while other magnetic materials will become magnetized and aggregate.

In addition, the problem using magnetite or maghemite nanoparticles in clinic is often limited by the biocompatibility and toxicity of these particles (Martin et al., 2008; Pisanic Ii et al., 2007). This happens from the body's defense system, the reticulo-endothelial system (RES), trying to remove these particles from the bloodstream as they pass through the liver, spleen and lymph nodes. The rapid removal of the iron oxide nanoparticles reduces their life-time. This is why it is necessary to produce nanocomposites with special surface modifications. The surface modification of the particles allows the water-insoluble drugs to be loaded and stored for a long time (Liong et al., 2008; Son et al., 2005). Despite some progress, the challenges in using surface modified magnetic iron oxide nanoparticles still exist. More work needs to be performed in the future.

5.3 Gas sensing

Gas detection with high sensitivity and selectivity is essential for controlling industrial, waste, and vehicle emissions, household activity and environmental monitoring. In the past decades, many sensor devices have been developed for various gases such as CO, CO_2, O_2, O_3, H_2, NH_3 and SO_2, as well as various organic vapors e.g., benzene, methanol, ethanol, amines and isopropanol (Jimenez-Cadena et al. 2007). Although semiconducting oxides have been quite useful as gas sensors, the operation at high temperatures often limits their functionality and applications. This has prompted the exploration of new materials that may offer higher sensing and selective capabilities than traditional ones.

Nanostructured metal oxides are one of the most commonly used materials for gas sensing because of the semiconductors make them possible for the electrical conductivity change when the surrounding atmosphere changes. Additionally, nanosized metal oxides exhibit high ratios of surface to volume, which favors the adsorption of gases on the particle surface, and hence increases the sensitivity in detection.

Iron oxide nanoparticles have shown good sensing capabilities toward hydrocarbon gases, CO and alcohols (Jimenez-Cadena et al., 2007; Han et al., 1996, 1999, and 2001). The studies by Zhang et al. (1996) and Tao et al. (1999) showed that γ-Fe_2O_3 nanosensors exhibited good sensitivity and selectivity to a range of hydrocarbon gases such as LPG, petrol and C_2H_2 at 380 °C, but poor sensitivity to H_2 and CO. However, Nakatani and Matsuoka (1983) together with Lee and Choi (1990) reported that the γ-Fe_2O_3-based sensors exhibited good sensitivity to H_2. This suggests that the gas-sensing characteristics of a nanosensor are related to its preparation process.

The sensitivity of iron oxide-based nanosensors can be improved by various doping schemes as well as by changing the sensing material structure. For example, the thin film type sensors tend to exhibit higher sensitivity than bulk material sensor(s) (Mohapatra and Anand, 2010). Tao and co-workers (1999) have studied the sensing characteristics of Y_2O_3-doped γ-Fe_2O_3 towards hydrocarbon gases, H_2 and CO and found that the addition of Y_2O_3 to γ-Fe_2O_3 resulted in a little difference in the sensitivity and selectivity compared with those made of pure γ-Fe_2O_3. Neri et al.(2002) have assessed the gas-sensing properties of Zn-doped Fe_2O_3 thin films prepared by liquid phase deposition method. They observed that the addition of metal Zn can increase the sensitivity of the Fe_2O_3 thin film to NO_2 below 250 °C.

5.4 Catalyst

A catalyst can attract atoms and/or molecules, and then change the surface conductivity and other properties. Different from sensing material, the catalyst often converts itself into a different species through a chemical reaction. The iron oxides (hematite and magnetite) have been applied in industry to produce chemicals with high efficiencies, such as ammonia (Haber process) and hydrocarbons (Fischer-Tropsch process) (Teja and Koh, 2009). It is expected that the nanoparticles with high surface areas can perform much better to enhance the chemical reaction rates than that of bulk states. For hematite, its thermal-dynamically stable structure allows it for high temperature oxidation catalysis (Sivula et al., 2010).

The catalysis effect can also be enhanced by coupling metal nanoparticles on the surface (Jiang and Yu 2009; Zhong et al., 2007). Jiang et al (2009) have reported the synthesis of Pd/α-Fe_2O_3 nanocomposites at ambient conditions, which displayed superior low-temperature catalytic activity toward CO oxidation to the pure α-Fe_2O_3 nanoparticles. It was proposed that the enhanced catalytic activity was due to the reaction between oxygen adsorbed on the reduced sites of the support (Fe^{2+}) and CO adsorbed on Pd at the metal-oxide interface, as shown in **Fig. 6**.

By using gold deposited iron oxide materials as a catalyst material, the oxidation and hydrogenation reaction of many organic compounds can be performed at much lower temperatures (Kung et al., 2007; Herzing et al., 2008; Lenz et al., 2009; Scirè et al., 2008). For example, Al-Sayari and co-workers (2007) have shown the dependence of the catalytic performance of Au/Fe_2O_3 catalyst that the non-calcined Au/Fe_2O_3 catalyst exhibited a high activity when pH≥ 5, whereas the activity of calcined Au/Fe_2O_3 catalyst was not influenced by the preparation conditions. Furthermore, the authors also noted that the catalytic activity of Fe_2O_3 toward CO oxidation was considerably lower than that of the Au/Fe_2O_3 catalyst.

Maghemite and magnetite/carbon composites have been found to be good catalysts for reducing the concentration of undesirable nitrogen in acrylonitrile-butadiene-styrene (ABS) degradation oil (Brebu et al., 2001), whereas hematite can be used as a photocatalyst for the degradation of chlorophenol and azo dyes (Bandara et al., 2007), as well as a support material for gold in catalysts for the oxidation of carbon monoxide (CO) at low temperatures (Zhong et al., 2007).

The challenge of catalysis research being the reaction mechanism for these systems are still yet to be confirmed or explained, especially for the metal oxide/gold systems (Astruc et al., 2005). The reaction can be compared from titanium oxide/gold. The rutile phase of titania provides a support for gold, in which CO will convert mostly along the perimeter between the titania and gold (Haruta, 2002). In other studies, it was proposed that the nature of the support material has much greater influence on the reactive properties of the deposited nanoparticle, because the active and selective sites are formed by negative gold particles (Milone et al., 2007).

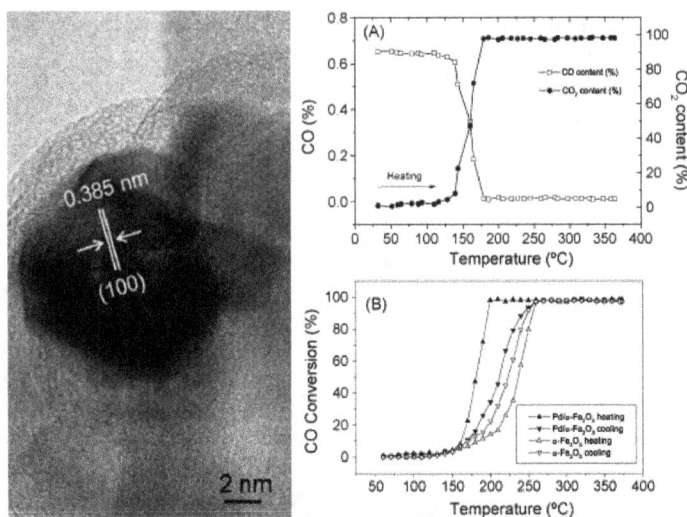

Fig. 6. HRTEM image of Pd particles binding on the surface of iron oxide, in which the lattice distance is ~0.385 nm, corresponding to Pd{1 0 0} planes. Catalytic activity of Pd/α-Fe₂O₃ nanocomposites showing the dependence of CO oxidation (A); (B) comparison of the catalytic activity of nanoparticles with and without doped palladium. Reprinted with permission from (Jiang and Yu, 2009).

6. Theoretical simulations

Beyond physical phenomena, theoretical methods have been developed and widely used to understand electronic, structure and forces of nanostructures (Cohen et al., 2008; Freund and Pacchioni, 2008; Hafner et al., 2006; Carter, 2008). Specifically, molecular dynamics (MD) method can be used for calculating interaction energies between surface modifiers and the modified matters, density functional theory (DFT) for binding energies, and Monte Carlo (MC) method for equilibrium properties (e.g., free energy, phase equilibrium) of particles. These methods have allowed researchers to understand and explain the growth mechanisms, structure, and functionalities of nanostructures (Hafner et al., 2006).

6.1 Molecular dynamics

MD simulation has been widely used for the study of the molecular behaviours in liquids and solids, examining material properties, and designing new materials, particularly for nanoparticles and nanocomposites. The MD method allows one to predict the time evolution of a system of interacting particles (atoms or molecules) and estimate relevant physicochemical properties. Specifically, it can calculate and simulate the interaction energies among atoms/molecules, which can help understand atomic positions, velocities, and forces. Thus, the macroscopic properties (e.g., pressure, energy, heat capacities) can be derived by means of statistical mechanics.

In our recent work, the MD method was used to explain the interactions between various goethite surfaces and surfactants of the nanorods. The simulation results of the side wall ($xy0$) surfaces with six different surfactants have been reported (Yue et al. 2010, 2011). The positively charged surfactants, CTAB (**Fig. 7**) and tetraethylammonium chloride (TEAC), were found to interact greatly with the side wall ($xy0$) of the nanorod, while the polymeric polyethylene glycol (PEG) and polyvinylpyrrolidone (PVP) and anionic surfactants (AOT) and Sodium Dodecyl Sulfate (SDS) were not suitable because of the low interaction energies among the surfaces. This is caused by the differences in the active sites on different surfaces (Kim et al. 2007). The ratios of iron and oxygen can vary greatly for different surfaces, in which the packing and exposure of atoms along a particular crystal plane will therefore determine the strength of adsorbed surface molecules. The simulation could provide quantitative information toward the interaction between surfactants and goethite surface(s), and hence understand the particle formation and growth mechanisms.

Through a similar MD simulation, the adsorption of minerals has been explored. Kerisit et al. (2006) simulated the interactions for electrolyte solutions to determine the surface properties of monovalent ions, such as NaCl, CsCl, and CsF on the (100) goethite surface. The calculations showed a structured interfacial region is in the first 15 Å on the surface. The structure of the mineral surface will also affect the arrangement and orientation of the water molecules, and hence the diffusive properties and distribution of the ionic species. In comparison, the adsorption of sodium ions is stronger than cesium ions because the former can occupy an interstitial site of mineral(s) due to smaller size.

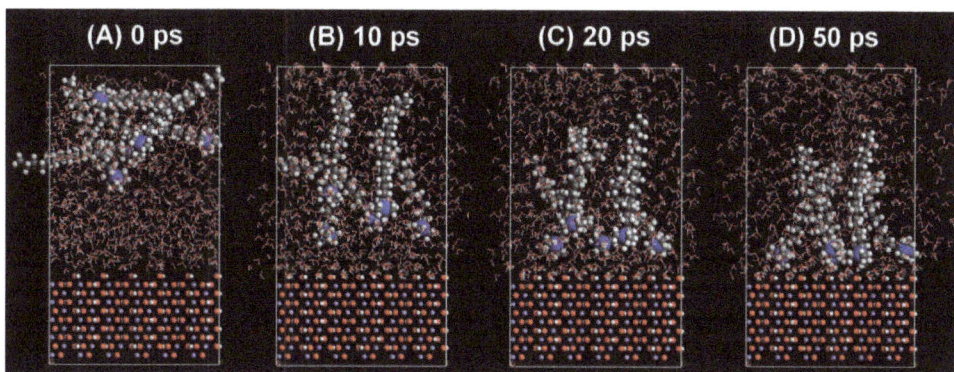

Fig. 7. MD simulation of CTAB molecular adsorption on the goethite crystal (010) surface at different time: (A) 0 ps, (B) 10 ps, (C) 20 ps, and (D) 50 ps. Reprinted with permission from (Yue et al., 2010).

Similarly, MD simulation was also employed to explain the growth mechanisms of akaganéite nanorods (Yue et al., 2011), as shown in **Fig. 8**, in which the atomic concentration profiles of various anions on different crystalline surfaces were compared. With the assistance of experimental techniques such as transmission electron microscopy (TEM), energy dispersive spectroscopy (EDS), and x-ray diffraction (XRD), the role of chloride ions in the lattice structure and forming β-FeOOH rodlike structure was determined. The analysis showed that the chloride ions were a small size, as well as having an intermediate interaction on the tunnel structure of the (001) surface, while the tight packing of the (100) and (110) surfaces does not allow interaction with any ions. The information was useful for the development of the simulation model, which explained the filling of the tunnel structure along (001) direction.

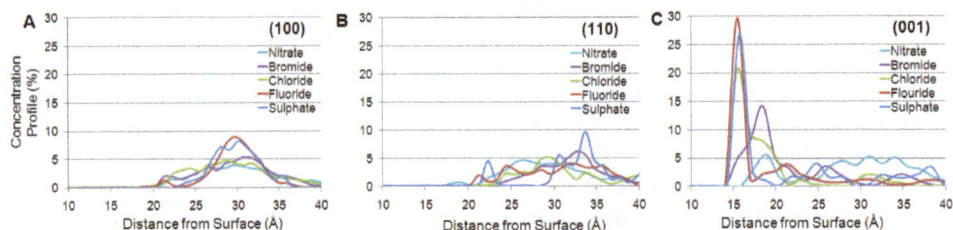

Fig. 8. The concentration profiles of various anions on the crystal surface of akaganéite nanorods: (A) (100); (B) (110); and (C) (001) plane.
Reprinted with permission from (Yue et al., 2011).

This MD method is used not only for small organic molecules but also for metallic nanoclusters. In our recent work (Yue et al., 2011), the $Fe_3O_4(111)$ surface modified with various surfactants, polymers, and silica, followed by the deposition of a Au nanoparticle was simulated by MD method). The results show the dynamic motion of the molecules on the $Fe_3O_4(111)$ surface, followed by the encapsulation of the Au nanoparticle surface. Through an analysis of the concentration profile, it reveals that $-NH_2$ groups within the molecule(s) are useful for attracting gold atoms, as shown in **Fig. 9**. Moreover, one-dimensional chainlike molecules allow higher flexibility to move toward the Au surface compared with three-dimensional structure (amorphous or polymerized silica)

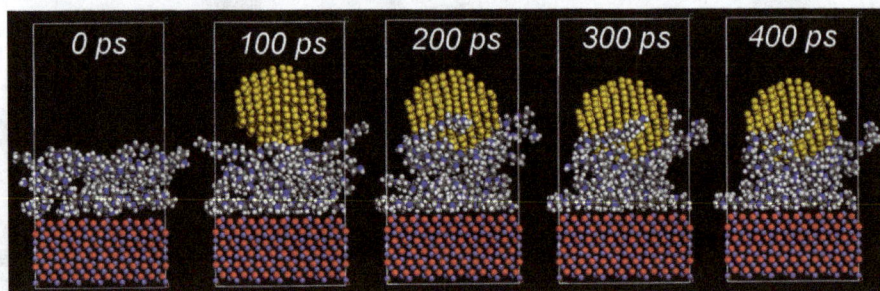

Fig. 9. Snapshots of PEI coating onto the surface of $Fe_3O_4(111)$ and the addition of a AuNP at various times. Reprinted with permission from (Yue et al., 2011).

This theoretical method is available for predicting the interaction energies and adsorption sites of molecules on the iron oxides surfaces. Aquino et al.(2006) simulated various molecules such as water, acetic acid, acetate, 2,4-dichlorophenoxyacetic acid, and benzene on the goethite (110) surface. The results show that two OH types, hydroxo and μ-hydroxo, were able to bend and act as proton acceptors, while the third type, μ3-hydroxo, acts only as proton donor due to its more pronounced rigidity.

However, MD is a classical simulation method which uses parameterized potentials (or forcefields), which cannot quantify electronic information of nanostructures. This method is limited to its accuracy, although the results can be obtained within a realistic period of time and larger length scales (Rustad et al., 2003; Zeng et al., 2008).

6.2 Density functional theory

DFT is another powerful simulation technique for understanding atom/molecular binder energies. The calculation is performed by using approximation method to simplify the Schrödinger's equation (Lado-Touriño and Tsobnang, 2000).

Many DFT studies have emphasized on the structural, electronic, catalytic, and magnetic properties of metal-oxide, such as Fe_2O_3, and Al_2O_3 (Alvarez-Ramirez et al., 2004; Ma et al., 2006; Mason et al., 2009; Rohrbach et al., 2004; Rollmann et al., 2004; Zhong et al., 2008; Mason et al., 2010). It has been extended into other systems, e.g., carbon nanotubes or graphene (Li et al., 2010; Chattaraj et al., 2009), transition metals (Cramer et al., 2009), semiconductors (Jin et al., 2011), and metals (e.g., Pd, Au, Cu) (Yang et al., 2007).

For example, Wong et al.(2011) demonstrated that the electronic and geometric structure of different metald (M = Au, Pt, Pd, or Ru) bilayers particularly on the α-Fe_2O_3(0001) support surface (**Fig. 10**). The analysis shows that the synergistic effect depends on the localized electron gain, electron transfer from Fe atoms to the dz^2 orbital of the metal bilayer, and interfacial metallic/ionic bonding. These effects were most pronounced for surfaces modified with Pt or Ru, while the Au bilayer is the most stable due to its low α-Fe_2O_3 lattice deformation and minimal surface of Fe atom spin quenching. Tuning the Ru bilayer can provide an optimal balance of these factors, and hence enhance the catalytic activity.

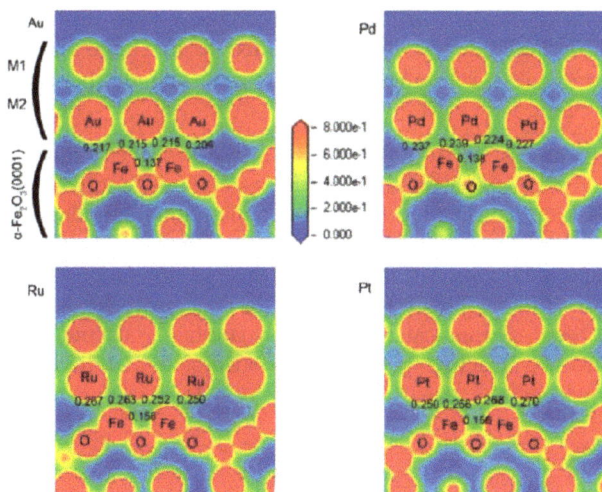

Fig. 10. Electron density contour maps of M/α-Fe_2O_3(0001) interfaces, where M = Ru, Pd, Au, and Pt, respectively, and the electron density is in the range 0.0–0.8 eV/Å^3. Reprinted with permission from (Wong et al., 2011).

Despite some success, the DFT method still has limitations in accurately describing the van der Waals interactions, phonon dispersion, spin-and space-degenerate states, strongly conjugated π systems, localization and delocalization errors for band gaps. Moreover, the DFT is difficult to solve the problems related to long range interactions and dispersion forces for complex biological systems. So far, the development of DFT technique is still demanded.

Besides DFT and MD simulations, Monte Carlo (MC) method, a stochastic method, has been employed to generate a statistical or probabilistic model for understanding particular systems. The MC method can be used to predict the crystalline structure of β-FeOOH (Kwon et al. 2006). By combination of quantitative X-ray structural analysis, the MC simulation has been used for characterizing the atomic-scale structure with and without chromium atoms. The results showed that the β-FeOOH particles containing chromium is distorted, while the particles without chromium is similar to its ideal structure. The combination of the experimental and MC simulation method can distinguish the differences between FeO_6 and CrO_6 octahedral units. However, this MC method can only provide information on equilibrium properties (e.g., free energy, phase equilibrium), but limited to the non-equilibrium systems.

7. Summary

This Chapter briefly overviews some experimental methods (hydrothermal, co-precipitate and microemulsion methods) used for the synthesis and surface modifications of low-dimensional iron oxide nanostructures with desirable functional properties (gas sensing, catalytic, magnetic, and biochemical properties), and a few theoretical simulation techniques (MD, DFT, and MC) for fundamental understandings. However, the challenges still exist. Experimentally, one of the big challenges is how to produce iron oxide nanostructures with desired characteristics (shape, size, and surface properties) for target applications. Theoretically, DFT and MD simulations are limited to the large-scale calculations (e.g., mesoscopic structure with size range of 0.1–10 μm) due to the current restraints in computational capability.

To overcome the limitations, the development of simple, cost-saving, and effective strategies for iron oxide and other nanostructures with desirable functional properties is highly demanded. For the computational modelings and simulation methods, much work needs to be performed in two directions: (i) to develop new and improved simulation techniques for large time and length scales; and (ii) to integrate diverse simulation techniques (DFT, MD, MC and others) on different levels together to form a powerful tool for exploring the structural, dynamic, and mechanical properties of nanomaterials and nanosystems. This is crucial to predict process–structure–property relationships in material design, optimization, and manufacturing.

8. Acknowledgement

We gratefully acknowledge the financial support of the Australia Research Council (ARC) the ARC Centres of Excellence for Functional Nanomaterials and ARC projects. The authors acknowledge access to the UNSW node of the Australian Microscopy & Microanalysis Research Facility (AMMRF).

9. References

Al-Sayari, S., Carley, A.F., Taylor, S.H. & Hutchings, G.J. (2007). Au/ZnO and Au/Fe$_2$O$_3$ catalysts for CO oxidation at ambient temperature: comments on the effect of synthesis conditions on the preparation of high activity catalysts prepared by coprecipitation. *Top. Catal.*, 44: 123-128.

Almeida, T.P., Fay, M., Zhu, Y. and Brown, P.D. (2009). Process Map for the Hydrothermal Synthesis of α-Fe$_2$O$_3$ Nanorods. *J. Phys. Chem. C*, 113(43): 18689-18698.

Alvarez-Ramirez, F., Martinez-Magadan, J.M., Gomes, J.R.B., & Illas, F. (2004), On the geometric structure of the (0001) hematite surface, *Surface Sci.*, 558, 4-14.

Aquino, A.J.A., Tunega, D., Haberhauer, G., Gerzabek, M.H. and Lischka, H. (2006). Quantum Chemical Adsorption Studies on the (110) Surface of the Mineral Goethite. *J. Phys. Chem. C*, 111(2): 877-885.

Astruc, D., Lu, F. and Aranzaes, J.R. (2005). Nanoparticles as Recyclable Catalysts: The Frontier between Homogeneous and Heterogeneous Catalysis. *Angew. Chem. Int. Ed.*, 44(48): 7852-7872.

Babic, M., Horák, D., Trchová, M., Jendelová, P., Glogarová, K.i., Lesný, P., Herynek, V., Hájek, M. and Syková, E. (2008). Poly(l-lysine)-Modified Iron Oxide Nanoparticles for Stem Cell Labeling. *Bioconjugate Chem.*, 19(3): 740-750.

Bandara, J., Klehm, U. and Kiwi, J. (2007). Raschig rings-Fe$_2$O$_3$ composite photocatalyst activate in the degradation of 4-chlorophenol and Orange II under daylight irradiation. *Appl. Catal. B*, 76(1-2): 73-81.

Boguslavsky, Y. & Margel, S. (2008). Synthesis and characterization of poly(divinylbenzene)-coated magnetic iron oxide nanoparticles as precursor for the formation of air-stable carbon-coated iron crystalline nanoparticles. *J. Colloid and Interface Sci.*, 317(1): 101-114.

Bomati-Miguel, O., Rebolledo, A.F. & Tartaj, P. (2008). Controlled formation of porous magnetic nanorods via a liquid/liquid solvothermal method. *Chem. Commun.*, (35): 4168-4170.

Brebu, M., Uddin, M.A., Muto, A., Sakata, Y. & Vasile, C. (2001). Catalytic degradation of acrylonitrile–butadiene–styrene into Fuel Oil 1. The effect of iron oxides on the distribution of nitrogen-containing compounds. *Energy & Fuels*, 15(3): 559-564.

Carter, E.A. (2008). Challenges in modeling materials properties without experimental input. *Science*, 321(5890): 800-803.

Catherine, C.B. & Adam, S.G.C. (2003). Functionalisation of magnetic nanoparticles for applications in biomedicine. *J. Phys. D: Appl. Phys.*, 36(13): R198.

Chattaraj, P.K. (2009). Chemical reactivity theory: a density functional view, pp. 9781420065435.

Chen, Z.P., Zhang, Y., Zhang, S., Xia, J.G., Liu, J.W., Xu, K. & Gu, N. (2008). Preparation and characterization of water-soluble monodisperse magnetic iron oxide nanoparticles via surface double-exchange with DMSA. *Colloids Surf. A*, 316(1-3): 210-216.

Chibowski, S., Patkowski, J. & Grzadka, E. (2009). Adsorption of polyethyleneimine and polymethacrylic acid onto synthesized hematite. *J. Colloid and Interface Sci.*, 329(1): 1-10.

Chin, A.B. and Yaacob, I.I. (2007). Synthesis and characterization of magnetic iron oxide nanoparticles via w/o microemulsion and Massart's procedure. *J. Mater. Process. Tech.*, 191(1-3): 235-237.

Cohen, A.J., Mori-Sánchez, P. & Yang, W. (2008). Insights into current limitations of density functional theory. *Science,* 321(5890): 792-794.

Cornell, R.M., Schuwertmann, U. *The iron oxides: structure, properties, reactions, occurrences, and uses,* Wiley-VCH, Weinheim, 2003.

Cramer, C.J. & Truhlar, D.G. (2009), Density functional theory for transition metals and transition metal chemistry. *Phys. Chem. Chem. Phys.* 11: 10757-10816.

Deng, Y.-H., Wang, C.-C., Hu, J.-H., Yang, W.-L. & Fu, S.-K. (2005). Investigation of formation of silica-coated magnetite nanoparticles via sol-gel approach. *Colloids Surf. A,* 262(1-3): 87-93.

Di Marco, M., Guilbert, I., Port, M., Robic, C., Couvreur, P. and Dubernet, C. (2007). Colloidal stability of ultrasmall superparamagnetic iron oxide (USPIO) particles with different coatings. *Int. J. Pharmaceutics,* 331(2): 197-203, 0378-5173.

Dong, W.T., Zhu, C.S. (2002). Use of Ethylene oxide in the sol–gel synthesis of α-Fe$_2$O$_3$ nanoparticles from Fe (III) Salt. *J. Mater. Chem.,* 12 (6): 1676–1683.

Erogbogbo, F., Yong, K.T., Hu, R., Law, W.C., Ding, H., Chang, C.W., Prasad, P.N. & Swihart, M.T. (2010). Biocompatible magnetofluorescent probes: luminescent silicon quantum dots coupled with superparamagnetic iron(III) oxide. *ACS Nano,* 4: 5131-5138.

Foy, S.P., Manthe, R.L., Foy, S.T., Dimitrijevic, S., Krishnamurthy, N. & Labhasetwar, V. (2010). Optical imaging and magnetic field targeting of magnetic nanoparticles in tumors. *ACS Nano,* 4(9): 5217-5224.

Freund, H.-J. & Pacchioni, G. (2008). Oxide ultra-thin films on metals: new materials for the design of supported metal catalysts. *Chem. Soc. Rev.,* 37(10): 2224-2242.

Garcia, K.E., Barrero, C.A, Morales, A.L. & Greneche, J.M. (2009). Magnetic structure of synthetic akaganeite: A review of Mössbauer data. *Rev. Fac. Ing. Univ. Antioquia,* 49: 185-191.

González-Carreño, T., Morales M.P., Gracia, M. & Serna, C.J. (1993). Preparation of uniform amma-Fe$_2$O$_3$ particles with nanometer size by spray pyrolysis. *Mater. Lett.,* 18: 151-155.

Guo, Z., Park, S., Wei, S., Pereira, T., Moldovan, M., BKarki, A., Young, D. P. & Hahn, H. T. (2007). Flexible high-loading particle-reinforced polyurethane magnetic nanocomposite fabrication through particle-surface-initiated polymerization. *Nanotech.* 18: 335704.

Hafner, J., Wolverton, C. & Ceder, G. (2006). Toward computational materials design: the impact of density functional theory on materials research. *MRS Bulletin,* 31: 659-668

Han, J.S., Yu, A.B., He, F.J. & Yao, T (1996). A study of the gas sensitivity of alpha-Fe$_2$O$_3$ sensors to CO and CH$_4$. *J. Mater. Sci. Lett.* 15: 434-436.

Han, J.S., Bredow, T., Davey, D.E., Yu, A.B. & Mulcahy, D.E. (2001). The effect of Al addition on the gas sensing properties of Fe$_2$O$_3$-based sensors. *Sens. and Actuators B* 75: 18-23.

Han, J.S., Davey, D.E., Mulcahy, D.E. & Yu, A.B. (1999). An investigation of gas response of alpha-Fe$_2$O$_3$(Sn)-based gas sensor. *Sens. and Actuators B* 61: 83-91.

Haruta, M. (2002). Catalysis of gold nanoparticles deposited on metal oxides. *CATTECH,* 6(3): 102-115.

Herzing, A.A., Kiely, C.J., Carley, A.F., Landon, P. & Hutchings, G.J. (2008). Identification of active gold nanoclusters on iron oxide supports for CO oxidation. *Science*, 321(5894): 1331-1335.

Hyeon, T., Lee, S.S., Park, J., Chung, Y. & Na, H.B. (2001). Synthesis of highly crystalline and monodisperse maghemite nanocrystallites without a size-selection Process. *J. Am. Chem. Soc.*, 123(51): 12798-12801.

Iida, H., Takayanagi, K., Nakanishi, T. & Osaka, T. (2007). Synthesis of Fe_3O_4 nanoparticles with various sizes and magnetic properties by controlled hydrolysis. *J. Colloid and Interface Sci.*, 314(1): 274-280.

Ilani, S., Rosenfeld, A. & Dvorachek, M. (1999). Mineralogy and chemistry of a Roman Remedy from Judea, Israel. *J. Archaeological Sci.*, 26(11): 1323-1326.

Im, S.H., Herricks, T., Lee, Y.T. & Xia, Y. (2005). Synthesis and characterization of monodisperse silica colloids loaded with superparamagnetic iron oxide nanoparticles. *Chem. Phys. Lett.*, 401(1-3): 19-23.

Ito, A., Shinkai, M., Honda, H. & Kobayashi, T. (2005). Medical application of functionalized magnetic nanoparticles. *J. Biosci. and Bioeng.*, 100(1): 1-11.

Jia, C.-J., Sun, L.-D., Yan, Z.-G., You, L.-P., Luo, F., Han, X.-D., Pang, Y.-C., Zhang, Z. & Yan, C.-H. (2005). Single-crystalline iron oxide nanotubes. *Angew. Chem. Int. Ed.*, 44(28): 4328-4333.

Jiang, X.C., Yu, A.B., Yang, W.R., Ding, Y., Xu, C. & Lam, S. (2010). Synthesis and growth of hematite nanodiscs through a facile hydrothermal approach. *J. Nanopart. Res.*, 12: 877-893.

Jiang, X.C. &Yu, A.B. (2009). Synthesis of $Pd/alpha-Fe_2O_3$ nanocomposites for catalytic CO oxidation. *J. Mater. Process. Tech.*, 209(9): 4558-4562.

Jimenez-Cadena, G., Riu, J. & Rius, F.X. (2007). Gas sensors based on nanostructured materials. *Analyst*, 132(11): 1083-1099.

Jin, D., Wang, W., Rahman, A., Lizhen, J., Zhang, H., Li, H., He, P., & Bao, S. (2011), Study on the interface between the organic and inorganic semiconductors, *Appl. Surface Sci.* 257: 4994-4999.

Kandori, K., Yamoto, Y. & Ishikawa, T. (2005). Effects of vinyl series polymers on the formation of hematite particles in a forced hydrolysis reaction. *J. Colloid and Interface Sci.*, 283(2): 432-439.

Kerisit, S., Ilton, E.S. & Parker, S.C. (2006). Molecular Dynamics simulations of electrolyte solutions at the (100) goethite surface. *J. Phys. Chem. B*, 110(41): 20491-20501.

Kim, D.K., Mikhaylova, M., Zhang, Y. & Muhammed, M. (2003). Protective coating of superparamagnetic iron oxide nanoparticles. *Chem. Mater.*, 15(8): 1617-1627.

Kim, H.-G., Kim, D.-W., Oh, C., Park, S.-H. & Oh, S.-G. (2007). Preparation of rod-type ferric oxyhydroxide particles by forced hydrolysis in the presence of a cationic surfactant. *J. Ceramic Process Res.*, 8(3): 172-176.

Kung, M.C., Davis, R.J. & Kung, H.H. (2007). Understanding Au-Catalyzed low-temperature CO oxidation. *J. Phys. Chem. C*, 111(32): 11767-11775.

Kwon, S.K., Suzuki, S., Saito, M., Kamimura, T., Miyuki, H. & Waseda, Y. (2006). Atomic-scale structure of beta-FeOOH containing chromium by anomalous X-ray scattering coupled with reverse Monte Carlo simulation. *Corrosion Sci.*, 48(6): 1571-1584.

Lado-Touriño, I. & Tsobnang, F. (2000). Using computational approaches to model hematite surfaces. *Computational Mater. Sci.*, 17(2-4): 243-248.

Lattuada, M. & Hatton, T.A. (2006). Functionalization of monodisperse magnetic nanoparticles. *Langmuir*, 23(4): 2158-2168.

Laurent, S., Forge, D., Port, M., Roch, A., Robic, C., Eliat, L.V. & Muller, R.N. (2008). Magnetic iron oxide nanoparticles: Synthesis, stabilization, vectorization physicochemical characterizations and biological applications. *Chem. Rev.*, 108: 2064-2110.

Lee, D.-D. & Choi, D.-H. (1990). Thick-film hydrocarbon gas sensors. *Sens. Actuators B*, 1(1-6): 231-235.

Lee, J., Isobe, T. & Senna, M. (1996). Preparation of ultrafine Fe$_3$O$_4$ particles by precipitation in the presence of PVA at high pH. *J. Colloid Interface Sci.*, 177(2): 490-494.

Lee, Y., Lee, J., Bae, C.J., Park, J.G., Noh, H.J., Park, J.H. & Hyeon, T. (2005). Large-scale synthesis of uniform and crystalline magnetite nanoparticles using reverse micelles as nanoreactors under reflux conditions. *Adv. Funct. Mater.*, 15(3): 503-509.

Lenz, J., Campo, B.C., Alvarez, M. & Volpe, M.A. (2009). Liquid phase hydrogenation of [alpha],[beta]-unsaturated aldehydes over gold supported on iron oxides. *J. Catal.*, 267(1): 50-56.

Li, Y., Zhou, Z., Yu, G., Chen, W., & Chen, Z. (2010), CO catalytic oxidation on iron-embedded graphene: computational quest for low-cost nanocatalysts, *J Phys. Chem. C*, 114: 6250-6254.

Li, Z., Chen, H., Bao, H. & Gao, M. (2004). One-pot reaction to synthesize water-soluble magnetite nanocrystals. *Chem. Mater.*, 16(8): 1391-1393.

Li, Z., Lai, X., Wang, H., Mao, D., Xing, C. & Wang, D. (2009). Direct hydrothermal synthesis of single-crystalline hematite nanorods assisted by 1,2-propanediamine. *Nanotechnology*, 20: 1-9.

Liong, M., Lu, J., Kovochich, M., Xia, T., Ruehm, S.G., Nel, A.E., Tamanoi, F. & Zink, J.I. (2008). Multifunctional inorganic nanoparticles for imaging, targeting, and drug delivery. *ACS Nano*, 2(5): 889-896.

Liu, H., Wei, Y., Li, P., Zhang, Y. & Sun, Y. (2007). Catalytic synthesis of nanosized hematite particles in solution. *Mater. Chem. Phys.*, 102(1): 1-6.

López-Quintela, M.A. & Rivas, J. (1993). Chemical Reactions in Microemulsions: A Powerful Method to Obtain Ultrafine Particles. *J. Colloid and Interface. Sci.*, 158(2): 446-451.

Lu, A.H., Salabas, E. & Schüth, F. (2007). Magnetic Nanoparticles: Synthesis, Protection, Functionalization, and Application. *Angew. Chem. Int. Ed.*, 46(8): 1222-1244.

Lu. J., Yang. S., Ng, K.M., Su, C.H.,Yeh, C.S., Wu, Y.N. & Shieh, D.B. (2006). Solid-State synthesis of monocrystalline iron oxide nanoparticle-based ferrofluid suitable for magnetic resonance imaging contrast application. *Nanotechnology*, 17: 5812-5820.

Lyon, J.L., Fleming, D.A., Stone, M.B., Schiffer, P. & Williams, M.E. (2004). Synthesis of Fe oxide core/Au shell nanoparticles by iterative hydroxylamine seeding. *Nano Lett.*, 4(4): 719-723.

Ma, X.Y., Liu, L., Jin, J.J., Stair, P.C., & Ellis, D.E. (2006), Experimental and theoretical studies of adsorption of CH$_3$ center dot on alpha-Fe$_2$O$_3$(0001) surfaces. *Surf. Sci.*, 600: 2874-85.

Mahmoudi, M., Simchi, A., Milani, A.S. & Stroeve, P. (2009). Cell toxicity of superparamagnetic iron oxide nanoparticles. *J. Colloid and Interface Sci.*, 336(2): 510-518,0021-9797.

Mammeri, F., Bourhis, E.L., Rozes, L. & Sanchez, C. (2005). Mechanical properties of hybrid organic-inorganic materials. *J. Mater. Chem.*, 15(35-36): 3787-3811.

Martin, A.L., Bernas, L.M., Rutt, B.K., Foster, P.J. & Gillies, E.R. (2008). Enhanced cell pptake of superparamagnetic iron oxide nanoparticles functionalized with dendritic guanidines. *Bioconjugate Chem.*, 19(12): 2375-2384.

Mason, S.E., Iceman, C.R., Tanwar, K.S., Trainor, T.P., & Chaka, A.M. (2009). Pb(II) adsorption on isostructural hydrated alumina and hematite (0001) surfaces: A DFT study, *J. Phys. Chem. C*, 113: 2159-2170.

Massart, R. (1981). Preparation of aqueous magnetic liquids in alkaline and acidic media. *IEEE Trans Magn MAG*, I7: 1247-1248.

Milone, C., Crisafulli, C., Ingoglia, R., Schipilliti, L. & Galvagno, S. (2007). A comparative study on the selective hydrogenation of [alpha],[beta] unsaturated aldehyde and ketone to unsaturated alcohols on Au supported catalysts. *Catal. Today*, 122: 341-351.

Mohapatra, M. & Anand, S. (2010), Synthesis and applications of nano-structured iron xxides/hydroxides – A Review. *Int. J. Eng. Sci. and Tech.*, 2(8): 127-146.

Mohapatra, S., Pramanik, N., Mukherjee, S., Ghosh, S. & Pramanik, P. (2007). A simple synthesis of amine-derivatised superparamagnetic iron oxide nanoparticles for bioapplications. *J. Mater. Sci.*, 42(17): 7566-7574.

Molday, R.S. & Mackenzie, D. (1982). *J. Immun. Methods*, 52: 353-367.

Morais, P.C., Oliveira, A.C., Tronconi, A.L., Goetze, T. & Buske, N. (2003). Photoacoustic spectroscopy: a promising technique to investigate magnetic fluids. *IEEE Trans. Magn.*, 39(5): 2654-2656.

Morber, J.R, Ding, Y., Haluska, M. S., Li, Y., Liu, J. P., Wang, Z. L. & Snyder, R. L. (2006). *J. Phys. Chem. B*, 110: 21672.

Mornet, S., Vasseur, S., Grasset, F. & Duguet, E. (2004). Magnetic nanoparticle design for medical diagnosis and therapy. *J. Mater. Chem.*, 14(14): 2161-2175.

Morrissey, J. & Guerinot, M.L. (2009). Iron uptake and transport in plants: The good, the bad, and the ionome. *Chem. Rev.*, 109(10): 4553-4567.

Nakatani, Y. & Matsuoka, M. (1983). Some Electrical Properties of γ-Fe_2O_3 Ceramics. *Jpn. J. Appl. Phys.*, 22: 232-239.

Neri, G., Bonavita, A., Galvagno, S., Siciliano, P. & Capone, S. (2002). CO and NO_2 sensing properties of doped-Fe_2O_3 thin films prepared by LPD. *Sens. Actuators B*, 82(1): 40-47,0925-4005.

Pascal C., Pascal J.L., Favier F., Elidrissi Moubtassim M.L. & Payen C. (1999). Electrochemical synthesis for the control of γ-Fe_2O_3 nanoparticle size. Morphology, microstructure, and magnetic behavior. *Chem. Mater.* 11: 141-147.

Pedro, T. Morales, M. del P., Veintenillas-verdaguer, S., Gonzalez-carreno, T. & Serma, C.J. (2003). The preparation of magnetic nanoparticles for applications in biomedicine. *J. Phys. D: Appl. Phys.*, 36(13): R182.

Pinho, S.L.C., Pereira, G.A., Voisin, P., Kassem, J., Bouchaud, V.r., Etienne, L., Peters, J.A., Carlos, L., Mornet, S.p., Geraldes, C.F.G.C., Rocha, J.O. & Delville, M.-H. (2010). Fine tuning of the relaxometry of γ-Fe_2O_3@SiO_2 nanoparticles by tweaking the silica coating thickness. *ACS Nano*, 4(9): 5339-5349.

Pisanic Ii, T.R., Blackwell, J.D., Shubayev, V.I., Fiñones, R.R. & Jin, S. (2007). Nanotoxicity of iron oxide nanoparticle internalization in growing neurons. *Biomater.* 28(16): 2572-2581.

Rockenberger, J., Scher, E.C. & Alivisatos, A.P. (1999). A new nonhydrolytic single-precursor approach to surfactant-capped nanocrystals of transition metal oxides. *J. Am. Chem. Soc.,* 121(49): 11595-11596.

Rohrbach, A., Hafner, J., & Kresse, G. (2004). Ab initio study of the (0001) surfaces of hematite and chromia: Influence of strong electronic correlations. *Phys. Rev. B,* 70: 125426.

Rollmann, G., Rohrbach, A., Entel, P. & Hafner, J. (2004), First-principles calculation of the structure and magnetic phases of hematite. *Phys. Rev. B,* 69: 165107.

Sahu, K.K., Rath, C., Mishra, N.C. & Anand, S. (1997). Microstructural and magnetic studies on hydrothermally prepared hematite. *J. Colloid Interface Sci.,* 185(2): 402-410.

Scirè, S., Crisafulli, C., Minicò, S., Condorelli, G.G. & Di Mauro, A. (2008). Selective oxidation of CO in H_2-rich stream over gold/iron oxide: An insight on the effect of catalyst pretreatment. *J. Molecular Catal. A: Chem.* 284(1-2): 24-32.

Sivula, K., Zboril, R., Le Formal, F., Robert, R., Weidenkaff, A., Tucek, J., Frydrych, J. & Gratzel, M. (2010). Photoelectrochemical water splitting with mesoporous hematite prepared by a solution-based colloidal approach. *J. Am. Chem. Soc.,* 132(21): 7436-7444.

Soler, M.A, Alcantara, G.B., Soares, F.Q., Viali, W.R., Sartoratto, P.P., Fernandez, J.R., da Silva, S.W., Garg, V.K., Oliveira, A.C. & Morais, P.C. (2007). Study of molecular surface coating on the stability of maghemite nanoparticles. *Surf. Sci.,* 601(18): 3921-3925.

Son, S.J., Reichel, J., He, B., Schuchman, M. & Lee, S.B. (2005). Magnetic nanotubes for magnetic-field-assisted bioseparation, biointeraction, and drug delivery. *J. Am. Chem. Soc.,* 127(20): 7316-7317.

Sousa, M.H., Rubim, J.C., Sobrinho, P.G. & Tourinho, F.A. (2001). Biocompatible magnetic fluid precursors based on aspartic and glutamic acid modified maghemite nanostructures. *J. Magn. Magn. Mater.,* 225(1-2): 67-72.

Sun, Q., Reddy, Marquez, M., Jena, P., Gonzalez, C. & Wang, Q. (2007). Theoretical Study on Gold-Coated Iron Oxide Nanostructure: Magnetism and Bioselectivity for Amino Acids. *J. Phys. Chem. C,* 111(11): 4159-4163.

Sun, S., Zeng, H., Robinson, D.B., Raoux, S., Rice, P.M., Wang, S.X. & Li, G. (2003). Monodisperse MFe_2O_4 (M = Fe, Co, Mn) nanoparticles. *J. Am. Cer. Soc.,* 126(1): 273-279.

Sun, S. & Zeng, H. (2004). Size-controlled synthesis of magnetite nanoparticles. *J. Am. Cer. Soc.,* 124: 8204-8205.

Tao, S., Liu, X., Chu, X. & Shen, Y. (1999). Preparation and properties of [gamma-Fe_2O_3 and Y_2O_3 doped gamma-Fe_2O_3 by a sol-gel process. *Sens. Actuators B,* 61(1-3): 33-38.

Tartaj, P., González-Carreño, T. & Serna, C.J. (2001). Single-step nanoengineering of silica coated maghemite hollow spheres with tunable magnetic properties. *Adv. Mater.,* 13(21): 1620-1624.

Teja, A.S. & Koh, P.-Y. Synthesis, properties, and applications of magnetic iron oxide nanoparticles. *Progress in Crystal Growth and Characterization of Mater.,* 55(1-2): 22-45.

Zhang, T., Luo, H, Zeng, H, Zhang, R. & Shen, Y.(1996). Synthesis and gas-sensing characteristics of high thermostability γ-Fe$_2$O$_3$ powder. *Sens. Actuators B*, 32: 181-184.

Tong, S., Hou, S., Zheng, Z., Zhou, J. & Bao, G. (2010). Coating optimization of superparamagnetic iron oxide nanoparticles for high T2 relaxivity. *Nano Lett.*, 10: 1530-6984.

Tristão, J., Ardisson, J., Sansiviero, M. & Lago, R. (2010). Reduction of hematite with ethanol to produce magnetic nanoparticles of Fe$_3$O$_4$ coated with carbon. *Hyperfine Interactions*, 195(1): 15-19.

Tristão, J.C., Silva, A.A., Ardisson, J.D. & Lago, R. (2009). Magnetic nanoparticles based on iron coated carbon produced from the reaction of Fe$_2$O$_3$ with CH$_4$: a Mössbauer study. LACAME 2008. J. Desimoni, C. P. Ramos, B. Arcondo, F. D. Saccone and R. C. Mercader, Springer Berlin Heidelberg: 21-25.

Varanda, L.C., Morales, M.P., Jafelicci, J.M. &Serna, C.J. (2002). Monodispersed spindle-type goethite nanoparticles from FeIII solutions. *J. Mater. Chem.*, 12(12): 3649-3653.

Vidal-Vidal, J., Rivas, J. & López-Quintela, M.A. (2006). Synthesis of monodispersed maghemite nanoparticles by the microemulsion method. *Colloids Surf. A*, 288: 44-51.

Vijayakumar, R., Koltypin, Y., Xu, X.N., Yeshurun, Y., Gedanken, A. & Felner, I. (2001). Fabrication of magnetite nanorods by ultrasonic irradiation. *J. Appl. Phys.*, 89, 6324-28.

Wang, H., Brandl, D.W., Le, F., Nordlander, P. & Halas, N.J. (2006). Nanorice: A hybrid plasmonic nanostructure. *Nano Lett.*, 6(4): 827-832.

Wang, L., Luo, J., Maye, M., Fan, Q., Rendeng, Q., Engelhard, M.E., Wang, C., Lin, Y. & Zhong, C.J. (2005). Iron oxide–gold core–shell nanoparticles and thin film assembly. *J. Mater. Chem.*, 15: 1821-1832

Wang, Z., Guo, H., Yu, Y. & He, N. (2006). Synthesis and characterization of a novel magnetic carrier with its composition of Fe$_3$O$_4$/carbon using hydrothermal reaction. *J. Magn. Magn. Mater.*, 302(2): 397-404.

Weckler, B. & Lutz, H.D. (1998) Lattice vibration spectra. Part XCV. Infrared spectroscopic studies on the iron oxide hydroxides goethite ([alpha]), akaganéite ([beta]), lepidocrocite ([gamma]), and feroxyhite ([delta]). *European J. Solid State and Inorg. Chem.*, 35(8-9): 531-544.

Wong, K., Zeng, Q.H. &Yu, A.B. (2011). Electronic structure of metal (M = Au, Pt, Pd, or Ru) bilayer modified α-Fe$_2$O$_3$(0001) surfaces. *J. Phys. Chem. C*, 115(11): 4656-4663.

Wu, W., He, Q. & Jiang, C. (2008). Magnetic iron oxide nanoparticles: Synthesis and surface functionalization strategies. *Nanoscale Res. Lett.*, 3: 397-415

Xu, Z., Hou, Y. & Sun, S. (2007). Magnetic core/shell Fe$_3$O$_4$/Au and Fe$_3$O$_4$/Au/Ag nanoparticles with tunable plasmonic properties. *J. Am. Chem. Soc.*, 129(28): 8698-8699.

Yang, Z., Lu, Z., Luo, G. & Hermansson, K. (2007), Oxygen vacancy formation energy at the Pd/CeO$_2$(111) interface. *Phys. Lett. A*, 369: 132-139.

Ye, Q.-L., Kozuka, Y., Yoshikawa, H., Awaga, K., Bandow, S. & Iijima, S. (2007). Effects of the unique shape of submicron magnetite hollow spheres on magnetic properties and domain states. *Physical Review B*, 75(22): 224404

Yue, J., Jiang, X.C. & Yu, A.B. (2010). Molecular dynamics study on the growth mechanism of goethite ([alpha]-FeOOH) nanorods. *Solid State Sciences*, 13(1): 263-270.

Yue, J., Jiang, X.C. & Yu, A.B. (2011). Experimental and theoretical study on the β-FeOOH nanorods: growth and conversion. *J. Nanoparticle Res.*, 13(9): 3961-3974.

Yue, J., Jiang, X.C., Zeng, Q. & Yu, A.B. (2010). Experimental and numerical study of cetyltrimethylammonium bromide (CTAB)-directed synthesis of goethite nanorods. *Solid State Sciences*, 12(7): 1152-1159.

Yue, J., Jiang, X.C. & Yu, A.B. (2011). Molecular Dynamics Study on Au/ Fe₃O₄ Nanocomposites and Their Surface Function toward Amino Acids. *J. Phys. Chem. B.*, 115:11693-11699.

Zeng, Q.H., Yu, A.B. & Lu, G.Q. (2008). Multiscale modeling and simulation of polymer nanocomposites. *Progress in Polymer Sci.*, 33(2): 191-269.

Zhai, Y., Zhai, J., Wang, Y., Guo, S., Ren, W. & Dong, S. (2009). Fabrication of iron oxide core/shell submicrometer spheres with nanoscale surface roughness for efficient surface-enhanced Raman scattering. *J. Phys. Chem. C* 113(17): 7009-7014.

Zhang, J., Liu, X., Guo, X., Wu, S. & Wang, S. (2010). A general approach to fabricate diverse noble metal (Au, Pt, Ag, Pt/Au)/Fe₂O₃ hybrid nanomaterials. *Chem.–Eur. J.*, 16(27): 8108-8116.

Zhang, S., Niu, H., Hu, Z., Cai, Y. & Shi, Y. (2010). Preparation of carbon coated Fe₃O₄ nanoparticles and their application for solid-phase extraction of polycyclic aromatic hydrocarbons from environmental water samples. *J. Chromatography A*, 1217(29): 4757-4764.

Zhang, W.M., Wu, X.L., Hu, J.S., Guo, Y.G. & Wan, L.J. (2008). Carbon coated Fe₃O₄ nanospindles as a superior anode material for lithium-ion batteries. *Adv. Funct. Mater.*, 18(24): 3941-3946.

Zhang, Z., Zhang, Q., Xu, L. & Xia, Y.-B. (2007). Preparation of nanometer γ-Fe₂O₃ by an electrochemical method in non-aqueous medium and reaction dynamics. *Synthesis and Reactivity in Inorg. Metal-Org. and Nano-Metal Chem.*, 37: 53–56.

Zhong, J. & Adams, J.B. (2008). Adhesive metal transfer at the Al(111)/a-Fe₂O₃(0001) interface: a study with ab initio molecular dynamics, modelling and simulation. *Mater. Sci. and Eng.*, 16: 085001(9 pages).

Zhong, Z., Ho, J., Teo, J., Shen, S. & Gedanken, A. (2007). Synthesis of porous α-Fe₂O₃ nanorods and deposition of very small gold particles in the pores for catalytic oxidation of CO. *Chem. Mater.*, 19(19): 4776-4782.

Zhong, Z., Lin, J., Teh, S.P., Teo, J. & Dautzenberg, F.M. (2007). A rapid and efficient method to deposit gold particles onto catalyst supports and its application for CO oxidation at low temperatures. *Adv. Funct. Mater.*, 17(8): 1402-1408.

Part 2

Testing Technology

7

Nanoscale Electrodeposition of Copper on an AFM Tip and Its Morphological Investigations

Udit Surya Mohanty, S. Y. Chen and Kwang-Lung Lin
Department of Materials Science and Engineering,
National Cheng Kung University
Tainan,
R.O.C, Taiwan

1. Introduction

Electrocrystallisation processes occurring at electrochemical solid/liquid interfaces have attracted the interest of many researchers from both fundamental and applied viewpoints. After the pioneering works of Max Volmer at the beginning of the last century (Volmer, 1934a, 1939b), the processes of electrocrystallisation have been the subject of numerous intensive studies, the results of which have been documented in several books (Bockris & Razumney 1967; Budevski, et al., 1996; Fischer,1954). The electrochemical method offers several advantages over vapour deposition techniques such as molecular beam epitaxy for depositing nanoscale superlattices. Additional technological advantages over the vapour deposition techniques consist in the relatively low processing temperature and the high selectivity. The low processing temperatures minimizes interdiffusion whereas the high selectivity of electrocrystallisation process allows uniform modification of surfaces and structures with complicated profiles. Phase formation and crystal growth phenomena are the most common morphological parameters observed in many technological important cathodic and anodic electrochemical reactions. One of the most frequently studied electrocrystallisation process is the cathodic metal deposition on foreign and native substrates from electrolytes containing complex metal ions (Fleischmann & Thirsk, 1963; Milchev, 2002; Paunovic & Schlesinger, 2006). Some of the typical cited examples are electrocrystallisation of Ag from Ag^+ containing electrolytes (Budevski et al., 1980; Fischer, 1969) and the electrodeposition of Cu (Budevski, 1983; Danilov et al., 1994; Hozzle et al., 1995; Michhailova et al. , 1993) which has recently become technologically important for the fabrication of Cu interconnects on integrated circuit chips (Andricacos et al., 1998; Oskam et al. 1998). Since the electrodeposition of metals is a process of great technological importance, a large number of studies have been carried out to understand the mechanism of electrodeposition of metals on conducting surfaces by employing a variety of electrochemical and spectroscopic techniques (Andricacos, 1999; Markovic & Ross, 1993). The conventional electrochemical methods such as cyclic voltammetry, impedance spectroscopy have been used to assess the mechanism and kinetics of metal electrocrystallisation. These techniques however provide information on the whole surface.

To fully understand the process, it is essential to obtain structural information on the substrate and the deposit in pm to the atomic level. Although techniques such as electron microscopy and optical microscopy have been employed to examine the morphology of the substrate and the metal deposit, they can be used only for the ex situ examination. The discovery of scanning tunnelling microscopy and atomic force microscopy (STM and AFM) offered new exciting possibilties for in situ studies of the electrocrystallisation phenomenon down to an atomic level (Binning & Rohrer, 1982; Lustenberger et al., 1988, Sonnenfeld & Hannsma, 1986). The application of these techniques in the last two decades has revolutionized the experimental work in this field and led to significant progress in the understanding of the atomistic aspects of the electrocrystallisation process (Gewirth & Siegenthaler, 1995; Staikov et al., 1994; Stegenthaler, 1992). These processes range from measuring the lateral force using a cantilever tip, measuring magnetic force, electrostatic force, Kelvin potential to the determination of surface conductivity. The invention of scanning probe microscopy (SPM) also provoked a rapid development of the modern nanoscience and nanotechnology dealing with nanoscale structures and objects including single atoms and molecules. Over, the years, many other types of scanning probe microscopic techniques have evolved from the base concept of AFM. Electrochemical fabrication of metal nanostructures has been reported using SPM-based lithography, typically by tip induced electrochemical deposition of metal ions transferred by the STM or AFM tip to the surface. (Allongne, 1995; Benenz et al., 2002). Many studies have been directed towards in situ STM and AFM imaging of metal underdeposition (Hachiya et al.1991; Li et al., 2001) and bulk deposition (Nichols et al., 1992; Yau et al. 1991). Since STM can only be applied to observe conductive surfaces, the existence of anodic oxide films as well as the space charge layer in the depletion condition makes the STM measurements of semiconductor electrodes relatively difficult (Batina & Nichols, 1992). On the other hand, AFM can image even non-conducting surfaces and electrochemical processes on the tip, which causes serious problems in the STM measurements in electrolyte solutions. AFM is also found to be more useful than STM in studying the electrode surface in situ. AFM works the same way as STM and can transfer materials from tip to substrate at a biased voltage. As AFM tips are normally made of silicon or silicon nitide, metallic materials have to be coated onto the AFM tip in order to make the deposition happen. Once it is coated with metals, it becomes no different from a STM tip, and deposition takes place under high electric field. AS AFM can work in liquid solution, it is possible to initiate electrochemical deposition using an AFM tip. Also the electrochemical reaction rate at the interface can be controlled by application of an external potential to the substrate. In particular, the amount of deposit and the kinetics of the metal deposition onto the surface can be controlled. Its because the electrochemical process is sensitive to the surface properties, in situ local deposition of metal can be made selective by tuning the surface characteristics. Copper has been electrochemically deposited onto GaAs surface by immersing the AFM tip into a mixture of $CuSO_4 + H_2SO_4$ solution (Carlsson et al., 1990).

The electrodeposition of Cu is strongly dependent on the structure of the substrate, applied potential and concentration of Cu^{2+} ions in the precursor solution. The nanoscale electrodeposition of metal in nanopatterned alkanethiol-modified Au (III) has been reported (Gewirth & Sigenthaler, 1995). Although the interaction between the tip and the sample enhanced Cu deposition on the surface, Cu did not deposit uniformly in the area scanned, but only deposited at the edges of the scanned area as well as in defects in the alkanethiol (self assembled monolayer) (SAM). The authors suggested the physical and electrostatic inhibition by the tip, or the diffusion of Cu ions to the area under the tip, even on a bare Au (III) surface. In one of the studies (Koinuma & Uosaki, 1994), AFM and scaling analysis have been employed to investigate the effect of current density, temperature and levelling agent on the morphology of electrolytically produced copper. However, very little information is available on the morphology of the nanostructures formed by the electrodeposition of copper on the AFM tip. In the present chapter an effort has been made to examine various modes of nucleation and growth of copper deposits on an AFM tip of 80 nm. Electron beam lithography techniques have been employed to facilitate selective electrodeposition of copper on the nanosize AFM tip in the presence of photoresist named poly (methyl methacyrlate).

2. Experimental method

In the present study an AFM probe NSC/18 /Ti-Pt was used which consisted of a Si base coated by films of Ti (1st layer, 15 nm) and Pt (2ndlayer) 10 nm. The Cantilevers employed were 230 μm in length and 40 μm in diameter with pyramidal tips of diameter 80 nm. The tip height ranged from 20 to 25 μm. The schematic diagram of the uncoated AFM probe containing the tip is shown in Fig. 1a. The magnified image is demonstrated in Fig. 1b. One of the AFM probes acted as a anode and the other as a cathode. Two connectors were placed on the AFM probes to facilitate the current flow between the power supplier and the probes, as illustrated in Fig. 2. For the preparation of a connector, a Cu plate of diameter 3.4 mm and length 12 mm was first plated with electroless Au (Fig. 3). The following electrolytic composition was used in the study: 2g/L KAu $(CN)_2$; 75 g/L $NH4Cl$; 50 g/L $6H_5Na3O_7 \cdot 2H2O$; and 10 g/L $NaH_2PO_2.H2O$. The pH of the electrolytic solution was maintained at 7 and the solution temperature was kept at 92°C.

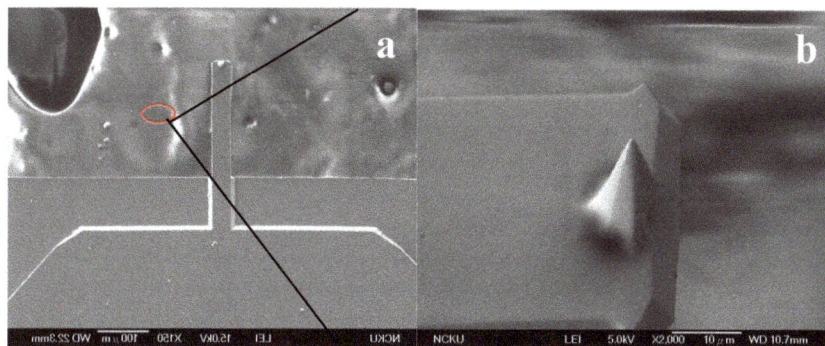

Fig. 1. SEM micrograph illustrating the schematic representation of an uncoated AFM probe

The layer of electroless gold on the copper plate provided good adhesion. The Cu plate coated with electroless Au was connected to a Cu wire as shown in Fig 3 by means of mechanical force. Finally, the connector was electroplated with a hard layer of Au to provide good abrasion ability and a thicker Au layer. The Si holders and the connectors were fixed together and were placed in a rectangular cell containing the electrolyte of composition 80 g/L $CuSO_4$ and 200 mL/L H_2SO_4. The power supply was fixed in the range of 10 µA to 100 A. Because the AFM probes cannot bear large amounts of current, a large electrode system consisting of Pt anode and Cu cathode was used as shown in Fig. 4.

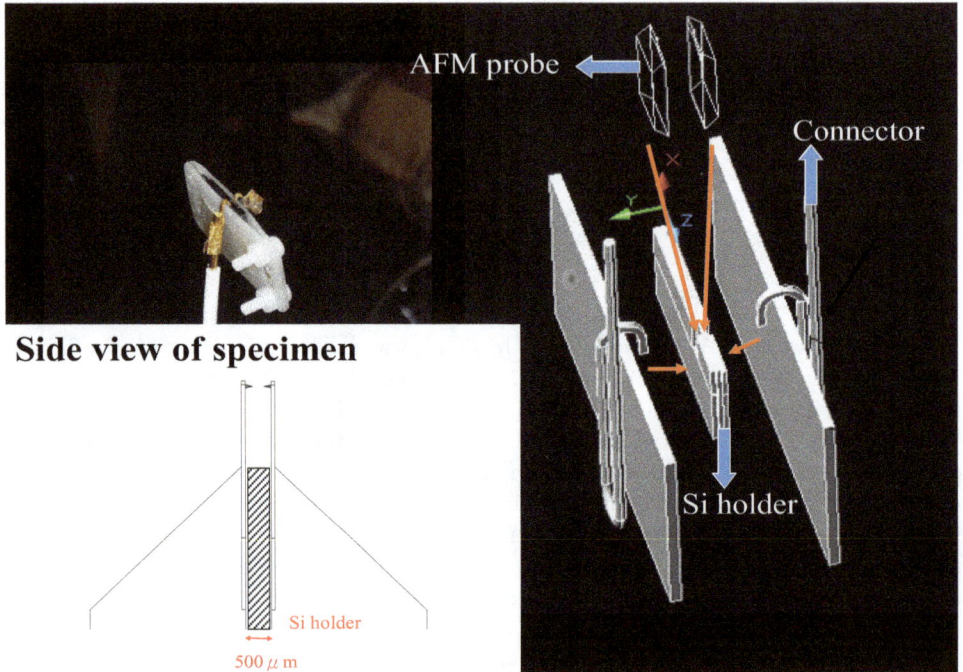

Fig. 2. Schematic diagram of the AFM probe placed on the Si holder

The cathode and anode were placed in parallel connection with the AFM probes. The current from the power supply was controlled between 10 and 0.1 A to provide a stable current between the AFM probes. Electron-beam lithography (EBL) techniques were used in our present work.

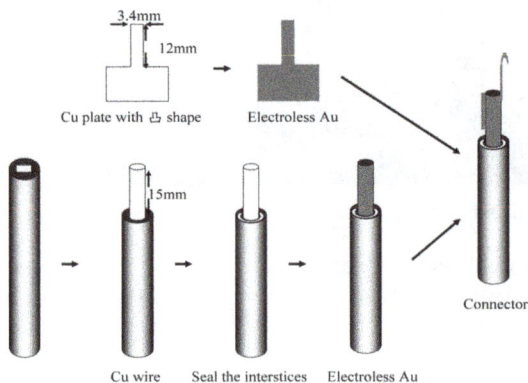

Fig. 3. Schematic diagram of the connector used in the electrodeposition process

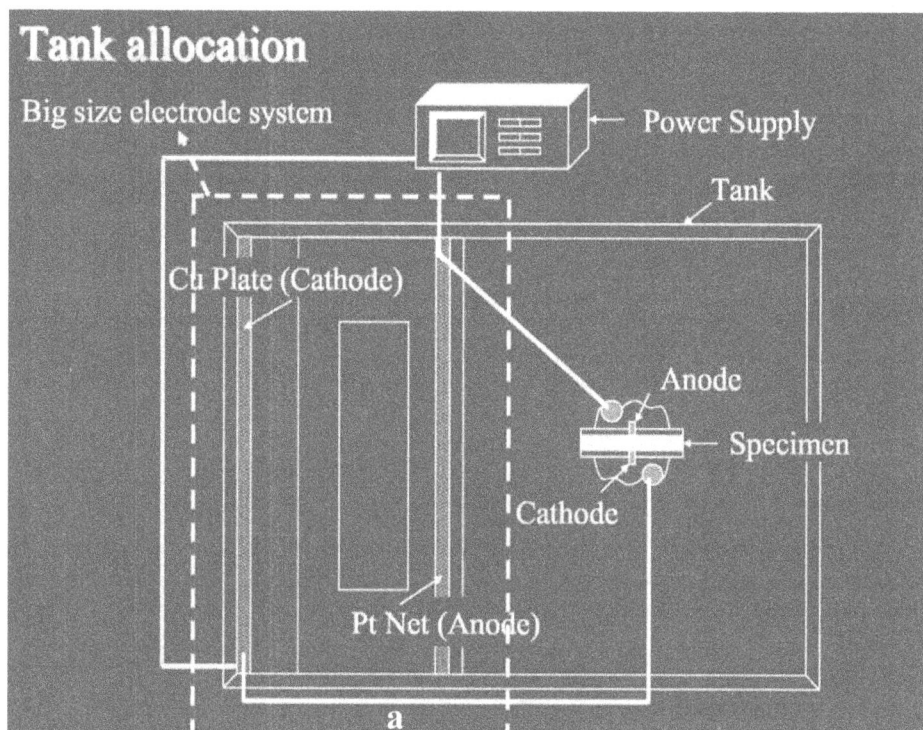

Fig. 4. Schematic diagram demonstrating the electrodeposition of copper on the AFM probe.

The samples were patterned using a JEOL 6400 thermionic emission scanning electron microscope equipped with the lithography software Elphy Quantum. The polymer used for EBL studies was poly_methyl methacrylate (PMMA). PMMA was the standard positive e-beam resist dissolved in a casting solvent anisole. The PMMA solution was spin-coated onto the AFM probe at a rotating speed of 1000 rpm for 60 s. Then baking was performed at 220°C to harden the film and to remove the remaining solvent. The EBL system employed a focused electron beam which moved across the sample to selectively expose a pattern in the resist previously designed with the system's built-in computer-assisted design tools.The open area of the AFM tip was selectively exposed to the high energy- beam electrons. The sample was then immersed in the developer solution (3:1 methyl isobutyl ketone: isopropyl alcohol developer) for 30 sec to selectively remove the resist from the exposed areas, whereas the unexposed resist remained insoluble in the developer. The process thus left a patterned resist mask on the sample that could be used for further processing. Finally, Cu was electrochemically deposited on the AFM tip.

2.1 Effect of various electrodeposition parameters

2.1.1 Effect of current density on copper electrodeposition

Scanning Electron Microscopy (SEM) was used to investigate the morphologies of the copper deposits nucleated on the AFM probe. SEM micrograph for Cu deposition on the investigated AFM probe for a plating time of 300 s and a current density of 0.03 A/dm^2 is shown in Fig. 5a . The secondary electron image (SEI) and back scattered electron image (BEI) are displayed in Figs. 5a and 5b. The figure reveals that only a slight amount of copper is electrodeposited on the AFM probe. Further increase in current density to 0.3 A /dm^2 enhanced the copper deposition on the AFM probe however the deposits observed are non-uniform and discontinuous (Fig.6a). The magnified image is seen in Fig. 6b. Furthermore, a gradual increase in the current density to 0.6 A /dm^2 results in uniform deposition of copper on the probe (Fig.7), nevertheless, no copper deposition is noticed on the AFM tip. Similar observations have been reported (Seah et al., 1998). They visualised this morphology on the basis of the fact that formation of more nucleation sites promoted uniform grain growth. In the present study, the formation of uniform copper deposits on the AFM probe could be attributed to the enhanced mass transfer of copper ions with the increase in current density. Litearture reports (Chang, 2001) describe that increase in plating current density increased the surface roughness and reduced the grain size of copper films due to an increase of plating overpotential. Several other researchers have demonstrated that the polarization overpotential increased with increasing the plating current density leading to high copper nucleation rate (Takahashi & Gross, 1999a, 2000 b; Tean et al., 2003; Teh et al., 2001).

The difficulty in depositing Cu ions on the AFM tip arises due to the local increase of the ion concentration in the electrolyte around the tip, which makes the effective local Nernst potential for deposition at the surface underneath the AFM tip more positive.

Fig. 5. SEM image obtained after copper deposition on the AFM tip for plating time of 300 s and current density of 0.03 A/dm^2 (a) SEI image (b) BEI image

Since, the standard electrode potential (ψ^e) of Cu^{2+} [ψ^e (Cu^{2+} + /Cu = +0.337 V) is larger than zero (Fu et al., 1990), from the theoretical point of view, the more positive the ψ^e value, the more easier it is for the reduction of metal ions, and the more negative the ψ^e value, the more difficult it is to reduce the metal ions. Our results suggest that the copper ions can be reduced to copper atoms more easily on the surface underneath the AFM tip. It might be possible that the effective Nernst potential which is required to initiate nucleation on the

AFM tip might be larger due to the deposition overvoltages. Also, the diameter of the AFM tip, which is around 80 nm, might induce high overpotential for deposition of copper on the tip.

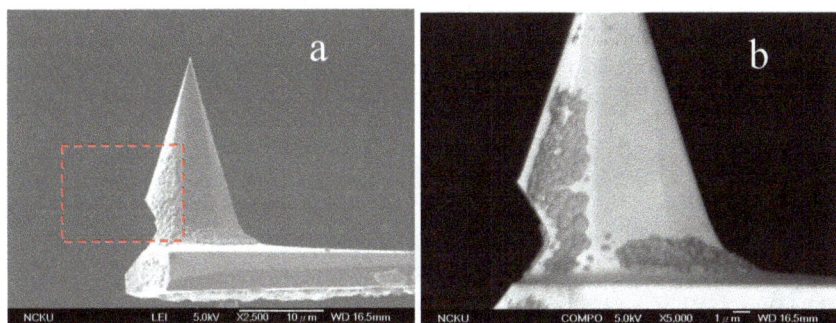

Fig. 6. SEM image obtained after copper deposition on the AFM tip for plating time of 300 s and current density of 0.3 A/dm² (a) SEI image (b) Magnified image (Lin, 2008)

The above results can also be explained on the basis of two reaction schemes which govern the Cu electrodeposition process on the AFM probe: one is the electrode surface reaction and the other one is the Cu^{2+} diffusion from the electrolyte solution to the electrode surface.

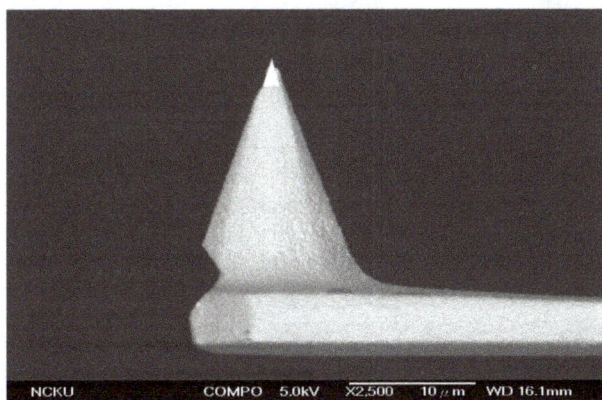

Fig. 7. SEM image obtained after copper deposition on the AFM tip for plating time of 300 s and current density of 0.6 A/dm² (Lin, 2008)

Polarization occurs when the rate of Cu^{2+} supply from the electrolyte solution is not faster than the rate of reaction at the electrode surface. The film morphology is primarily dependent on the degree of polarization (Seah et al., 1999).Thus higher polarization would make electrodeposition slower resulting in a smoother film. Since the effect of increasing current density is to increase the electrode surface reaction, a faster surface reaction makes Cu^{2+} undersupplied from the electrolyte solution. Hence, the polarization is higher and smoother film morphology is observed. Nevertheless, when the applied current density is greater than the limiting current density, it is impossible for the electrode to gain any Cu ions from the electrolyte solution; thereby leading to an increase in the Cu film surface roughness.

2.1.2 Effect of plating time on copper electrodeposition

The effect of different plating times during copper electrodeposition on the AFM probe is investigated. The plating time was varied from 5 to 900 s for different current densities. The SEM micrograph in Figs. 8-11 illustrates the morphology of copper deposit formed under current density of 0.3 A/dm² and various plating times namely 5, 60, 300, 540 s respectively. The results reveal a random distribution of copper crystals on the cantilever with no trace of copper deposits on the AFM tip. This morphology clearly suggests the case of instantaneous nucleation.

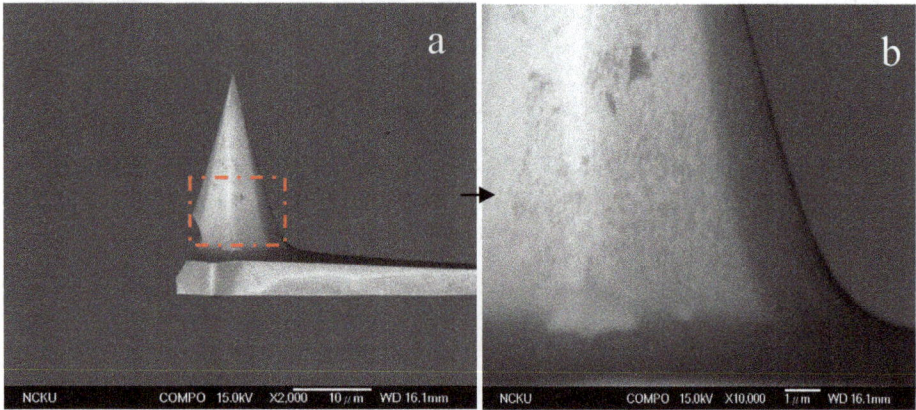

Fig. 8. SEM micrographs illustrating Cu deposition on the AFM tip at a current density of 0.3 A/dm² and plating time of 5 s (a) 2000 T (b) 10000 T magnification of the marked area in red

As instantaneous nucleation corresponds to a slow growth of nuclei on a small number of active sites, all activated at the same time. It can be noted from the SEM images displayed in Fig. 8-11, that in most of the samples the nuclei may be nucleated almost simultaneously, as confirmed by their similar size. In other words i.e at high nucleation rates (instantaneous nucleation), all nuclei are formed immediately after imposition of the potential and grow at the same rate. As a result, they are all of the same age and their number remains constant.

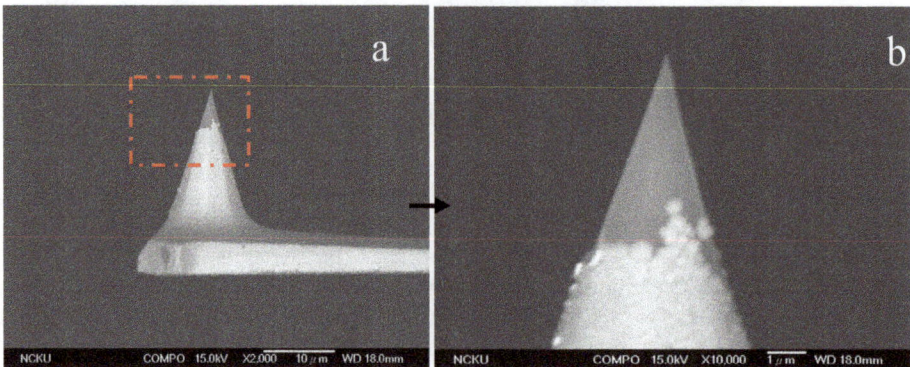

Fig. 9. SEM micrographs illustrating Cu deposition on the AFM tip at a current density of 0.3 A/dm² and plating time of 60 s (a) 2000 T (b) 10000 T magnification of the marked area in red

The mode of instantaneous nucleation is described by the following equation involving the first-order kinetics law (Budevski et al., 1996; Milchev, 1997)

$$N = N_0[1 - \exp(-At)] \tag{1}$$

where N is the number of sites converted into nuclei at time t and A is the nucleation rate constant, N_0 is the respective saturation value. Nucleation does not occur simultaneously over the entire cathode surface and a diameter distribution for the crystallites ensues. When A is very high, $N \equiv N_0$, all surface sites are converted immediately into nuclei and the nucleation is said to be instantaneous. The nonhomogeneity and overgrowth of the Cu deposits may be due to the existence of low nucleation overpotential in the area beneath the tip. At low overpotentials, the nucleation is described well by the model of instantaneous nucleation for reasonably long time scales (Kelber et al., 2006)

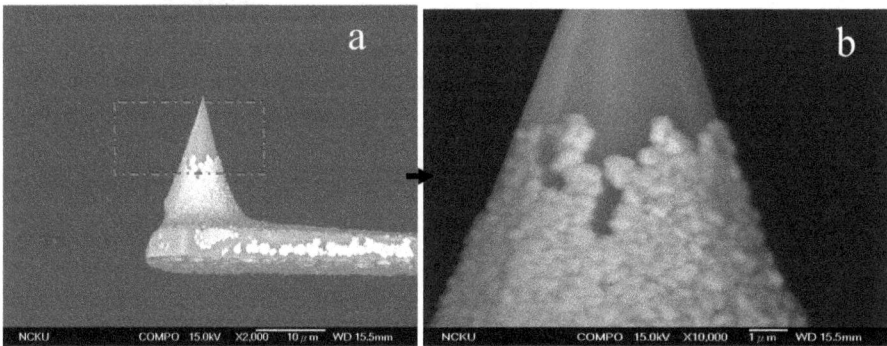

Fig. 10. SEM micrographs illustrating Cu deposition on the AFM tip at a current density of 0.3 A/dm² and plating time of 300 s (a) 2000 T (b) 10000 T magnification of the marked area in red

However, the morphology of copper deposits formed under current density of 0.6 A/dm² and plating time of 900 s were found to be totally different. The copper layer on the AFM probe also shows resemblance to a candle base (Fig. 12), and a thicker layer of copper deposits are grown on the whole of the AFM probe containing the tip. Also, on the basis of instantaneous nucleation model, It has been reported (Thirsk & Harrison, 1972) that under the diffusion controlled three-dimensional growth, the cathodic current density is proportional to $t^{1/2}$.

The growth of copper layer also takes place slowly and farther away from the tip. Also it can be noticed that the growth rate on the side of tip is faster than on the tip (Fig. 12). From the results it could be established that higher current density and higher plating time increases the mass transfer of Cu^{2+} ions in the open area beneath the tip, thereby enhancing the rate of Cu deposition between the open area and the tip. The variance of the thickness of copper deposits on the tip and its surrounding area might be attributed to the nanoscale dimension of the AFM tip as compared to the whole of the AFM probe. Literature reports reveal (Seah et al., 1999) that in case of nanocrystalline electrodeposited Cu the pinhole number-density necessary for full coverage on the substrate can be reduced by increasing the current density. However, abnormal crystallite growth-leading to the formation of bimodal grain

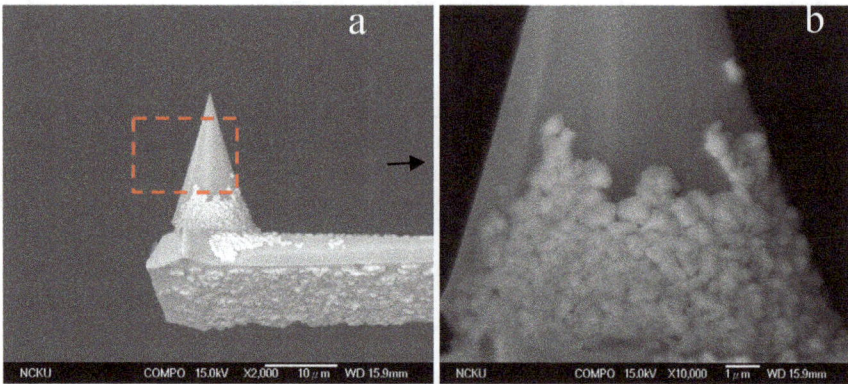

Fig. 11. SEM micrographs illustrating Cu deposition on the AFM tip at a current density of 0.3 A/dm^2 and plating time of 540 s (a) 2000 T (b) 10000 T magnification of the marked area in red

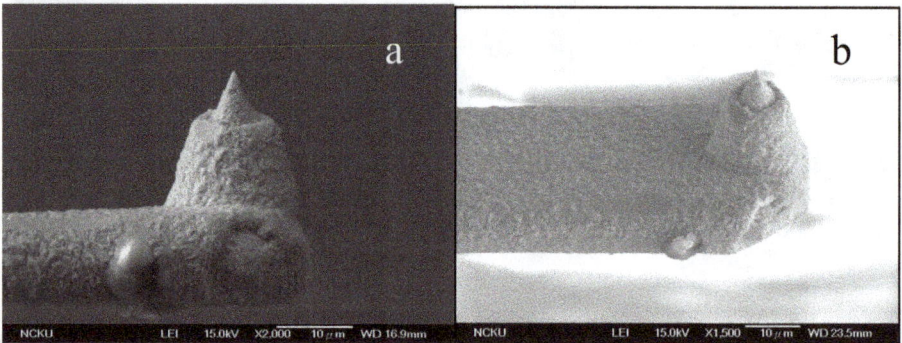

Fig. 12. SEM micrographs illustrating Cu deposition on the AFM tip at a current density of 0.6 A/dm^2 and plating time of 900 s (a) SEI image (b) image taken at 35^0 tilt (Lin, 2008)

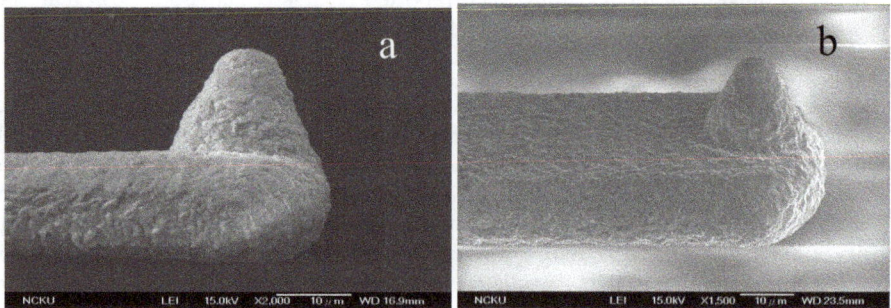

Fig. 13. SEM micrographs illustrating Cu deposition on the AFM tip at a current density of 0.6 A/dm^2 and plating time of 1200 s (a) SEI image (b) image taken at 35^0 tilt (Lin, 2008)

structures- can be suppressed by increasing the electrodeposition current density. In our present case, the crystal growth variation is seen on the open area below the tip and the tip itself. The morphology observed in Fig. 12 is a case of progressive nucleation followed by growth.

As nucleation progresses, the nuclei begins to overlap. Each nucleus is defined by its own diffusion zone through which copper diffuses, thus representing the mass-supply mechanism for continuation of growth. Progressive nucleation corresponds to fast growth of nuclei on many active sites all activated during the course of electroreduction (Pardave et al., 2000). Fig.12b. shows the SEM micrograph for copper deposition at 0.6 A/dm^2 and 900s taken at a tilted angle of 35^0. Further increase in the plating time to 1200 s for similar current density resulted in an entirely different morphology from the micrograph shown in Fig. 12. The AFM probe containing the nanoscale AFM tip seems to be entirely covered with copper deposits and also a significant increase in growth and thickness of the deposits are observed in Figs. 13a. Fig. 13b represents the SEM image tilted at an angle of 45^0 for clear depiction of the copper deposition on the AFM probe. The copper deposition process on the AFM probe proceeds through instantaneous and progressive nucleation modes for different values of current density. The mechanisms for instantaneous and progressive nucleation modes are described below.

Once nucleation begins, crystals growth may be determined by the rate of charge- transfer or diffusion process. Simple equations have been described (Harrison & Thirsk, 1971) for two- or three dimensional nucleation and crystal growth processes occurring on a foreign substrate for charge transfer control reactions.

For two-dimensional (2D) instantaneous nucleation and cylindrical growth, current is described by

$$i = 2zF\text{лh } N_0k^2_{\ 2D}{}^t \,/\, \rho \exp{(\text{-л } M^2 \, N_0k_2{}^2{}_D \, t^2)} \,/\, \rho^2 \qquad (2)$$

And for 2D progressive nucleation

$$i = z \, F\text{лhMK}_2{}^2{}_DA_2Dt^2 \,/\, \rho \exp{(\text{ -л } M^2 \, k_2{}^2{}_D A_2D \, t^3 \,/\, 3\rho^2)} \qquad (3)$$

where k_{2D} represents the lateral growth rate constants (mol cm^{-2} s^{-1}), h is the layer height in cm, N_0 represents the total number of active centers (cm^{-2}), A_{2D} the nucleation rate (nuclei cm^{-2} s^{-1}), M is the atomic weight (g mol^{-1}) and ρ the density (g cm^{-3}) of the deposit. For these type of mechanisms the current usually increases and then decreases to zero when the surface gets completely covered by two dimensional crystals However, for three dimensional (3D) instantaneous nucleation and growth, the current is depicted by the following equations below.

$$i = z \, F \, K' \, [1 - \exp{(\text{-л } M^2k^2 \, N_0 \, t^2 \,/\rho^2)}\,] \qquad (4)$$

and for 3D progressive nucleation:

$$i = zFK' \, [\, 1\text{- } \exp{(\text{ -л}M^2 \, k^2A_{3D} \, t^3 \,/\, 3\rho^2)}\,] \qquad (5)$$

Where k and k' signify the lateral and vertical growth rate constants (mol cm^2 s^{-1}) and A_{3D} the nucleation rate (nuclei cm^2 s^{-1}). Hence nucleation and growth phenomena are affected by many factors i.e a combination of 2D and 3D growth (Abyaneh & Fleischmann, 1981; Creus et al., 1992), the death and rebirth of nuclei (Abyaneh & Fleischmann, 1981) and the secondary three dimensional (3D) growth on top of the first growth layers (Abyaneh et al., 1982).

2.2 Electron Beam lithography studies

EBL (Electron-Beam lithography) technique followed soon after the development of the scanning electron microscope (SEM) in 1955 (Smith, 1955) and is one of the earliest processes used for IC fabrication (Buck, 1957). To date, EBL is widely exploited to produce structures in the sub-100 nm range (Allee et al., 1991; Matsui et al., 1989; Sun et al., 2005). Also, as compared with photolithography, the lateral resolution achieved by EBL is higher because the beam of electrons can be focused to produce probe size as small as 1 nm. More over, electrons do not suffer from optical thin-film interference. For ICs, where at present low beam energy and thick conventional resists are employed; electron scattering is the most important factor whereas for nanolithography, which utilizes high beam energy and thin resists, secondary electron emission is the most dominant factor. The resolution of EBL is also dependent on the chemical nature of the resist. Recently, new class of resists such as organic self-assembled mono layers (SAMs) has been developed to fabricate structures below 10 nm (Golzhauser et al., 2000; Lercel, 1996) Currently, electron beam lithography is used principally in support of the integrated circuit industry, where it has three niche markets. The first is in maskmaking, typically the chrome-on glass masks used by optical lithography tools. It is the preferred technique for masks because of its flexibility in providing rapid turnaround of a finished part described only by a computer. The ability to meet stringent line width control and pattern placement specifications, on the order of 50 nm each, is a remarkable achievement.

2.2.1 Principle of EBL

The principle of pattern transfer based on EBL consists of several process steps. The process steps are essentially the same as those used for photolithography, except that the pattern on the resist is formed by scanning directly the focused particle beam across the surface. The lithographic sequence usually begins with coating of substrates with a positive or negative resist. Positive resists such as poly (methyl-methacrylate) (PMMA) used in the present chapter become more soluble in a developing solvent after exposure because the radiation causes local bond breakages and thus chain scission. This causes the exposed regions containing material of lower mean molecular weight to dissolve after the development. Nevertheless, negative resists become less soluble in solvent after exposure because cross-linking of polymer chains occurs. If in case, a region of a negative resist-covered film is exposed, only the exposed region will be covered by the resist after development. Subsequently, the resist-free parts of the substrate can be selectively coated with metal or etched before removal of the unexposed resist thus leaving the desired patterns at the surface. Fabrication of metallic nanostructures has been widely explored using conventional EBL and lift off techniques. However, this top-down approach cannot be employed for the fabrication of high aspect ratio vertical structures since gradual accumulation of materials at the top of the resist blocks and closes the opening of the structures during the evaporation of metal. Electrodeposition of metals into the holes formed in presence of PMMA resist is a convenient alternative to solve this problem (Simon et al., 1997). The fabrication of dense ultra-small magnetic arrays by filling nanoholes with electrodeposited Ni has been demonstrated (Xu et al., 1995).

2.2.2 EBL induced Copper deposition

Electron beam lithography technique is used in the present study to enable selective electrodeposition of Cu on the AFM tip and the open area beneath it. The selective electrodeposition of Cu on n-type Si (111) surfaces covered with organic monolayers by using e-beam lithographic techniques has been reported (Balaur et al., 2004). Selective copper deposition on e-beam patterned alkane and biphenylthiols has been reported (Kalten Poth et al., 2002) at suitable deposition potentials. 1-octadecanethiol (ODT) was used as a "positive template" leading to copper deposition only on the irradiated parts, 1,1'-biphenyl-4-thiol (BPT) on the other hand acted as a "negative template," where the irradiated and cross-linked biphenyl layer exhibited a blocking behavior, allowing copper deposition on the non-irradiated parts. In the present study, the open area of the nanosize AFM tip was selectively exposed to the e-beam. It is noticed that copper electrodeposition occurs on the exposed area of the AFM tip. For the copper electrodeposition process, the current density applied was 0.6 A/dm², and the electrodeposition time was varied from 300 to 2400s.

Fig. 14. SEM micrograph demonstrating Cu deposition on the AFM tip after EBL treatment under current density of 0.6 A/dm² and electrodeposition time of 300s (a) BEI image (b) Image taken at the tip site (c) Exposure area site

SEM micrographs for copper electrodeposition on the AFM tip and the open area beneath it for various deposition times (i.e 300, 600, 1200, 2400 s) and current density of 0.6 A/dm^2 are presented in Figs. 14-17.These SEM micrographs were taken after exposure to the electron beam. In Figs. 14 (a) – (c) the micrographs for copper deposition on the AFM tip under current density of 0.6 A dm^{-2} and electrodeposition time of 300 s are clearly depicted. Copper deposition is found to be minimum and non-uniform in these images. Further increase in the electrodeposition time to 600 s for the similar current density and exposure to the e-beam increases the amount of copper deposits on the nanosize AFM tip and the open area beneath it (Fig. 15 a). SEM micrographs in Fig. 15b and 15 c refers to the magnified images of the AFM tip and the exposed site.

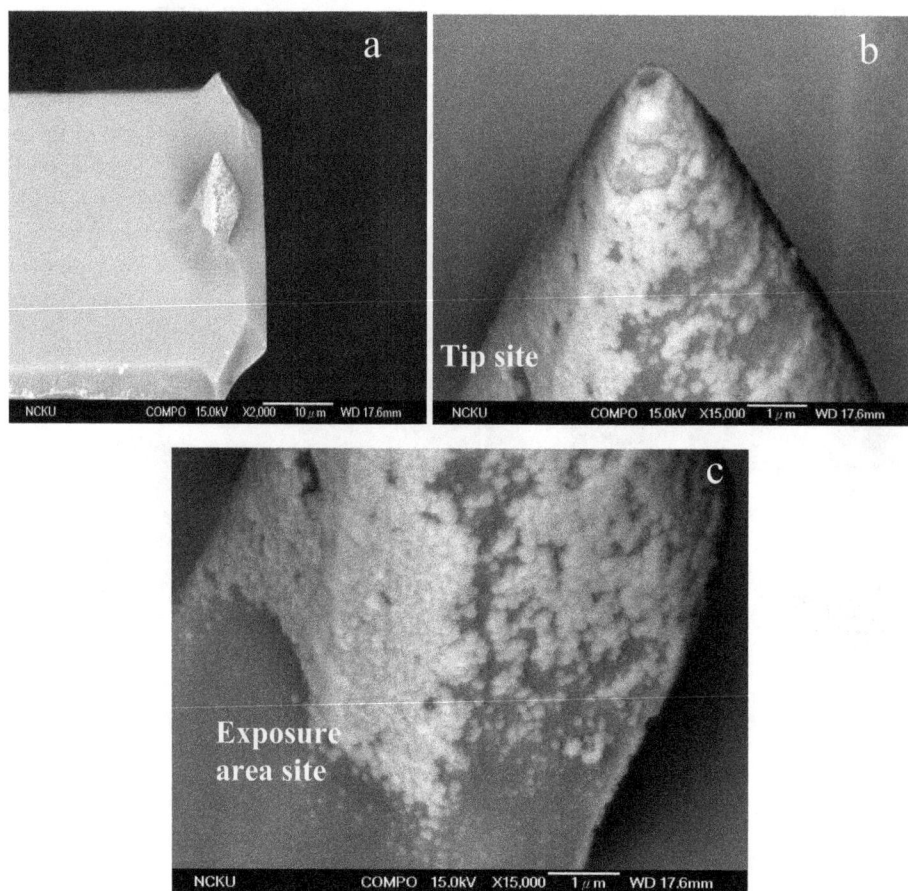

Fig. 15. SEM micrograph demonstrating Cu deposition on the AFM tip after EBL treatmentunder current density of 0.6 A/dm^2 and electrodeposition time of 600s (a) BEI image (b) Image taken at the tip site (c) Exposure area site

The micrographs reveal that copper deposition is not uniform in the open area beneath the AFM tip. However a significant change in the morphology of copper deposits is observed when the electrodeposition time was increased to 1200 s. SEM micrograph in Fig. 16a shows that an uniform layer of copper is deposited on the AFM tip and the open area beneath the tip. These results indicate that the exposure of the tip to the high energy electron beam might have facilitated the electrodeposition of copper on the tip. The micrographs in Fig. 16 b reveal that some copper is being deposited on the edges of the cantilever. This is because the PMMA layers on the edges are found to be thinner than on the platform. Those places are not exposed to the e-beam; therefore the developer could dissolve the PMMA layer on the edges and hence copper deposition took place on the edges. The overpotential required to deposit copper on the edges is lower than on the AFM tip.

Fig. 16. SEM micrograph demonstrating Cu deposition on the AFM tip after EBL treatment under current density of 0.6 A/dm² and electrodeposition time of 1200s (a) SEI image (b) Magnified to 5000 T (c) Tip site (d) Cantilever site

Theoretically it has been established (West, 1971) that deposition at low overpotentials is dominated by surface diffusion; hence nucleation and growth occur primarily at step edges and dislocations (Winand, 1975). Fig. 17 illustrates the morphology of copper deposition on the nanosize AFM tip obtained under current density of 0.6 A/ dm² and electrodeposition time of 2400 s and after exposure to the e-beam. The micrograph in Fig.17a distinctly shows that copper is deposited on the AFM tip and a very thick growth of copper deposits is seen on the open area beneath the tip. From the series of micrographs obtained at different electrodeposition times and current density of 0.6 A /dm², it is noticed that copper gets deposited both on the AFM tip and the open area beneath it, the most uniform deposition seen at 2400 s of electrodeposition time.

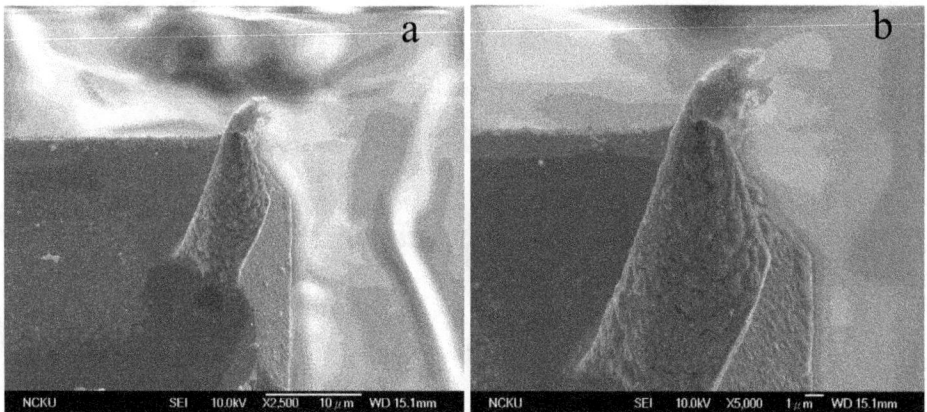

Fig. 17. SEM micrograph demonstrating Cu deposition on the AFM tip after EBL treatment under current density of 0.6 A/dm² and electrodeposition time of 2400s (a) SEI image (b) Magnified to 5000 T

The PMMA coated on the AFM tip becomes more soluble in a developing solvent after exposure to the e-beam because the radiation causes local bond breakages and thus chain scission (Djenizian et al., 2006) as mentioned above. It could be clearly seen from Fig. 17b that the unexposed areas below the AFM tip remain covered with PMMA. However, the resist free parts of the AFM tip are selectively coated with copper. Reports on the selective electrodeposition of Cu (Balaur et al., 2004) on n-type Si (1 1 1) surfaces covered with organic monolayers and e-beam modified using e-beam lithographic techniques have also been established. Copper was electrochemically deposited in the e-beam modified regions and the selectivity of the deposition of copper in these regions was strongly dependent on the applied e-beam dose. The selective deposition of copper on the nanosize AFM tip can be described on the basis of Volmer-Weber approach which states that higher numbers of activation sites are triggered with a higher overvoltage. In the Volmer-Weber model, nucleation and growth are strongly potential dependent. At low cathodic potentials, only a few sites are involved because the energy level is not sufficient whereas at high cathodic voltages more initiation sites contribute to the nucleation process. It implies that at low overpotentials the crystallites have to grow extremely large to reach coalescence and form a homogenous deposit. In the present study, higher overpotential existing on the AFM tip might have increased the number of activation sites, leading to the preferential deposition of copper on exposure to the e-beam.

3. Conclusion

The investigations made in this chapter have highlighted electrodeposition as an attractive approach for the preparation of nanostructured materials. Copper electrodeposition on a nanosize AFM tip of diameter 80 nm was established by varying the magnitude of current densities with electrodeposition time and vice versa. Significant changes in the morphology of copper deposits were observed with changes in the above parameters. Morphological investigations by SEM revealed that a nonuniform layer of copper was formed on the open areas surrounding the tip and the AFM probe; however, deposition of copper on the AFM tip could not be achieved in the absence of photoresist. Electron beam lithography technique facilitated the formation of copper deposits on the nanosize AFM tip of diameter 80 nm in the presence of PMMA. Copper was electrochemically deposited on the e-beam modified regions of the AFM probe at a current density of 0.6 A/dm² with electrodeposition times ranging from 300 to 2400 s. The most uniform deposition on the AFM tip was noticed after EBL treatment under current density of 0.6 Adm⁻² and electrodeposition time of 2400 s.

4. Acknowledgment

The authors acknowledge financial support of this study from the National Science Council of China under NSC 94-2811-E-006-021. The Department of Materials Science and Engineering, National Cheng Kung University assisted in meeting the publication costs of this article.

5. References

Abyaneh, M.Y.; Hendrikx. J.; Visscher, W. & Barendrecht. E (1982). Studies of Electroplating using an EQCM. I. Copper and Silver on Gold. *Journal of Electrochemical Society,* Vol. 129, No.12 (December 1982) pp.2654-2659., ISSN 0013-4651.

Abyaneh, M.Y.& Fleischmann, M. (1981), The Electrocrystallisation of Nickel. *Journal of Electroanalytical Chemistry*, Vol. 119, No. 187, pp. 197, ISSN 1572-6657.

Andricacos, P. C. ; Uzoh, C.; Dukovic, J. O.; Horkans, J. ; Deligianni, H (1998) Damascene copper electroplating for chip interconnections *IBM J. Res. Develop* Vol 42 (April 2010) No.5, 567-574, ISSN:0018-846.

Andricacos, P. C (1999) Copper on Chip Interconnections –A Breakthrough in Electrodeposition to make better chips, *Interface* 1999, 8,32.

Allee, D. R.; Umbach, C.P. & Broers, A.N. (1991). Direct nanometer scale patterning of SiO_2 with electron- beam irradiation. Journal of Vaccuum Science and Technology B, Vol. 9, No.16 (November 1991)pp.2838-2841, ISSN 1071-1023.

Allongne, P.A. (1995) in Advances in Electrochemical Science and Engineering, (eds : Gerischer, H. & Tobias, C. W), ISBN 3-527-28273-4,Weinheim, Germany.

Balaur, E. ; Djenizian, T.; Boukherroub, R.; Chazalviel, J. N. ; Ozanam, F. & Schmuki, P. Electroplating: an alternative transfer technology in the 20nm range. Electrochemistry Communications. Vol.6, No.2, (February 2004), pp. 153-157, ISSN 1366-2481.

Batina, K. & Nichols, D.M. (1992). An in situ scanning tunneling microscopy study of the initial stages of bulk copper deposition on gold(100): the rim effect. *Langmuir*, Vol.8, No.10 (Oct 1992) pp.2572-2576, ISSN 0743-7463.

Benenz, P.; Xiao, X. Y.& Baltruschat, H. (2002). Tip-Induced Nanostructuring of a Clean and Ethene –modified Pt (111) Electrode with Cu. *Journal of Physical Chemistry B*, Vol. 106, No. 14, (March 2002) pp 3673-3680, ISSN 1089-5647.

Binning, G. & Rohrer, H. (1982). Scanning Tunneling Microscopy. *Helvetica Physica Acta*, Vol. 55, pp.726-735, ISSN 0036-8075.

Bockris J. O.M. & Razumney, G. A. (1967) *Fundamental Aspects of Electrocrystallization*, Plenum Press, New York.

Buck, D.A. & Shoulders. K (1957), In Proceedings Eastern Joint Computer Conference (ATTE, New Yoork), p.55.

Budevski, E. ; Staikov, G. & Lorenz, W. J. (1996) *Electrochemical Phase Formation and Growth – An Introduction to the Initial Stages of MetalDeposition*, ISBN 3527294228978352729 4220,VCH, Weinheim.

Budevski, E.; Bostanov, V. & Staikov, G. (1980) *Annu. Rev. Mater. Sci.* 10, 85.

Budevski E. (1983) *Comprehensive Treatise of Electrochemistry*, Vol. 7 (Eds: B. E.Conway, J. O'M. Bockris, E. Yeager, S. U. M. Kahn, R. E. White , Plenum Press, New York, p.399.

Budevski, E. .; Staikov, G. & Lorentz, W. (1996). *An Introduction to Initial Stages of MetalDeposition*, p. 4, Wiley VCH, ISBN 3527294228, Weinheim, Germany.

Carlsson, P.; Holmstriom, B. ; Kita H. & Uosaki, V. (1990). Novel application of scanning tunneling microscopy — tip current voltammetry of n-GaAs and p-GaP in electrolyte solution.*Surface Science*, Vol. 237, No.1-3 (November 1990) pp. 280 -290, ISSN 0039-6028.

Chang, S.C. ; Shieh, J.M. ; Lin, K.C.; Dai, B.T. ; Wang, T.C.; Chen, C.F. ; Feng, M.S. ; Li, Y.H. & Lu, C.P. (2001). Investigations of effects of bias polarization and chemical parameters on morphology and filling capability of 130 nm damascene electroplated copper. *Journal of Vacuum Science and Technology B.* Vol.19 , No.3 (May 2001) pp. 767-774, ISSN 1071-1023.

Creus ,H. A.; Carro, P.; Gonzalez, S.; Salvarezza, R.C & Arvia, A.J. (1992). Electrochemical kinetics and growth modes of silver deposits on polyfaceted platinum spherical electrodes Electrochmica Acta, Vol. 37, No. 12, (September 1992), pp. 2215-2227, ISSN 0013-4686.

Danilov, A. I., Molodkina, E. B & Polukarov Yu. M. (1994) *Russ. J.Electrochem.* 1994, 30, 674.

Djenizian, T. & Schmuki, P. (2006) Electron beam lithographic techniques and electrochemical reactions for the micro- and nanostructuring of surfaces under extreme conditions. *Journal of Electroceramics*, Vol. 16, No.1, (Feb 2006), pp.9-14, ISSN 1385-3449.

Fischer, H. (1954) *Elektrolytische Abscheidungund Elektrokristallisation von Metallen*,Springer, Berlin

Fischer, H. (1969) Electrocrystallisation of Metals under ideal and real conditions. *Angew. Chem. Int. Ed.Engl.* , 8, 108-119, ISSN 1521-3773.

Fleischmann, M. & Thirsk, H. R. (1963). *Advances in Electrochemistry and Electrochemical Engineering, Vol. 3, P. Delahay (Ed.)*, Wiley, New York.

Fu, C. X.; Shen, W. X. & Yao T. T. (1990) *Physical Chemistry*, 4th ed., p. 603, Higher Education Press, Beijing.

Gewirth, A.A. & Siegenthaler (1995), Nanoscale Probes of the Solid/ Liquid Interface, NATO ASI series E: Applied sciences vol. 288 , 334 pp, ISBN 0-7923-3454-X, Kluwer Academic Publishers, Dordrecht/ Boston/ London.

Gileadi, E. & Tsionsky, V. Studies of Electroplating Using an EQCM. I. Copper and Silver on Gold. *Journal of Electrochemical Soc.*iety. vol. 147, No.2 (Feb 2000) pp. 567-574, ISSN 0013-4651.

Golzhauser, A.; Geyer, W.; Stadler, V.; Eck, W.; Grunze, M.; Edinger, K.; Weimann, Th.& Hinz. P (2000). Nanoscale patterning of self-assembled monolayers with electrons. *Journal of Vacuum Science and Technology. B*, Vol. 18, No.6 (November 2000) 3414-3418, ISSN 1071-1023.

Hachiya, T.; Honbo, H. & Iteya, K (1991). Detaled Underpotential Deposition of Copper on Gold (111) in aqueous-Solutions. *Journal of Electroanalytical Chemistry*, Vol. 315, pp.275-291, ISSN 1572-6657.

Hozzle, M. H.; Zwing, V & Kolb D. M (1995). The influence of steps on the deposition of Cu on Au (III) Electrochimica Acta , Vol 40, No. 10,pp 1237-1247, ISSN 0013-4686 (95) 00055-0.

Harrison, J.A. & Thirsk, H. R. (1971), A guide to the study of electrode kinetics, In :*Electroanalytical Chemistry*, A. J. Bard, (Ed.), Vol5, 67, Marcel Dekker, London.

KaltenPoth, G.; Volkel, B.; Nottbohm, C.T.; Golzhauser.A & Buck, M. (2002). Electrode modification by electron-induced patterning of self-assembled monolayers. *Journal of Vacuum Science and Technology B*, Vol.20 , No.6 (November 2002), pp 2734-2738, ISSN 1071-1023.

Kelber, J..; Rudenja, J. &. Bjelkevig, C. (2006) Electrodeposition of copper on Ru (0 0 0 1) in sulfuric acid solution: Growth kinetics and nucleation behavior. *Electrochimica Acta*, Vol.51, No.15, (April 2006) pp.3086-3090, ISSN 0013-4686.

Koinuma, M. & Uosaki, K (1994) In situ observation of anodic dissolution process of n-GaAs in HCl solution by electrochemical atomic force microscope. *Journal of Vaccum Science and Technology B*, Vol.12, pp 1543-1546, ISSN 1071-1023.

Lercel, M. J.; Craighead, H. G.; Parikh, A.N.; Seshadri, K. & Allana, D. L. (1996), Sub-10 nm lithography with self-assembled monolayers. Applied Physics Letters, Vol. 68, No.11, pp.1504-1506, ISSN 0003-6951.

Lin, K.L.; Chen, S.Y. & Mohanty, U.S (2008) Effect of Current density and Plating Time on the Morphology of Copper Deposits on an AFM tip. Journal of Electrochemical Society. Vol. 155, No. 4, pp. D251-D255, ISSN 0013-4651.

Li, Y.; Maynor, B.W. & Lu. J (2001). Electrochemical AFM "Dip-Pen" Nanolithography. .*Journal of American Chemical Society*, Vol. 123, No.9 (March 201) 2105-2106, ISSN 0002-7863.

Lustenberger, P.; Rohrer, H; Christoph, R & Stegenhaler, H. (1988) Journal of Electroanalytical Chemistry, Vol. 243, pp.225.

Matsui, S.; Ichihashi, T .& Mito, M. (1989). Electron beam induced selective etching and deposition technology. *Journal of Vacuum Science and Technology B*, Vol. 7, No.15 September 1989) pp.1182-1190, ISSN 1071-1023 .

Markovic, N.M.; & Ross, P.N, Jr (1993). Electrodeposition of copper on Pt (111) and Pt (100) single crystal surfaces, *Journal of Vacuum Science and Technology A*, Vol. 11, No.4, pp. 2225-2232, ISSN 0734-2101.

Michailova, E. ; Vitanova, I.; Stoychev, D. & Milchev, A. (1993). Initial stages of Copper Electrodeposition in presence of organic additives. *Electrochimica Acta* 1993, 38, 2455-2458, ISSN 0013-4686 (93) 85116-G.

Milchev, A.. (1997). Electrochemical alloy formation—theory of progressive and instantaneous nucleation without overlap. *Electrochimica Acta*. Vol. 42, No.10, pp. 1533-1536, ISSN 0013-4686.

Milchev, A (2002) *Electrocrystallization:Fundamentals of Nucleation and Growth*, ISBN 1-4020-7090-X, Kluwer Academic Publishers,Boston/Dordrecht/London.

Nichols, R. J.; Beckman.; Mayer, H.; Batina, N & Kolb, D.M (1992). An in situ scanning tunnelling microscopy study of bulk copper deposition and the influence of an organic additive. *Journal of Electroanalytical Chemistry*, Vol. 330, No.1-2 (July 1992) pp. 381-394, ISSN 1572-6657.

Oskam, G.; Long, J. G.; Natarajan, A. ; Searson, P. C (1998) Electrochemical deposition of metals onto silicon. *J. Phys. D: Appl. Phys.* Vol. 31, No.16 (August 1998), pp.1927-1949, ISSN 0022-3727.

Pardave, M. P.; Ramirez, M. T. ; Gonzalez, I.; & Scharfiker, B. R. (2000). *J. Electrochem. Soc.*, 147, 567 (200)

Paunovic, M. & Schlesinger, M. (2006) *Fundamentals of Electrochemical Deposition*, ISBN 978-0-471-71221-3, Wiley-Interscience, NewYork.

Seah, C. H.; Mridha, S; Chan, L.H (1998), Growth morphology of electroplated copper: effect of seed material and current density, IITC, 98-158, IEEE, ISSN 0-7803-4285-2/98.

Seah, C.H.; Mridha, S & Chan, L.H (1999), Fabrication of D.C.-plated nanocrystalline copper electrodeposits. *Journal of Materials Processing Technology*, Vol. 89-90 , (May 1999), pp.432-436, ISSN 0924-0136.

Simon, G.; HaghiriGosnet, A.M.; Carcenac, F. & Lannois, H (1997), Electroplating: An alternative transfer technology in the 20nm range. *Microelectronics Engineering*, Vol. 35, No. 1-4 (February 1997), pp.51-54, ISSN 0167-9317.

Smith, K.C. A & Oatley, C.W (1955). Brazilian Journal of Applied Physics, Vol. 6, pp.391.

Sonnenfeld, R. & Hannsma, P.K (1986) Atomic Resolution Microscopy in Water, *Science*, Vol. 232, No.4747, pp.211-213,, ISSN 0036-8075.

Staikov, G.; Juttner, K.; Lorenz, W. J & Budevski, E (1994). Metal depsoition in the nanometer range.,*Electrochimica Acta*, Vol. 39, No.8-9, (June 1994), pp. 1019-1029, ISSN 0013-4686.

Stegenthaler, H (1992). in Scanning Tunneling Microscopy II, Springer Ser. Sur. Sci, Vol. 28(eds: Ewisendanger, R.; Guntherodt, H. J, Springer, Berlin.

Sun, S.; Chong, K. S. L and Leggett, G.J (2005). Photopatterning of self-assembled monolayers at 244 nm and applications to the fabrication of functional microstructures and nanostructures, *Nanotechnology*, 16, pp.1798-1808, ISSN 0957-4484.

Takahashi, K.M. & Gross, M.E. (1999). Transport Phenomena That Control Electroplated Copper Filling of Submicron Vias and Trenches. *Journal of Electrochemical Society*. Vol. 146, No.12 (December 1999) pp. 4499-4503,.

Takahashi, K.M. (2000). Electroplating Copper onto Resistive Barrier Films. *Journal of Electrochemical Society*. Vol. 147, No.4 (April 2000) pp 1414- 1417, ISSN 0013-4651.

Tan, M. & Harb, J.N. (2003) Additive Behavior during Copper Electrodeposition in Solution Containing Cl-, PEG, and SPS *Journal of Electrochemical Soc.*iety. Vol.150, pp C420-C425, ISSN 0013-4651.

Teh, W.H.; Koh, L.T.; Chen, S.M.; Xie, J.; Li, C.Y. & Foo, P.D. (2001) Study of microstructure and resistivity evolution for electroplated copper films at near-room temperature. *Microelectronics Journal*, Vol 32, No.7 (July 2001), pp. 579-585, ISSN 0026-2692.

Thirsk, H. R. & Harrison, J. A. (1972), A guide to the study of electrode kinetics, ISBN 0126877505, London, New York, Academic Press.

Volmer, M (1934) *Das Elektrolytische Kristallwachstum, Hermannet Cie*, Paris.

Volmer, M (1939) *Kinetik der Phasenbildung, Steinkopf*, Dresden.

West, J. M. (1971). *Electrodeposition and Corrosion Process*, pp. 204, 2nd Edition, ISBN 978-04422093525, Van Nostrand Reinhold, New York.

Winand, R (1975). Electrocrystallisation of Copper. *Transactions of the Institution of Mining and Metallurgy. Section C, Mineral processing and extractive metallurgy*, Vol.84,pp.67-75, ISSN 0371-8553.

Xu, W.; Wong, J.; Cheng, C.C.; Johnson, R. & Scherer.A. (1995), Fabrication of ultrasmall magnets by electroplating . *Journal of Vacuum Science and Technology, B*. Vol. 13, No.6 (November 1995) pp.2372-2375, ISSN 1071-1023.

Yau, S.L.; Gao, X; Chang, S.C; Schardt, B.C & Weaver , M. J.(1991) Atomic-resolution scanning tunneling microscopy and infrared spectroscopy as combined in situ probes of electrochemical adlayer structure: carbon monoxide on rhodium (111). *Journal of American Chemical Society*, Vol. 113, No.16, pp. 6049-6056, ISSN 0002-7863.

8

Iron Oxide Nanoparticles Imaging Tracking by MR Advanced Techniques: Dual-Contrast Approaches

Shengyong Wu
Medical Imaging Institute of Tianjin
China

1. Introduction

Recently a number of imaging modalities have been presented for cellular imaging including magnetic resonance imaging (MRI), optical imaging, and positron emission tomography (PET) based on the background of growing demand for molecular imaging to noninvasively and longitudinally visualize cell migration and track transplanted cells in vivo, also to monitor cell biodistribution. Cellular MRI, with its superb ability of resolving soft tissue anatomies in three-dimensions (3D) with high spatial resolution in comparison to other modalities, is particularly important as a noninvasive tool to provide unique information on the dynamics of cell migration *in vivo* (Modo, 2005; Arbab, 2008a; Zhang, 2008).

In vivo MRI of cells is very useful for studying tumors, inflammation, stem cell therapy, and immune response, etc. Cells labeled with commercially available iron oxide nanoparticles (iron particles) can be imaged for weeks with MRI. The labeling procedure does not exhibit any alteration to cell viability or function (Bulte, 2004; Oude Engberink, 2007). Superparamagnetic iron oxides (SPIO) and ultra-small superparamagnetic iron oxide (USPIO) particles are commercial MR contrast agents for cell labeling due to their biocompatibility and strong effects upon T_2 and T_2^* relaxation. Several labeling methods have been developed to incorporate sufficient quantities of iron into cells. Cellular MRI has now been widely used for tracking transplanted iron-labeled therapeutic cells in vivo (Bulte, 2004; Oude Engberink, 2007). The technique has recently been introduced into the clinic (de Vries, 2005). The effect from iron particles is seen as hypointensity or negative-contrast on T_2- and T_2^*-weighted images because of the shortening of T_2 and T_2^* relaxation times. However, concerns have been raised that the negative-contrast could be non-specific and difficult to differentiate from signal hypo-intensities resulting from susceptibility artifacts (i.e. from the presence of air or other field inhomogeneities), flow related signal losses, and calcification. Therefore, several positive-contrast and even dual-contrast imaging techniques have recently been developed for tracking iron-labeled cells. Dual-contrast imaging effectively permits detection of the presence of iron-labeled cells with both negative- and positive-contrast within a single image. This chapter illustrates negative- and positive-contrast MR techniques for tracking iron-labeled cells. Particular attention was paid to recently developed positive-contrast cell tracking techniques, the status of dual-contrast approaches of new MRI pulse sequences and image postprocessing techniques and their perspectives. The new advanced technology in imaging contrast of iron oxide NPs on multimodal platform will also be introduced.

2. Negative-contrast MRI techniques

Cellular MRI is a newly emerging field of MR research that allows the "non-invasive, quantitative, and repetitive imaging of targeted macromolecules and biological processes in living organisms" (Herschman, 2003). Cellular MRI requires that cells are labeled with MR contrast agent to make them distinct from the surrounding tissues. Iron oxide nanoparticles are regarded as the most extensively applied contrast agent in cell imaging and cell tracking studies based on the fact of their strong negative contrast effect, biocompatibility, variety in core size and coating surface, as well as ease of detection at microscopic level (Muja, 2009). SPIO and USPIO are currently the predominant MRI contrast agents. The description of the physical and chemical properties of SPIO and USPIO can be found in recent reviews (Herschman, 2003; Thorek, 2006; Muja, 2009). The sizes of monocrystalline iron oxide nanoparticles (MIONs) $\approx$ 3 nm in diameter, USPIO particles $\approx$ 15-30 nm, SPIO particles $\approx$ 60-180 nm and micron sized iron oxide particles (MPIOs) can be as large as 10 μm (Shapiro, 2005). Some of the SPIO and USPIO agents, such as Endorem (SPIO, Guebert), Ferumoxides (SPIO, Berlex) and Resovist (USPIO, Schering), are already approved by the Food and Drug Administration (FDA) and are extensively used for imaging of the liver, central nervous system (CNS) and lymphatic system (Arbab, 2004b; Helmberger, 2005; Manninger, 2005), etc. Cationic transfection agents such as poly-L-lysine or the FDA-approved protamine sulfate are used to increase labeling efficiency *in vitro*. SPIO particles may decrease T_2^* by magnetic susceptibility effect and T_2 by dipole-dipole interaction or scalar effect between protons and magnetic centre. A large magnetisation difference occurs as a result of the nonhomogeneous distribution of superparamagnetic particles, which gives rise to local field gradients that accelerate the loss of phase coherence of the spins contributing to the MR signal. Iron-labeled cells cause significant signal dephasing due to the magnetic field inhomogeneity induced in water molecules near the cell such that iron-labeled cells were visualized as signal voids on T_2 and T_2^* weighted images (negative-contrast MR imaging). Negative-contrast techniques are the most commonly used approach for the detection of the SPIO-labeled cells.

While cell-based therapies have attracted well attention as novel therapeutics for the treatment of so many kinds of diseases, investigations (Zhang, 2005; Heyn, 2005, 2006) have showed that single, living, highly phagocytic large cells, such as macrophages, or human endothelial cells can be tracked over time in MRI using a 3.0 T even 1.5 T scanner. As an example of stem cell-based studies, investigators (Anderson, 2005) demonstrated that MRI of iron-labeled stem cells was directly identified in neovasculature of a glioma model. The cells were labeled using the ferumoxides/poly-L-lysine complex in vitro and the labeled cells were then injected in the model, and their migration toward and incorporation into the tumor neovasculature was visualized in vivo with negative-contrast MRI. Other studies have shown that ferumoxides-TA labeled human MSCs will home to liver (Arbab, 2004a), tumors (Khakoo, 2006), or heart (Kraitchman, 2005), illustrated at negative-contrast imaging with MR scan and confirmed at histologic evaluation. A group (Zhu, 2006) labeled neural stem cells (NSCs) obtained from patients with traumatic brain injury then performed intracerebral injections of either ferumoxide-labeled or unlabeled cells around the injured tissue of them as the first study in the field of noninvasive imaging of stem cell treatment of brain injury, and their serial MRI about 7-10 weeks demonstrated that stem-cell engraftment and migration after implantation can be detected noninvasively with the use of MRI.

Also, in an early study (Kircher, 2003a), a highly derivatized cross-linked iron oxide (CLIO) nanoparticle was used to efficiently label cytotoxic T lymphocytes (CTLs) for in vivo tracking of the injected cells to melanoma cell line at near single-cell resolution, with MRI and optimized the labeling protocol (three-dimensional nature of the calculated T_2 maps), showing no cytotoxic and not influencing cell behavior or effector function. Despite the fact that the high spatial resolution given by MRI provides accurate evaluation of morphology of lymphoid organ, the sensitivity and ability to quantify MR data is still limited when compared with nuclear medicine based techniques. For MR cell tracking to be clinically useful, it should be defined for the detection limits of the MR method which will be utilized. The related clinical studies with 3.0 T scanners suggest that negative-contrast techniques possibly detect 150,000 Feridex labeled cells after directly injected into the lymph nodes of patients (de Vries, 2005). Another recent example of study by Laboratory for Gene Transcript Targeting, Imaging and Repair in Massachusetts General Hospital demonstrated that functionalization allows SPIO nanoparticles to be targeted, and it showed that their phosphorothioate-modified DNA probes linked to SPIO could be used to identify differential gene expression due to amphetamine exposure with high reliability using the calculation of rate of signal reduction (R_2*) in T_2*-weighted MR images (Liu, 2009). There are also extensive published works with detailed descriptions of many aspects of labeled cells for detection with negative-contrast MRI (Ferrucci, 1990; Bulte, 2004b; Hsiao, 2007; Gonzalez-Lare, 2009). Those and many of other preclinical studies have provided evidences for the potential translation of iron oxide NPs labeling and cellular MR imaging to the clinic applications.

An important property of USPIO is its ability to shorten T_1 and T_2 relaxation times (Small, 1993; Li, 2005). USPIO-labeled cells can be tracked in T_1 and T_2/T_2* weighted images, which should increase the accuracy and the specificity for detection of the labeled cells (Kelloff, 2005), such as in imaging assessment on angiogenesis of tumor (Niu, 2011), atherosclerotic plaques (Metz, 2011), or arthritis (Lefevre, 2011). USPIO nanoparticles recently have shown potential in the imaging of molecular biomarkers, such as integrins that are heterodimeric transmembrane glycoproteins, a family of adhesion molecules playing a major role in angiogenesis and tumor metastasis (Chen, 2009; Tan, 2011).

Much of the progress in detecting individual iron-labeled cells has achieved from improvements in contrast agent design that increases targeting and intracellular uptake properties (Cerdan, 1989; Weissleder, 1990; Bulte, 2001; Zhao, 2002). Improvements in MR hardware and pulse sequence design also have played an important role during recent progress in this area of research. Although negative-contrast MRI has shown promise as a means to visualize labeled cells (Hogemann, 2003; Heyn, 2005), some remaining issues may hamper its wide applications: (1) it is difficult to distinguish the signal voids of labeled cells from those of complex background tissue signals; (2) With the resulting signal void as the means for detection, partial-volume effects are often severe and go far beyond the real cell size; (3) it is difficult to discriminate iron-induced susceptibility changes from those caused by other susceptibility artifacts due to i.e. air/tissue interfaces, or peri-vascular effects.

3. Positive-contrast and dual-contrast MRI techniques

The "white-marker imaging" positive-contrast mechanism was introduced by Seppenwoolde et al. in 2003 (Seppenwoolde, 2003). Since then, several groups have developed positive-contrast or dual-contrast pulse sequences for tracking iron-labeled cells *in vitro* and *in vivo* (Table 1).

3.1 Gradient-dephasing technique: "white-marker" imaging

"White-marker" imaging was initially presented to create positive-contrast around paramagnetic intravascular device markers used in magnetic-resonance-based interventional procedures (Seppenwoolde, 2003). The gradient-dephasing technique uses a slice gradient to dephase the background water signal followed by an incomplete gradient rephasing pulse which was exploited for the depiction and tracking of paramagnetic susceptibility markers. Local magnetic field inhomogeneities were selectively visualized with positive-contrast, such as those created by iron-labeled cells for "white-marker" imaging. Advanced methods were developed to separate magnetic susceptibility effects from partial volume effects in "white marker" imaging in order to avoid compromising the identification of magnetic structures (Seppenwoolde, 2007). However, this method is only sensitive to macroscopic field inhomogeneities caused by paramagnetic material, to a volume surrounding the paramagnetic material that is free of other field variations (Zurkiya, 2006).

A similar gradient depashing technique termed gradient echo acquisition for superparamagnetic particles (GRASP), by dephasing of the background signal, has been used to detect positive-contrast from superparamagnetic particles based on the phenomena that the z-rephasing gradient is reduced so that dipolar fields generated by the cells are rephased and positive signal can be observed (Mani, 2006a), also to image ferritin deposition in a rabbit model of carotid injury with relatively low concentrations of iron oxides at 1.5 T MR scanner (Mani, 2006b). The GRASP technique was used to successfully image low concentrations of ferumoxides (0.05 mM Fe corresponding to 2.8 μg Fe/mL) and ferritin (5 μg Fe/mL) in gel phantoms (Mani, 2006). GRASP "white-marker" imaging has several advantages including ease of implementation, high sensitivity, no influence on positive signal due to both B_0 and B_1 field inhomogeneities, and fast acquisition with various TE values. The feasibility of GRASP was tested to aid in dynamically tracking stem cells in a mouse model of myocardial infraction (Mani, 2008). Using T_2^*-GRE and GRASP techniques at 9.4 T scanner, iron-labeled embryonic stem cells were visualized in the border zone of infarcted mice at 24 hours, and 1 week following implantation. The positive signal in areas containing iron-loaded stem cells corresponded precisely with the signal loss detected within images produced with conventional GRE sequences. Regions that contained iron-labeled cells were confirmed by histology (Mani, 2008). The presence of the signal loss because of iron-labeled cells would have been difficult to detect on T_2^*-weighted images without using the positive-contrast sequence. The region of the myocardium containing the iron-labeled cells was clearly visible when both GRASP and T_2^*-weighted techniques (dual contrast imaging) were applied. Dual-contrast effects act to extend the signal change well beyond the location of the particle or

MR sequences	Contrast agents	Experimental conditions	Biological target	Application and Results
gradient-dephasing technique & GRASP	Ferritin	*In vitro* and *in vivo*	Endogeneous ferritin	Crush injured rabbit carotid arteries
				Myocardial infraction
	Ferumoxides	*In vitro* and *in vivo*	Embryonic stem cell-derived cardiac precursor cell	
				Injected into the hind limb of mouse
	Ferumoxides	*In vitro* and *in vivo*	Embryonic stem cell line TL-1	
off-resonance (OR) method	Ferumoxides	*In vitro* and *in vivo*	SPIO-luc-mouse embryonic stem cell	Injection into hindlimbs of mouse
Off-resonance saturation	mMION/ SPPM	Gel phantom/ *in vivo*	the $_v\beta_3$-expressing microvasculature	molecular imaging of cancer
IRON technique	MION-47	*In vivo*	Macrophage	Atherosclerotic plaque
				MR lymphography
	MION-47	*In vivo*	Macrophage	
SR-SPSP sequence	Ferumoxides	*In vitro* and *in vivo*	Human bone marrow stromal cells	Injection into the hind legs of mouse
FLAPS sequence	Ferumoxides	*In vitro* and *in vivo*	GFP-R3230Ac cell line	Injection into the hind legs of rat
UTE imaging	Ferumoxides	*In vitro* and *in vivo*	G6 glioma cells	Implanted cellular imaging
SWEET sequence	Ferumoxides	*in vivo*	Human epidermal carcinoma cells	Visualization of magnetically labeled tumor cells

Note: GRASP, superparamagnetic particles/susceptibility; IRON, oxide nanoparticles–resonant water suppression; SR-SPSP, self-refocused spatial-spectral; FLAPS, fast low-angle positive contrast steady-state free precession; UTE, ultrashort echo-time

Table 1. Summary of Previously Published Studies of Positive- and Dual-contrast Techniques

cell itself. This form of signal amplification increases sensitivity in detecting the labeled cells within a complex image background. With the use of signal amplification, potential future applications of (U)SPIO include 'doping' of therapeutic cell preparations with a small fraction of labeled cells, to allow cell tracking without altering the majority of the cells. This would allow for better delineation and identification of labeled cells with both techniques. The challenge for both techniques is the difficulty associated with attempting to quantify the concentration of the labeled cells in vivo because of the susceptibility artifact produced via the iron particles.

Generally, to resolve issues associated with volume averaging and other artifacts that may limit the clinical utility of MRI to detect iron labeled cells (especially in tissues other than the brain), GRASP technique has been developed to differentiate between the signal generated by the cells and signal loss cause by various artifacts (Mani, 2006, 2008), and to specifically avoid the signal loss generated by the iron laden cells to be confused with signal caused by other sources (motion, perivascular effects, coil inhomogeneities, etc.). In the recent study (Briley-Saebo, 2010), the GRASP sequence was also used to both detect and confirm the presence of the Feridex labeled dendritic cells (DCs) in the draining lymph nodes of nude mice 24 h after footpad injection. The results showed the possibility to detect and longitudinally track ex vivo human DC vaccines in the spleen of mice for up to 2 weeks, with greater lymphoid targeting observed following i.v. injection, relative to subcutaneous foot-pad injection; also showed good correlation between in vivo R_2^* values on a 9.47 Tesla dedicated mouse scanner and Feridex concentration, with detection limits of 3.2% observed for the spleen. But investigators didn't detect the Feridex labeled cells within the liver and spleen using the GRASP sequence while they indicated that, the dipole effects would be limited and signal enhancement would not be observed when the iron particles being homogenously distributed over a large volume (such as the liver or spleen). They further demonstrated the values of nodes the white marker sequence, GRASP, in accurate detection and identification of Feridex labeled DCs in superficial lymph, and indicated that the appropriate utilization of animals models and MR validated imaging strategies might allow for the optimization of human DC vaccine therapies and improved therapeutic success, whereas white marker sequences maybe most effective when the iron laden cells being compartmentalized within a limited volume (such as in lymph nodes, tumors, or myocardium). On the basis of a recent report (Sigovan, 2011) of the feasibility study on a positive contrast technique, GRASP at a relatively high field 4.7 T, for a novel superparamagnetic nanosystem designed for tumor treatment under MRI monitoring, investigators found that the magnetic nanoparticles for drug delivery can be detected using positive contrast, and suggested that the combined negative and susceptibility methods allow good quality images of large magnetic particles and offer their follow-up for theranostic applications.

3.2 Off-resonance Imaging (ORI)

Off-resonance MRI approaches have also been developed to produce positive-contrast. With this method, a spectrally selective radio frequency (SSRF) pulse was used to excite only the susceptibility-shifted, or 'off-resonance', water signals (Cunningham, 2005; Foltz, 2006), at the frequency shift induced by the iron particles. Since only the off-resonance signal due to iron particles are excited and refocused, the background on-resonance signal is largely eliminated.

Iron-labeled mouse embryonic stem cells were imaged as positive-contrast through suppression of background tissue with these off-resonance methods (Suzuki, 2008). A spin-echo sequence was used with million-fold (120 dB) suppression of on-resonance water by matching the profiles of a 90° excitation and a 180° refocusing pulse. The positive-contrast signal from the volume of cells was affected by how well the excitation profile was defined. The method is therefore inherently limited by the complication associated with unwanted magnetization from the regions that suffer from chemical shift or susceptibility-related artifacts (e.g., from fat/lipid present in the region of interest and/or imperfect B_0 shimming, due to air/tissue interfaces, etc.) (Farrar, 2008). Although ORI techniques are being increasingly used to image iron oxide imaging agents such as MION, the diagnostic accuracy, linearity, and field dependence of ORI have not been fully characterized. After the sensitivity, specificity, and linearity of ORI were examined as a function of both MION concentration and magnetic field strength (4.7 and 14 T), and MION phantoms with and without an air interface as well as MION uptake in a mouse model of healing myocardial infarction were imaged, the linear relationship between MION-induced resonance shifts and with MION concentration were illustrated, whereas T_2 showed comparable to the TE and then decreasing after increasing initially with MION concentration and the ORI signal/sensitivity being highly non-linear. Improved specificity of ORI in distinguishing MION-induced resonance shifts and linearity can be expected at lower fields (4.7 T, on-resonance water linewidths 15 Hz) with on-resonance water linewidths decreased, air-induced resonance shifts reduced, and longer T_2 values observed, thus ORI will be likely optimized at low fields with very short TEs choosing and with moderate MION concentrations. Off-resonance approaches generate positive contrast but have a lower sensitivity than T_2^*-weighted imaging and are more complex to perform at high field strengths. Superparamagnetic iron-oxide nanoparticles become saturated above 0.5 Tesla and thus have equal sensitivity at clinical field strengths (1.5–3.0 T) and at the higher field strengths often used in preclinical studies (Sosnovik, 2009).

An alternative off-resonance technique termed inversion-recovery with on-resonant water suppression (IRON) sequence was proposed by a serial studies from one lab (Stuber, 2005, 2007). The IRON method used a spectrally-selective saturation pre- pulse to suppress the signal originating from on-resonant protons in the background tissue while preserving the signal from off-resonant spins in proximity to the iron particles. However, since the size of the signal-enhanced region is dependent on the bandwidth of the water suppression pulse, this scheme requires extra steps to adjust the center frequency and bandwidth of the pre-pulse to locate the exact site proximal to the cells. IRON sequence has been successfully applied for in vivo tracking of iron-loaded stem cells (Stuber, 2007).

The utility of IRON method combined with injection of the long-circulating MION-47 has been recently evaluated by investigators in Johns Hopkins University School of Medicine (Korosoglou, 2008a) for developing a novel contrast-enhanced MR angiography technique. One important aspect of the study was fat suppression for the IRON sequence with an initial radiofrequency pulse offset by 440 Hz at 3.0 T, and with spin inversion, to cause zero longitudinal magnetization of the targeted species for the radiofrequency pulses (105° for fat, 100° for water), which obviously shortened the subsequent recovery time. The usage of MION-47 allowed acquisition of multiple image sets over a 1- or 2-day period with high spatial resolution.

IRON techniques with a commercially available MION-47 were recently employed to detect macrophage-rich atherosclerotic plaques in a rabbit model of atherosclerosis (Korosoglou, 2008b), in which pre-contrast imaging was performed in 7 Watanabe rabbits and 4 control New Zealand rabbits, and post-contrast imaging was repeated on days 1 and 3 after intravenous injection of MION-47. A second injection was performed on day 3 after imaging and post-contrast imaging performed again on day 6. There was a significant increase in signal intensity within aortic atherosclerotic plaques following administration of MION-47 (48% increase on day 3 and 72% increase on day 6) versus hypointensity (negative-contrast) in conventional MR images, but no enhancement was seen in control rabbits that lacked atherosclerosis. The positive-contrast regions corresponded to regions demonstrating deposition of iron particles within macrophage-rich atherosclerotic plaques. These findings not only validated that MION-47 is a successful imaging agent for macrophage-rich atherosclerosis, but also suggested that positive-contrast IRON MRI can be applied to the general class of iron oxide particles. This is significant as USPIO-enhanced MR imaging has been previously studied in human (Trivedi, 2006); enabling IRON MRI sequences to be directly applied to patient care.

Korosoglou et al. also investigated the utility of IRON techniques and MION-47 to create positive-contrast MR-lymphography (Korosoglou, 2008c). After six rabbits received a single bolus injection of 80 mmol Fe/kg MION-47, MRI was performed at baseline, 1 day, and 3 days using conventional T_1- and T_2^*-weighted sequences and IRON. On T_2^*-weighted images, as expected, signal attenuation was observed in areas of para-aortic lymph nodes after MION-47 injection. However, using IRON the para-aortic lymph nodes exhibited very high contrast enhancement, which remained 3 days after injection. IRON in conjunction with iron particles can be therefore used to perform positive-contrast MR-lymphography, particularly 3 days after injection of the contrast agent, when signal is no longer visible within blood vessels. The proposed method may have the potential as an adjunct for nodal staging in cancer screening.

Iron-labeled radioembolization microspheres were visualized for in vivo tracking during trans-catheter delivery to VX2 liver tumors in a rabbit model (Gupta, 2008). The study was performed for real-time observation of microsphere delivery with dual-contrast techniques. The results showed significant changes in post-injection contrast-to-noise ratio (CNR) values from those of pre-injection at positions of microsphere deposition with both negative- and positive-contrast.

The off-resonance MRI method possesses some advantages including no need for dephasing gradients or saturation pulses, high suppression efficiency, and flexible selection of the excited frequency band to encompass spins in the vicinity of the iron particles without fat tissue off-resonance. This technique, however, was not slice-selective such that it can result in interference from insufficiently suppressed background signals or less background signal with regions of greater susceptibility excluded from the selected slice. This technique can also cause less on-resonant signal to be suppressed, has less flexibility in RF pulse design, and can lead to less erroneous off-resonant signal detection in a multi-slice manner with individually shimmed slices (Zurkiya, 2006).

The off-resonance saturation method has been developed by Zurkiya and Hu, in which water protons are imaged with and without the presence of an off-resonance saturation pulse (Zurkiya, 2006). This method relies on diffusion-mediated saturation transfer to reduce the on-resonance MRI signal due to the off-resonance saturation (ORS) pulse, similar to chemical exchange saturation transfer techniques (Ward, 2000). This approach has been verified that greatly improved tumor detection accuracy over the conventional T_2^*-weighted methods because of its ability to turn "ON" the contrast of superparamagnetic polymeric micelles (SPPM) nanoparticles (Khemtong, 2009). SPPM nanoparticles encoded with cyclic (RGDfK) ligand (arginine-glycine-aspartic acid), cRGD, were able to target the $\alpha_v\beta_3$-expressing microvasculature in A549 non-small cell lung tumor xenografts in mice. The results suggest that the combination of ORS imaging with cancer-targeted SPPM nanoparticles will show promise in detecting biochemical markers at early stages of non–small cell lung tumor development, and could further enhance the sensitivity of contrast and provide new opportunities in imaging biomarkers setting of *in vivo* tumor target.

The study (Zurkiya, 2008) transfected cells with genes from magnetotactic bacteria (i.e., MagA) under doxycycline-regulated gene expression, resulting in the intracellular production of iron oxide nanoparticles similar to synthetic SPION. MagA-expressing cells could be visualized by MRI after transplantation in the mouse brain after 5 d of induction with doxycycline. The generalized implementation of these techniques as treatment strategies in stem cell tracking needs to be explored. Investigators have recently inserted magnetic reporter genes into cells. After the expression of iron storage proteins formed stored iron then MRI can be used to detect it. Another transgene reporter, an adenoviral vector carrying a transgene for light- and heavy-chain ferritin protein to transfect cells has been shown that they could be detected by in vivo magnetic resonance imaging (Genove, 2005).

Balchandani et al. recently developed a self-refocused spatial-spectral (SR-SPSP) pulse, which is successful in creating positive-contrast images of SPIO-labeled cells (Balchandani, 2009). This pulse can enable slice-selective, spin-echo imaging of off-resonant spins without an increase in TE, which is essentially a phase-matched 90° SPSP pulse and a 180° SPSP pulse combined into one pulse. This results in a considerably shorter TE than possible with two separate pulses. The simultaneous spatial and spectral selectivity allows the imaging of off-resonant spins while selecting a single slice. The SR-SPSP pulse is also suitable for any application requiring spatial and spectral selectivity, such as tracking metallic devices or replacing standard pulses in MR spectroscopic imaging sequences. More recently a novel combination of off-resonance (ORI) positive-contrast MRI and $T(2\rho)$ relaxation in the rotating frame (ORI-$T_{2\rho}$ method) for positive-contrast MR imaging of USPIO in a mouse model of burn trauma and infection with *Pseudomonas aeruginosa (PA)*, was also reported to have direct implications in the longitudinal noninvasive monitoring of infection, and show promise in testing the new-developed anti-infective compounds (Andronesi, 2010). The same group also reported that ORI-$T_{2\rho}$ method proved to have slightly higher sensitivity than ORI, and MR imaging clearly showed migration and accumulation of labeled MSCs to the burn area which can be confirmed by histology staining for iron labeled cells (Righi, 2010).

3.3 Fast low angle steady-state free precession (FLAPS) sequence

FLAPS imaging has been proposed for time-efficient acquisition of off-resonance positive-contrast images (Dharmakumar, 2007). The technique takes advantage of the unique spectral response of the steady-state free precession (SSFP) signal to achieve signal enhancement from off-resonant spins while suppressing signal from on-resonant spins at relatively small flip angles (Dharmakumar, 2006). Besides the positive-contrast generated by the weakly off-resonant spins, the spins in and around the core of the local magnetic susceptibility (LMS)-shifting media (such as labeled cells) experience large deviations from the central frequency leading to intra-voxel dephasing that was observed as negative-contrast in FLAPS images. So this technique has the capability to identify the presence of labeled cells with both negative- and positive-contrast within a single image.

Zhang et al. recently investigated the feasibility of imaging iron-labeled green fluorescent protein (GFP)-expressing cells with the dual-contrast method and compared its measurements with traditional negative-contrast technique (Zhang, 2009). The GFP-cell was incubated for 24 hours using 20 mg Fe/mL concentration of SPIO and USPIO nanoparticles. The labeled cells were imaged using the FLAPS technique, and FLAPS images with positive-contrast were compared with negative-contrast T_2*-weighted images. The results demonstrated that SPIO and USPIO labeling of GFP cells had no effect on cell function or GFP expression, and the labeled cells were observed as a narrow band of signal enhancement surrounding signal voids in FLAPS images. Positive- and negative-contrast images were both valuable for visualizing labeled GFP-cells. MRI of labeled cells with GFP expression holds great potential for monitoring the temporal and spatial migration of gene markers and cells, and enhances our understanding of cell- and gene-based therapeutic strategies. These findings suggested that the dual-contrast nature of the FLAPS approach offers significant advantages to the field of cellular MRI. A highly sought feature of cellular imaging is the quantification of labeled cells. Past studies have shown that it may be possible to define a relation between number of cells and MR transverse relaxation time constants (apparent T_2 or T_2*). However, since the specificity of the labeled cells is often compromised in GRE images, it is often difficult to use the time constant thus derived as a reliable metric to quantify the number of cells. These previous FLAPS investigations showed that local contrast was exponentially related to the number of cells. Furthermore, the dual-contrast filter, using an image metric that is analogous to local contrast, can provide additional quantitative information regarding those regions containing the labeled cells. This technique still could be limited by the magnetic perturbations around MNPs. A careful investigation of how the output of dual-contrast image filters can be used to derive quantitative information regarding the concentration of labeled cells from *in vivo* images has been demonstrated (Dharmakumar, 2009).

3.4 Ultra short echo time methods

It has been introduced that ultrashort echo-time (UTE) imaging had capability of imaging materials with extremely short T_2 and very fast signal decay (Robson, 2006; Rahmer, 2009), and did as a new and promising approach that allowed the detection of short-T_2 signal components, such as tendons, ligaments, menisci, periosteum, and cortical bone before signals within these tissues decay to a level where they were not observable with conventional spin echo pulse sequences. Due to the very short TE (on the order of 1/10 ms) used for UTE imaging, only negligible T_2 decay occurs before sampling, and consequently high signal from the short-T_2 components can be obtained. Coolen et al. reported that MRI parameters could be optimized for positive-contrast detection of iron-oxide labeled cells using double-echo Ultra-short echo time (d-UTE) sequences (Coolen, 2007). During these studies, there was a linear correlation between signal intensity and concentration USPIO labeled cells. Another group found that the enhancement due to the presence of short T_2 USPIO accumulation generally agreed with signal loss within GRE images during ex vivo MR of aorta atherosclerotic rabbit (Crowe, 2005).

Liu et al. recently measured ultrashort T_2^* relaxation in tissues containing a focal area of SPIO nanoparticle-labeled cells. MRI experiments in phantoms and rats with iron-labeled tumors demonstrated that these cells can be detected even at ultrashort T_2^* down to 1 ms or less (Liu, 2009). The authors suggested that combining ultrashort T_2^* relaxometry with the multiple gradient echo T_2^* mapping techniques should improve the ability to measure the relaxation of tissues with high densities of implanted iron- labeled cells. In another investigation, T1-weighted positive contrast enhancement from SPIO particles was achieved from the UTE imaging then this sequence, taking advantage of the unique effect of MNPs on relaxation time domain, was also examined to validate its positive contrast imaging capability of "probe" targeting to U87MG human glioblastoma cells through an SPIO conjugated RDG with high affinity to the cells overexpressing integrin $\alpha_v\beta_3$ (Zhang, 2011). So the study was regarded as providing a dual contrast imaging method from UTE technique plus T_2-weighted TSE images in its application of molecular imaging of glioma with potential quantification of SPIO nanoparticles suggested by previously published report (Liu, 2009).

The more recent study (Girard, 2011) showed that both contrast mechanisms of optimizing T_1 contrast from UTE technique with conventional T_2^* contrast of SPIO, even an extra subtraction of a later echo signal from the UTE signal, could be powerful both in improving the specificity by providing long T_2^* background suppression and increasing detection sensitivity, in molecular imaging application of tumor-targeted IONPs in vivo. A hybrid sequence, PETRA (Pointwise Encoding Time reduction with Radial Acquisition) (Grodzki, 2011), combined the features of single point imaging with radial projection imaging with no need of hardware changes, to show shorter encoding times over the whole k-space and to enable higher resolution for tissue with very short T_2, compared to the UTE sequence, so that it could avoids problems derived from the UTE but with good image quality and might improve e.g. orthopedic MR imaging as well as MR-PET attenuation correction. A 3D imaging technique (Seevinck, 2011) from the group in University Medical Center Utrecht, The Netherlands, applying center-out RAdial Sampling with Off-Resonance reception (co-RASOR) by the using of UTE technique (for the minimization of subvoxel dephasing at locations with high magnetic field gradients in the vicinity of the magnetized objects), and a hard, nonselective RF block pulse and radial sampling of k-space, was also presented to depict and accurately localize small paramagnetic objects with high positive contrast but ideally without background signal.

3.5 Others new MRI pulse sequences and image postprocessing techniques

Several other new sequences were reported on positive- and dual-contrast methods of MR cell tracking. Kim et al. recently developed simple means of detecting iron-labeled cells by using susceptibility weighted echo-time encoding technique (SWEET) (Kim, 2006). The subtraction of two sets of image volumes acquired at slightly-shifted echo time generates positive-contrast at the cell position. In a more recent study, the SWEET method was employed to selectively enhance the effect of the magnetic susceptibility caused by SPIO-labeled KB cells (KB cell is a cell line derived from a human carcinoma of the nasopharynx, used as an assay for antineoplastic agent). It was also demonstrated that this method could be used to visualize SPIO-labeled KB cells and their tumor formation in mice for at least a 2-week period (Kim, 2009).

Dual-contrast images can also be achieved by applying T_2*-weighted imaging combined with different post-processing techniques from the magnetic field map (Ward, 2000; Zurkiya, 2006). A susceptibility gradient mapping (SGM) technique has been recently developed, in which a color map of 3D susceptibility-gradient vector for every voxel is generated with calculated echo-shifts, and the map presents a 3D form of a positive-contrast images (Dahnke, 2008; Liu, 2008). Hyperintensities of SGM were seen in areas surrounding the 1×106 ferumoxides/protamine sulfate complex labeled flank C6 glioma cells of experimental rat model. The sensitivity of the method was compared to white-marker and IRON positive-contrast methods for visualizing the proliferation of tumor cells for labeled tumors that were approximately 5mm (small), 10 mm (medium) and 20 mm (large) in diameter along the largest dimension (Liu, 2008). The number of positive voxels detected around small and medium tumors was significantly greater with the SGM technique than those with the other two techniques, while similar as the "white-marker" technique for large tumors that could not be visualized with the IRON technique. The SGM is a post-processing technique and its positive-contrast images can be derived directly from the T_2*-weighted images without requiring dedicated positive-contrast pulse sequences, thereby it can provide the flexibility to display susceptibility gradients or suppress susceptibility artifacts in specific directions; not like the "white marker" or IRON techniques that require specialized pulse sequence designs and extra scans in addition to those obtained for conventional anatomic imaging. With SGM the hyperintense regions on positive-contrast images originating from SPIO labeled cells can be easily differentiated from other signal voids in T_2 or T_2*-weighted images.

The phase gradient mapping (PGM) techniques have recently developed independently by two groups, one related derived phase gradient maps from standard phase images also including a phase unwrapping procedure to assist the analysis and characterization of object-induced macroscopic phase perturbations (Bakker, 2008); another one utilized fast Fourier transform (FFT) to form phase gradients and develop positive contrast maps by the use of PGM but without need of phase unwrapping, so as to be appropriate technique for the visualization of magnetic nanoparticulate system (Langley, 2011; Zhao, 2011). By the method introduced recently of dual contrast with therapeutic iron nanoparticles at 4.7 T scanner (Sigovan, 2011), or postprocessing methods, with the measure of the T_2*, an efficient estimation of nanoparticle concentration can be made (Langley, 2011). Applications of two kind of approaches, the traditional relaxometry method and model-based method, have demonstrated that, besides the detection of SPIO nanoparticles by positive contrast methods, quantification of the SPIO concentration also play important role in clinical evaluation of results from different treatments with monitoring cellular therapies, and the former derives from the signal decay associated with areas containing contrast SPIO particles (Kuhlpeter, 2007; Rad, 2007; Liu, 2009), assuming that the rate varies linearly with contrast agent concentration; the later derives from the formation of magnetic field by SPIO-containing region (Dixon, 2009).

3.6 T_1 & T_2 (T_2^*) multi-contrast for cell tracking

As introduced in as earlier as 1990s, it is possible to achieve positive contrast and dual contrast with superparamagnetic particles by employing T_1- and/or T_2-weighted sequences (Canet, 1993; Chambon, 1993; Small, 1993). Although most earlier clinical trials with magnetic nanoparticles as contrast agents were evaluated almost exclusively on T_2-w fast spin echo (FSE) and T_2^*-w gradient echo (GRE) sequences, and the strong T_1 contrast enhancement effect of magnetic nanoparticles has rarely been used in clinical and molecular imaging (Reimer, 1995; Yamamoto, 1995; Tang, 1999), the effect of SPIO or USPIO on proton relaxation is not confined to T_2 and T_2^* effect. They should be considered to influence T_1 relaxivity with increased SI on T_1-w GRE sequences at low concentrations. For in vivo imaging application of MNPs, optimal combination of negative and positive contrast methods is still under evaluation.

Superparamagnetic iron oxide particles (SPIO) were used shortly after gadolinium-chelate magnetic resonance (MR) contrast agent as well known, while USPIO being the strong T_2 relaxivity that produces negative contrast also a high T_1 relaxivity with an increase in SI on T_1-weighted images (Small, 1993), so that a biphasic imaging sequence protocol (only immediate postadministration and 20-24 hr delayed images) in the *in vivo* study allowed visualization of the dynamic enhancement patterns of both normal tissue and potentially tumor based on early T_1-shortening effects produced by intravascular USPIO particulate agent (BMS 180549, previously AMI-227) and marked T_1-shortening produced following agent uptake by liver and spleen, as well as showed markedly less T_2-shortening at 20-24 hr within both liver and spleen.

The more recent investigation (Zhang, 2011) demonstrated that an appropriate SPIO core size and concentration range was paid much attention to obtain positive contrast with UTE imaging, and this technique could be used with the receptor targeted SPIO molecular imaging probe so as to provide an opportunity for monitoring cancer cells with overexpression of integrin $\alpha_v\beta_3$ in addition to negative contrast by the approach of T_2 relaxometry mapping.

Investigators recently synthesized a biocompatible water-dispersible Fe_3O_4–SiO_2–Gd–DTPA–RGD nanoparticle with r_1 relaxivity of 4.2 $mm^{-1}s^{-1}$ and r_2 relaxivity of 17.4 $mm^{-1}s^{-1}$ at the Gd/Fe molar ratio of 0.3:1, indicating the potential to use this multifunctional agent for dual-contrast MR imaging of tumor cells over-expressing high-affinity $\alpha_v\beta_3$ integrin *in vitro* and *in vivo* (Yang, 2011).

4. Imaging contrast of IRON-labeled cell on multimodular platform

MRI can be commonly used to set up a kind of nanomedicine platform for applications of multimodality probe to obtain information about concomitant anatomic, chemical, and physiological features of body. This kind of approach has been found under the background that, the nanomedicine platform could capitalize on the availability of specific probes, while achieving an theranostic (integrated diagnostic and therapeutic) design to allow for the visualization of therapeutic efficacy by noninvasive imaging methods such as MRI (Guthi, 2010), for example, in the field of tumor imaging researches, the combination of diagnostic capability with therapeutic intervention is critical to address the challenges of cancer heterogeneity and adaptive resistance, also molecular diagnosis by imaging is important to verify the cancer biomarkers in the tumor tissue and to guide target-specific therapy. It has been thought that ideal multimodality imaging probes enhance capabilities from complementary imaging modalities to enable both noninvasive and invasive molecular imaging (e.g, via probes with MRI and NIR fluorescence reporter capabilities) and to facilitate verification of disease detection and deliver additional evidences for the pathology (eg, probes with reporter capabilities for both positron emission tomography and MRI) (Kircher, 2003b; Lee, 2008). As for the establishment and utilizations of multimodular platform, such as optical and multimodality molecular imaging; multifunctional PET/MRI contrast agent; focused ultrasound/magnetic nanoparticle targeting delivery; design magnetic nanoparticles, etc, some topics are beyond of the scope of this chapter, and some good review papers have already published, so readers are recommended to check them (Jaffer, 2009; Chomoucka, 2010; Liu, 2010; Veiseh, 2010).

Guthi et al. recently introduced a multifunctional methoxy-terminated PEG-b-PDLLA micelle system that was encoded with a lung cancer-targeting peptide (LCP) and loaded with SPIO together with doxorubicin for MR imaging and therapeutic delivery in their *in vitro* study of a lung cancer (Guthi, 2010), they presented a significantly increased cell targeting, micelle uptake, superb T_2 relaxivity for ultrasensitive MR detection and cell cytotoxicity in $\alpha_v\beta_6$-expressing lung cancer cells, with confocal laser scanning microscopy of Doxo fluorescence also used to study the targeting specificity of LCP-encoded micelles to $\alpha_v\beta_6$-expressing H2009 over the $\alpha_v\beta_6$-negative H460 cells. The same micelles were previously conjugated with a cRGD ligand that can target $\alpha_v\beta_3$ integrins on tumor endothelial (SLK) cells (Nasongkla, 2006), illustrating growth inhibition of tumor SLK cells with ultrasensitive detection by MRI. The same lab in University of Texas Southwestern Medical Center at Dallas has previously demonstrated a multi-functional micelle design that allows for the vascular targeting of tumor endothelial cells, MRI ultrasensitivity, and controlled release of doxorubicin (Doxo) for therapeutic drug delivery (Nasongkla, 2006; Khemtong, 2009). Investigators (Guthi, 2010) found that SPIO-clustered polymeric micelle design has considerably decreased the MR detection limit to subnanomolar concentrations (< nM) of micelles through the increased T_2 relaxivity and high loading of SPIO per micelle particle; suggested that, on that multifunctional platform, the application of positive contrast imaging, such as ORS, could further enhance the contrast sensitivity and allow for the *in vivo* imaging of tumor-specific markers.

The proposed approaches of dual imaging (e.g. with CLIO modified with a NIR fluorophore, therapeutic siRNA sequences, and a cell penetrating peptide for cancer) Medarova, 2007), even multi-modular imaging (e.g. with triple functional iron oxide nanoparticles) (Xie, 2010) demonstrate potential for the creation of targeted multifunctional nanomedicine platforms.

5. Perspectives

There is an increasing interest in using cellular MRI to monitor behavior and physiologic functions of iron-labeled cells in vivo. Iron particles provide good MR probing capabilities and some of these agents are currently available for clinical applications. Based on the fact that iron particles exhibit unique nanoscale properties of super-paramagnetism and have the potential to be utilized as excellent probes for cellular imaging and molecular imaging, several MR techniques have recently been proposed to increase the detection sensitivity for image contrast generated with iron-labeled cells, including negative-, positive- and dual-contrast methods for visualization of iron-labeled cells in vitro and in vivo.

The hyperintense regions on positive-contrast images originating from iron-labeled cells can be easily differentiated from other signal voids on T_2 or T_2^*-weighted images, therefore providing a greater degree of certainty in the determination of labeled cells. Moreover, the hyperintensities appeared to illustrate a greater sensitivity than the dark spots on regular MR images. Because positive-contrast imaging approaches do not provide sufficient anatomical information, it is necessary to combine positive-contrast techniques with conventional gradient echo or spin echo imaging, to achieve dual-contrast. Also, the combinined gadolinium and SPIO-enhanced imaging in a 'dual contrast' MRI could be the more accurate technique for the detection of rntities, especially of tumors. Additionally, some new applications of agents for MR imaging have been tested so as to obtain dual-contrast agents for noninvasive imaging studies. Dual-contrast MRI techniques for in vivo cell tracking will add to the growing armamentarium for preclinical cellular MR imaging and further demonstrate the value and diagnostic power of molecular MR imaging, and multifunctional iron oxide nanoparticles together with MRI will have unique advantages with diagnostic and therapeutic capabilities. Simutaneously, the "concept" of dual-contrast imaging can be expaned into imaging evaluation on the platform of dual-modality (or even multimodal approach) including the simultaneous MRI-PET of new method for functional and morphological imaging with blooming perspectives for further development.

While much progress has been made to date, many challenges still face cellular MRI approaches aimed at assessing the migration, homing and function of transplanted therapeutic iron-labeled cells in vivo. For cellular MRI techniques to be successful, the combined expertise of basic scientists, clinicians and representatives from industry will undoubtedly be essential.

6. References

Anderson, S., Glod, J., Arbab, A., Noel, M., Ashari, P., Fine, H., & Frank, J. (2005). Noninvasive MR Imaging of Magnetically Labeled Stem Cells to Directly Identify Neovasculature in a Glioma Model. *Blood*, Vol. 105, No. 1, (August 2004), pp. 420-425, ISSN 0006-4971

Andronesi, O., Mintzopoulos, D., Righi, V., Psychogios, N., Kesarwani, M., He, J., Yasuhara, S., Dai, G., Rahme, L., & Tzika, A. (2010). Combined Off-resonance Imaging and T_2 Relaxation in The Rotating Frame for Positive Contrast MR Imaging of Infection in a Murine Burn Model. *J Magn Reson Imaging*, Vol. 32, No. 5, (November 2010), pp. 1172-1183, ISSN 1053-1807

Arbab, A., Jordan, E., Wilson, L., Yocum, G., Lewis, B., & Frank, J. (2004a). In Vivo Trafficking and Targeted Delivery of Magnetically Labeled Stem Cells. *Hum Gene Ther*, Vol.15, No. 4, (July 2004), pp. 351–360, ISSN 1043-0342

Arbab, A., Yocum, G., Kalish, H., Jordan, E., Anderson, S., Khakoo, A., Read, E., & Frank, J. (2004b). Efficient Magnetic Cell Labeling with Protamine Sulfate Complexed to Ferumoxides for Cellular MRI. *Blood*, Vol. 104, No. 4, (April 2004), pp. 1217-1223, ISSN 1053-1807

Arbab, A., & Frank, J. (2008). Cellular MRI and Its Role in Stem Cell Therapy. *Regen Med*, Vol. 3, No. 2, (March 2008), pp. 199-215, ISSN 1746-0751

Bakker, C., de Leeuw, H., Vincken, K., Vonken, E., & Hendrikse, J. (2008). Phase Gradient Mapping as an Aid in The Analysis of Object-induced and System-related Phase Perturbations in MRI. *Phys. Med. Biol.*, Vol. 53, No. 18, (September 2008), pp. N349–N358, ISSN 0031-9155

Balchandani, P., Yamada, M., Pauly, J., Yang, P., & Spielman, D. (2009). Self-refocused Spatial-spectral Pulse for Positive Contrast Imaging of Cells Labeled with SPIO Nanoparticles. *Magn Reson Med*, Vol. 62, No. 1, (May 2009), pp. 183-192, ISSN 0740-3194

Briley-Saebo, K., Leboeuf, M., Dickson, S., Mani, V., Fayad, Z., Palucka, A., Banchereau, J., & Merad, M. (2010). Longitudinal Tracking of Human Dendritic Cells in Murine Models Using Magnetic Resonance Imaging. *Magn Reson Med*, Vol. 64, No. 5, (October 2010), pp. 1510-1519, ISSN 0740-3194

Bulte, J., Trevor Douglas, T., Witwer, B., Zhang. S, Strable, E., Lewis, B., Zywicke, H., Miller, B., van Gelderen, P., Moskowitz, B., Duncan, I., & Frank, J. (2001). Magnetodendrimers Allow Endosomal Magnetic Labeling and In Vivo Tracking of Stem Cells. *Nature Biotechnology*, Vol. 19, No. 12, (December 2001), pp. 1141–1147, ISSN 1087-0156

Bulte, J., & Kraitchman, D. (2004a). Iron Oxide MR Contrast Agents for Molecular and Cellular Imaging. *NMR Biomed*, Vol. 17, No. 7, (November 2004), pp. 484–499, ISSN 0952-3480

Bulte, J., & Kraitchman, D. (2004b). Monitoring Cell Therapy Using Iron Oxide MR Contrast Agent. *Curr Pharm Biotechnol*, Vol. 5, No. 6, (December 2004) , pp. 567-584, ISSN 1389-2010

Canet, E., Revel, D., Forrat, R., Baldy-Porcher, C., de Lorgeril, M., Sebbag, L., Vallee, J., Didier, D., & Amiel, M. (1993). Superparamagnetic Iron Oxide Particles and Positive Enhancement for Myocardial Perfusion Studies Assessed by Subsecond T1-weighted MRI. *Magn Reson Imaging*, Vol. 11, No. 8, (March 2004), pp. 1139–1145, 0730-725X

Cerdan S, Lötscher HR, Künnecke B, Seelig J. (1989). Monoclonal Antibodycoated Magnetite Particles as Contrast Agents in Magnetic Resonance Imaging of Tumors. *Magn Reson Med*, Vol. 12, No. 2, (November 2005), pp. 151–163, ISSN 0740-3194

Chambon, C., Clement, O., Blanche, R., Schouman-Claeys, E., & Frija, G. (1993). Superparamagnetic Iron Oxides as Positive MR Contrast Agents: In Vitro and In Vivo Evidence. *Magn Reson Imaging*, Vol. 11, No. 4, (March 2004), pp. 509–519, ISSN 0730-725X

Chen, K., Xie, J., Xu, H., Behera, D., Michalski, M., Biswal, S., Wang, A., & Chen, X. (2009). Triblock Copolymer Coated Iron Oxide Nanoparticle Conjugate for Tumor Integrin

Targeting. *Biomaterials*, Vol. 30, No. 36, (September 2009), pp. 6912–6919, ISSN 0142-9612

Chomoucka, J., Drbohlavova, J., Huska, D., Adam, V., Kizek, R., & Hubalek, J. (2010). Magnetic Nanoparticles and Targeted Drug Delivering. *Pharmacol Res*, Vol. 62, No. 2, (Febuary 2010), pp. 144-149, ISSN 1043-6618

Coolen, B., Lee, P., Shuter, B., & Golay, X. (2007). Optimized MRI Parameters for Positive Contrast Detection of Iron-oxide Labeled Cells Using Double-echo Ultra-short Echo Time (d-UTE) Sequences, *Proceedings of Intl Soc Magn Reson Med*, ISSN 1545-44362005, Berlin, Germany, May 2007

Crowe, L., Wang, Y., Gatehouse P., Tessier, J., Waterton, J., Robert, P., Bydder, G., & Firmin, D. (2005). Ex vivo MR Imaging of Atherosclerotic Rabbit Aorta Labelled with USPIO – Enhancement of Iron Loaded Regions in UTE Imaging, *Proceedings of Intl Soc Magn Reson Med*, ISSN 1545-4436, Miami Beach, Florida, USA, May 2005

Cunningham, C., Arai, T., Yang, P., McConnell, M., Pauly, J., & Conolly, S. (2005). Positive Contrast Magnetic Resonance Imaging of Cells Labeled with Magnetic Nanoparticles. *Magn Reson Med*, Vol. 53, No. 5, (April 2005), pp. 999–1005, ISSN 0740-3194

Dahnke, H., Liu, W., Herzka, D., Frank, J., & Schaeffter, T. (2008). Susceptibility Gradient Mapping (SGM): A New Postprocessing Method for Positive Contrast Generation Applied to Superparamagnetic Iron Oxide Particle (SPIO)-labeled Cells. *Magn Reson Med*, Vol. 60, No. 3, (August 2008), pp. 595–603, ISSN 0740-3194

de Vries, I., Lesterhuis, W., Barentsz, J., Verdijk, P., van Krieken, J., Boerman, O., Oyen, W., Bonenkamp, J., Boezeman, J., Adema, G., Bulte, J., Scheenen, T., Punt, C., Heerschap, A., & Figdor, C. (2005). Magnetic Resonance Tracking of Dendritic Cells in Melanoma Patients for Monitoring of Cellular Therapy. *Nature Biotechnology*, Vol. 23, No. 11, (October 2005), pp. 1407–1413, ISSN 1087-0156

Dharmakumar, R., Koktzoglou, I., & Li, D. (2006). Generating Positive Contrast from Off-Resonant Spins with Steady-state Free Precession Magnetic Resonance Imaging: Theory and Proof-of-principle Experiments. *Phys. Med. Biol.*, Vol. 51, No. 17, (August 2006), pp. 4201-4215, ISSN 0031-9155

Dharmakumar, R., Koktzoglou, I., & Li, D. (2007). Factors Influencing Fast Low Angle Positive Contrast Steady-state Free Precession (FLAPS) Magnetic Resonance Imaging. *Phys. Med. Biol.*, Vol. 52, No. 11, (May 2007), pp. 3261-3273, ISSN 0031-9155

Dharmakumar, R., Zhang, Z., Koktzoglou, I., Tsaftaris, S., & Li, D. (2009). Dual-contrast Cellular Magnetic Resonance Imaging. *Mol Imaging*, Vol. 8, No. 5, (October 2009), pp. 254-263, ISSN 1535-3508

Dixon, W., Blezek, D., Lowery. L, Meyer, D., Kulkarni, A., Bales, B., Petko, D., Foo, T. (2009). Estimating Amounts of Iron Oxide from Gradient Echo Images. *Magn Reson Med*, Vol. 61, No. 5, (Febuary 2009), pp. 1132–1136, ISSN 0740-3194

Farrar, C., Dai, G., Novikov, M., Rosenzweig, A., Weissleder, R., Rosen, B., Sosnovik, D. (2008). Impact of Field Strength and Iron Oxide Nanoparticle Concentration on The Linearity and Diagnostic Accuracy of Off-resonance Imaging. *NMR Biomed*, Vol. 21, No. 5, (October 2007), pp. 453-463, ISSN 0952-3480

Ferrucci, J., & Stark, D. (1990). Iron Oxide-enhanced MR Imaging of The Liver and Spleen: Review of the First 5 years. *AJR Am J Roentgenol*, Vol. 155, No. 5, (November 1990), pp. 943-950, ISSN 0361-803X

Foltz, W., Cunningham, C., Mutsaers, A., Conolly, S., Stewart, D., & Dick, A. (2006). Positive-contrast Imaging in the Rabbit Hind-limb of Transplanted Cells Bearing Endocytosed Superparamagnetic Beads. *J Cardiovasc Magn Reson*, Vol. 8, No. 6, (November 2006), pp. 817-823, ISSN 1097-6647

Frank, J., Anderson, S., Kalsih, H., Jordan, E., Lewis, B., Yocum, G., & Arbab, A. (2004). Methods For Magnetically Labeling Stem and Other Cells for Detection by In Vivo Magnetic Resonance Imaging. *Cytotherapy*, Vol. 6, No. 6, (November 2004), pp. 621-625, ISSN 1465-3249

Genove, G., DeMarco, U., Xu, H., Goins, W., & Ahrens, E. (2005). A New Transgene Reporter for In Vivo Magnetic Resonance Imaging. *Nat Med*, Vol. 11, No. 4, (April 2005), pp. 450-454, ISSN 1078-8956

Girard, O., Du, J., Agemy, L., Sugahara, K., Kotamraju, V., Ruoslahti, E., Bydder, G., & Mattrey, R. (2011). Optimization of Iron Oxide Nanoparticle Detection Using Ultrashort Echo Time Pulse Sequences: Comparison of T1, T2*, and Synergistic T1 − T2* Contrast Mechanisms. *Magn Reson Med*, Vol. 65, No. 6, (Febuary 2011), pp. 1649-1660, ISSN 0740-3194

Gonzalez-Lara, L., Xu, X., Hofstetrova, K., Pniak, A., Brown, A., & Foster, P. (2009). In Vivo Magnetic Resonance Imaging of Spinal Cord Injury in The Mouse. *J Neurotrauma*, Vol. 26, No. 5, (May 2009), pp. 753-762, ISSN 0897-7151

Grodzki, D., Jakob, P., & Heismann, B. (2011). Ultrashort Echo Time Imaging Using Pointwise Encoding Time Reduction with Radial Acquisition (PETRA). *Magn Reson Med*, Article first published online : 30 JUN 2011, ISSN 0740-3194

Gupta, T., Virmani, S., Neidt, T., Szolc-Kowalska, B., Sato, K., Ryu, R., Lewandowski, R., Gates, V., Woloschak, G., Salem, R., Omary, R., & Larson, A. (2008). MR Tracking of Iron-labeled Glass Radioembolization Microspheres during Transcatheter Delivery to Rabbit VX2 Liver Tumors: Feasibility Study. *Radiology*, Vol. 249, No. 3, (October 2008), pp. 845-854, ISSN 0033-8419

Guthi, J., Yang, S., Huang, G., Li, S., Khemtong, C., Kessinger, C., Peyton, M., Minna, J., Brown, K., & Gao, J. (2010). MRI-visible Micellar Nanomedicine for Targeted Drug Delivery to lung cancer cells. *Mol Pharm*, Vol. 7, No. 1, (Febrary 2010), pp. 32-40, ISSN 1543-8384

Helmberger, T., & Semelka, R. (2001). New Contrast Agents for Imaging The Liver. *Magn Reson Imaging Clin N Am*, Vol. 9, No. 4, (November 2001), pp. 745-766, ISSN 1064-9689

Herschman, H. (2003). Molecular Imaging: Looking At Problems, Seeing Solutions. *Science*, Vol. 302, No. 5645, (October 2003), pp. 605-608, ISSN 0036-8075

Heyn, C., Bowen, C., Rutt, B., & Foster, P. (2005). Detection Threshold of Single SPIO-labeled Cells with FIESTA. *Magn Reson Med*, Vol. 53, No. 2, (Febuary 2005), pp. 312-320, ISSN 0740-3194

Heyn, C., Ronald, J., Mackenzie, L., MacDonald, I., Chambers, A., Rutt, B., Foster, P. (2006). In Vivo Magnetic Resonance Imaging of Single Cells in Mouse Brain with Optical Validation. *Magn Reson Med*, Vol. 55, No. 1, (January 2006), pp. 23-29, ISSN 0740-3194

Högemann-Savellano, D., Bos, E., Blondet, C., Sato, F., Abe, T., Josephson, L., Weissleder, R., Gaudet, J., Sgroi, D., Peters, P., & Basilion, J. (2003). The Transferrin Receptor: a Potential Molceular Imaging Marker for Human Cancer. *Neoplasia*, Vol. 5, No. 6, (November 2003), pp. 495- 506, ISSN 1522-8002

Hsiao, J., Tai, M., Chu, H., Chen, S., Li, H., Lai, D., Hsieh, S., Wang, J., & Liu, H. (2007). Magnetic Nanoparticle Labeling of Mesenchymal Stem Cells Without Transfection Agent: Cellular Behavior and Capability of Detection with Clinical 1.5T Magnetic Resonance at The Single Cell Level. *Magn Reson Med*, Vol. 58, No. 4, (October 2006), pp. 717–724, ISSN 0740-3194

Jaffer, F., Libby, P., & Weissleder, R. (2009). Optical and Multimodality Molecular Imaging: Insights Into Atherosclerosis. *Arterioscler Thromb Vasc Biol.*, Vol. 29, No. 7, (July 2009), pp. 1017-1024. ISSN 1079-5642

Kelloff, G., Krohn, K., Larson, S., Weissleder, R., Mankoff, D., Hoffman, J., Link, J., Guyton, K., Eckelman, W., Scher, H., O'Shaughnessy, J., Cheson, B., Sigman, C., Tatum, J., Mills, G., Sullivan, D., & Woodcock, J. (2005). The Progress and Promise of Molecular Imaging Probes in Oncologic Drug Develpoment. *Clin Cancer Research*, Vol. 11, No. 22, (November 2005), pp. 7967-7985, ISSN 1078-0432

Khakoo, A., Pati, S., Anderson, S., Reid, W., Elshal, M., Rovira, I., Nguyen, A., Malide, D., Combs, C., Hall, G., Zhang, J., Raffeld, M., Rogers, T., Stetler-Stevenson, W., Frank, J., Reitz, M., & Finkel, T. (2006). Human Mesenchymal Stem Cells Exert Potent Antitumorigenic Effects in A Model of Kaposi's Sarcoma. *J Exp Med*, Vol. 203, No. 5, pp. 1235–1247, ISSN 0022-1007

Khemtong, C., Kessinger, C., Ren, J., Bey, E., Yang, S., Guthi, J., Boothman, D., Sherry, A., & Gao, J. (2009). In Vivo Off-resonance Saturation Magnetic Resonance Imaging of Alphavbeta3-targeted Superparamagnetic Nanoparticles. *Cancer Res*, Vol. 69, No. 4, (Febrary 2009), pp. 1651-1658, ISSN 0008-5472

Kim, Y., Bae, K., Lee, Y., Park, T., Yoo, S., & Park, H. (2006). Positive-Contrast Cellular MR Imaging Using Susceptibility Weighted Echo-time Encoding Technique (SWEET), *World Congress on Medical Physics and Biomedical Engineering*, ISBN 9783540368397, Coex Seoul, Korea, August 2006

Kim, Y., Bae, K., Yoo, S., Park, T., & Park, H. (2009). Positive Contrast Visualization for Cellular Magnetic Resonance Imaging Using Susceptibility-weighted Echo-time Encoding. *Magn Reson Imaging*, Vol. 27, No. 5, (December 2008), pp. 601-610, 0730-725X

Kircher, M., Allport, J., Graves, E., Love, V., Josephson, L., Lichtman, A., & Weissleder, R. (2003a). In Vivo High Resolution Three-dimensional Imaging of Antigen-specific Cytotoxic T-lymphocyte Trafficking to Tumors. *Cancer Res*, Vol. 63, No. 20, (October 2003), pp. 6838-6846, ISSN 0008-5472

Kircher, M., Mahmood, U., King, R., Weissleder, R., & Josephson, L. (2003b). A Multimodal Nanoparticle for Preoperative Magnetic Resonance Imaging and Intraoperative Optical Brain Tumor Delineation. *Cancer Res*, Vol. 63, No. 23, (Deceber 2003), pp. 8122–8125, ISSN 0008-5472

Korosoglou, G., Shah, S., Vonken, E., Gilson, W., Schär, M., Tang, L., Kraitchman, D., Boston, R., Sosnovik, D., Weiss, R., Weissleder, R., & Stuber, M. (2008a). Off Resonance Angiography: A New Method to Depict Vessels − phantom and Rabbit Studies. *Radiology*, Vol. 249, No. 2, (September 2008), pp. 501–509, ISSN 0033-8419

Korosoglou, G., Weiss, R., Kedziorek, D., Walczak, P., Gilson, W., Schär, M., Tang, L., Kraitchman, D., Boston, R., Sosnovik, D., Weiss, R., Bulte, J., Weissleder, R., & Stuber, M. (2008b). Noninvasive Detection of Macrophage-rich Atherosclerotic Plaque in Hyperlipidemic Rabbits Using "Positive contrast" Magnetic Resonance Imaging. *J Am Coll Cardiol*, Vol. 52, No. 6, (August 2008), pp. 483–491, ISSN 0735-1097

Korosoglou, G., Tang, L., Kedziorek, D., Cosby, K., Gilson, W., Vonken, E., Schär, M., Sosnovik, D., Kraitchman, D., Weiss, R., Weissleder, R., & Stuber, M. (2008c). Positive Contrast MR-lymphography Using Inversion Recovery with ON-resonant Water Suppression (IRON). *J Magn Reson Imaging*, Vol. 27, No. 5, (May 2008), pp. 1175-1180, ISSN 1053-1807

Kraitchman, D., Tatsumi, M., Gilson, W., Ishimori, T., Kedziorek, D., Walczak, P., Segars, W., Chen, H., Fritzges, D., Izbudak, I., Young, R., Marcelino, M., Pittenger, M., Solaiyappan, M., Boston, R., Tsui, B., Wahl, R., & Bulte, J. (2005). Dynamic Imaging of Allogeneic Mesenchymal Stem Cells Trafficking to Myocardial Infarction. *Circulation*, Vol. 112, No. 10, (August 2005), pp. 1451–1461. ISSN 0009-7322

Kuhlpeter, R., Dahnke, H., Matuszewski, L., Persigehl, T., von Wallbrunn, A., Allkemper, T., Heindel, W., Schaeffter, T., & Bremer, C. (2007). R2 and R2 Mapping for Sensing Cell-Bound Superparamagnetic Nanoparticles. *Radiology*, Vol. 245, No. 2, (September 2007), pp. 449–457, ISSN 0033-8419

Langley, J., Liu, W., Jordan, E., Frank, J., & Zhao, Q. (2011). Quantification of SPIO Nanoparticles In Vivo Using the Finite Perturber Method. *Magn Reson Med*, Vol. 65, No. 5, (November 2010), pp. 1461-1469, ISSN 0740-3194

Lee, H., Li, Z., Chen, K., Hsu, A., Xu, C., Xie, J., Sun, S., & Chen, X. (2008). PET/MRI Dual-Modality Tumor Imaging Using Arginine-glycine-aspartic (RGD)-conjugated Radiolabeled Iron Oxide Nanoparticles. *J Nucl Med*, Vol. 49, No. 8, (July 2008), 1371–1379. ISSN 0161-5505

Lefevre S, Ruimy D, Jehl F, Neuville A, Robert P, Sordet C, Ehlinger M, Dietemann JL, Bierry G. (2011). Septic Arthritis: Monitoring with USPIO-enhanced Macrophage MR Imaging. *Radiology*, Vol. 258, No. 3, (March 2011), pp. 722-728, ISSN 0033-8419

Li, W,, Tutton, S,, Vu, A., Pierchala, L., Li, B., Lewis, J., Prasad, P., & Edelman, R. (2005) First-pass Contrast-enhanced Magnetic Resonance Angiography in Humans Using Ferumoxytol, A Novel Ultrasmall Superparamagnetic Iron Oxide (USPIO)-based Blood Pool Agent. *J Magn Reson Imaging*, Vol. 1, No. 5, (January 2005), pp. 46-52, ISSN 1053-1807

Liu, C., Ren, J., Yang, J., Liu, C., Mandeville, J., Rosen, B., Bhide, P., Yanagawa, Y., & Liu, P. (2009). Dna-based Mri Probes for Specic Detection of Chronic Exposure to Amphetamine in Living Brains. *The Journal of Neuroscience*, Vol. 29, No. 34, (August 2009), pp. 10663-10670, ISSN 0270-6474

Liu, H., Hua, M., Yang, H., Huang, C., Chu, P., Wu, J., Tseng, I., Wang, J., Yen, T., Chen, P., & Wei, K. (2010). Magnetic Resonance Monitoring of Focused Ultrasound/Magnetic Nanoparticle Targeting Delivery of Therapeutic Agents to the Brain. *Proc Natl Acad Sci U S A.*, Vol. 107, No. 34, (August 2010), pp. 15205-15210, ISSN 0027-8424

Liu, W., Dahnke, H., Jordan, E., Schaeffter, T., & Frank, J. (2008). In Vivo MRI Using Positive-contrast Techniques in Detection of Cells Labeled with Superparamagnetic

Iron Oxide Nanoparticles. *NMR Biomed*, Vol. 21, No. 3, (March 2008), pp. 242-250, ISSN 0952-3480

Liu, W., Dahnke, H., Rahmer, J., Jordan, E., & Frank, J. (2009). Ultrashort T2* Relaxometry for Quantitation of Highly Concentrated Superparamagnetic Iron Oxide (SPIO) Nanoparticle Labeled Cells. *Magn Reson Med*, Vol. 61, No. 4, (April 2009), pp. 761–766, ISSN 0740-3194

Mani, V., Briley-Saebo, K., Hyafil, F., & Fayad, Z. (2006a). Feasibility of In Vivo Identification of Edogenous Ferritin with Positive Contrast MRI in Rabbit Carotid Crush Injury Using GRASP. *Magn Reson Med*, Vol. 56, No. 5, (November 2006), pp. 1096–1106, ISSN 0740-3194

Mani, V., Briley-Saebo, K., Itskovich, V., Samber, D., & Fayad, Z. (2006b). Gradient Echo Acquisition for Superparamagnetic Particles with Positive Contrast (GRASP): Sequence Characterization in Membrane and Glass Superparamagnetic Iron Oxide Phantoms at 1.5 T and 3 T. *Magn Reson Med*, Vol. 55, No. 1, (January 2006), pp. 126–135, ISSN 0740-3194

Mani, V., Adler, E., Briley-Saebo, K., Bystrup, A., Fuster, V., Keller, G., & Fayad, Z. (2008). Serial In Vivo Positive Contrast MRI of Iron Oxide-labeled Embryonic Stem Cell Derived Cardiac Precursor Cells in A Mouse Model of Myocardial Infarction. *Magn Reson Med*, Vol. 60, No. 1, (July 2008), pp. 73–81, ISSN 0740-3194

Manninger, S., Muldoon, L., Nesbit, G., Murillo, T., Jacobs, P., & Neuwelt, E. (2005). An Exploratory Study of Ferumoxtran-10 Nanoparticles as a Blood-brain barrier Imaging Agent Targeting Phagocytic Cells in CNS Inflammatory Lesions. *AJNR Am J Neuroradiol.*, Vol. 26, No. 9, (October 2005), pp. 2290–2300.

Medarova, Z., Pham, W., Farrar, C., Petkova, V., & Moore, A. (2007). In Vivo Imaging of siRNA Delivery and Silencing in Tumors. *Nat Med*, Vol. 13, No. 3, (Febrary 2007), pp. 372–377, ISSN 1078-8956

Metz, S., Beer, A., Settles, M., Pelisek, J., Botnar, R., Rummeny, E., & Heider, P. (2011). Characterization of Carotid Artery Plaques with USPIO-enhanced MRI: Assessment of Inflammation and Vascularity as In Vivo Imaging Biomarkers for Plaque Vulnerability. *Int J Cardiovasc Imaging*, Vol. 27, No. 6, (October 2010), pp. 901-912, ISSN 1569-5794

Modo, M., Hoehn, M., & Bulte, J. (2005). Cellular MR imaging. *Mol Imaging.*, Vol. 4, No. 3, (July 2005), pp. 143-164, ISSN 1535-3508

Muja, N., & Bulte, J. (2009). Magnetic Resonance Imaging of Cells in Experimental Disease Models. *Prog Nucl Magn Reson Spectrosc.*, Vol. 55, No. 1, (July 2009), pp. 61-77, ISSN 0079-6565

Nasongkla, N., Bey, E., Ren, J., Ai, H., Khemtong, C., Guthi, J., Chin, S., Sherry, A., Boothman, D., & Gao, J. (2006). Multifunctional Polymeric Micelles as Cancer-targeted, MRI-ultrasensitive Drug Delivery Systems. *Nano Lett.*, Vol. 6, No. 11, (November 2006), pp. 2427–2430, ISSN 1530-6984

Niu, G., & Chen, X. (2011). Why Integrin as A Primary Target for Imaging and Therapy. *Theranostics*, Vol. 1, No. 1, (January 2011), pp. 30–47, ISSN 1838-7640

Oude Engberink, R., van der Pol, S., Döpp, E., de Vries, H., & Blezer, E. (2007). Comparison of SPIO and USPIO for In Vitro Labeling of Human Monocytes: MR Detection and Cell Function. *Radiology*, Vol. 243, No. 2, (May 2007), pp. 467–474, ISSN 0033-8419

Politi, L., Bacigaluppi, M., Brambilla, E., Cadioli, M., Falini, A., Comi, G., Scotti, G., Martino, G., & Pluchino, S. (2007). Magnetic-resonancebased Tracking and Quantification of Intravenously Injected Neural Stem Cell Accumulation in The Brains of Mice with Experimental Multiple Sclerosis. *Stem Cells*, Vol. 25, No. 10, (Jun 2007), pp. 2583–2592, ISSN 0250-6793

Rad, A., Arbab, A., Iskander, A., Jiang, Q., & Soltanian-Zadeh, H. (2007). Quantification of Superparamagnetic Iron Oxide (SPIO)-labeled Cells Using MRI. *J Magn Reson Imaging*, Vol. 26, No. 2, (August 2007), pp 366-374, ISSN 1053-1807

Rahmer, J., Bornert, P., & Dries, S. Assessment of Anterior Cruciate Ligament Reconstruction Using 3D Ultrashort Echo-time MR Imaging. (2009). *J Magn Reson Imaging*, Vol. 29, No. 2, (Febrary 2009), pp 443-448, ISSN 1053-1807

Reimer, P., Rummeny, E., Daldrup, H., Balzer, T., Tombach, B., Berns, T., & Peters, P. (1995). Clinical Results with Resovist: A Phase 2 Clinical Trial. *Radiology*, Vol. 195, No. 2, (May 1995), pp. 489-496, ISSN 0033-8419

Righi, V., Andronesi, O., Mintzopoulos, D., Fichman, A., & Tzika, A. (2010). Molecular MR Imaging of Labeled Stem Cells in A Mouse Burn Model In Vivo, *Proceedings of Intl Soc Magn Reson Med*, ISSN 1545-4436, Stockholm, Sweden, May 2010

Robson, M., & Bydder, G. (2006). Clinical Ultrashort Echo Time Imaging of Bone and Other Connective Tissues. *NMR Biomed*, Vol. 19, No. 7, (November 2006), pp. 765-780, ISSN 0952-3480

Seevinck, P., de Leeuw, H., Bos, C., & Bakker, C. (2011). Highly Localized Positive Contrast of Small Paramagnetic Objects Using 3D Center-out Radial Sampling with Off-resonance Reception. *Magn Reson Med*, Vol. 65, No. 1, (January 2008), pp. 146–156, ISSN 0740-3194

Seppenwoolde, J., Viergever, M., & Bakker, C. (2003). Passive Tracking Exploiting Local Signal Conservation: The White Marker Phenomenon. *Magn Reson Med*, Vol. 50, No. 4, (October 2003), pp. 784–790, ISSN 0740-3194

Seppenwoolde, J., Vincken, K., & Bakker, C. (2007). White-marker Imaging-Separating Magnetic Susceptibility Effects from Partial Volume Effects. *Magn Reson Med*, Vol. 58, No. 3, (September 2007), pp. 605–609, ISSN 0740-3194

Shapiro, E., Skrtic, S., & Koretsky, A. (2005). Sizing It Up: Cellular Mri Using Micron-sized Iron Oxide Particles. *Magn Reson Med*, Vol. 53, No. 2, (Febuary 2005), pp. 329–338, ISSN 0740-3194

Sigovan, M., Hamoudeh, M., Al Faraj, A., Charpigny, D., Fessi, H., & Canet-Soulas, E. (2011). Positive Contrast with Therapeutic Iron Nanoparticles at 4.7 T. *MAGMA*, Vol. 24, No. 5, (May 2011), pp. 259-265, ISSN 0968-5243

Small, W., Nelson, R., & Bernardino, M. (1993). Dual-contrast Enhancement of Both T1- and T2-weighted Sequences Using Ultra Small Superparamagnetic Iron Oxide. *Magn Reson Med*, Vol. 11, No. 5, (May 1993), pp. 645–654, ISSN 0740-3194

Sosnovik, D. (2009). Molecular Imaging of Myocardial Injury: A Magnetofluorescent Approach. *Curr Cardiovasc Imaging Rep.*, Vol. 2, No. 1, (Febuary 2009), pp. 33–39, ISSN 1941-9066

Stuber, M., Gilson, W., Schaer, M., Bulte, J., & Kraitchman, D. (2005). Shedding Light on the Dark Spot with IRON - A Method That Generates Positive Contrast in The Presence of Superparamagnetic Nanoparticles, *Proceedings of Intl Soc Magn Reson Med*, ISSN 1545-4436, Miami Beach, Florida, USA, May 2005

Stuber, M., Gilson, W., Schär, M., Kedziorek, D., Hofmann, L., Shah, S., Vonken, E., Bulte, J., & Kraitchman, D. (2007). Positive Contrast Visualization of Iron Oxide-labeled Stem Cells Using Inversion-recovery with ON-resonant Water Suppression (IRON). *Magn Reson Med*, Vol. 58, No. 5, (November 2007), pp. 1072–1077, ISSN 0740-3194

Suzuki, Y., Cunningham, C., Noguchi, K., Chen, I., Weissman, I., Yeung, A., Robbins, R., & Yang, P. (2008). In Vivo Serial Evaluation of Superparamagnetic Iron-oxide Labeled Stem Cells by Off-resonance Positive Contrast. *Magn Reson Med*, Vol. 60, No. 6, (December 2008), pp. 1269-1275, ISSN 0740-3194

Tan, M., & Lu, Z. (2011). Integrin Targeted MR Imaging. *Theranostics*, Vol. 1, No. 1, (January 2011), pp. 30–47, ISSN 1838-7640.

Tang, Y., Yamashita, Y., Arakawa, A., Namimoto, T., Mitsuzaki, K., Abe, Y., Katahira, K., & Takahashi, M. (1999). Detection of Hepatocellular Carcinoma Arising in Cirrhotic Livers: Comparison of Gadolinium- and Ferumoxides Enhanced MR Imaging. *AJR Am J Roentgenol*, Vol. 172, No. 6, pp. 1547-1554, ISSN 0361-803X

Thorek, D., Chen, A., Czupryna, J., & Tsourkas, A. (2006). Superparamagnetic Iron Oxide Nanoparticle Probes for Molecular Imaging. *Ann Biomed Eng*, Vol. 34, No. 1, (January 2006), pp. 23-38, ISSN 0090-6964

Trivedi, R., Mallawarachi, C., U-King-Im, J., Graves, M., Horsley, J., Goddard, M., Brown, A., Wang, L., Kirkpatrick, P., Brown, J., & Gillard, J. (2006). Identifying Inflamed Carotid Plaques Using In Vivo USPIO-enhanced MR Imaging to Label Plaque Macrophages. *Arterioscler Thromb Vasc Biol*, Vol. 26, No. 7, (July 2006), pp. 1601–1606, ISSN 1079-5642

Veiseh, O., Gunn, J., & Zhang, M. (2010). Design and Fabrication of Magnetic Nanoparticles for Targeted Drug Delivery and Imaging. *Adv Drug Deliv Rev*, Vol. 62, No. 3, (November 2009), pp. 284-304, ISSN 0169-409X

Ward, K., Aletras, A., & Balaban, R. (2000). A New Class of Contrast Agents for MRI Based on Proton Chemical Exchange Dependent Saturation Transfer (CEST). *J Magn Reson.*, Vol. 143, No. 1, (March 2000), pp. 79-87, ISSN 1090-7807

Weissleder, R., Elizondo, G., Wittenberg, J., Rabito, C., Bengele, H., & Josephson, L. (1990). Ultrasmall Superparamagnetic Iron Oxide: Characterization of a New Class of Contrast Agents for MR Imaging. *Radiology*, Vol. 175, No. 2, (May 1990), pp. 489-493, ISSN 0033-8419

Xie, J., Chen, K., Huang, J., Lee, S., Wang, J., Gao, J., Li, X., & Chen, X. (2010). PET/NIRF/MRI Triple Functional Iron Oxide Nanoparticles. *Biomaterials*, Vol. 31, No. 11, (April 2010), pp. 3016-3022, ISSN 0142-9612

Yamamoto, H., Yamashita, Y. Yoshimatsu, S., Baba, Y., Hatanaka, Y., Murakami, R., Nishiharu, T., Takahashi, M., Higashida, Y., & Moribe, N. (1995). Hepatocellular Carcinoma in Cirrhotic Livers: Detection with Unenhanced and Iron Oxide-enhanced MR Imaging. *Radiology*, Vol. 195, No. 1, (April 1995), pp. 106-112, ISSN 0033-8419

Yang, H., Zhuang, Y., Sun, Y., Dai, A., Shi, X., Wu, D., Li, F., Hu, H., & Yang, S. Targeted Dual-ontrast T(1)- and T(2)-weighted Magnetic Resonance Imaging of Tumors Using Multifunctional Gadolinium-labeled Superparamagnetic Iron Oxide Nanoparticles. *Biomaterials*, Vol. 32, No. 20, (March 2011), pp. 4584–4593, ISSN 0142-9612

Zhang, L., Zhong, X., Wang, L., Chen, H., Wang, Y., Yeh, J., Yang, L., & Mao, H. T1-weighted Ultrashort Echo Time Method for Positive Contrast Imaging of Magnetic Nanoparticles and Cancer Cells bound with the Targeted Nanoparticles. *J Magn Reson Imaging*, Vol. 33, No. 1, (January 2011), pp. 194-202, ISSN 1053-1807

Zhang, Z., Dharmakumar, R., Mascheri, N., Fan, Z., Wu, S., & Li, D. (2009). Comparison of SPIO and USPIO Cell Labeling for Tracking GFP Gene Marker with Negative and Positive Contrast MRI. *Mol Imaging*, Vol. 8, No. 3, (October 2009), pp. 148-155, ISSN 1535-3508

Zhang, Z., Mascheri, N., Dharmakumar, R., & Li, D. (2008). Cellular Magnetic Resonance Imaging: Potential for Use in Assessing Aspects of Cardiovascular Disease. *Cytotherapy*, Vol. 10, No. 6, (March 2008), pp. 575-586, ISSN 1465-3249

Zhang, Z., van den Bos, E., Wielopolski, P., de Jong-Popijus, M., Bernsen, M., Duncker, D., & Krestin, G. (2005). In Vitro Imaging of Single Living Human Umbilical Vein Endothelial Cells with a Clinical 3.0-T MRI Scanner. *MAGMA*, Vol. 18, No. 4, (August 2005), pp. 175-185, ISSN 0968-5243

Zhao, M., Kircher, M., Josephson, L., & Weissleder, R. (2002). Differential Conjugation of Tat Peptide to Superparamagnetic Nanoparticles and Its Effect on Cellular Uptake. *Bioconjug Chem*, Vol. 13, No. 4, (July 2002), pp. 840–844.

Zhao, Q., Langley, J., Lee, S., & Liu, W. (2011). Positive Contrast Technique for the Detection and Quantification of Superparamagnetic Iron Oxide Nanoparticles in MRI. *NMR Biomed*, Vol. 24, No. 5, (October 2010), pp. 464–472, ISSN 0952-3480

Zhu, J., Zhou, L., & XingWu, F. (2006). Tracking Neural Stem Cells in Patients with Brain Trauma. *N Engl J Med.*, Vol. 30, No. 355, (November 2006), pp. 2376–2378, ISSN 0028-4793

Zurkiya, O., Chan, A., & Hu, X. (2008). MagA is Sufficient for Producing Magnetic Nanoparticles in Mammalian Cells, Making It an MRI Reporter. *Magn Reson Med*, Vol. 59, No. 6, (Jun 2008), pp. 1225–1231, ISSN 0740-3194

Zurkiya, O., & Hu, X. (2006). Off-resonance Saturation as a Means of Generating Contrast with Superparamagnetic Nanoparticles. *Magn Reson Med*, Vol. 56, No. 4, (October 2006), pp. 726–732, ISSN 0740-3194

9

New Trends on the Synthesis of Inorganic Nanoparticles Using Microemulsions as Confined Reaction Media

Margarita Sanchez-Dominguez[1], Carolina Aubery[2] and Conxita Solans[2]
[1]Centro de Investigación en Materiales Avanzados, S. C. (CIMAV), Unidad Monterrey;
GENES-Group of Embedded Nanomaterials for Energy Scavenging
[2]Instituto de Química Avanzada de Cataluña, Consejo Superior de Investigaciones
Científicas (IQAC-CSIC); CIBER en Biotecnología,
Biomateriales y Nanomedicina (CIBER-BBN)
[1]Mexico
[2]Spain

1. Introduction

The development of nanotechnology depends strongly on the advances in nanoparticle preparation. Nowadays, there are a number of technologies available for nanoparticle synthesis, from the gas phase techniques such as laser evaporation (Gaertner & Lydtin, 1994), sputtering, laser pyrolisis, flame atomization and flame spray pyrolisis (Kruis et al. 1998), etc, to the liquid phase techniques such as coprecipitation from homogeneous solutions and sol-gel reactions (Qiao et al. 2011), solvothermal processes (Gautam et al. 2002), sonochemical and cavitation processing (Suslick et al. 1996), and surfactant and polymer-templated synthesis (Holmberg, 2004). Amongst the surfactant-based approaches, the microemulsion reaction method is one of the most used techniques for the preparation of very small and nearly monodispersed nanoparticles. This method offers a series of advantages with respect to other methods, namely, the use of simple equipment, the possibility to prepare a great variety of materials with a high degree of particle size and composition control, the formation of nanoparticles with often crystalline structure and high specific surface area, and the use of soft conditions of synthesis, near ambient temperature and pressure. The traditional method is based on water-in-oil microemulsions (W/O), and it has been used for the preparation of metallic and other inorganic nanoparticles since the beginning of the 1980's (Boutonnet et al., 1982). The droplets of W/O microemulsions are conceived as tiny compartments or "nanoreactors". The main strategy for the synthesis of nanoparticles in W/O microemulsions consists in mixing two microemulsions, one containing the metallic precursor and another one the precipitating agent. Upon mixing, both reactants will contact each other due to droplets collisions and coalescence, and they will react to form precipitates of nanometric size (Figure 1). This precipitate will be confined to the interior of microemulsion droplets. Numerous investigations have been published about the use of W/O microemulsions for the preparation of a variety of nanomaterials, such as metallic and bimetallic nanoparticles, single metal oxide as well as mixed oxides, quantum dots, and even complex ceramic materials (Boutonnet et al., 1982; Destrée & Nagy, 2006; Eastoe et al. 2006; Holmberg, 2004; López-Quintela et al. 2004; Pileni 1997 and 2003). Materials synthesized in w/o microemulsions exhibit unique surface properties; for example, nano-catalysts prepared by this method show better performance (activity, selectivity) than those prepared by other methods (Boutonnet et al. 2008).

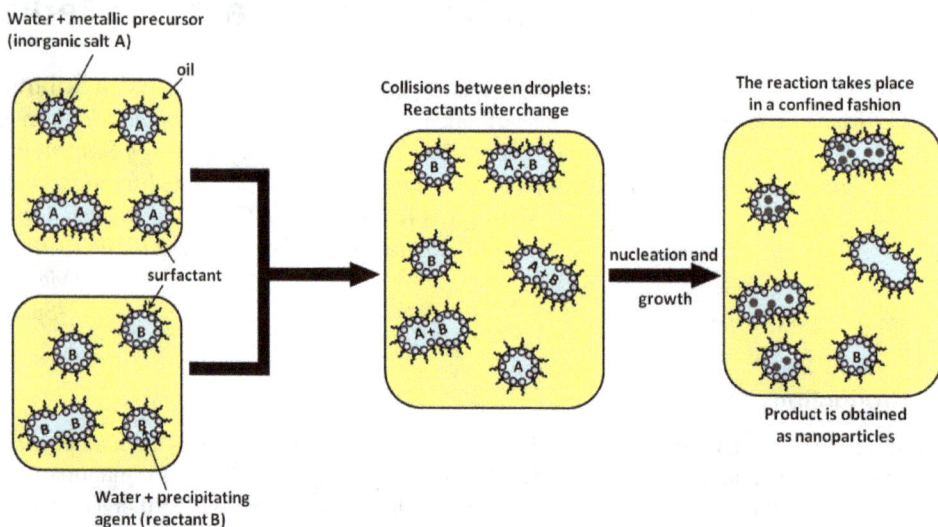

Fig. 1. Scheme of the w/o microemulsion reaction method for the synthesis of inorganic nanoparticles.

In spite of the superior properties and performance of nanoparticles obtained in w/o microemulsions, this method has not found good acceptance at the industrial level, mainly due to the employment of large amounts of oils (solvents) which represent the continuous and hence main component of these systems. In addition, most studies employ relatively low concentration of the metal precursors, leading to small yields of nanoparticles per microemulsion volume. These drawbacks affect negatively from the economic and ecologic point of view. It is the aim of this chapter to review the newest trends in the synthesis of inorganic nanoparticles using microemulsions as confined reaction media, with the objective to identify those alternatives or approaches that make this type of colloidal media more attractive for nanoparticle synthesis from the environmental , economic, technological, and scientific point of view. Some of those approaches are: the synthesis of advanced materials, such as mixed oxides and complex ceramics with nanocrystalline structure, core-shell particles, mixed materials with key nano-heterojunctions, etc, which may be difficult to obtain by other methods; optimization of microemulsion compositions, by making use of advanced phase behaviour knowledge; use of bicontinuous microemulsions in semi-continuous batches, and last but not least, a novel approach based on the use of oil-in-water microemulsions instead of w/o microemulsions as confined reaction media. An introductory section about the generalities and properties of microemulsion systems as well as on the use of microemulsions as reaction media for nanoparticle synthesis is first included. In addition, other aspects of nanoparticle synthesis are reviewed, such as study of reaction kinetics; influence of microemulsion dynamics on the characteristics of the obtained materials, as well as phase-transfer and isolation of nanoparticles from the microemulsion reaction media.

2. General properties and formation of microemulsions

2.1 Microemulsions: Definition and basic properties

Microemulsions are transparent and thermodynamically stable colloidal dispersions in which two liquids initially immiscible (typically water and oil) coexist in one phase due to the presence of a monolayer of surfactant molecules with balanced hydrophilic-lipophilic properties (Danielsson & Lindman, 1981). They are optically isotropic and transparent. In contrast to emulsions, for which formation requires a considerable energy input, microemulsions form spontaneously upon gentle components mixing, once composition and temperature conditions are right. Depending on the ratio of oil and water and on the hydrophilic-lipophilic balance (HLB) of the surfactant, microemulsions can exist as oil-swollen micelles dispersed in water (oil-in-water microemulsions, O/W), or water-swollen inverse micelles dispersed in oil (water-in-oil microemulsions, W/O); at intermediate compositions and HLBs, bicontinuous structures can exist. When a dispersed phase is present, it consists of droplets with a narrow size distribution in the order of 2-50 nm.

The formation of microemulsion depends on the surfactant type and structure, e.g. single hydrocarbon chain ionic surfactants require the incorporation of cosurfactant or electrolytes for microemulsion formation due to their high hydrophilic character; in contrast, double chain ionic surfactants and ethoxylated non-ionic surfactants may form microemulsions without cosurfactant. Lowering the interfacial tension between the oily and aqueous phase ($\gamma_{o/w}$) is the main role of the surfactant (or surfactant/cosurfactant mixture). The extremely low $\gamma_{o/w}$ (in the order of $10^{-2} - 10^{-3}$ mN m^{-1}) achieved is one of the main microemulsion characteristics: the decrease on $\gamma_{o/w}$ is caused by the surfactant, overcoming the surface energy term caused by the huge increase in interfacial area. In addition, the spontaneous dispersion of numerous water or oil droplets causes an entropy increase, yielding a thermodynamically stable system. The extremely low interfacial tension is decisive for microemulsion formation, and depends on the composition of the system (Kunieda & Friberg, 1981; Cross, 1987).

Microemulsions are dynamic systems, and it has been shown that droplet content exchange processes can occur in the order of millisecond time scales (Fletcher et al., 1987; Clarke et al., 1990). Collisions are produced due to constant Brownian motion of the droplets. When these collisions are sufficiently violent, the surfactant layer breaks up and the micellar exchange can be produced. It is thought that the micellar exchange process is characterized by an activation energy (E_a or energy barrier), which is affected by the flexibility or rigidity of the surfactant layer (Fletcher & Horsup, 1992; Lindman & Friberg, 1999}, in addition to diffusion processes (Fletcher et al., 1987).

2.2 Microemulsions and phase equilibria

Phase behavior studies by means of equilibrium phase diagrams of polar solvent/amphiphile/nonpolar solvent systems provide essential information on microemulsion formation and structure. In 1954, Winsor predicted four types of equilibria which was latter experimentally evidenced: i) Winsor I: oil-in-water (o/w) microemulsions are formed, and the surfactant-rich water phase coexists with the oil phase where surfactant is only present as monomers; ii) Winsor II: water-in-oil (w/o) microemulsions are formed and the surfactant-rich oil phase coexists with the surfactant-poor aqueous phase; iii) Winsor III (middle phase): a three-phase system where a bicontinuous middle-phase microemulsion (rich in surfactant) coexists with both excess water and oil phases; and iv) Winsor IV: a single-phase (isotropic) micellar solution (microemulsion), that forms upon addition of a sufficient quantity of amphiphile. Figure 2 shows how this equilibria can be affected by salinity (for ionic surfactants) or temperature (for non-ionic surfactants), and also illustrates the structural variability of microemulsions (O/W, W/O and bicontinuous (BC)).

Some typical equilibrium phase diagrams are shown in Figure 3 (Destrée & Nagy, 2006). In each of these diagrams L_2 denotes a region where one phase W/O microemulsions are formed. AOT (Sodium 2-ethylhexylsulfosuccinate) based systems are amongst the best characterized systems, and it has been found that the size of the inverse microemulsion droplets formed by this type of systems increases linearly with the amount of water added to the system (Pileni, 1998) and can increase from 4 nm to 18 nm with 0.1 M sodium AOT surfactant (water/AOT/isooctane). AOT based systems are probably the most used for the synthesis of inorganic nanoparticles in w/o microemulsions, for two reasons: good control of droplet size as explained above and the large microemulsion regions found in water/AOT/alkane systems, which give rise to a great deal of compositions available for nanoparticle synthesis. Systems based on cetyltrimethylammonium bromide (CTAB), usually combine this surfactant with alcohols such as hexanol as the oil phase. This alcohol can act as co-surfactant, adsorbing at the oil/water interface along with the surfactant. As shown in Figure 3 the microemulsion region of water/CTAB/hexanol system is relatively narrow, however, when shorter alcohols such as butanol are added as cosurfactant, the microemulsion regions are considerably enlarged (Košak et al., 2004).

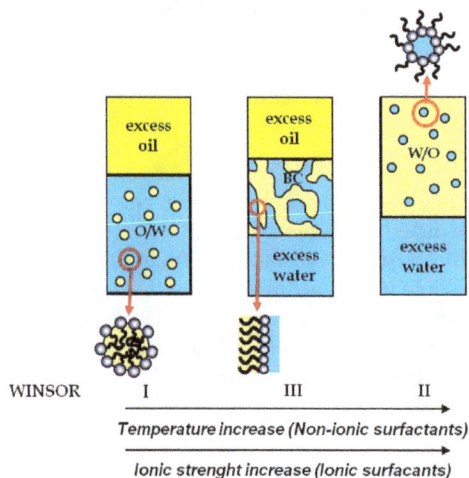

Fig. 2. Winsor classification of microemulsion equilibria. Microemulsion phase sequence as a function of temperature and salinity for non-ionic and ionic surfactants, respectively.

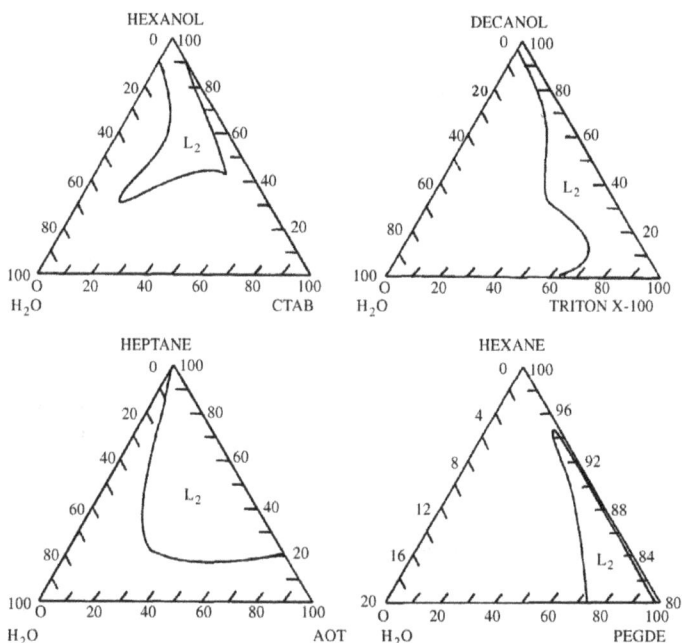

Fig. 3. Water / Nonionic surfactant / Oil pseudo bynary phase diagram, as a function of temperature. Reproduced with permission (Destrée & Nagy, 2006).

As for non-ionic surfactant – based systems, Triton® X-100 (octyl-phenol ethoxylate) is one of the most used, however, alkyl-phenol ethoxylate surfactants such as this one have a limited biodegradability. Their metabolites of degradation have low solubility and are toxic, for example, nonylphenol has been proven to be an endocrine disrupter (Jobling & Sumpter, 1993). On the other hand, aliphatic fatty alcohol ethoxylates such as PEGDE (penta(ethylene glycol) dodecyl ether) are more environmentally friendly; for nanoparticle synthesis, the technical-grade options are usually chosen due to their lower cost. A special feature of non-ionic surfactant systems is the sensitivity of their hydrophilic-lipophilic properties to temperature, and although sometimes this characteristic is seen as a drawback, the possibility for phase-behavior tuning can be used as an advantage for the formulation of non-ionic microemulsions. In addition, nonionic surfactants have a great capacity of hydration by their ethoxylated (EO) units; hence, an appropriate selection of surfactant, oil and precursor salts/precipitating agent concentration, in combination with the rich structural behavior that such a system may display as a function of temperature, can lead to highly optimized formulations in terms of aqueous phase uptake and hence reactants loading. A good premise to this behavior is the enormous efficiency boost in the formation of middle phase microemulsions by the use of block copolymer surfactants reported by Strey et al. (Jakobs et al., 1999).

2.3 Effect of precursor salts and additives on the phase behavior

Although nonionic microemulsion systems are mainly affected by temperature changes, the addition of electrolytes and cosurfactant can also produce shifts in the solubilization and

T_{HLB} (hydrophilic-lipophilic balanced temperature or phase inversion temperature) of the systems (Aramaki et al., 2001; Kunieda et al., 1995; Shinoda, 1968). The use of ionic surfactants may have some drawbacks, as usually the aqueous phase uptake of ionic microemulsions is reduced in the presence of precursor salts due to screening effects, and hence microemulsion regions become smaller (Liu et al., 2000; Gianakas et al., 2006). Additionally, complex species could interfere with particle growth by adsorption to their surface, and contaminations of ceramic nanoparticles with the surfactant counterions aer possible. Often, the effect of addition of precursor salts or precipitating agent on the phase behavior and structure of microemulsion systems is underlooked. Generally two microemulsions with a fixed water/surfactant ratio are prepared without taking into account the influence the added salt has on the size and the structure of the water droplets.

Recently, Stubenrauch et al (Magno et al., 2009) and Sanchez-Dominguez et al. (Aubery et al., 2011) have reported systematic studies on the effects of addition of reactants to nonionic microemulsion systems. It was shown (Magno et al., 2009) that, depending on the aqueous nature of nonionic microemulsion systems, and the salting-in or salting-out effect of the additives, both increase or decrease on the water solubilization could be obtained. The same group studied the effects of different salts on the water solubilization of ionic microemulsions of the system aqueous phase / AOT/butanol /decane (Stubenrauch et al., 2008). They found that depending on the type of precursor (salts of Pt, Bi, or Pb) or the reducing agent (NaBH$_4$), different behaviors can be obtained, and it was necessary to add different amounts of SDS and 1-butanol in order to keep both the w/o nature of the microemulsion droplets as well as their size (which was only assessed theoretically based on microemulsion composition).

In the studies by our group on the effects of addition of precursor salts and precipitating agent to the non-ionic microemulsion system aqueous solution / Synperonic® 13/6.5 / isooctane (Aubery et al., 2011), several factors were taken into account: phase behavior (pseudoternary phase diagrams at constant temperature), dynamicity (presence or absence of percolation in W/O structures, or formation of bicontinuous microemulsions), and droplet size (DLS). It was possible to obtain w/o microemulsions at a wide range of overlapping compositions for both precursor salts and precipitating agent. In fact, the microemulsion regions were considerably enlarged upon addition of precursors and precipitating agent; this behavior is contrary to what is typically obtained with ionic systems which have their microemulsion region reduced with addition of salts. It was difficult to obtain both type of microemulsions in either a non-percolated or percolated state; this was characterized extensively by conductivity, FT PGSE NMR and hydrophilic dye diffusion studies. When pseudobinary phase diagrams as a function of temperature were carried out, there were some compositions and temperatures at which both precursor salts and precipitating agent microemulsions were either percolated, not percolated or bicontinuous.

2.4 Dynamic processes

Among the dynamic processes in microemulsions, interactions of droplets components and droplet- droplet interactions must be taken into account (Fletcher et al, 1987; Fletcher & Horsup, 1992; Moulik & Paul, 1998). Concerning interactions of droplets components in a W/O nonionic microemulsion, a schematic representation is depicted in Figure 4. The example concerns an aqueous droplet stabilized by a mixture of surfactant and cosurfactant molecules. The aqueous domain is composed by bounded water (hydrating the surfactant/cosurfactant hydrophilic domains) and free water (forming the droplet core). The exchange process to reach microemulsion equilibria comprises: 1) exchange of water between the bounded and free state; 2) exchange of cosurfactants among the interfacial film, the continuous phase and the dispersed phase (depending on its solubility). If ionic species are solubilized in the aqueous solution, they exchange ions between the bounded and free water. The composition of the aqueous droplets, their concentration and the temperature are mayor factors defining further interactions between them.

Droplet-droplet interactions depend strongly on droplet concentration, solvent viscosity, temperature, rigidity or flexibility of interfacial layer, and interactions between surfactant tails (Capek, 2004; Lopez-Quintela, 2003). When water or oil droplet dispersions are present, the droplets continuously collide, break apart, aggregate and break apart giving rise to dynamic processes in microemulsions. These dynamic processes allows microemulsion droplets to continuously exchange their content in microsecond scales. The composition of the aqueous droplets, as explained above, has a great influence on droplet interactions. As an example, the interfacial layer plays an important role on the formation, stability and discrete nature of microemulsion droplets. The film rigidity has been observed to increase with the surfactant hydrocarbon chain, whereas it substantially decreases with cosurfactant addition. The surfactant packing capacity can be also affected by the ionic strength of the droplets (Aramaki et al., 2001; Kunieda et al., 1995). The increase of surfactant molecules in the layer is proportional to the rigidity of the micelles.

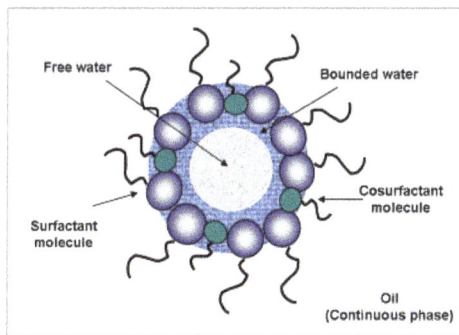

Fig. 4. Schematic representation of W/O microemulsion droplet. (Adapted with permission from Moulik, 1998).

Changes on microemulsion dynamics giving rise to structural transitions can be explained in terms of percolation. Figure 5 refers to a percolation process taking place in W/O microemulsions (Borkovec et al., 1988), as the oil to water ratio ϕ_o is varied. As observed in Figure 5, at high oil concentration, the fraction of water in discrete droplets increases with water composition (decreasing oil concentration) up to a concentration, where it drastically decreases. This concentration is called Percolation Concentration C_P. Although clusterization occurs below C_P (low water concentrations), these clusters remain finite in size respect to the bulk solution. C_P represents the concentration at which the first infinite cluster appears. Further increase on water concentration would lead to the disappearance of discrete water droplets to give rise to an increase of infinite water and oily domains, which are characteristic of bicontinuous microemulsions. Water percolation can also be induced by temperature, and is defined as the percolation temperature T_P.

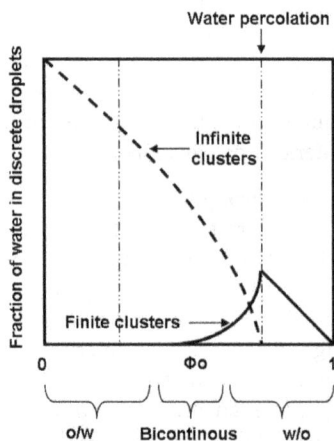

Fig. 5. Schematic representation of structural regimes of microemulsions caused by water percolation as a function of the relative amount of oil. Reprinted (adapted) with permission from Borkovec M., Eicke H.-F., Hammerich H., & Das Gupta, B. (1988) *J. Phys. Chem.*, 92, 1, pp. (206-211). Copyrigh (1988) American Chemical Society

3. The microemulsion reaction method: introduction and generalities

A brief description of the synthesis of nanoparticles in W/O microemulsions has already been given in Section 1, along with an explanatory figure (Figure 1). Colloidal nanoparticle formation is a complex process, which includes nucleation and growth steps -giving rise to nanoparticle formation- as well as eventual coagulation and flocculation.

3.1 Mechanism of nanoparticle formation

A model of particle precipitation in a homogeneous aqueous medium has been proposed by La Mer (La Mer & Dinegar, 1950). The model involves particle nucleation at short times. As soon as monomer formation takes place due to chemical reaction, its concentration increases up to the point of spontaneous nucleation, which occurs over a critical supersaturation concentration [C]$_C$. Afterwards, growth takes place (Figure 6). The growing step is mainly controlled by the diffusion of monomers in solution (C) onto the particles surface. Thus, C reaches a maximum and afterwards it begins to decrease. This decrease in monomer concentration is due to the growth of the particles by diffusion. In microemulsions, the number of nucleated sites is expected to be higher, comparing to homogeneous reactions, as illustrated in Figure 6. On the other hand, the diffusion controlled particle growth should occur at lower rate. Another model is based on the thermodynamic stabilization of the particles. In this model the particles are thermodynamically stabilized by the surfactant. The size of the particles remains constant when the precursor concentration and the size of the aqueous droplets vary. Nucleation occurs continuously during the nanoparticle formation.

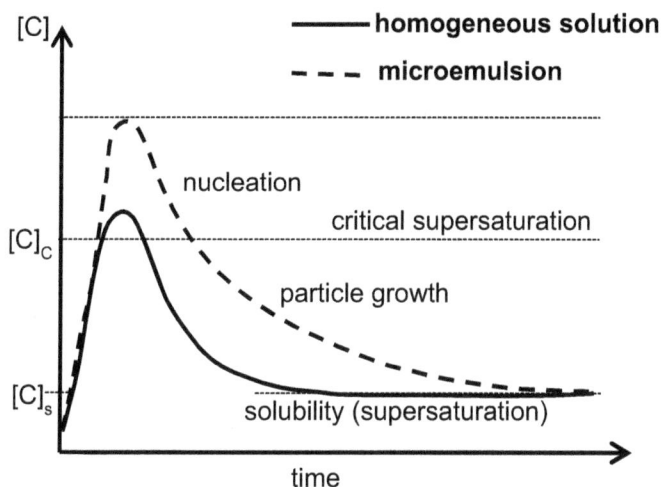

Fig. 6. Monomer concentration [C] as a function of time in microemulsions, compared to a homogeneous system. (Adapted from La Mer & Dinegar, 1950 and Schmidt, 1999).

3.2 Reaction kinetics

Although microemulsions as reaction media for the synthesis of inorganic nanoparticles have been extensively studied, the kinetics of these reactions is still not completely understood. As mentioned above, several types of nanoparticles have been synthesized using a variety of surfactant systems, and relationships between the nanoparticles characteristics and the microemulsion media are not straightforward due to the diversity of variables which can have an influence, and this may be closely related with complex kinetics. An effort to relate the surfactant media with the reaction kinetics was reviewed by Lopez-Quintela et al. (Lopez-Quintela et al., 2004), concerning both inorganic and organic syntheses in microemulsions. Few studies can be cited concerning the follow-up of reactions with time, due to the fast rate of microemulsion reactions. Some of these works are pointed out below:

1. Bandyopadhyaya et al. (Bandyopadhyaya et al., 1997) have modelized $CaCO_3$ formation in microemulsions by carbonation. A time-scale analysis was developed, resulting in a model of reaction kinetic that closely corresponded to results obtained experimentally.
2. Chew et al. (Chew et al., 1990) have studied the effect of alkanes in the formation of AgBr particles in ionic W/O microemulsions (using AOT as surfactant), where the transmittance of the reactions were followed with time with UV-Vis and Stopped-Flow Spectrophotometry. They have found an increase on reaction rate with the chain lenght of the alkane.
3. Curri et al. (Curri et al., 2000) studied the role of cosurfactant on the synthesis of CdS nanoclusters, using CTAB as surfactant. Stopped-Flow Spectrophotometry was used in order to compare a reaction using CTAB plus cosurfactant and other carried out using AOT. They have summarized two different cosurfactant effects: the influence of the

surfactant film flexibility on particle growth and the particles stabilization in solution, determined by the adsorption of cosurfactant onto the particle surface.

4. Lopez-Quintela et al. (De Dios et al., 2009) simulated the kinetics of nanoparticles formation in microemulsions. Simulations were carried out by comparing Ag, Ag-Au and Au formation with experimental data reported by Destrée and Nagy. (Destrée & Nagy, 2006). The detailed comprehension of the kinetics taking place in microemulsion reactions is limited by the experimental data in this direction. Hence, systematic studies focused on reaction rates are greatly encouraged in order to advance in this field.

3.3 Parameters influencing on nanoparticle synthesis

Although complete control of particle characteristics is still far from clear and direct, some results on this field can be pointed out as shown below.

Aqueous solution concentration. It have been described in several publications the particle size dependency with water:surfactant molar ratio (w_0). In general, it has been observed that, as increasing w_0, an increase on particle size is observed (Pileni, 1997; Lopez-Quintela, 2003). However, Cason et al. (Cason et al., 2001) have found that, with different w_0, it was possible to obtain constant particle size if the reaction time increases for the synthesis to get completed. They proposed that the growth of the particles is affected by w_0. It was considered that for low w_0 values, the aqueous solution is not enough to completely hydrate the polar groups of the surfactant and the counterion. As a consequence, the film rigidity is higher compared to higher w_0 values. This influences on the micellar exchange and, as a consequence, the growth rate decreases. Increasing w_0, the micelle rigidity decreases generating an increase in the growth rate up to a certain concentration, where further increase in w_0 simply causes reagent dilution, which causes a decrease in the growth rate. Some studies have indicated a decrease on particle size with w_0 (Bagwe & Khilar, 1997).

Reagent concentration: Particle size have been determined to be directly dependent on reagent concentration (Lopez-Quintela, 2003). An example is the work carried out by Destrée & Nagy (Destrée & Nagy, 2006). They have synthesized Pt nanoparticles, using different concentrations of K_2PtCl_4. An increase on particle diameter from 2 to 12 nm was obtained, by increasing the concentration of the precursor. On the other hand, an increase on the precipitating/precursor ratio generally causes a decrease on particle size (Lisiecki & Pileni, 2003). It is thought that increasing precipitating agent concentration, particle nucleation can be favored in a higher extent, which further grow simultaneously, resulting in particles with lower size and polydispersity.

Surfactant and cosurfactant: Studies in order to determine the effect of nonionic hydrophilic and lypophilic surfactant groups have been developed. As the lypophilic chain of the surfactant is longer, smaller particles are obtained due to the increased micellar rigidity Generally, the addition of cosurfactant causes an increase micellar exchange, due to the decrease in the interfacial film rigidity. It is thought that the increase in microemulsion droplet size is counteracted with the increase on surfactant film curvature, generating smaller particles than without cosurfactant (Lopez-Quintela et al., 2004).

Solvent: Some studies have shown that low weight oil molecules, with low molecular volumes, can penetrate in the sufactant hydrocarbon chains, increasing the film curvature and rigidity (Cason et al., 2001). This effect has been observed to produce micellar exchange decrease and, consequently, smaller particles are obtained.

Electrolytes: Some studies reveal the possible dependence of nanoparticle shape with electrolyte addition (Filankembo et al., 2003). Pileni (Pileni, 2003) has postulated that the selective ion or molecule adsorption over nanocrystal layers can affect their growth in certain directions, which could explain the apparent preference on certain particle shape.

Microemulsion structure: Some studies have claimed about the nanoparticle shape partial dependence on microemulsion structure, where the microemulsion media acts as a template. A particular example is the work carried out by Pileni (Pileni, 2001) on the preparation of Copper nanoparticles from microemulsions by varying the internal structure. Spherical water droplets resulted in spherical particles, water cilinders resulted in cylindrical copper nanocrystals (with spherical particles) and a mixture of W/O microemulsion with lamellar phase resulted in a mixture of particle shape such as spheres, cylinders, etc. It was found that the template was not the only parameter which controls the shape of nanocrystals. There are examples of nonexistent correlation between the microemulsion structure and the nanoparticles obtained, which supports the nanoparticle shape dependence on electrolyte adsorption (Chen & Lin, 2001).

Even though there is a diversity of studies carried out in order to relate nanoparticle characteristics with microemulsion properties, there is a gap in the effects of microemulsion dynamic behavior on nanoparticle characteristics, as systematic studies in this direction are scarce. The transport and micellar dynamics influence to some extent the nanoparticle formation, and it is important to take this into account in order to understand the basics of nanoparticles synthesis by this route. This type of studies may give rise to improvements on controlled nanoparticle characteristics.

4. Recent advances in the use of microemulsions as confined reaction media for the synthesis of inorganic nanoparticles

There have been a number of advances in different aspects of the synthesis of nanoparticles in microemulsions over the last four years. The main ones are: the use of other types of microemulsions for synthesis (O/W and bicontinuous microemulsions), the preparation of more complex architectures (core/shell and multishell, hybrid nanocrystals), the synthesis of more complex ceramics (spinels, perovskites, etc), modeling of reactions in microemulsions, and novel approaches for the separation of nanoparticles from the reaction mixtures. The most outstanding examples of each of these aspects are given below.

4.1 The use of other types of microemulsions for inorganic nanoparticle synthesis

One of the main drawbacks of the technique reviewed so far (synthesis in W/O microemulsions) and the main reason why it has not been generally accepted for production at the industrial scale, is the fact that these microemulsions employ large quantities of organic solvent, as well as its limited production capacity, since this is restricted to the amount of aqueous phase solubilized and the concentration of precursor which often cannot be that high due to interactions with the surfactant, as discussed in Section 2.3. Some research groups have been working in new approaches to overcome these drawbacks.

4.1.1 Synthesis in oil-in-water microemulsions

Our research group in collaboration with the group of Boutonnet, has developed a novel and straightforward approach based on O/W microemulsions (Sanchez-Dominguez et al., 2009). From a practical and environmental point of view, the possibility of preparing inorganic nanoparticles using O/W instead of W/O microemulsions may be highly advantageous, since the major (continuous) phase is water. The method consists in the use of organometallic precursors, dissolved in nanometer scale oil droplets of O/W microemulsions (Figure 7), and stabilized by a monolayer of hydrophilic surfactant. The first work reported as a proof of concept the synthesis of metallic (Pt, Pd, and Rh) as well as metal oxide (CeO_2) nanoparticles (Sanchez-Dominguez et al., 2009). Small (around 3 nm), nanocrystalline materials with a narrow size distribution were obtained (Figure 8).

It was followed by the synthesis of the following mesoporous nanocrystalline oxides: CeO_2, ZrO_2, $Ce_{0.5}Zr_{0.5}O_2$, and TiO_2 (Sanchez-Dominguez et al., 2010). Small particle size (3 nm), and high specific surface area (200-380 m^2 g^{-1}) was obtained for all materials. Nanocrystalline cubic CeO_2 and $Ce_{0.5}Zr_{0.5}O_2$ were obtained under soft conditions (35°C). The materials were evaluated as catalyst supports in the CO oxidation reaction by doping them with Au (2 wt%, impregnation technique). The resulting catalysts showed a high Au dispersion (HRTEM/EDX). These materials showed a good activity in CO oxidation at low temperature (T_{50} of 44°C for TiO_2). This study demonstrates the feasibility of this approach for the preparation of highly active catalysts.

In a more recent study by the same group (Tiseanu et al., 2011), Eu-doped luminescent CeO_2 nanocrystals were prepared by the same method. Several characterization techniques (X-ray diffraction, RAMAN spectroscopy, UV-Vis diffuse-reflectance, FTIR as well as time-resolved photoluminescence spectroscopy) were used to characterize the nanocrystals, and it was shown that there was a surface enrichment of Eu^{3+}, which diffused progressively to the inner Ceria sites upon calcination. Under excitation into the UV and visible spectral range, the calcined europium doped ceria nanocrystals display a variable emission spanning the orange-red wavelengths. A remarkable result was that the surface area of the powders remained as high as 120 m^2 g^{-1} even after calcination at 1000°C.

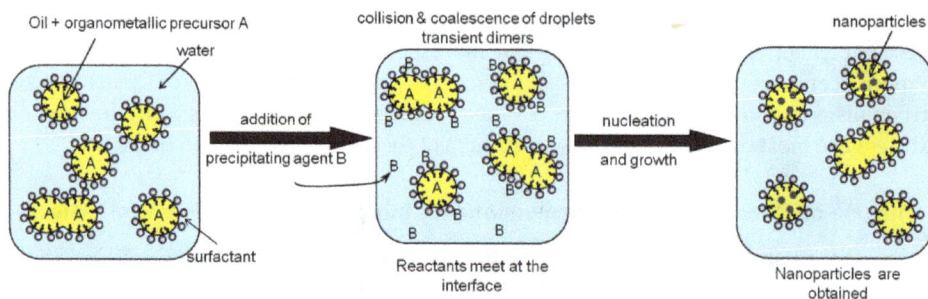

Fig. 7. TEM micrographs and related particle size distribution histograms of nanoparticles prepared in O/W microemulsions: (a) Pt, (b) Pd, (c) Rh and (d) CeO_2. Scale bar: 50 nm, except d (10 nm) and inset of d (5 nm). Reproduced with permission (Sanchez-Dominguez et al., 2009).

It should be pointed out that in all of these examples, only one microemulsion is used for synthesis, as opposed to what is typically needed with the W/O method (two microemulsions, one bearing the precursors and another one the precipitating agent in the aqueous phase). Since most precipitating agents are water-soluble, it means that it can be added directly to the microemulsion without affecting its O/W structure, and hence only one microemulsion, containing the organometallic precursor is prepared. Hence, the mechanism occurring in this approach is most likely different; possibly, it is an interfacial reaction. Modelization studies in conjunction with kinetic experiments need to be carried out in order to clarify this point. Considering these results, the perspectives of this novel O/W microemulsion reaction approach are very positive, and should complement the W/O microemulsion method, offering a greener alternative. Finally, it must be highlighted that the typical metal loading in the microemulsions reported, and hence the typical production capacity ranges from 2 to 5 grams of nanoparticles per kg of microemulsion, which is comparable and in some cases superior to typical metal loadings achieved in W/O microemulsions.

4.1.2 Synthesis in bicontinuous microemulsions

An interesting approach to boost the metal loading and hence the nanoparticle production capacity of microemulsions is the use of bicontinuous microemulsions. Lopez et al. (Esquivel et al., 2007; Loo et al., 2008) have reported this approach for the synthesis of magnetic nanoparticles. A microemulsion system based on cationic surfactants was used for the synthesis of a mixture of maghemite/magnetite nanoparticles, using bicontinuous microemulsions at 80°C, with 30-40 wt% of aqueous phase. They obtained small nanoparticles (8 nm) with a narrow size distribution, a nanocrystalline structure and superparamagnetic behavior. Furthermore, the yield of the reactions was as high as 1.16 g of product per 100 grams of microemulsion, which is rather high compared to what can be obtained in most w/o microemulsion systems (0.05 - 0.2 grams per 100 g microemulsion).

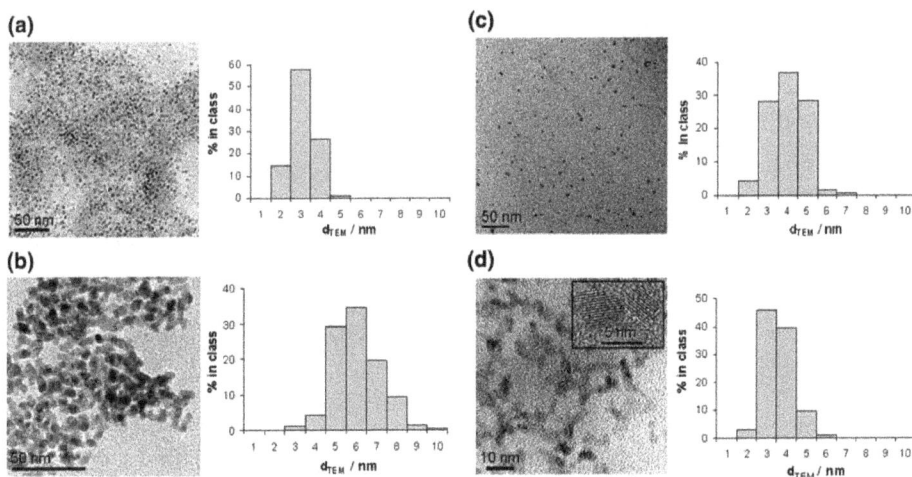

Fig. 8. TEM micrographs and related particle size distribution histograms of nanoparticles prepared in O/W microemulsions: (a) Pt, (b) Pd, (c) Rh and (d) CeO_2. Scale bar: 50 nm, except d (10 nm) and inset of d (5 nm). Reproduced with permission (Sanchez-Dominguez et al., 2009).

The same research group reported recently the synthesis of silver nanoparticles by the same approach (Reyes et al., 2010; Sosa et al., 2010), by using a microemulsion system based on AOT/SDS as the surfactant system and toluene as the oil. Depending on the surfactant: oil ratio, the authors found the formation of only globular nanoparticles or a mixture of interconnected, worm-like structures plus globular nanoparticles. The reaction yields for these materials was also remarkably high (up to 1.4 g of silver nanoparticles per 100 g of microemulsion). In all of these works, only one microemulsion was necessary for nanoparticle preparation, as the precipitating agent was added directly, as an aqueous solution, to the microemulsion containing the metallic precursors. This aspect also contributes to the greener quality of this approach as compared to the traditional W/O microemulsion reaction method.

4.1.3 Synthesis in microemulsions with an optimized aqueous phase uptake

In the work carried out by our group concerning a nonionic system (Aubery et al., 2011), large microemulsion regions were obtained when the reactants were incorporated, as mentioned in Section 2.3. Thanks to this high aqueous phase uptake and the overlap of microemulsion regions for both precursor salts and precipitating agent, synthesis of Mn-Zn ferrite nanoparticles could be carried out using a wide range of compositions. Futhermore, different scenarios were available for nanoparticle synthesis: W/O non-percolated, W/O percolated, and bicontinuous microemulsions. Differences were observed in the characteristics of the synthesized nanoparticles depending on the type of microemulsions used, and in all cases spinel nanocrystalline particles with superparamagnetic properties were obtained, directly in the microemulsion, without the need for calcination. The aqueous phase content ranged from 5 wt% to 50 wt%, which represents a boost in the production capacity. This study should encourage further research into optimized non-ionic microemulsion systems, since although the presence of salts affects their phase behavior, it does so in such a way that aqueous solubilization can be significantly increased at a certain temperature, which can be investigated by phase behavior studies.

4.2 Preparation of more complex architectures

In this regard, most of the studies concern core-shell studies, although some other structures include multiple core-shell particles, hollow spheres and nanowires and nanorods.

4.2.1 Core/shell nanoparticles

A large majority of the core-shell nanoparticles synthesized in W/O microemulsions contain silica, usually as the shell material. In the last few years, the W/O microemulsion approach has been gaining popularity over the well-known adaptation of the Stöber method (Nann & Mulvaney, 2004), for coating a diversity of nanoparticles with a silica shell. This is because it has been observed that the microemulsion method results in a better shell thickness control (Dong H, 2009), as compared to the adapted Stöber method, which is based on the sol-gel technique. It must be pointed out that in the majority of the studies, the core material was synthesized in a previous step, by a different method, usually hydrothermal or solvothermal techniques (Dong B. et al., 2009; Dong H et al., 2009; Qian et al., 2009; Vogt et al., 2010; Wang J. et al., 2010). Nevertheless, a very interesting point from these investigations is the strategy on how this silica shell is deposited onto the core; this sophisticated approach is probably the reason for the high control achieved. The core nanoparticles are usually functionalized for one or two purposes: one is in order to be very well dispersed in one of the microemulsion phases (the oily or the aqueous phase), the other is for very controlled deposition of silica via hydrolytic copolymerization with silanized molecules such as (3-aminopropyl)triethoxysilane (APTES), which were covalently linked to the core particles. By this approach, uniform CdTe@silica nanoparticles with a regular core – shell structure, 48±3 nm in diameter were obtained by Dong H. et al. (Dong H. et al., 2009). In their work, the initial core CdTe particles, synthesized by a hydrothermal method, were functionalized with thioglycolic acid, so they could be reacted with APTES and then dispersed in the aqueous phase of the microemulsion. The silica precursor, TEOS, was dissolved in the oily phase of the microemulsion (cyclohexane and octanol), and the silica shell was then formed by addition of ammonia. In the work by Dong B. et al (Dong B. et al., 2009), on the other hand, the core ZnS:Mn particles were functionalized with oleic acid and hence dispersed in the oil phase of the microemulsion, and the silica layer was deposited by reacting TEOS with ammonia in the W/O microemulsion containing the core particles dispersed in the oil. Figure 9 shows TEM results of these core-shell particles.

Fewer examples deal with the formation of core-shell nanoparticles in which both the core and the shell have been synthesized in a W/O microemulsion (Chung et al., 2011; Takenaka et al., 2007). Takenaka et al. prepared Ni nanoparticles in a W/O microemulsion, and afterwards TEOS and ammonia were added in order to form the silica layer. Core-shell nanoparticles with 20-50 nm diameter and a Ni shell (5 nm) were formed. For comparison, silica nanoparticles were prepared also in W/O microemulsions but the Ni nanoparticles were prepared by impregnation of these silica nanoparticles. Their catalytic activity in the partial oxidation of methane reaction was evaluated, and the core-shell nanoparticles had a better performance than the impregnated ones (Takenaka et al., 2007). On the other hand, Chung et al. prepared silica nanoparticles coated with a thin layer of CeO$_2$, and the material was also prepared in W/O microemulsions in a two-step procedure (Chung et al., 2011). This reaction turned out to be challenging as the formation of CeO$_2$ shell was competing with bulk precipitation. The problem was overcome by coupling two strategies:

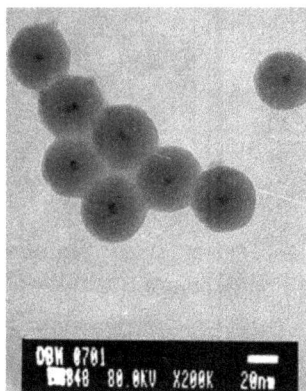

Fig. 9. TEM image of ZnS:Mn@silica nanoparticles with a core – shell structure. (Reproduced with permission, Dong B. et al. 2009).

functionalization of the surface of the core silica nanoparticles with an organoamine group, and step-wise, semi-batch addition of the second microemulsion containing the Ce precursor. In this way, the silica cores were homogenously coated with a CeO_2 shell.

As for core-shell nanoparticles made up of materials different from silica, the synthesis of both the core and the shell is usually carried out in W/O microemulsions, either in a two step process by preparing first the core and the later deposition of the shell (by adding more aqueous phase or more microemulsion comprising the second component), or both precursors are incorporated simultaneously, but the different reaction kinetics for each of the products results in a core-shell structure. The following core-shell nanomaterials can be listed: Pt@CeO$_2$ (Yeung & Tsang, 2009 and 2010), Co@Ag (Garcia-Torres et al., 2010), Fe$_2$O$_3$@Au (Iglesias-Silva et al., 2010), Ni@Au (Chiu et al., 2009), Ag@Polystyrene (Li et al., 2009), and CdS@TiO$_2$ (Ghows & Entezari, 2011). So far, the core@shell structures of these materials is not as well defined and controlled as that obtained with core@silica materials.

4.2.2 Hollow nanospheres

Jiang et al. have prepared hollow nanospheres of Ni (Jiang et al., 2010) and CuS (Jiang, 2011), by following an approach which resembles that reported by Sanchez-Dominguez et al. (Sanchez-Dominguez et al., 2009). Jiang et al. used an o/w microemulsion in which the precursor (a naphtenate), was dissolved in the oil phase (dimethylbenzene) of a water/SDS/butanol/dimethylbenzene microemulsion. The precipitating agent was added in the water phase. The authors explain that an interfacial reaction occurs, and hollow nanospheres of about 100-200 nm are formed (Figure 10, for Ni hollow spheres). These are made-up of smaller nanoparticles. One difference between Jiang's method and that reported by Sanchez-Dominguez et al is that in the former, the temperatures used for reaction are higher (85°C for Ni; post-synthesis hydrothermal treatment for CuS), whereas in the latter the temperatures used are near room temperature (25-35°C).

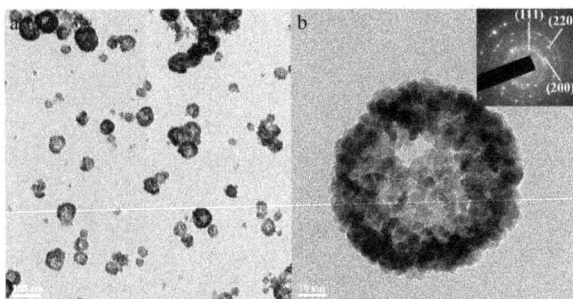

Fig. 10. TEM image of the Ni hollow spheres and (b) a single Ni hollow sphere. Inset: SAED pattern. Reproduced with permission (Jiang et al., 2010).

4.2.3 Nanowires, nanorods

Some works describe the formation of nanowire-like or nanorod structures. Usually, in order to obtain such high aspect ratio structures, it is necessary to carry out the synthesis at a relatively high temperature, or include a certain post-synthesis thermal treatment. Wang et al. synthesized single-crystalline ZnO nanowire bundles with a length of about 1 μm and a diameter of about 20–30 nm (Wang G. et al., 2010). The approach was by reacting zinc acetate with hydrazine in a w/o microemulsion based on water/dodecylbenzene sulfonic acid sodium salt /xylene. The reaction temperature is not mentioned, however for reflux of xylene is achieved around 140°C. The relatively high reaction temperature and the heating time is possibly the driving force for the growth of the nanowires, as different structures were obtained at shorter reaction times. Also, it must be pointed out that the precursor used, zinc acetate, is soluble in both water and the oil phase, which is an unusual approach.

Wu et al. synthesized nanowires of Zn/Co/Fe layered double hydroxides using a w/o microemulsion based on water/CTAB/n-hexanol n-hexane (Wu et al., 2010). In their approach, the sulfate salts were used as precursors, hence these were dissolved in the aqueous phase only, and urea was used as precipitant. The influence of reaction temperature, time, urea concentration and CTAB to water molar ratio on the structure and morphology of Zn/Co/Fe-layered double hydroxides was investigated. The possible reason for nanowire growth is the solvothermal treatment of the reaction mixture which was carried out in an autoclave at 80-180°C during 6-24 hours. The thermal treatment in the autoclave was a key factor for annealing and therefore obtaining both a crystalline structure and formation of high aspect ratio particles (nanowires).

4.3 Synthesis of complex ceramics

The w/o microemulsion reaction method has been used for the synthesis of complex ceramic nanoparticles such as perovskites, spinels, aluminates, and hexaferrites. Often, nanoparticles of precursors such as hydroxides or other amorphous compounds are synthesized in the microemulsions, and these are afterwards calcined at a certain temperature in order to obtain the desired crystalline structure.

He et al. synthesized nanoparticles of perovskite-type oxides $La_{0.8}Ce_{0.2}Cu_{0.4}Mn_{0.6}O_3$ and $La_{0.8}Ce_{0.2}Ag_{0.4}Mn_{0.6}O_3$ (He et al., 2007). The microemulsion used was CTAB/butanol/water /heptane, and for comparison purposes, the same materials were synthesized by the sol-gel technique. The precipitation of the precursors was carried out with NaOH for the microemulsion method, whereas citric acid was used for the sol-gel method. The particle size distribution was smaller and more uniform and the specific surface area was higher for the particles synthesized in microemulsions than those synthesized by sol-gel. Furthermore, the catalytic activity in the NO reduction by CO was evaluated. Performance of perovskites synthesized in microemulsion was superior than that of materials synthesized by sol-gel.

Gianakas et al. (Gianakas et al., 2007) reported the synthesis of spinel-type metal aluminates MAl_2O_4 sutwhere M=Mg, Co, o Zn using w/o and bicontinuous microemulsions. They carried out a very complete phase behavior study, which included pseudoternary phase diagrams for each precursor combination as well as the precipitating agent, ammonia. The microemulsion system was: aqueous solution/ CTAB/butanol/ octane. The spinel structure was achieved after calcination at 800°C. It was found that spinels synthesized in reverse microemulsions showed better surface and textural properties, as well as smaller particle size than spinel synthesized in bicontinuous microemulsions. As for catalytic activity, which was evaluated in the NO reduction by CO, the spinels synthesized in w/o microemulsions was slightly superior. Similar characteristic size was obtained by Wang et al. (Wang Y. et al.,

2007) for nanoparticles of manganese-doped barium aluminate $BaAl_{12}O_{19}$: Mn^{2+}; calcination at 1300°C was carried out in order to obtain the crystalline phase expected. The evaluation of photoluminescent properties of this material showed that this phosphor is a good candidate to replace Hg lamps.

Other good examples of ceramic materials obtained in w/o microemulsions include: barium hexaferrite ($BaFe_{12}O_{19}$) nanoparticles (Xu et al., 2007), tungsten oxide (WO_3) nanoparticles (Asim et al., 2007), and rutile TiO_2 nanoparticles (Keswani et al., 2010). In the last example, it is remarkable that the rutile phase was obtained at room temperature, without the need for thermal treatment, hence the size of the rutile nanocrystals remained as small as 4 nm.

4.4 Modeling of reactions in microemulsions

There have been a number of studies dealing with the theoretical aspects of nanoparticle formation by the microemulsion reaction method. Most of these studies use the Monte Carlo method. The studies carried out in the last four years are focused on several aspects: kinetics of nanoparticle formation (de Dios et al., 2009), formation of bimetallic nanoparticles (Tojo et al., 2009; Angelescu et al., 2010), droplet exchange (Niemann & Sundmacher, 2010), cluster coalescence (Kuriyedath et al., 2010), and core-shell nanoparticle formation (Viswanadh et al., 2007).

Kinetics of nanoparticle formation in microemulsion were studied for the Ag and Au nanoparticles using Monte Carlo simulations by de Dios et al. (de Dios et al., 2009). It was shown that, although the material interdroplet exchange depends primarily on the flexibility of surfactant film, a slow reaction rate leads to a more effective material interdroplet exchange for a given microemulsion. Two factors contribute to this result. Firstly, a slow reaction implies that autocatalytic growth takes place for a longer period of time, because there are available reactants. If the reaction is faster, the reactants are almost exhausted at early stages of the process. As a consequence, autocatalytic growth is only possible at the beginning. Secondly, a slow reaction rate implies the continuous production of seed nuclei, which can be exchanged between micelles due to their small size, allowing the coagulation of two nanoparticles. This exchange only takes place at early stages of the synthesis. Both factors, autocatalysis and ripening, favor the slow growth of the biggest nanoparticles leading to the production of larger particles when the reaction is slower.

With respect to the formation of bimetallic nanoparticles in microemulsions, the same research group (Tojo et al., 2009), carried out Monte Carlo studies in order to explain the different structures that can be obtained when bimetallic nanoparticles are synthesized in microemulsions. They observed that the difference in reduction rates of both metals is not the only parameter to determine metal segregation; the interdroplet channel size also plays an important role. The reduction rate difference determines nanoparticle structure only in two extreme cases: when both reactions take place at the same rate, a nanoalloy structure is always obtained. In contrast, if both reactions have very different rates, the nanoparticle shows a core-shell structure. However, in the large interval between both extreme cases, the nanoparticle structure is strongly dependent on the intermicellar exchange, which is mainly determined by the flexibility of the surfactant film around the microemulsion droplets. In a related study by Angelescu et al. (Angelescu et al., 2010), it was found that the bimetallic nanoparticle structure is mainly determined by the difference in the reduction rates of the two metal ions and the excess of reducing agent. An intermetallic structure is always obtained when both reduction reactions take place at about the same rate. When the metal ions have very different reduction potentials, a core-shell to intermetallic structure transition is found at increasing the excess of the reducing agent. An enhancement of the intermetallic structure at the expense of the core-shell, can be obtained either by decreasing the concentration of both metal salts or by increasing the interdroplet exchange rates. The results obtained by these studies has positive implications in the general formation of bimetallic nanoparticles with a given structure (core-shell or nano-alloy).

4.5 Novel approaches for the separation of nanoparticles from reaction mixtures

Often, the nanoparticles formed in a microemulsion are so well dispersed in the reaction media that some solvent has to be added in order to destabilize the microemulsion, which causes desorption of surfactant from the particles, which aggregate and precipitate, making their separation by centrifugation or filtration easier. Sometimes, during this aggressive process the nanoparticles end up so agglomerated that it is difficult to re-disperse them. Some novel and straightforward approaches have been proposed for an improved recovery or phase transfer of nanoparticles from microemulsion media.

Eastoe et al. (Hollamby & Eastoe, 2009; Myakonkaya et al., 2010, 2011; Nazar et al., 2011; Vesperinas et al., 2007) have proposed three approaches for nanoparticle recovery. One of them is based on the use of a photodestructible surfactant for microemulsion formation, and in the final step, irradiation with UV-light induces microemulsion destabilization and hence separation of Au nanoparticles (Vesperinas et al., 2007). In another approach, excess water is added at the end of the reaction, to the microemulsion containing the nanoparticles, inducing a change in phase behavior and hence microemulsion destabilization, followed by phase separation. Interestingly, by this approach, usually the nanoparticles remain in the oil phase, which can be diluted with organic solvents to form stable nanoparticle dispersions (Nazar et al., 2011). This method shows potential benefits for dispersion, storage, application, and recovery of NPs, with the great advantage that it is not necessary to add organic solvents for nanoparticle separation. In other approach by the same group, nanoparticle separation has been achieved by changing the solvent quality, for example, adding squalene to water/AOT/octane microemulsion containing Au nanoparticles (Myakonkaya et al., 2010).

Abecassis et al. have proposed nanoparticle separation by thermally inducing the phase separation of the microemulsion media (Abecassis et al., 2009). This was applied to the synthesis of Au NPs, which upon destabilization remained preferentially in the oil phase.

5. Conclusions and perspectives

It has been shown that the microemulsion reaction method is a versatile technique, useful for the controlled synthesis of a large variety of nanomaterials, from metals, metal oxides, ceramics, quantum dots, magnetic nanoparticles, etc. The method has now been extended to the synthesis of other types or architectures, such as core-shell, multishell, hollow spheres and nanowires, in addition to the traditional small globular particles. Although for about 25 years only w/o microemulsions were used for the synthesis of inorganic nanoparticles, in the last five years the use of o/w and bicontinuous microemulsions has also been developed, and their usefulness for the synthesis of a variety of nanomaterials has been demonstrated. These developments are greener than the traditional w/o microemulsion method, so it should contribute to an advance towards the industrial use of microemulsions for nanoparticle synthesis. Furthermore, there have been efforts towards boosting the metal loading in microemulsions, in order to increase their production capacity. The investigations on novel approaches for nanoparticle recovery should also be taken into account by more research groups for the improvement of nanoparticle quality and dispersability in different media. The new developments reviewed here should encourage the preparation of novel materials with different architectures, in order to respond quickly to the demands of Nanotechnology and Materials Science. It is hoped that this chapter is useful to students and researchers who start exploring the microemulsion reaction method for nanoparticle synthesis, as well as for those not new to the field but who are looking for the newest trends in this fascinating technique.

6. Acknowledgements

The authors acknowledge financial support by Ministerio Ciencia e Innovación (MICINN Spain, grant number CTQ2008-01979) and Generalitat de Catalunya (Agaur, grant number 2009SGR-961).

7. References

Abecassis B., Testard F., & Zemb T. (2009) Gold nanoparticle synthesis in worm-like catanionic micelles: Microstructure conservation and temperature induced recovery. *Soft Matter*, 5, 5, (March 2009), pp. (974-978).

Angelescu D. G., Magno L. M., & Stubenrauch C. (2010) Monte Carlo simulation of the size and composition of bimetallic nanoparticles synthesized in water in oil microemulsions. *J. Phys. Chem. C*, 114, 50, (Dec. 2010), pp. (22069-22078).

Aramaki K., Hayashi T., Katsuragi T., Ishitobi M., & Kunieda H. (2001) Effect of Adding an Amphiphilic Solubilization Improver, Sucrose Distearate, on the Solubilization Capacity of Nonionic Microemulsions, *J. Colloid Interface Sci.*, 236, 1, (April 2001), pp. (14-19).

Asim N., Radiman S., & bin Yarmo M. A. (2007) Synthesis of WO3 in nanoscale with the usage of sucrose ester microemulsion and CTAB micelle solution. *Mater. Lett.*, 61, 13, (May 2007), pp. (2652-2657).

Aubery C., Solans C., & Sanchez-Dominguez, M. (2011) Tuning high aqueous phase uptake in nonionic w/o microemulsions for the synthesis of Mn-Zn ferrite nanoparticles: phase behavior, characterization and nanoparticle synthesis. *Langmuir*, 27, 23, (Oct. 2011), pp. (14005–14013).

Bagwe R. P., & Khilar K. C. (1997) Effects of the Intermicellar Exchange Rate and Cations on the Size of Silver Chloride Nanoparticles Formed in Reverse Micelles of AOT. *Langmuir*, 13, 24, (Nov. 1997), pp. (6432-6438).

Bandyopadhyaya R., Kumar R., Gandhi K. S., & Ramkrishna, D. (1997) Modeling of Precipitation in Reverse Micellar Systems, *Langmuir*, 13, 14, (July 1997), pp. (3610-3620).

Borkovec M., Eicke H.-F., Hammerich H., & Das Gupta, B. (1988) Two Percolation Processes in Microemulsions. *J. Phys. Chem.*, 92, 1, (Jan 1988), pp. (206-211).

Boutonnet M., Kizzling J., & Stenius P. (1982) The preparation of monodisperse colloidal metal particles from microemulsions. *Colloids and Surfaces*, 5, 3, (November 1982), pp. (209-225)

Boutonnet M., Lögdberg S., & Svensson E. E. (2008) Recent developments in the application of nanoparticles prepared from w/o microemulsions in heterogeneous catalysis. *Curr. Opin. Colloid Interface Sci.*, 13, 4, (August 2008), pp. (270-286).

Capek, I. (2004) Preparation of metal nanoparticles in water-in-oil (wyo) microemulsions. *Adv. Colloid Interface Sci.* 2004, 110, 1-2, (June 2004), pp. (49-74)

Cason J. P., Miller M. E., Thompson J. B., & Roberts C. B. (2001) Solvent Effects on Copper Nanoparticle Growth Behavior in AOT Reverse Micelle Systems. *J. Phys. Chem. B.*, 105, 12, (March 2001), pp. (2297-2302).

Chen C. C., & Lin J. J. (2001) Controlled Growth of Cubic Cadmium Sulfide Nanoparticles Using Patterned Self-Assembled Monolayers as a Template. *Adv. Mater.*, 13, 2, (Jan, 2001), pp. (136-139).

Chew C. H., Can L. M., & Shah D. O. (1990) The Effect of Alkanes On The Formation Of Ultrafine Silver Bromide Particles In Ionic W/O Microemulsions. *J. Dispersion Sci. Technol.*, 11, 6, (Dec. 1990), pp. (593-609).

Chiu H. K., Chiang I. C., & Chen D. H. (2009) Synthesis of NiAu alloy and core-shell nanoparticles in water-in-oil microemulsions. *J. Nanopart. Res.*, 11, 5, (July 2009), pp. (1137-1144).

Chung S. H., Lee D. W., Kim M. S., & Lee K. Y. (2011) The synthesis of silica and silica-ceria, core-shell nanoparticles in a water-in-oil (W/O) microemulsion composed of heptane and water with the binary surfactants AOT and NP-5. J. Colloid Interface Sci., 355, 1, (March 2011), pp. (70-75).

Clarke S., Fletcher P. D. I., & Ye X. (1990) Langmuir, 6, 7, (July 1990), pp. (1301-1309).

Cross, J. (1987). Nonionic Surfactants: Chemical Analysis, Surfactant Science Series (1st edition), Marcel Decker Inc., ISBN-10: 0824776267, New York, NY, USA.

Curri M. L., Agostiano A., Manna L., Della Monica M., Catalano, M., Chiavarone L., Spagnolo V., & Lugara M. (2000) Synthesis and Characterization of CdS Nanoclusters in a Quaternary Microemulsion: the Role of the Cosurfactant *J. Phys. Chem. B*, 104, 35, (August 2000), pp. (8391-8397).

Cushing B. L., Kolesnichenko V. L., & O'Connor C. J. (2004) Recent Advances in the Liquid-Phase Syntheses of Inorganic Nanoparticles. *Chem. Rev.*, 104, 9, (Nov. 2004), pp. (3893-3946).

Danielsson I., and Lindman B. (1981) The definition of microemulsion. *Colloids and Surfaces*, 3, 4, (December 1981) pp. (391-392).

De Dios M., Barroso F., Tojo C., & Lopez-Quintela, M. A. (2009) Simulation of the kinetics of nanoparticle formation in microemulsions. *J Colloid Interface Sci.*, 333, 2, (May 2009), pp. (741-748).

Destrée C., & Nagy J. B. (2006) Mechanism of formation of inorganic and organic nanoparticles from microemulsions. *Adv. Colloid Interface Sci.*, 123-126, (November 2006), pp. (353-367).

Dong B., Cao L., Su G., Liu W., Qu H., & Jiang D. (2009) Synthesis and characterization of the water-soluble silica-coated ZnS:Mn nanoparticles as fluorescent sensor for Cu^{2+} ions. *J. Colloid Interface Sci.*, 339, 1, (Nov. 2009), pp. (78-82).

Dong H., Liu Y., Ye Z., Zhang W., Wang G., Liu Z., & Yuan J. (2009) Luminescent nanoparticles of silica-encapsulated cadmium-tellurium (CdTe) quantum dots with a core-shell structure: Preparation and characterization. Helvetica Chimica Acta, 92, 11, (Nov. 2009), pp.(2249-2256).

Eastoe J., Hollamby M. J., & Hudson L. (2006) Recent advances in nanoparticle synthesis with reversed micelles, *Adv. Colloid Interface Sci.*, 128-130, (December 2006), pp. (5-15).

Esquivel J., Facundo I. A., Treviño M. E., & López R. G. (2007) A novel method to prepare magnetic nanoparticles: precipitation in bicontinuous microemulsions. *J. Mater. Sci.*, 42, 21, (Nov. 2007), pp. (9015–9020)

Filankembo A., Giorgio S., Lisieki I., & Pileni, M. P. (2003) Is the Anion the Major Parameter in the Shape Control of Nanocrystals? *J. Phys. Chem. B*, 107, 30, (April 2003), pp. (7492-7500).

Fletcher P. D. I., Howe A. M., & Robinson B. H. (1987) J. Chem. Soc. Faraday Trans. 1, The kinetics of solubilisate exchange between water droplets of a water-in-oil microemulsion, 83, 4, (April 1987), pp. (985-1006).

Fletcher P. D. I., & Horsup D. I. (1992) Droplet dynamics in water-in-oil microemulsions and macroemulsions stabilised by non-ionic surfactants. Correlation of measured rates with monolayer bending elasticity. *J. Chem. Soc. Faraday Trans. 1*, 88, 6, (June 1992), pp. (855-864).

Gaertner, G.F., & Lydtin H. (1994) Review of ultrafine particle generation by laser ablation from solid targets in gas flows. *Nanostructured Mater.*, 4, 5, (September 1994), pp. (559-568).

Garcia-Torres J., Vallés E., & Gómez E. (2010) Synthesis and characterization of Co@Ag core-shell nanoparticles. *J. Nanopart. Res.*, 12, 6, (August 2010), pp. (2189-2199).

Gautam U. K., Ghosh M., Rajamathi M., & Seshadri R. (2002) Solvothermal routes to capped oxide and chalcogenide nanoparticles, *Pure Appl. Chem.*, 74, 9, (September 2002), pp. (1643-1649).

Ghows N., & Entezari M. H. (2011) Fast and easy synthesis of core-shell nanocrystal (CdS/TiO2) at low temperature by micro-emulsion under ultrasound. *Ultrason. Sonochem.*, 18, 2, (March 2011), pp. (629-634).

Giannakas A. E., Ladavos A. K., Armatas G. S., Petrakis D. E., & Pomonis P. (2006) Effect of composition on the conductivity of CTAB butanol octane nitrate salts ($Al(NO_3)_3$ + $Zn(NO_3)_2$) microemulsions and on the surface and textural properties of resulting spinels $ZnAl_2O_4$. *J. Appl. Surf. Sci.*, 252, 6, (Jan. 2006), pp. (2159-2170).

Giannakas A. E., Ladavos A. K., Armatas G. S., & Pomonis P. J. (2007) Surface properties, textural features and catalytic performance for NO + CO abatement of spinels MAl_2O_4 (M = Mg, Co and Zn) developed by reverse and bicontinuous microemulsion method. *Appl. Surface Sci.*, 253, 16, (June 2007), pp. (6969-6979).

He H., Liu M., Dai H., Qiu W., & Zi X. (2007) An investigation of NO/CO reaction over perovskite-type oxide $La_{0.8}Ce_{0.2}B_{0.4}Mn_{0.6}O_3$ (B = Cu or Ag) catalysts synthesized by reverse microemulsion. *Catal. Today*, 126, 3-4, (August 2007), pp. (290-295).

Hollamby M. J., Eastoe J., Chemelli A., Glatter O., Rogers S., Heenan R. K., & Grillo, I. (2010) Separation and purification of nanoparticles in a single step. *Langmuir*, 26, 10, (May 2010), pp. (6989-69949.

Holmberg K. (2004) Surfactant-templated nanomaterials synthesis. *J. Colloid Interface Sci.*, 274, 2, (June 2004), pp. (355-364).

Iglesias-Silva E., Vilas-Vilela J. L., López-Quintela M. A., Rivas J., Rodríguez M., & León L. M. (2010) Synthesis of gold-coated iron oxide nanoparticles. *J. Non-Cryst. Solids*, 356, 25-27, (June 2010), pp. (1233-1235).

Jakobs B., Sottman T., Strey R., Allgaier J., Willner L., & Richter D. (1999) Amphiphilic Block Copolymers as Efficiency Boosters for Microemulsions, *Langmuir*, 15, 20, (Sep. 1999), pp. (6707-6711).

Jiang D., Deng Y., Wang H., Shen B., Wu Y., Liu L., Zhong C., & Hu W. (2010) Fabrication of Nickel hollow spheres by microemulsion-template-interface reaction route. *Mater. Lett.* 64, 6, (March 2010), pp. (746-7489).

Jiang D., Hu W., Wang H., Shen B., & Deng Y. (2011) Microemulsion template synthesis of copper sulfide hollow spheres at room temperature. *Colloids Surface A: Physicochem. Eng. Aspects*, 384, 1-3, (July 2011), pp. (228-232).

Jobling S., & Sumpter J. P. (1993) Detergent components in sewage effluent are weakly oestrogenic to fish: An in vitro study using rainbow trout (Oncorhynchus mykiss) hepatocytes. *Aquatic Toxicology.* 27, 3-4, (Dec. 1993), pp. (361-372).

Keswani R. K., Ghodke H., Sarkar D., Khilar K. C., & Srinivasa, R. S. (2010) Room temperature synthesis of titanium dioxide nanoparticles of different phases in water in oil microemulsion. *Colloids Surface A: Physicochem. Eng. Aspects*, 369, 1-3, (Oct. 2010), pp. (75-81).

Košak A., Makovec D., & Drofenik M. (2004) The preparation of MnZn-ferrite nanoparticles in a water/CTAB, 1-butanol/1-hexanol reverse microemulsion, *Phys. Stat. Sol. C*, 1, 12, (Dec. 2004), pp. (3521-3524)

Kruis F. E., Fissan H., & Peled A. (1998) Synthesis of nanoparticles in the gas phase for electronic, optical and magnetic applications—a review. *J. Aerosol Sci.* 29, 5-6, (June 1998), pp. (511-535).

Kunieda H., & Friberg S.E. (1981) Critical Phenomena in a Surfactant/Water/Oil System. Basic Study on the Correlation between Solubilization, Microemulsion, and Ultralow Interfacial Tensions. *Bull. Chem. Soc. Jpn*, 54, 4, (April, 1981), pp (1010-1014).

Kunieda, H.; Nakano, A.; Pes, M. A. (1995) Effect of Oil on the Solubilization in Microemulsion Systems Including Nonionic Surfactant Mixtures, *Langmuir*, 11, 9, (Sep. 1995), pp. (3302-3306).

Kuriyedath S. R., Kostova B., Kevrekidis I. G., & Mountziaris T. J. (2010) Lattice Monte Carlo simulation of cluster coalescence kinetics with application to template-assisted synthesis of quantum dots. *Ind. Eng. Chem. Res.* 49, 21, (Nov. 2010), pp. (10442-104499.

La Mer V., & Dinegar R. (1950) Theory, Production and Mechanism of Monodisperse Hydrosols. *J. Am.Chem. Soc.*, 72, 11, (Nov. 1950), pp. (4847-4854).

Li Y. C., Wang C. P., Hu P. F., & Liu, X. J. (2009) A method for synthesizing the core (Ag)/shell (PSt) composite nanoparticles. *Mater. Lett.*, 63, 20, (August 2009), pp. (1659-1661).

Lindman B., & Friberg S. E.(1999) Chapter 1: Microemulsions: A Historical Overview, In: *Handbook of Microemulsion Science and Technology*, Eds. Kumar, P.; Mittal, K.L., pp. (1-12), Marcel Dekker, Inc., ISBN: 0-8247-1979-4, New York, NY, USA.

Lisiecki I., & Pileni M. P. (2003) Synthesis of Well-Defined and Low Size Distribution Cobalt Nanocrystals: The Limited Influence of Reverse Micelles, *Langmuir*, 19, 22, (Sep. 2003), pp. (9486-9489).

Liu C., Rondinone A. J., & Zhang Z. J. (2000) Synthesis of magnetic spinel ferrite $CoFe_2O_4$ anoparticles from ferric salt and characterization of the size-dependent superparamagnetic properties. *Pure Appl. Chem.*, 72, 1-2, (Jan. 2000), pp. (37-45).

Loo A. L., Pineda M. G., Saade H., Treviño M. E., & Lopez, R.G. (2008) Synthesis of magnetic nanoparticles in bicontinuous microemulsions. Effect of surfactant concentration. *J. Mater. Sci.*, 43, 10, (May 2008), pp. (3649–3654).

López-Quintela M. A. (2003) Synthesis of nanomaterials in microemulsions: formation mechanisms and growth control. *Curr. Opin. Colloid Interface Sci.*, 8, 2 (June 2003), pp. (137-144).

López-Quintela M. A., Tojo C., Blanco M. C., García Rio L., & Leis J. R. (2004) Microemulsion dynamics and reactions in microemulsions. *Curr. Opin. Colloid Interface Sci.*, 9, 3-4 (November 2004), pp. (264-278).

Magno M., Angelescu D. G, & Stubenrauch, C. (2009) Phase diagrams of non-ionic microemulsions containing reducing agents and metal salts as bases for the synthesis of bimetallic nanoparticles, *Colloids Surfaces A: Physicochem. Eng. Aspects*, 348, 1-3, (Sep 2009), pp. (116-123)

Moulik S. P., & Paul B. K.(1998) Structure, dynamics and transport properties of microemulsions, *Adv. Colloid Interface Sci.*, 78, 2, (Sep. 1998), pp. (99-195).

Myakonkaya O., Deniau B., Eastoe J., Rogers S. E., Ghigo A., Hollamby M., Vesperinas A., Sankar M., Taylor S. H., Bartley J. K., & Hutchings G. J. (2010) Recycling nanocatalysts by tuning solvent quality. *J. Colloid Interface Sci.*, 350, 2, (Oct. 2010), pp (443-446).

Myakonkaya O., Eastoe J., Mutch K. J., & Grillo I. (2011) Polymer-induced recovery of nanoparticles from microemulsions. *Phys. Chem. Chem. Phys.*, 13, 8, (Sep. 2010), pp. (3059-3063).

Nann T., & Mulvaney P. (2004) Single Quantum Dots in Spherical Silica Particles. *Angew. Chem. Int. Ed.*, 43, 40, (Oct. 2004), pp. (5393–5396).

Nazar M. F., Myakonkaya O., Shah, S. S., & Eastoe J. (2011) Separating nanoparticles from microemulsions. *J. Colloid Interface Sci.*, 354, 2, (Feb. 2011), pp. (624-629).

Niemann B., & Sundmacher K. (2010) Nanoparticle precipitation in microemulsions: Population balance model and identification of bivariate droplet exchange kernel. *J. Colloid Interface Sci.*, 342, 2, (Feb. 2010), pp. (361-371).

Pileni M. P., Zemb T., & Petit C. (1985) Solubilization by reverse micelles: Solute localization and structure perturbation. *Chem. Phys. Lett.* 118, 4, (August 1985), pp(414–420).

Pileni M. P. (1997) Nanosized Particles Made in Colloidal Assemblies. *Langmuir*, 13, 13, (June 1997), pp. (3266-3276).

Pileni M.P. (1998) Fabrication and Properties of Nanosized Material Made by Using Colloidal Assemblies as Templates, *Cryst. Res. Technol.*, 33, 7-8 (Dec. 1998), pp. (1155-1186)

Pileni M. P. (2001) Mesostructured Fluids in Oil-Rich Regions: Structural and Templating Approaches, *Langmuir*, 17, 24, (Nov. 2001), pp. (7476-7486).

Pileni M. P. (2003) The role of soft colloidal templates in controlling the size and shape of inorganic nanocrystals. *Nat. Mater.*, 2, 3 (March 2003), pp. (145-150).

Qian L. P., Yuan D., Yi G. S., & Chow G. M. (2009) Critical shell thickness and emission enhancement of $NaYF_4$:Yb,Er/$NaYF_4$/silica core/shell/shell nanoparticles. J. Mater. Res., 24, 12, (Jan 2009), pp. (3559-3568).

Qiao, S. Z., Liu J., & Lu G. Q. M. (2011) Chapter 21: Synthetic Chemistry of Nanomaterials, In: *Modern Inorganic Synthetic Chemistry*, Eds. Xu, R.; Pang, W.; Huo, Q., pp. (479-506) Elsevier, ISBN 13: 978-0-444-53599-3, Amsterdam, The Netherlands.

Reyes P. Y., Espinoza J. A., Treviño M. E., Saade H., & López R. G. (2010) Synthesis of Silver Nanoparticles by Precipitation in Bicontinuous Microemulsions. J. *Nanomaterials*, Volume 2010, Article ID 948941, 7 pages. doi:10.1155/2010/948941

Sanchez-Dominguez M., Boutonnet M., & Solans C. (2009) A novel approach to metal and metal oxide nanoparticle synthesis: the oil-in-water microemulsion reaction method. J. *Nanopart. Res.*, 11, 7, (Oct. 2009), pp. 1823-1829

Sanchez-Dominguez M., Liotta L. F., Di Carlo G., Pantaleo G., Venezia A. M., Solans C., & Boutonnet M. (2010) Synthesis of CeO_2, ZrO_2, $Ce_{0.5}Zr_{0.5}O_2$, and TiO_2 nanoparticles by a novel oil-in-water microemulsion reaction method and their use as catalyst support for CO oxidation. *Catal. Today*, 158, 1-2, (Dec. 2010), pp. (35-43)

Schmidt J., Guesdon C., & Schomäcker R. (1999). Engineering aspects of preparation of nanocrystalline particles in microemulsions. J. *Nanopart.Res.*, 1, 2, (June 1999), pp.(267-276).

Shinoda K. (1968) Proceedings of 5th Int. Congr. Detergency, Barcelona, Spain, September 1968, p. 275.

Sosa Y. D., Rabelero M., Treviño M. E., Saade H., & López R. G. (2010) High-Yield Synthesis of Silver Nanoparticles by Precipitation in a High-Aqueous Phase Content Reverse Microemulsion, J. *Nanomaterials*, Volume 2010, Article ID 392572, 6 pages doi:10.1155/2010/392572

Stubenrauch C., Wielpütz T., Sottmann T., Roychowdhury C., & Di Salvo F. J. (2008) Microemulsions as templates for the synthesis of metallic nanoparticles. *Colloids Surfaces A: Physicochem. Eng. Aspects*, 317, 1-3, (March 2008), pp. (328–338)

Suslick K. S., Hyeon T., & Fang M. (1996) Nanostructured Materials Generated by High-Intensity Ultrasound: Sonochemical Synthesis and Catalytic Studies. *Chem. Mater.* 8, 8, (August 1996), pp. (2172-2179).

Takenaka S., Umebayashi H., Tanabe E., Matsune H., & Kishida, M. (2007) Specific performance of silica-coated Ni catalysts for the partial oxidation of methane to synthesis gas. J. Catal. 245, 2, (Jan. 2007), pp. (392-400).

Tiseanu C., Parvulescu V. I., Boutonnet M., Cojocaru B., Primus P. A., Teodorescu C. M., Solans C., & Sanchez-Dominguez M. (2011) Surface versus volume effects in luminescent ceria nanocrystals synthesized by an oil-in-water microemulsion method. Phys. Chem. Chem. Phys., 13, 38, (August 2011), pp. (17135–17145)

Tojo C., De Dios M., & López-Quintela M. A. (2009) On the structure of bimetallic nanoparticles synthesized in microemulsions. J. *Phys. Chem. C*, 113, 44, (Nov. 2009), pp. (19145-19154).

Vesperinas A., Eastoe J., Jackson S., & Wyatt P. (2007) Light-induced flocculation of gold nanoparticles. *Chem. Commun.*, 43, 38, (August 2007), pp. (3912-3914).

Viswanadh B., Tikku S., & Khilar K. C. (2007) Modeling core-shell nanoparticle formation using three reactive microemulsions. *Colloids Surfaces A: Physicochem. Eng. Aspects*, 298, 3, (May 2007), pp. (149-157).

Vogt C., Toprak M. S., Muhammed M., Laurent S., Bridot J. L., & Müller R. N. (2010) High quality and tunable silica shell–magnetic core nanoparticles, *J. Nanopart. Res.*, 12, 4, (May 2010), pp. (1137-1147).

Wang G., Deng Y., & Guo L. (2010) Single-crystalline ZnO nanowire bundles: Synthesis, mechanism and their application in dielectric composites. *Chem. Eur. J.*, 16, 33, (Sep. 2010), pp. (10220-10225).

Wang, J., Tsuzuki T., Sun L., & Wang X. (2010) Reverse microemulsion-mediated synthesis of SiO_2-coated ZnO composite nanoparticles: Multiple cores with tunable shell thickness. *ACS Appl. Mater. Interfaces*, 2, 4, (March 2010), pp. (957-960).

Wang Y. H., & Li F. (2007) Synthesis of $BaAl_{12}O_{19}$:Mn^{2+} nanophosphors by a reverse microemulsion method and its photoluminescence properties under VUV excitation. *J. Lumin.*, 122-123, (Jan-April 2007), pp. (866-868).

Wu H., Jiao Q., Zhao Y., Huang S., Li X., Liu H., & Zhou M. (2010) Synthesis of Zn/Co/Fe-layered double hydroxide nanowires with controllable morphology in a water-in-oil microemulsion. *Mater. Charact.*, 61, 2, (Feb. 2010), pp. (227-232).

Yeung C. M. Y., & Tsang S. C. (2009) Noble metal core-ceria shell catalysts for water-gas shift reaction. J. Phys. Chem. C, 113, 15, (April 2009), pp. (6074-6087).

Yeung C. M. Y., & Tsang S. C. (2010) Some optimization in preparing core-shell Pt-ceria catalysts for water gas shift reaction. *J. Mol. Catal. A: Chem.*, 322, 1-2, (May 2010), pp. (17-25).

Xu P., Han X., & Wang, M. (2007) Synthesis and magnetic properties of $BaFe_{12}O_{19}$ hexaferrite nanoparticles by a reverse microemulsion technique. *J. Phys. Chem. C*, 111, 16, (April 2007), pp. (5866-5870).

10

Laser-Combined STM and Related Techniques for the Analysis of Nanoparticles/Clusters

Hidemi Shigekawa[1], Shoji Yoshida[1],
Masamichi Yoshimura[2] and Yutaka Mera[3]
[1]University of Tsukuba
[2]Toyota Technological Institute
[3]University of Tokyo
Japan

1. Introduction

Nanoscale particles and clusters have been attracting considerable attention from researchers and engineers from fundamental and practical viewpoints owing to their high potential for providing an extremely wide range of functional characteristics compared with ordinary solid materials, such as chemical reactivity and electrical, magnetic, optical and mechanical properties [1-5]. In fact, nanoparticles with novel functions have been realized in various fields including catalysis, biology, plasmonics, electronic devices, magnetism and so forth, on the basis of their wide range of properties. The modification of nanoparticle surfaces is producing further advances in the development of functions including those of composite materials.

For the further development of novel functions based on nanoparticles/clusters and to optimize their use, it is essential to understand the physics and chemistry of such materials in relation to their macroscopic functions. However, because nanoparticles/clusters are generally defined as particles with diameters of 1-100 nm (1-10 nm in the field of nanotechnology), conventional analysis techniques are considered to average the information of nanostructures over the ensemble, limiting the understanding of individual characteristics. Furthermore, using conventional methods, analysis of the effect of local structures in each element such as atomic-scale defects, which are considered to determine the overall characteristics of small materials, is difficult. Therefore, the introduction of new methods for the analysis of these highly functional small materials is eagerly awaited.

Scanning tunneling microscopy (STM) is one of the most promising techniques for such purposes. The characteristics of materials can be obtained at the atomic scale not only for their surface but also for their inner structures including the transient dynamics. Furthermore, external perturbation such as by thermal, mechanical or electromagnetic excitation enables advanced measurements. Among the various STM techniques useful for the study of nanoparticles/clusters, STM in combination with optical technologies, which enables probing of the response of local electronic structures to optical treatment, is an interesting approach for considering the future applications of such materials. On the other hand, sample preparation is an important factor in the analysis of nanoscale materials. Tunneling current and bias voltage can be used to modify target materials to obtain a deeper understanding of their characteristic properties. In addition, the STM tip plays an essential role in measurement. A Ag tip, for example, is used to enhance the effect of local excitation, and a carbon nanotube (CNT) tip is an excellent probe for observing fine nanoscale structures.

In this chapter, we review and discuss the STM-based techniques developed in combination with optical technologies and their application to the analysis of nanoscale particles and clusters.

2. Laser-combined STM and related techniques

2.1 Probing methods

In this section, we discuss probing methods of STM combined with optical technologies.

2.1.1 Photoabsorption spectroscopy

Photoabsorption spectroscopy is a major branch of optical spectroscopy used to explore the electronic states of materials. Photoabsorption spectroscopy using STM (STM-PAS) provides a high spatial resolution in STM and high spectral accuracy of optical spectroscopy compared with scanning tunneling spectroscopy (STS). The detected signal in STM-PAS is the current flowing through the STM tip in response to the modulation of spectroscopic light. The simplest STM-PAS scheme is based on a lock-in (LI) technique with the intensity modulation of light while the wavelength is swept. Spatially resolved photoabsorption spectroscopy using STM was first demonstrated with a resolution of ∼50 nm [6]. It was then shown that STM-PAS enables the nanometer-scale imaging of isolated subsurface defects in semiconductors through the absorption spectra associated with the defects [7].

Here we introduce two advanced STM-PAS schemes.

2.1.1.1 Fourier transform STM-PAS

The simplest STM-PAS scheme has two inherent technical problems. The first is the spurious spectra that are often generated by temporal instabilities or positional drift of the STM tip, which cause the tunneling current to change with time because acquisition for LI detection requires a long time to sweep the wavelength of spectroscopic light. The second problem is an undesirable excess component in the photomodulated tip current due to the photothermal expansion of the tip material [8].

Fourier transform STM-PAS (STM-PAS-FT) based on the Fourier transformation technique was devised to solve such technical problems [9]. This scheme is essentially the same as that used in Fourier transform near-infrared spectroscopy (FT-NIR). Multiple lights modulated with different frequencies corresponding to their wavelengths generate a tip current with an interferogram caused by the superposition of current components modulated at the different frequencies. The photoabsorption spectrum is computed from the interferogram via Fourier transformation. In this case, the photothermal expansion is suppressed by the simultaneous illumination of multiplexed lights.

Figure 1(a) shows a schematic diagram of the experimental setup of STM-PAS-FT. Figure 1(b) shows a typical photoabsorption spectrum obtained around the band gap of Si by the STM-PAS-FT scheme (a) and the LI scheme (b). Although it took about 100 min to acquire the spectrum for a single sweep of photon energy from 0.68 to 1.55 eV in the LI scheme, it took only 16 min to obtain a. high quality STM-FT-NIR spectrum in the range of 0.25–1.85 eV (1s for each scan and 1000 scans totally) [9]. Compared with the long acquisition time for the LI spectrum, that for a single STM-FT-NIR spectrum is much shorter (1 s), which enables us to avoid acquiring spurious spectra.

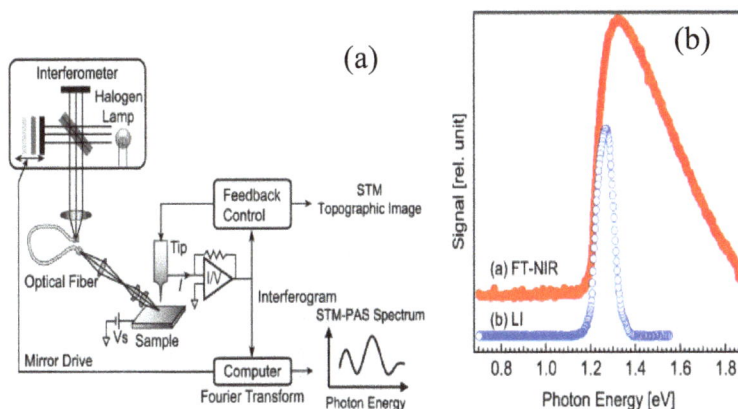

Fig. 1. (a) Experimental setup of Fourier transform near-infrared measurement. (b) STM-FT-NIR spectrum and LI spectrum of a Si substrate obtained at 94 K. The acquisition time was 16 min for FT-NIR measurement (1s for each scan and 1000 scans totally) and 100 min for LI measurement (1 scan). The decrease in intensity above 1.3 eV in the LI spectrum is due to the cutoff filter used in the grating monochromator. [9]*

Figure 2 shows an STM topographic image (a) and a two-dimensional map (b) of Si signals integrated from 1.3 to 1.5 eV in the STM-PAS-FT spectra recorded in the framed area in (a) (8 spectra were measured and averaged at each pixel) [9]. The samples were hemispherical

Fig. 2. (a) STM image of GeSn nanodots/Si obtained at 98 K. (b) Map of STM-PAS-FT signal integrated from 1.3 to 1.5 eV for the spectrum obtained from the framed area in (a). [9]*

Ge$_{1-x}$Sn$_x$ (x=0.1) nanodots epitaxially grown on Si substrates with an ultrathin SiO$_2$ film. The deposition of Ge and Sn was controlled to 24 monolayers to grow nanodots with diameters of ~40 nm. After the samples were annealed at 770 K, the surface was terminated with atomic hydrogen to suppress surface states. The region of bright contrast in (b) matches the region without nanodots in (a) reasonably well, i.e., the expected part of the Si substrates. The contrast indicates the spatial resolution of STM-PAS-FT to be ~10 nm.

Figure 3(a) shows a set of photoabsorption spectra obtained on different GeSn nanodots with various lateral diameters [10]. The peak indicated in each spectrum by an arrow is observed at an energy lower than the gap energy of Si (~1.2 eV) and exhibits a clear blue shift with decreasing dot size, suggesting that the signal is induced by optical transitions between discrete levels in the quantum dots. The photoabsorption energy of a spherical nanodot with radius R is given by

$$E_{abs} = E_{bulk} + \frac{\hbar^2 \pi^2}{2\mu R^2} - 1.786\frac{e^2}{\varepsilon R} - 0.248\frac{e^2}{8\pi\varepsilon a_B^*} \tag{1}$$

Here, $a_B^* \equiv 4\pi\hbar^2\varepsilon / \mu e^2$ represents the exciton Bohr radius, $\hbar$ is Planck's constant, ε is the dielectric constant of the nanodot material, e is the electron charge and μ is the reduced mass of carriers. The first term E_{bulk} is the bandgap energy of the bulk crystal [11]. The solid curve in Fig. 3 shows the theoretical curve calculated from Eq. (1). The energy position of the

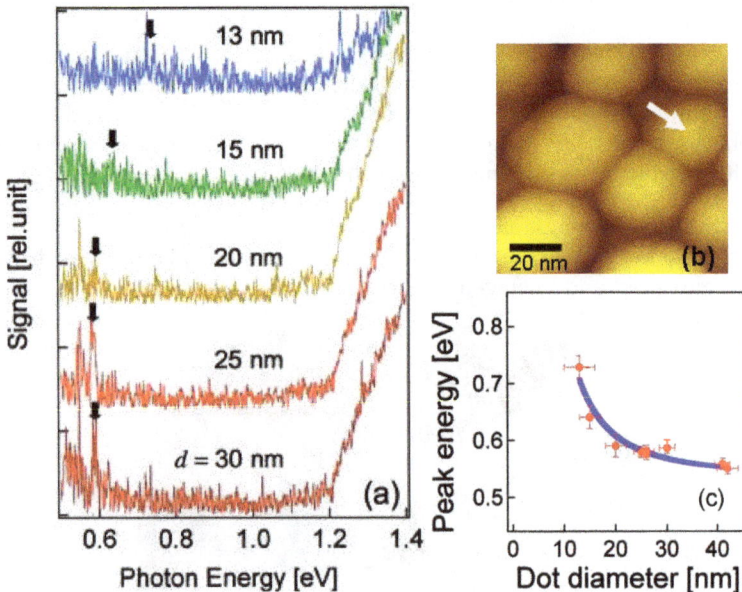

Fig. 3. (a) STM-PAS-FT spectra obtained from individual GeSn nanodots of various sizes. (b) Typical topographic STM image of Ge$_{1-x}$Sn$_x$ nanodots. The white arrow indicates the position where the spectrum for d=30 nm was acquired. (c) Peak energy in (a) as a function of the lateral diameter of the nanodots. The solid curve is calculated from Eq. (1). [10]

peak is in good agreement with the optical transition energy between discrete levels theoretically predicted by the size dependence due to the quantum confinement effect.

2.1.1.2 Electric field modulation spectroscopy

Conventional electric field modulation spectroscopy (EFMS) techniques, such as electroreflectance and photoreflectance, are established tools used for the accurate measurement of interband transition energies in semiconductors [12]. The principle of EFMS is based on the fact that applying an electric field to a semiconducting material causes an oscillatory change in the optical absorption coefficient α depending on the wavelength, i.e., the Franz-Keldysh effect [13]. The spectral line shape of EFMS is closely related to energy derivatives of the unperturbed dielectric function, and represents features corresponding to interband transitions. By combining EMFS with STM (STM-EFMS), we can perform EFMS with nanometer spatial resolution [14].

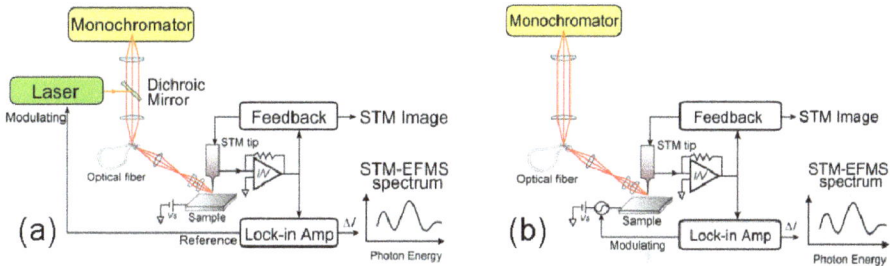

Fig. 4. Schematic illustrations of STM-EFMS measurement: (a) OM scheme, (b) BM scheme.[15]*

Figure 4 illustrates schematics of STM-EFMS measurements using two different schemes for electric field modulation: (a) optical modulation (OM) and (b) bias modulation (BM). In OM, electric field modulation is achieved by an SPV periodically induced by chopped light illumination with energy above the bandgap of the sample from a diode laser. In BM, electric field modulation is achieved by applying a modulated bias voltage to the sample, which directly modulates the tip-induced band bending beneath the tip. In STM-EFMS, the change in α is detected as a change in the STM tip current, which is synchronized with the modulation of the electric field.

Figure 5 shows a typical STM-EFMS spectrum (solid curve) and the spectrum obtained by the conventional EFMS method (dashed curve). The STM-EFMS spectrum reproduces the main features of the band structure. Two distinct structures observed at photon energies approximately $h\nu$=1.41 and 1.78 eV are ascribed to the interband absorption edge of GaAs and the spin split-off band absorption, respectively. It was demonstrated that the spatial resolution of STM-EFMS measurements was of nanometer scale [15, 16]. A typical STM topographic image of a β-FeSi$_2$ nanodot sample is shown in Fig. 6(a). The sample was epitaxially grown on an n-Si(111) substrate covered with an ultrathin SiO$_2$ film. After the growth, the surface was terminated with atomic hydrogen. Figure 6(b) shows STM-EFMS spectra at 96 K obtained by the two schemes in an energy range lower than the absorption edge of Si [15].

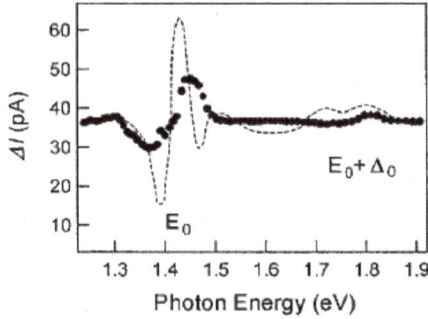

Fig. 5. STM-EFMS spectrum obtained for a perfect (110) cleaved surface of GaAs. The broken curve indicates the EFMS spectrum measured by the conventional electroreflectance method applied to a macroscopic GaAs sample. E_0 and $E_0+\Delta_0$ denote the interband absorption edge of GaAs and to the spin split-off band absorption, respectively. [14]*

For epitaxially grown sufficiently strained β-FeSi$_2$ nanodots on Si, bandgap crossover, i.e., change from indirect band to direct band, is theoretically predicted [18], which, however, has not been confirmed experimentally despite that the mechanism is of great importance for application. The two spectra exhibit a common feature from 0.72 to 0.76 eV. The energy positions of the signals, 0.72–0.76 eV, closely match the absorption thresholds detected by macroscopic measurements of the photoabsorption coefficient for a bulk β-FeSi$_2$ crystal at 100 K [17] and the energy threshold is attributed to optical transitions across the indirect bandgap. Therefore, these findings strongly indicate that the β-FeSi$_2$ nanodot sample examined was an indirect-gap semiconductor, instead of the theoretical prediction.

Using the STM-EFMS scheme, the band structure of individual nanodots can be explored with high accuracy.

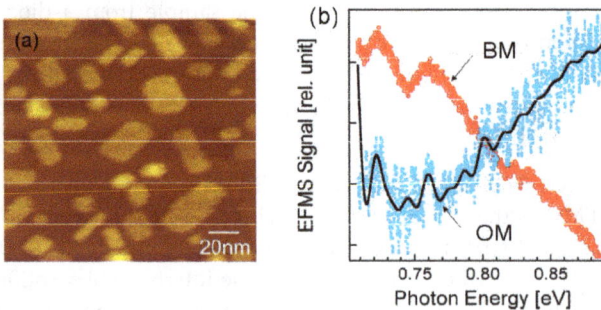

Fig. 6. (a) Typical STM topographic image of β-FeSi$_2$ nanodot sample. Bright contrasts with heights of 5-10 nm are H-terminated β-FeSi$_2$ nanodots grown on Si(111). (b) STM-EFMS spectra measured on β-FeSi$_2$ nanodots by optical modulation (OM) and bias modulation (BM). The common features near 0.72 – 0.76 eV agree well with the absorption thresholds detected by macroscopic measurements of the photoabsorption coefficient in a bulk β-FeSi$_2$ crystal. [15]*

2.1.2 Light-emission spectroscopy

When carriers are injected from an STM tip to a sample, light emission (LE) is induced in some cases. STM luminescence spectroscopy is a measurement scheme in which the emitted light is collected to explore the local electronic properties of materials (Fig. 7) [19-22].

The mechanism of photon emission depends on the process of measurement such as plasmon polariton (SPP) excitation in conductive samples, carrier recombination in semiconductor samples and the HOMO-LUMO transition in molecular samples. Information on molecular vibrations can be obtained by analyzing the spectrum [22], such as by inelastic tunneling spectroscopy [23], which may be used for the analysis of composite nanoparticles/clusters combined with organic materials.

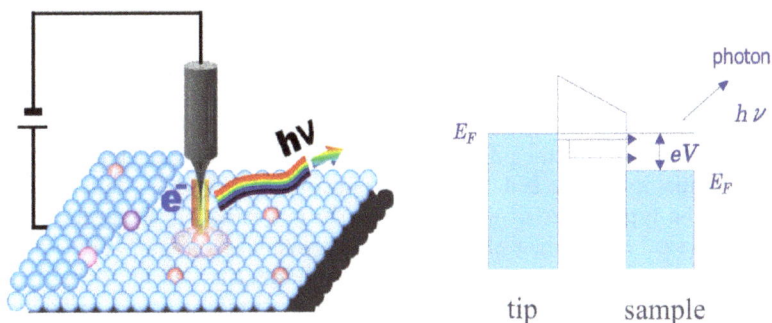

Fig. 7. Schematic illustrations of LE-STM setup and basic mechanism.

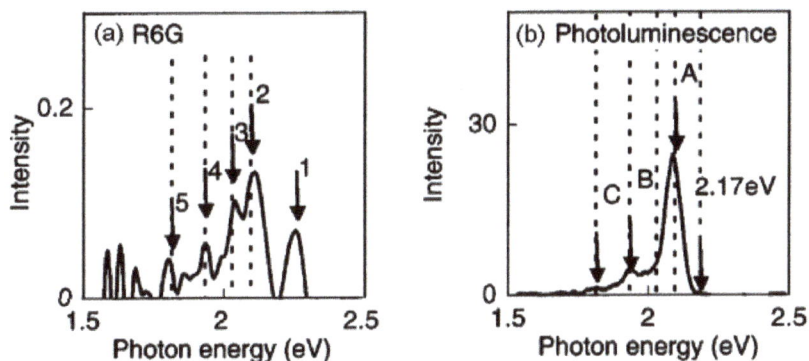

Fig. 8. (a) is the STM light emission spectrum of a single R6G molecule. (b) Photoluminescence (PL) spectrum of R6G on HOPG. The cutoff of the PL spectrum at 2.17 eV is due to the short-wavelength-cutoff filter inserted in the collection optics. [24]*

From the distribution of emission intensity on a sample surface (photon map), we can investigate the geometry of the electronic structures of the sample. The photon map also enables us to estimate the transport properties of minority carriers by considering the

---

* Reprinted with permission from each reference. Copyright American Institute of Physics.

diffusion length. When a local spectrum is analyzed as a photon map, more detailed information such as the distribution of elements can be obtained. However, STM luminescence spectra are affected by various factors other than sample properties, such as the tip shape, tip material and the characteristics of the substrate used for the experiment; thus, careful analysis is necessary to determine the physical properties of the target material from the STM luminescence spectra. In the case of organic materials, damages due to carrier injection must be avoided.

Figure 8(a) shows the STM luminescence spectrum obtained from a single rhodamine 6G (R6G) molecule on an HOPG surface. The features of the spectrum are in good agreement with the photoluminescence spectrum of a layer of rhodamine molecules on HOPG (Fig. 8(b)) [24].

2.1.3 Photoexcitation spectroscopy

Dynamical processes have often been studied by a laser pump-probe method where a pump pulse excites a sample and a subsequently arriving probe pulse with a delay time of t_d is used to track its temporal evolution [25, 26]. The temporal resolution attainable in such experiments is limited only by the pulse width, which is generally in the femtosecond range. However, the spatial resolution is determined by the optical diffraction limit, which is large compared with the typical size of materials and devices currently being developed, and therefore, the physical observables are obscured by ensemble averaging. Thus, high spatial resolution in pump-probe experiments would provide new insights into nanoscale structures and materials and unveil a rich variety of dynamical features of light-sensitive phenomena in unexplored regimes such as charge transfer, phase transitions, electronic transitions, carrier or spin transport and quantum coherence.

In contrast, STM easily provides atomic-scale spatial resolution despite its low temporal resolution (typically worse than 100 kHz) [27-31]. Therefore, if the tunneling process directly produced by optical excitation can be measured, high temporal and spatial sensitivity can be simultaneously achieved with the atomic-scale resolution of STM [32-41]. A promising setup for achieving this is pulse-pair-excited STM (PPX-STM), in which, in analogy with pump-probe experiments, a sequence of paired laser pulses with a certain delay time t_d excites the sample surface beneath the STM tip, and the tunneling current I is measured as a function of t_d. To detect a faint time-resolved tunneling current with a high signal-to-noise ratio, the rectangular modulation of delay time with a pulse-picking procedure is used (shaken-pulse-pair excited STM: SPPX-STM), enabling the spatial mapping of time-resolved tunneling current [33].

Figure 9 shows the setup of SPPX-STM. When paired optical pulses arrive at a sample beneath the STM tip, they generate pulses of raw tunneling current I^*, reflecting the excitation and relaxation of the physical properties of the sample. If these current pulses decay rapidly compared with the time scale of the STM preamplifier bandwidth, they can be temporally averaged in the preamplifier but cannot be detected directly in the signal I. Even in this case, the relaxation dynamics can be probed through the t_d dependence of I. When t_d is sufficiently long, paired optical pulses with the same intensity independently induce two current pulses with the same I^*. In contrast, when t_d is short and the second pulse illuminates the sample in the excited state induced by the first pulse, the second current pulse may have a different magnitude, depending on t_d. A typical process that can be observed using this mechanism is absorption bleaching in semiconductors; when the carriers excited by the first optical pulse remain in the excited state, the absorption of the second optical pulse is suppressed. In such a case, the current I^* induced by the second current pulse decreases depending on t_d, reflecting the decay of the excited carriers excitation by the first-pulse. Signal I also depends on t_d, because the magnitude difference of the second current pulse changes the temporally averaged value of the tunneling current. Accordingly, the relaxation dynamics of the excited carriers of the target material, namely, the decay of carrier density after excitation by the first optical pulse, can be probed by STM at the resolution of the pulse width, that is, in the femtosecond range.

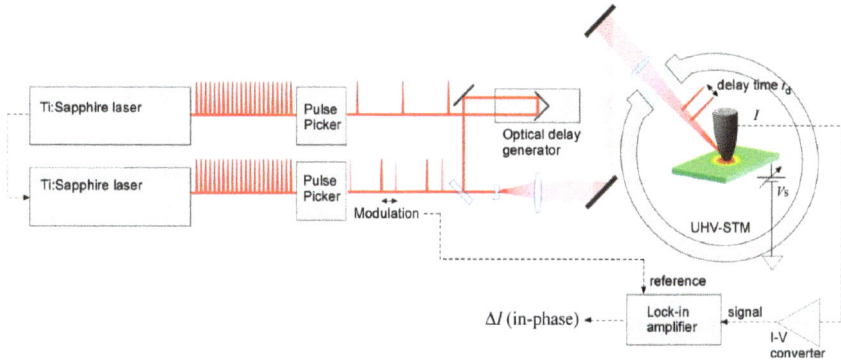

Fig. 9. Schematic illustration of SPPX-STM

In SPPX-STM the sophisticated control of delay-time generation and modulation with a pulse-picking procedure is essential. This enables the probing of nanometer-scale structures with a wide range of relaxation lifetimes. Using the pulse-picking method, a large and discrete modification of t_d can be realized by changing the selection of pulses that transmit the pulse pickers, which is suitable for modulating t_d in SPPX-STM. In this method, the delay time dependence of the tunneling current, $\Delta I(t_d) \equiv I(t_d) - I(\infty)$, is accurately probed with a high acquisition rate, where $I(\infty)$ is the tunneling current for a delay time sufficiently long for the excited state to be relaxed. Accordingly, SPPX-STM has made it possible to visualize the carrier dynamics in nanometer-scale structures with a wide range of relaxation lifetimes. Figure 10 shows the capability of wide timescale measurement.

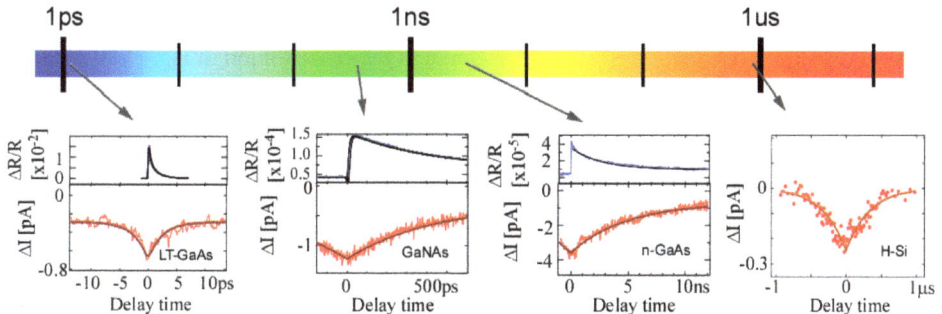

Fig. 10. SPPX-STM signals obtained for various samples. Upper spectra of LT-GaAs, GaNAs and n-GaAs were obtained by optical pump-probe method (R: reflectivity of probe pulse).

As an example, SPPX-STM has been applied to the analysis of carrier dynamics in a Co nanoparticle/GaAs(110) system. When Co is deposited on a GaAs, nanoparticles are formed (Fig. 11(a)). In this system, photoexcited minority carriers (holes) captured at the surface are recombined with electrons tunneling from the STM tip via the gap states formed by Co as shown in Fig. 11(c). This is considered to be enhanced by the existence of gap states at the Co nanoparticle sites. Understanding such a charge transport mechanism through nanoparticles is of great importance not only for the development of nanoscale electronic devices but also for their application to the finer control of chemical reactions in catalysis.

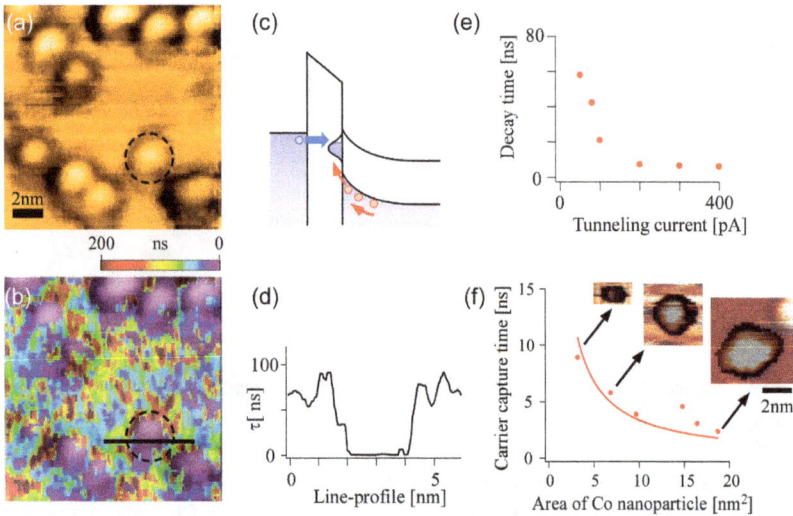

Fig. 11. (a) STM image of Co/GaAs, (b) 2D map of decay constant, (c) schematic model of recombination at gap states, (d) cross section along the line in (b), (e) decay constant as a function of tunneling current, (f) decay constant as a function of Co particle size.

Figure 11(b) shows the overlap of the STM image in Fig. 11(a) with the map of the decay constant obtained over the surface. The two-dimensional (2D) map of the decay constant shown in color scale indicates the decay constant of the photoinduced carrier density at each point. The positional agreement is good. As shown in the cross section in Fig. 11(d) obtained along the line in Fig. 3(b), the decay is rapid in the Co regions. In such regions, photoinduced holes trapped at the surface are recombined with electrons tunneling from the STM tip at the gap states; thus, there are two limitations in this process: the tunneling current and hole-capture rate. When the tunneling current is sufficient, the hole-capture rate becomes the limiting factor of the recombination process. Figure 11(e) shows the relation between the decay constant and tunneling current. As expected, the decay constant decreases with increasing tunneling current and has a saturated value of 6.9 ns, which corresponds to the hole-capture rate of this system. The decay process should depend on the gap-state density. Figure 11(f) shows the decay constant as a function of the Co nanoparticle size. The time constant increases with decreasing nanoparticle size as expected.

In SPPX-STM, the nonlinear interference between the excitations is essential, which depends on the material we measure. In SPPX-STM applied to a semiconductor, tip-induced band bending and surface photovoltage play important roles in the measurement. However, in general, such as dipole formation, charge transfer, changes in conductance, and vibration that causes the change in the tip-sample distance are possible mechanisms for producing SPPX-STM signals. Therefore, SPPX-STM enables the nanoscale probing of transient dynamics over a wide range of time scales, simultaneously with the observation of local structures by STM.

Another promising technique is STM combined with synchrotron radiation (SR-STM), which probes core-level photoemission, enabling the identification of atomic species of the target materials [42]. The spatial resolution has been improved to ~10 nm, and therefore in the near future, in addition to the analysis of isolated nanostructures, probing of the inner structures of targets may become possible.

2.2 Manipulation for fine measurement

The tunneling current and bias voltage in STM, which are the basic parameters of STM measurement, can be used for the modification of target materials. Probing, for example, the effect of atomic-scale defects on local electronic structures enables the clarification of the fundamental mechanism in each element and its relation to macroscopic functions. For nanoparticles/clusters, such effects are essential for determining the characteristic properties of their total systems.

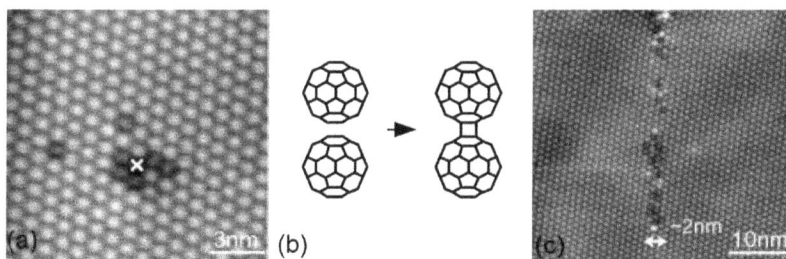

Fig. 12. (a) STM image of current-injection-induced polymerized C_{60} molecules (dark contrasts), (b) schematic of C_{60} polymerization, (c) nanoscale patterning of polymerized C_{60} molecules formed by scanning the tip along the longitudinal direction. [43][†]

Figure 12(a) shows an STM image of a C_{60} crystalline film within a thickness of several monolayers grown on an HOPG surfaces in ultrahigh vacuum (UHV). Before acquiring the image, electrons were injected at the point indicated by a cross in Fig. 12(a) at a sample bias voltage of Vs=+4.2 V. The dark contrasts around the point represent intracluster structures with a stripe pattern, suggesting the frozen rotation of C_{60} molecules despite the room temperature. Namely, the dark sites are C_{60} molecules polymerized with molecules in the underlayer. The polymerization was induced by the injection of low-energy electrons from the STM probe tip. Figure 12(c) shows a line structure consisting of polymerized C_{60} clusters confined in a width as small as ~2 nm, which is a good example of nanoscale electron-beam

---

† Copyright The Japan Society of Applied Physics.

patterning [43]. When a template such as nanoscale cavity is used, individual C_{60} molecules are stabilized in each cavity even at room temperature. Manipulation of a single C_{60} molecule using STM tunneling current was successfully carried out (Fig.13) [44.].

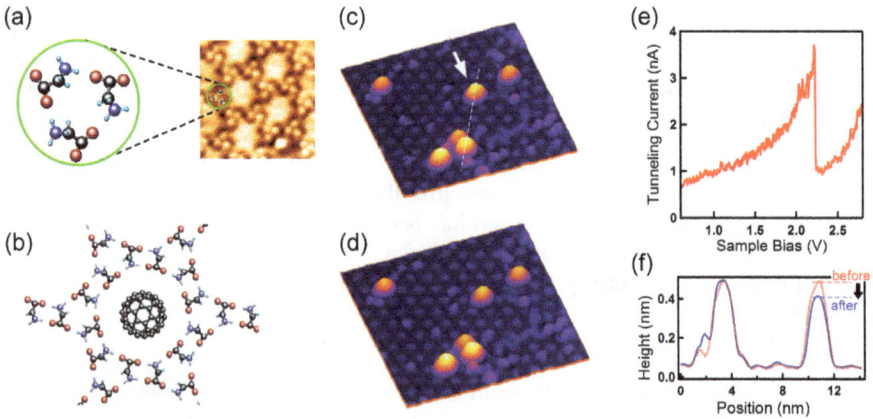

Fig. 13. (a) STM image of glycine-nanocavity array (template). (b) Schematic illustration of C_{60} molecule stabilized in a nanocavity. STM images of C_{60} molecules stabilized by a glycine template before (c) and after (d) the injection of tunnel current on the molecule indicated by arrow. (e) Change in tunnel current upon manipulation. (f) Cross sections along the line in (a).

Figures 14(b) and (c) show topographic STM images of a single-walled carbon nanotube (SWNT). After acquiring the image in Fig. 14(b), the STM tip was fixed at the position marked in Fig. 14(b), and a tunneling current of 0.1 nA with 7.0 V bias voltage was injected. Figure 13(c) shows the defect generated at the probed site. The finite flat LDOS around the Fermi level shown in Fig. 14(c), which was measured before defect generation, indicates that

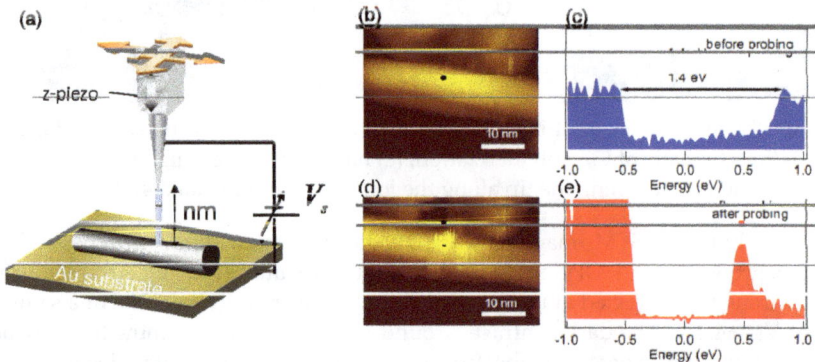

Fig. 14. (a) Schematic illustration of tunnel current injection. (b) STM image of an SWNT acquired with Vs=1.0 V and It=0.1 nA at 95 K. (c) dI/dV vs bias-voltage curve, obtained at the position marked in (b), exhibiting the features of a metallic SWNT characterized by a finite flat LDOS in the first van Hove gap. (d) STM image acquired after current injection at the marked position. (e) dI/dV vs bias-voltage curve, obtained at the position marked in (d), exhibiting a HOMO-LUMO gap of ~0.7 eV. [45]

the tube was initially metallic. Figure 14(e) shows the LDOS measured at the defect, which is characterized by a HOMO-LUMO gap that opened across the Fermi level. The HOMO-LUMO gap was observed to be over 2 nm along the long axis of the CNT and is considered to act as a barrier to carrier transport along the metallic SWNT. This result indicates that we can modify the local electronic properties of a single cluster in a controlled manner using the STM modification technique [45, 46].

Another method of manipulation is mechanical deformation of clusters by an STM tip. For example, the change in the HOMO-LUMO gap of C_{60} molecule due to deformation was observed thorough the measurement of tunneling current under the compression of the molecule by STM tip [47].

Combination of STM manipulations with optical techniques enables further analysis of nanoscale materials.

2.3 Probe technology

For the STM measurement of nanoscale materials, choosing the most suitable STM tip depending on the specific experiment is important. In this section, silver tips for optical measurement, glass-coated tips for photoemission measurements, molecular tip for chemical analysis and CNT probes for high resolution imaging will be described.

2.3.1 Insulator-coated metal tips for SR-STM

STM combined with a synchrotron radiation light source (SR-STM) has attracted considerable attention owing to the possibility of elemental analysis at nanometer resolution by detecting the core-level electrons of surface atoms. The fabrication of a tip coated with an insulating thin film is the key to achieving high spatial resolution by reducing the photoinduced current impinging to the side wall of the tip [48, 49]. For example, a W tip was coated with glass except for the region less than 5 μm from the tip apex using a focused ion beam (FIB) technique [48]. Using this state-of-the-art STM tip, the photoinduced current was dramatically reduced by a factor of ~40 compared with that of an untreated tip. Recently, using this tip in combination with the Lock-in (LI) detection method, a spatial resolution of as high as ~10 nm was demonstrated on checkerboard-patterned Ni and Fe samples [42].

2.3.2 Silver tips for TERS and STML

Tip-enhanced Raman spectroscopy (TERS) [50] is a promising method of chemical analysis at the nanometer level. Under external illumination, a sharp tip is used to create a localized light source and excite a specimen surface. According to classical electromagnetic theory, a sharp metal tip is suitable for enhancing the Raman scattering of nearby molecules. It is known that Ag produces greater enhancement than Au in the visible range because the imaginary part of its permittivity is much smaller. The silver tip is also used in STM-induced luminescence (STML), where STML intensities are enhanced by about one order of magnitude compared with those obtained using tungsten tips [51]. There have been many reports on the fabrication of Ag tips by electrochemical etching with various electrolytes such as a mixture of perchloric acid and ethanol [52]. Using such a tip, single-molecule tip-enhanced Raman spectra from brilliant cresyl blue (BCB) sub-monolayers deposited on a flat

Au surface were obtained [53]. A highly enhanced electric field was created in the gap of 1 nm between the tip and sample. For STM imaging, the tip apex should be free of oxidation or contaminants. Atomically resolved STM imaging and STML spectra with a high signal-to-noise ratio are obtained using an electrochemically etched Ag tip followed by tip cleaning by Ar ion sputtering in UHV [54].

2.3.3 Molecular tips

Carboxyl-terminated SWNTs from solution phases can be attached onto Au tips through self-assembed monolayers for using in STM [55. 56]. In addition to the high-resolution imaging of molecules (such as diether) on a surface, these CNT tips enable chemically selective observation due to electron tunneling through hydrogen-bond interactions between the atached molecule and carboxyl groups at SWNTs. The differentiation of DNA bases and chiral recognition on a single-molecule basis have also been demonstrated using molecular tips [56]. In a similar way, voltage-induced chemical contrast in an STM image was reported using chemically modified tips with hydrogen-bond donors [57]. Moreover, molecular orbitals of metal phthalocyanines on metal surfaces have been clearly imaged with an O_2-functionalized STM tip, where the observations were supported by theoretical calculations [58].

2.3.4 CNT probes

CNTs are one of the most intriguing materials in nanotechnology [59]. A CNT has a one-dimensional cylindrical structure with sistinct physical characteristics such as a small diameter, high aspect ratio, high stiffness, high conductivity and so forth. In view of the shape and electric conductivity required for a high-resolution STM tip, these properties of CNTs make them ideal as a tip material for probing extremely small objects such as nanoclusters. [60]. Mechanical attachment, direct growth and dielectrophoresis are methods employed to fabricate CNT-STM tips. A single CNT tip can be prepared by mechanical attachment, which enables the high-resolution ghost-free imaging of nanoclusters. However, this method is time-consuming, and the other methods (direct growth and dielectrophoresis) are more suitable for the mass production of CNT tips.

(a) Mechanical attachment method

Following the first approach to fabricating CNT probes under an optical microscope [61]. a more sophisticated method was proposed, where the attachment of a CNT onto a probe is performed in two independent precise stages under SEM observation, where beam-deposited amorphous carbon is used as a glue [62]. Although the thus mechanically prepared CNT-STM tips exhibited atomic resolution on Au(111) reconstruction, the cleaning process of CNTs by heating in UHV was necessary for stable observation. [63]. A metal coating method has been proposed for improving the electric conductivity between a CNT and the supporting metal tip [64]. Automotive exhaust catalysts consist of metal nanoclusters supported on metal oxide surfaces. Since catalytic activity can be altered by controlling the size of the nanoclusters because of its strong size dependence, the precise characterization of metal nanoclusters is essential. High-resolution UHV-STM images of size-selected Pt(n) (n=4,7-10,15) clusters deposited on TiO_2(110)-(1x1) surfaces were obtained using a CNT tip (Fig. 15) [65]. Clusters of Pt(7) (Fig.15(b)) and smaller were oriented flat on the surface with a planar structure, and a planar-to-three-dimensional transition was observed at n=8 (Fig.15(c)). Color scale shows the structures of the top layers. Individual Pt atoms are clearly identified, especially for (c) Pt(8) and (d) Pt(9), indicating that the geometry of the clusters is atomically resolved and the details can be analyzed with a CNT tip.

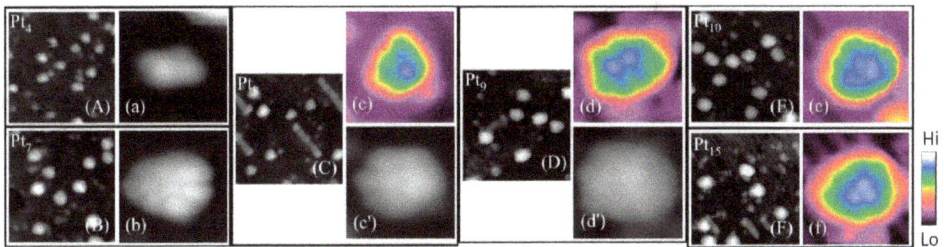

Fig. 15. STM images of $TiO_2(110)$ surface after deposition of size-selected Pt_n^+ (n = 4,7–10,15) cluster ions. Images with uppercase letters are 20×20 nm^2 views and those with lowercase letters are 3.5×3.5 nm^2 views of a cluster on the same surface. TiO_2 surface after the deposition of [(A), (a)] Pt_4^+, [(B)(b)] Pt_7^+, [(C)(c)(c')] Pt_8^+, [(D)(d)(d')] Pt_9^+, [(E)(e)] Pt_{10}^+, [(F)(f)] Pt_{15}^+ [65].

(b) Direct growth method

This method is suitable for the mass production of CNT probes. Chemical vapor deposition (CVD) is commonly used for the synthesis of CNTs [66]. Tips fabricated by the direct growth method sometimes consist of numerous CNTs, and selective growth at the apex is required for stable operation of the tip. For this purpose, several methods of pinpointing catalysis have been reported [67, 68]. The growth direction of CNTs is also important in measurements. Plasma CVD is suitable for controlling the alignment of CNTs [69]. By optimizing the reaction at the sharp apex, CNT probes can be directly grown on the apex of a tungsten probe without reducing its sharpness, as shown in Fig.16 [70]. Thin films of Fe or Co (20-30 nm) are used as a catalyst, and the growth of CNTs with a diameter of ~40 nm has been observed. Because the magnetic nanoparticles are located at the tip of CNTs, this type of probe can also be utilized to study the magnetic properties of nanoclusters with higher spatial resolution [71]. For spin-polarized STM measurement, a magnetic coating of Fe (10-20 monolayers) on a cleaned tungsten tip is conventionally used [72], which may be improved using a CNT tip.

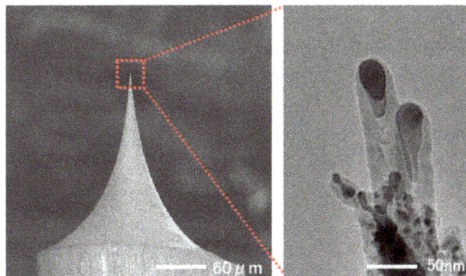

Fig. 16. SEM and TEM images of a grown CNT-STM probe. Two CNTs are grown on the apex of the STM tip. Black contrast corresponds to metal particles used as catalyst.

(c) Dielectrophoresis

When an alternating electric field (~MHz) is applied between asymmetric electrodes (for example, a metal probe and counter plane electrode) immersed in CNT solution (where solvent is water, alcohol, dichloroethane, etc.), the CNTs are polarized and become attached to the probe (dielectrophoresis). [73]. The high-yield synthesis of conductive CNT tips for the multiprobe microscope [74-76] was reported using the dielectrophoresis method [77]. After Pt-Ir coating, such a tip were successfully applied for electronic transport measurement by multiprobe STM using the four-terminal method.

3. Summary

Laser-combined STM and related techniques have been reviewed and discussed focusing on the analysis of nanoscale particles and clusters. The addition of optical technologies to STM provides new approaches to the study of nanoscale-material physics and chemistry. Near-field optical microscopy (NSOM) and other techniques [78-86], which have not been discussed in this chapter, are expected to play complementary roles in understanding and developing the physics and chemistry of new nanoparticles/clusters for realizing novel functional devices.

4. References

[1] M. Haruta, Catal. Today 36, 153 (1997).
[2] P. Jena and A. W. Castleman (Eds), Nanoclusters, A Bridge across Disciplines, Elsevier (2011).
[3] J. A. Alonso, Structure and Properties of Atomic Nanoclusters, Imperial College Press (2006).
[4] J. P. Liu, E. Fullerton, O. Gutfleish and D. J. Sellmyer (Eds), Nanoscale Magnetic Materials and Applications, Springer (2009).
[5] F. J. Owens and C. P. Poole (Eds), The Physics and Chemistry of Nanosolids, Wiley (2008).
[6] J. M. R. Weaver, L. M. Wapita, and H. K. Wickramasinghe, Nature 342 (1989) 783.
[7] A. Hida, Y. Mera, and K. Maeda, Appl. Phys. Lett. 78 (2001) 3190.
[8] S. Grafstrom, P. Schuller, J. Kowalski, and R. Neumann, J. Appl. Phys. 83 (1998) 3453.
[9] N. Naruse, Y. Mera, Y. Fukuzawa, Y. Nakamura, M. Ichikawa, and K. Maeda, J. Appl. Phys., 102 (2007) 114301.
[10] N. Naruse, Y. Mera, Y. Nakamura, M. Ichikawa, and K. Maeda, Appl. Phys. Lett. 94 (2009) 093104.
[11] Y. Kayanuma, Phys. Rev. B 44 (1991) 13085.
[12] D. E. Aspnes, in Handbook on Semiconductors, edited by M. Balkanski (North-Holland, Amsterdam, 1980) Vol.2, p. 109.
[13] D. E. Aspnes, Phys. Rev. 147 (1969) 554.
[14] A. Hida, Y. Mera, and K. Maeda, Appl. Phys. Lett. 78 (2001) 3029.
[15] N. Naruse, Y. Mera, Y. Nakamura, M. Ichikawa, and K. Maeda, J. Appl. Phys. 104 (2008) 074321.
[16] A. Hida, Y. Mera, and K. Maeda, Physica B 308-310 (2001) 1145.
[17] H. Udono et al., Thin Solid Films 461 (2004) 182.
[18] L. Miglio, V. Meregalli, and O. Jepsen, Appl. Phys. Lett. 75 (1999) 385.
[19] C. Thirstrup, M. Sakurai K. Stokbro, and M. Aono, Phys. Rev. Lett. 82 (1999) 1241.

[20] C. Chen, C. A. Bobisch, and W. Ho, Science 325 (2009) 981.
[21] A. Okada, K. Kanazawa, K. Hayashi, N. Okawa, T. Kurita, O. Takeuchi and H. Shigekawa Appl. Phys. Exp. 3 (2010) 015201.
[22] C. Chen, P. Chu, C. A. Bobisch, D. L. Mills, and W. Ho, Phys. Rev. Lett. 105 (2010) 217402.
[23] Y. Sainoo, Y. Kim, T. Okawa, T. Komeda, H. Shigekawa, and M. Kawai, Phys. Rev. Lett. 95 (2005) 246102.
[24] Y. Uehara and S. Ushioda, Appl. Phys. Lett. 86 (2005) 181905.
[25] A. Othonos, J. Appl. Phys. 83 (1998) 1789 and references therein.
[26] J. Shah, Ultrafast Spectroscopy of Semiconductors and Semiconductor Nanostructures (Berlin: Springer 1999).
[27] H. Mamin, H. Birk, P. Wimmer and D. Rugar, J. Appl. Phys. 75 (1994) 161.
[28] J. Wintterlin, J. Trost, S. Renisch, R. Schuster, T. Zambelli and G. Ertl, Surf. Sci. 394 (1997) 159.
[29] M. Rost et al., Rev. Sci. Instrum. 76 (2005) 053710.
[30] L. Petersen et al., Rev. Sci. Instrum. 72 (2001) 1438.
[31] U. Kemiktarak, T. Ndukum, K. Schwab, K. CandEkinci, Nature 450 (2007) 85.
[32] Y. Terada, S. Yoshida, O. Takeuchi and H. Shigekawa, J. Phys. Condens. Matter 22 (2010) 264008
[33] Y. Terada, S. Yoshida, O. Takeuchi and H. Shigeakawa: Nature photonics, 4 (2010) 869.
[34] H. Shigekawa, S. Yoshida, O. Takeuchi, M. Aoyama, Y. Terada, H. Kondo and H Oigawa, Thin Solid Films, 516 (2008) 2348.
[35] Y. Terada, M. Aoyama, H. Kondo, A. Taninaka, O. Takeuchi and H. Shigekawa, Nanotechnology 18 (2007) 044028.
[36] Y. Terada, S. Yoshida, O. Takeuchi and H. Shigekawa, Advances in Optical Technologies 2011 (2011) 510186.
[37] S. Yoshida, Y. Terada, R. Oshima, O. Takeuchi and H. Shigekawa, Nanoscale 2012 (2012), DOI:10.1039/C2NR11551D.
[38] O. Takeuchi, M. Aoyama, R. Oshima, Y. Okada, H. Oigawa, N. Sano, H. Shigekawa, R. Morita and M. Yamashita, Appl. Phys. Lett. 85 (2004) 3268.
[39] O. Takeuchi, R. Morita, M. Yamashita and H. Shigekawa: Jpn. J. Appl. Phys. 41 (2002) 4994.
[40] H. Shigekawa, O. Takeuchi, Y. Terada and S. Yoshida: Handbook of Nanophysics, Edited by Klaus D. Sattler Taylor & Francis (2010), vol. 6, Principles and Methods.
[41] M. Yamashita, H. Shigekawa and R. Morita (Eds), Mono-Cycle Photonics and Optical Scanning Tunneling Microscopy-Route to Femtosecond Angstrom Technology- (Springer, 2005).
[42] T. Okuda et al., Phys. Rev. Lett. 102 (2009) 105503.
[43] Y. Nakamura, Y. Mera and K. Maeda, Jpn. J. Appl. Phys. 44 (2005) L1373.
[44] K. Kanazawa, A. Taninaka, H. Huang, N Nishimura, S. Yoshida, O. Takeuchi, and H. Shigekawa, Chem. Commun. 47 (2011) 11312.
[45] K. Yamada, H. Sato, T. Komaguchi, Y. Mera, and K. Maeda, Appl. Phys. Lett. 94 (2009) 253103.
[46] M.Berthe, S.Yoshida, Y.Ebine, K.Kanazawa, A.Okada, A.Taninaka, O.Takeuchi, N. Fukui, H. Shinohara, S.Suzuki, K.Sumitomo, Y. Kobayashi, B.Grandidier, D.Stievenard and H.Shigekawa, Nano Lett. 7 (12) (2007) 3623-3627.
[47] C. Joachim, J. K. Gimzewski, R. R. Schlittler, and C. Chavy, Phys. Rev. Lett., 74 (1995) 2102.
[48] K. Akiyama et al., Rev. Sci. Instrum. 76 (2005) 083711.

[49] A. Saito et al., Surf. Sci. 601 (2007) 5294.
[50] R. M. Stockle, Y. D. Suh, V. Deckert and R. Zenobi, Chem. Phys. Lett. 318 (2000) 131.
[51] R. Berndt, J. K. Gimzewski, and P. Johansson, Phys. Rev. Lett. 71 (1993) 3493.
[52] M. Iwami, Y. Uehara and S. Ushioda, Rev. Sci. Instrum. 69 (1998) 4010.
[53] W. H. Zhang, B. S. Yeo, T. Schmid, R. Zenobi, J. Phys. Chem. C111 (2007) 1733.
[54] C. Zhang et al., Rev. Sci. Instrum. 82 (2011) 083101.
[55] T. Nishino, T. Ito and Y. Umezawa, Anal. Chem. 74 (2002) 4275.
[56] T. Nishino and Y. Umezawa, Anal. Sci. 26 (2010) 1023.
[57] D. Gingery and P. Bühlmann, Surf. Sci. 605 (2011) 1099.
[58] Z. Cheng et al., Nano Res. 4 (2011) 523.
[59] S. Iijima, Nature, 354, 56 (1991).
[60] J. M. Marulanda, 2011, *Electronic Properties of Carbon Nanotubes*, InTech.
[61] H. Dai, J. H. Hafner, A. G. Rinzler, D. T. Colbert, and R. E. Smalley, Nature 384 (1996) 147.
[62] S. Akita et al., J. Phys D: Appl. Phys. 32 (1999) 1044.
[63] W. Mizutani, N. Choi, T. Uchihashi and H. Tokumoto, Jpn. J. Appl. Phys. 40 (2001) 4328.
[64] T.Ikuno et al., Jpn. J. Appl. Phys. 43 (2004) L644.
[65] N. Isomura, X. Wu, and Y. Watanabe, J. Chem Phys. 131(16) (2009) 164707.
[66] W. Wongwiriyapan et al., Jpn. J. Appl. Phys. 45 (2006) 1880.
[67] C. L. Cheung, J. H. Hafner, and C. M. Lieber, PNAS 97 (2000) 3809.
[68] I. T. Clark, G. Rius, Y. Matsuoka, and M. Yoshimura, J. Vac. Sci. Technol. B28 (2010) 1148.
[69] M. Yoshimura, S. Jo and K. Ueda, Jpn. J. Appl. Phys. 42 (7B) (2003) 4841.
[70] K. Tanaka, M. Yoshimura, and K. Ueda, e-J. Surf. Sci. Nanotech. 4 (2006) 276.
[71] K. Tanaka, M. Yoshimura and K. Ueda, J. Nanomaterials 2009 (2009) 147204.
[72] J. E. Bickel, et al., Phys. Rev. B84 (2011) 054454.
[73] K. Ueda, M. Yoshimura, M. Ishikawa, T. Nagamura, Japan Patent: 3557589
[74] I. Shiraki, F. Tanabe, R. Hobara, T. Nagao, and S. Hasegawa, Surf. Sci. 493 (2001) 633.
[75] J. Onoe , T. Nakayama, M. Aono, and T. Hara, Appl. Phys. Lett. 82 (2003) 595.
[76] M. Ishikawa, M. Yoshimura and K. Ueda, Jpn. J. Appl. Phys. 44 (2005) 1502.
[77] H. Konishi, Y. Murata, W. Wongwiriyapan, M. Kishida, K. Tomita, K. Motoyoshi, S. Honda, M. Katayama, S. Yoshimoto, K. Kubo, R. Hobara, I. Matsuda, S. Hasegawa and M. Yoshimura, Rev. Sci. Instr. 78 (2007) 013703.
[78] Y. Terada, S. Yoshida, A. Okubo, K. Kanazawa, M. Xu, O. Takeuchi and H. Shigekawa, Nano Lett. 8 (11), (2008) 3577-3581.
[79] S. Yoshida, Y. Kanitani, O. Takeuchi and H. Shigekawa, Appl. Phys. Lett. 92 (2008) 102105.
[80] S. Yoshida, Y. Kanitani, R. Oshima, Y. Okada, O. Takeuchi and H. Shigekawa, Phys. Rev. Lett. 98 (2007) 026802.
[81] A. Hagen et al., *Phys. Rev. Lett.* 95 (2005) 197401.
[82] H. Watanabe, Y. Ishida, N. Hayazawa, Y. Inouye and S. Kawata, Phys. Rev. B 69 (1004) 1.
[83] S. Yasuda, T. Nakamura, M. Matsumoto and H. Shigekawa, J. Am. Chem. Soc. 125 (2003) 16430.
[84] D. Futaba, R. Morita, M. Yamashita, S. Tomiyama and H. Shigekawa, Appl. Phys. Lett. 83 (2003) 2333.
[85] S. Kawata, M. Ohtsu, M. Irie (Eds), Nano-Optics, Springer (2002).
[86] S. Grafstorm (o: umlaut), J. Appl. Phys. 91 (2002) 1717.
[87] S. Loth, M. Etzkorn, C. Lutz, D. Eigler and A. Heinrich, Science 24 (2010) 1628.

11

Phase Separations in Mixtures of a Nanoparticle and a Liquid Crystal

Akihiko Matsuyama

Department of Bioscience and Bioinformatics, Kyusyu Institute of Technology
Japan

1. Introduction

Liquid crystal suspensions including various micro- and nano-colloidal particles have recently been received great attention for many practical applications such as nanosensors and devices, etc. When large colloidal particles of micronscale are dispersed in a uniform nematic liquid crystal phase, the colloidal particles disturb a long-range orientational order of the nematic phase. For a strong anchoring between the colloidal surface and a liquid crystal, different defect structures such as hedgehogs or Saturn rings can appear around a single colloidal particle, due to strong director deformations.(Fukuda, 2009; Skarabot et.al., 2008; Stark, 2001) Experiments have also shown two-dimensional crystalline structures of colloidal particles.(Loudet et. al., 2004; Musevic et. al., 2006; Nazarenko et. al., 2001; Pouling et.al., 1997; Yada et. al., 2004; Zapotocky et.al., 1999) On the other hand, under a weak surface anchoring between the colloidal surface and a liquid crystal, the coupling to the orientational elasticity of the liquid crystals tends to expel the colloidal particles and the suspension shows a phase separation into an almost pure nematic phase coexisting with a colloidal rich phase.(Anderson et.al., 2001; Pouling et. al., 1994) Such phase separations induced by a nematic ordering have also been discussed in flexible polymers dispersed in a nematic liquid crystal.(Chiu & Kyu, 1999; Das & Ray, 2005; Dubaut et.al., 1980; Matsuyama & Kato, 1996; Shen & Kyu, 1995)

If the colloidal particles are ~1-10nm in diameter, these "nanoparticles" are too small to distort the nematic director and defects do not form. In this case, the system can show a homogeneous single phase or phase separations,(Anderson et.al., 2001; Anderson & Terentjev, 2001; Caggioni et. al., 2005; Meeker et. al., 2000; Yamamoto & Tanaka, 2001) depending on the interaction between a colloidal particle and a liquid crystal. Although the theoretical progress on the description of a director around colloidal particles with strong anchoring conditions has been noticeable,(Araki & Tanaka, 2004; Fukuda & Yokoyama, 2005; Kuksenok et.al., 1996; Lubensky et. al., 1998; Yamamoto, 2001) little theoretical work exists in phase separations.(Popa-Nita et. al., 2006; Pouling et. al., 1994)

In this chapter, we focus on nanoparticles dispersed in liquid crystals and discuss phase separations and phase behaviors in mixtures of a nanoparticle and a liquid crystal. It is mainly based on authors' original theoretical works obtained within recent years. The nanoparticles have a variety in the shape such as spherical and rodlike. In this chapter, we focus on (1) mixtures of a liquid crystal and a spherical nanoparticle and (2) mixtures of a liquid crystal and a rodlike nanoparticle, such as carbon nanotube. The topics are currently interested in the advanced fields of nanoparticles and fundamental sciences.

When the nanoparticles are dispersed in isotropic solvents, the system may show phase separations, or sodification, between a liquid and a crystalline phase, depending on temperature and concentration, etc. These phase separations are induced by a balance between steric repulsions and attractive dispersion forces. However, the nature of phase separations of nanoparticles dispersed in liquid crystalline solvents is quite difference. The key point is ordering of nanoparticles induced by liquid crystalline ordering. Depending on the interaction between nanoparticles and liquid crystals, we have a variety of phase separations.

The aim of this chapter is to introduce such a new kind of phase separations. We review recent mean field theories to describe phase separations (or phase diagrams) in mixtures of a nanoparticle and a liquid crystal and summarize the variety of phase separations in such nanoparticle dispersions, where liquid crystalline ordering (nematic and smectic A phases) and nanoparticle ordering compete. In Section 2, We discuss spherical nanoparticles dispersed in liquid crystals. Nanotubes dispersed in liquid crystals are discussed in Section 3. The effects of external forces, such as magnetic and electric fields, on the phase behaviors are also discussed in Section 3.

2. Spherical nanoparticles dispersed in liquid crystals

2.1 Ferroelectric nanoparticles dispersed in nematic liquid crystals

Small nanoparticles do not significantly perturb the nematic director. However, it has been recently discovered that ferroelectric nanoparticles can greatly enhance the physical properties of nematic liquid crystals. Recent experimental(Copic et.al., 2007; Li et. al., 2006) and theoretical(Kralj et.al., 2008; Lopatina & Selinger, 2009) studies have shown that low concentrations of ferroelectric nanoparticles ($BaTiO_3$) increase the orientational order of a liquid crystal and increase the nematic-isotropic transition temperature, due to the coupling between the ferroelectric nanoparticle with electric dipole moment and the orientational order of liquid crystals.(Lopatina & Selinger, 2009)

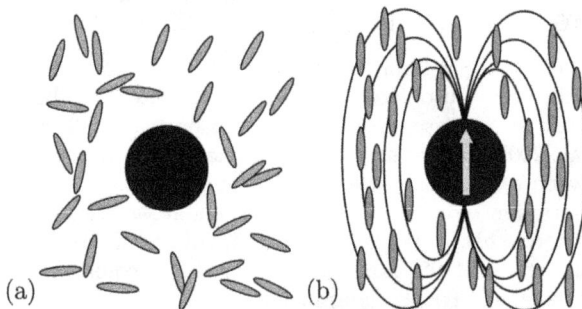

Fig. 1. Nanoparticles dispersed in liquid crystal. (a) Nanoparticle with no electric dipole moment, in an isotropic phase. (b) Ferroelectric particle with dipole moment, which produces an electric field that interacts with orientational order of the nematic phase. Reproduced with permission from (Lopatina & Selinger, 2009) . Copyright 2009 American Physical Society.

As shown in Fig. 1, the orientational distribution of the nanoparticle dipole moment interacts with the orientational order of liquid crystals and stabilizes the nematic phase. The electric field generated by the nanoparticle interacts with the order parameter of the liquid crystal through the free energy

$$F_{int} = -\frac{\Delta\epsilon\rho_{NP}p^2}{180\pi\epsilon_0\epsilon^2 R^3}S_{LC}S_{NP},$$ (1)

where $\Delta\epsilon$ is the dielectric anisotropy of the aligned liquid crystal, ρ_{NP} is the number density of nanoparticles, and p is the electric dipole moment, S_{LC} (S_{NP}) is the scalar orientational order parameters of the liquid crystals (nanoparticles). This free energy can predict the enhancement in the isotropic-nematic transition temperature and in the response to an applied electric field. The attractive interaction between the liquid crystal and the nanoparticle through the order parameters is important to understand the phase behaviors. The next section we consider the free energy to describe the phase separations.

2.2 Phase ordering in mixtures of a spherical nanoparticle and a liquid crystal

We consider mixtures of a spherical nanoparticle and a liquid crystal. By taking into account the ordering of liquid crystals and nanoparticles, we can expect six possible phases in this mixture. Figure 2 shows the schematically illustrated six phases. The isotropic (I) phase means both liquid crystals and nanoparticles have no positional and orientational order. In the nematic (N) phase, liquid crystals have an orientational order, while nanoparticles have no positional order. Similarly, in the smectic A (A) phase, liquid crystals have a smectic A order, while nanoparticles have no positional order. When the concentration of nanoparticles is high, we may have a crystalline (C) phase of nanoparticles dispersed in an isotopic matrix of liquid crystals. We can also expect a nematic-crystal (NC) phase and a smectic A-crystal (AC) phase, where nanoparticles form a crystalline structure dispersed in a nematic and a smectic A matrix of liquid crystals. To describe these phases, depending on temperature and concentration, we take into account three scalar order parameters: an orientational order parameter for a nematic phase, one-dimensional translational order parameter for a smectic A phase, and a translational order parameter for a crystalline phase of nanoparticles.

2.3 Free energy of mixtures of a spherical nanoparticle and a liquid crystal

We consider a binary mixture of N_c spherical nano-colloidal particles of the diameter R_c and N_r low-molecular weight liquid crystal molecules (liquid crystals) of the length l and the diameter d. The volume of the liquid crystal and that of the nanoparticle are given by $v_r = (\pi/4)d^2l$ and $v_c = (\pi/6)R_c^3$, respectively. Let $\rho_r(\mathbf{u}, \mathbf{r})$ and $\rho_c(\mathbf{r})$ be the number density of liquid crystals and colloidal particles with an orientation $\mathbf{u}$ (or its solid angle Ω) at a position $\mathbf{r}$, respectively. The free energy F of the dispersion at the level of second virial approximation is given by(Matsuyama & Hirashima, 2008a; Matsuyama, 2010a)

$$\begin{aligned}
\beta F/V = &\int \rho_r(\mathbf{r}, \mathbf{u})\left[\beta\mu_r^\circ + \ln\rho_r(\mathbf{r}, \mathbf{u}) - 1\right]d\mathbf{r}d\Omega \\
&+ \int \rho_c(\mathbf{r})\left[\beta\mu_c^\circ + \ln\rho_c(\mathbf{r}) - 1\right]d\mathbf{r} \\
&+ \frac{1}{2}\iint \rho_r(\mathbf{r}, \mathbf{u})\rho_r(\mathbf{r}', \mathbf{u}')\beta_{rr}(\mathbf{r}, \mathbf{u}; \mathbf{r}', \mathbf{u}')d\mathbf{r}d\Omega d\mathbf{r}'d\Omega', \\
&+ \frac{1}{2}\iint \rho_c(\mathbf{r})\rho_c(\mathbf{r}')\beta_{cc}(\mathbf{r}; \mathbf{r}')d\mathbf{r}d\mathbf{r}', \\
&+ \iint \rho_c(\mathbf{r})\rho_r(\mathbf{r}', \mathbf{u}')\beta_{cr}(\mathbf{r}; \mathbf{r}', \mathbf{u}')d\mathbf{r}d\mathbf{r}'d\Omega',
\end{aligned}$$ (2)

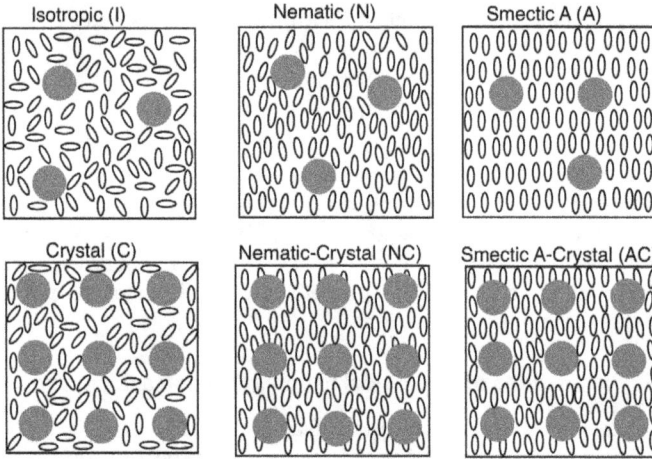

Fig. 2. Ordering of nanoparticles dispersed in liquid crystals. We here take into account three scalar order parameters: an orientational order parameter for a nematic phase, one-dimensional translational order parameter for a smectic A phase, and a translational order parameter for a crystalline phase of nanoparticles.

where $d\Omega$ is the solid angle, μ_i° is the standard chemical potential of a particle $i, j (= r, c)$, $\beta \equiv 1/k_B T$; T is the absolute temperature, k_B is the Boltzmann constant, $\beta_{ij} \equiv 1 - \exp[-\beta u_{ij}]$ is the Mayer-Mayer function, and u_{ij} is the interaction energy between two particles i and j.

Let $f_r(\mathbf{r}, \mathbf{u})$ be the distribution function of liquid crystals and then the density can be expressed as

$$\rho_r(\mathbf{r}, \mathbf{u}) = c_r f_r(\mathbf{r}, \mathbf{u}), \tag{3}$$

where $c_r \equiv N_r/V$ is the average number density of liquid crystals. We here consider a nematic and a smectic A phase of liquid crystals and use the decoupled approximation(Kventsel et.al., 1985) for the distribution function:

$$f_r(\mathbf{r}, \mathbf{u}) = f_r(\tilde{z}_r) f_r(\mathbf{u}), \tag{4}$$

where $\tilde{z}_r \equiv z/l$, l is the average distance between smectic layers, $f_r(\tilde{z}_r)$ is the translational distribution function of liquid crystals for a smectic A phase, and $f_r(\mathbf{u})$ is the orientational distribution function of liquid crystals for a nematic phase. Similarly, using the translational distribution function $f_c(\mathbf{r})$ of nanoparticles, the density of nanoparticles can be expressed as

$$\rho_c(\mathbf{r}) = c_c f_c(\mathbf{r}), \tag{5}$$

where $c_c \equiv N_c/V$ is the average density of nanoparticles. The total number N_r of liquid crystals and N_c of nanoparticles must be conserved and then we have the normalization conditions:

$$\iint \rho_r(\mathbf{r}, \mathbf{u}) d\mathbf{r} d\Omega = N_r/V, \tag{6}$$

and

$$\int \rho_c(\mathbf{r}) d\mathbf{r} = N_c/V. \tag{7}$$

The orientational order parameter S of a nematic phase is given by(Maier & Saupe, 1958)

$$S = \int P_2(\cos\theta) f_r(\theta) d\Omega, \tag{8}$$

where $P_2(\cos\theta) \equiv 3(\cos^2\theta - 1/3)/2$. The translational order parameter σ_s of a smectic A phase is given by(McMillan, 1971)

$$\sigma_s = \int_0^1 \cos(2\pi\tilde{z}_r) f_r(\tilde{z}_r) d\tilde{z}_r. \tag{9}$$

In the McMillan's model,(McMillan, 1971) the order parameter for the smectic A phase is given by $\langle P_2(\cos(\theta)) \cos(2\pi\tilde{z}_r) \rangle$. In Eq. (4), we have used the decoupled approximation: $\langle P_2(\cos(\theta)) \cos(2\pi\tilde{z}_r) \rangle = S\sigma_s$. It has been reported that the decoupled model for the smectic A phase is in quantitative agreement with the original McMillan's theory.(Kventsel et.al., 1985) In the decoupled model, the smectic A phase is defined by $S \neq 0$ and $\sigma_s > 0$.

For a crystalline phase, we here consider a face-centered cubic (fcc) structure of nanoparticles for example. The translational order parameter for a fcc crystalline phase can be calculated by(Kirkwood & Monroe, 1941)

$$\sigma_f = \int_0^1\int_0^1\int_0^1 \cos(2\pi\tilde{x}) \cos(2\pi\tilde{y}) \cos(2\pi\tilde{z}) f_c(\tilde{\mathbf{r}}) d\tilde{\mathbf{r}}, \tag{10}$$

where L is the lattice size of a fcc crystal and we define $\tilde{x} \equiv x/L$, $\tilde{y} \equiv y/L$, $\tilde{z} \equiv z/L$, and $d\tilde{\mathbf{r}} \equiv d\tilde{x}d\tilde{y}d\tilde{z}$. It is possible to consider the other crystalline structure such as a body-centered cubic and a simple cubic, etc.(Matsuyama, 2006a;b)

When the interaction between liquid crystals is a short-range attractive interaction, the anisotropic part of the interaction can be given by Fourier components of the potential:(McMillan, 1971)

$$\beta_{rr} \simeq -(v_r l/d)\nu S(1 + \gamma\sigma_s \cos(2\pi z/l)) P_2(\cos\theta) \tag{11}$$

where we have retained the lowest order of the Fourier components. The $\nu(\equiv U_a/k_B T)$ is the orientational dependent (Maier-Saupe) interaction parameter between liquid crystals(Maier & Saupe, 1958) and the γ shows the dimensionless interaction of a smectic phase(Matsuyama & Kato, 1998; McMillan, 1971). According to the McMillan theory, the parameter γ is given by $\gamma = 2\exp[-(r_0/l)^2]$, which can vary between 0 and 2, and increases with increasing the chain length of alkyl end-chains of a liquid crystal. The smectic condensation is more favored for larger values of γ. For the anisotropic interaction between nanoparticles in a fcc crystalline phase, the anisotropic part of the interaction can be given by expanding β_{cc} at the lowest order of the Fourier components:(Kirkwood & Monroe, 1941)

$$\beta_{cc} \simeq -(v_c R_c/d)g\sigma_f \cos(2\pi\tilde{x}) \cos(2\pi\tilde{y}) \cos(2\pi\tilde{z}), \tag{12}$$

where the coefficient β_{cc} is proportional to the total surface area $(v_c R_c)$ of two particles. The parameter $g(\equiv -\beta f_0)$ is the dimensionless interaction parameter between nanoparticles, where the interaction energy f_0 consists of an entropic and enthalpic terms. In this paper, we only consider short-range interactions between particles. The long-range interaction, due to the presence of surface charges, is not taken into account. Similarly to Eq. (11), the anisotropic interaction between a nanoparticle and a liquid crystal in a nematic and a smecticA phase is given by

$$\beta_{cr} \simeq \frac{1}{2}(v_c l/R_c)\omega S\left(1 + \omega_1 \sigma_s \cos(2\pi z/l)\right) P_2(\cos\theta), \tag{13}$$

where β_{cr} is proportional to the surface area (v_c/R_c) of a nanoparticle(Stark, 1999). The $\omega \equiv w_0/k_B T$ shows the dimensionless interaction parameter between a liquid crystal and a particle surface. When $\omega > 0$, or repulsive interaction between a liquid crystal and a nanoparticle, doping nanoparticles disturb the orientational ordering of liquid crystals, or the orientational elasticity of the liquid crystals tends to expel the particles to be lower the elastic free energy of a nematic phase(Pouling et. al., 1994). In the mean field level, the elastic distortion cost of a director is taken into account in the order of ωS^2. The negative values of $\omega(< 0)$ indicate the attractive interactions between a liquid crystal and a nanoparticle and the particles tend to disperse into a liquid crystalline matrix as indicated in Fig. 1. The last term $\omega_1(> 0)$ is the coupling between a smectic liquid crystal and a colloidal surface.

We here assume the system is incompressible. Let $\phi_r = v_r N_r/V$ and $\phi_c = v_c N_c/V$ be the volume fraction of a liquid crystal and a nano-colloidal particle, respectively. Using the axial ratio $n_r(\equiv l/d)$ of a liquid crystal and $n_c \equiv R_c/d$, the volume of a particle is given by $v_r = a^3 n_r$ and $v_r \simeq (an_c)^3$, where we define $a^3 \equiv (\pi/4)d^3$. To describe phase behaviors of the incompressible blends, we calculate the free energy of mixing for the binary mixtures of a liquid crystal and a nanoparticle:

$$\Delta F = F(N_c, N_r) - F(N_c, 0) - F(0, N_r), \tag{14}$$

where the $F(N_c, 0)$ and $F(0, N_r)$ are the reference free energy of the pure nanoparticles and the pure liquid crystal in an isotropic phase, respectively. Substituting Eqs. (11)-(13) into (14), the mixing free energy is given by

$$\Delta F = F_{mix} + F_c + F_{nem} + F_{sm} + F_{anc}, \tag{15}$$

where the each term is given as following.

The first term in Eq. (15) shows the free energy for mixing of colloids and liquid crystals in the isotropic phase:

$$a^3 \beta F_{mix}/V = \frac{\phi_r}{n_r} \ln \phi_p + \frac{\phi_c}{n_c^3} \ln \phi_c + \chi \eta_c \phi_r, \tag{16}$$

where the first and the second terms in Eq. (16) correspond to the entropy of isotropic mixing for liquid crystals and colloidal particles, respectively. We here have added the third term which shows the isotropic interaction parameter $\chi \equiv U_0/k_B T$ related to the dispersion force between a colloidal particle and a liquid crystal, where U_0 is the interaction energy between a colloid and a liquid crystal in an isotropic state. A positive χ denotes that the colloid-liquid crystal contacts are less favored compared with the colloid-colloid and liquid crystal-liquid crystal contacts. This interaction parameter is well known as the Flory-Huggins parameter in polymer solutions.(Flory, 1953) For a colloidal particle, its surface only can interact with the surrounding solvents and so the probability for the colloid-liquid crystal contact is proportional to $\eta_c \phi_r$, where the η_c is the surface fraction of colloidal particles and is given by

$$\eta_c \equiv \frac{a^2 n_c^3 N_c}{a^2(n_r N_r + n_c^3 N_c)} = \frac{\phi_c}{n_c}. \tag{17}$$

Then the dispersion interaction due to the mixing is given by $\chi \eta_c \phi_r$. On increasing the diameter n_c of a colloid, the interaction term decreases with a fixed ϕ_c. Eq. (16) corresponds to the extended Flory-Huggins free energy for the isotropic mixtures of a liquid crystal, whose the number of segments is n_r, and a colloidal particle, whose the number of segments is n_c^3. The second term in Eq. (15) shows the free energy for a crystalline ordering of colloidal particles:

$$a^3 \beta F_c / V = \frac{\phi_c}{n_c^3} \int_0^1 \int_0^1 \int_0^1 f(\tilde{r}) \ln f(\tilde{r}) d\tilde{r}$$

$$-\frac{1}{2} g \eta_c^2 \sigma_f^2, \tag{18}$$

where the first term in Eq. (18) shows the entropy loss due to the crystalline ordering. When the colloidal particles have no positional order, we have the distribution function $f(\tilde{r}) = 1$ and the free energy (F_c) becomes zero. The third term in Eq. (15) shows the free energy for nematic ordering of liquid crystals:

$$a^3 \beta F_{nem} / V = \frac{\phi_r}{n_r} \int f_r(\theta) \ln 4\pi f_r(\theta) d\Omega$$

$$-\frac{1}{2} v \phi_r^2 S^2, \tag{19}$$

where the first term in Eq. (19) shows the entropy change due to the nematic ordering. The forth term in Eq. (15) shows the free energy for smectic A ordering of liquid crystals:

$$a^3 \beta F_{sm} / V = \frac{\phi_r}{n_r} \int_0^1 f_r(\tilde{z}_r) \ln f_r(\tilde{z}_r) d\tilde{z}_r$$

$$-\frac{1}{2} v \gamma \phi_r^2 (S \sigma_s)^2. \tag{20}$$

The last term in Eq. (15) shows the anchoring interaction between a colloidal surface and a liquid crystal:

$$a^3 \beta F_{anc} / V = \frac{\omega}{2} \eta_c \phi_r \left(S^2 + \omega_1 (S \sigma_s)^2 \right). \tag{21}$$

In a thermal equilibrium state, the distribution functions of nanoparticles and liquid crystals are determined by minimizing the free energy (15) with respect to these functions: $(\delta F / \delta f_c(\tilde{r}))_{\{f_r(\theta), f_r(\tilde{z}_r)\}} = 0$, $(\delta F / \delta f_r(\theta))_{\{f_c(\tilde{r}), f_r(\tilde{z}_r)\}} = 0$, and $(\delta F / \delta f_z(\tilde{z}_r))_{\{f_c(\tilde{r}), f_r(\theta)\}} = 0$. The order parameters S, σ_s, and σ_f can be determined by Eqs. (8), (9), and (10), respectively. Using these distribution functions and order parameters, we can calculate the free energy of our systems. The chemical potentials of a nanoparticle and a liquid crystal can be obtained from this free energy.

2.4 Phase transitions in nanoparticle/liquid crystal mixtures

In a mixture of a nanoparticle and a liquid crystal, we have some phase transitions, depending on temperature and concentration(Matsuyama, 2009).

One is the nematic-isotropic phase transition of this mixture. The nematic-isotropic transition (NIT) temperature is given by

$$\tau_{NI} = T/T_{NI}^{\circ} = 1 - (1 + \alpha_a/n_c)\phi_c, \tag{22}$$

where T_{NI}° shows the NIT temperature of a pure liquid crystal. We here define the ratio α_a between the anchoring strength (w) and the nematic interaction (v):

$$\alpha_a \equiv w/v. \tag{23}$$

The value of α_a shows the anchoring strength. The negative sign represents attractive interaction between a nanoparticle and a liquid crystal and thus the nanoparticles tend to disperse in a liquid crystalline matrix. On the other hand, the positive sign represents the repulsive interaction and the liquid crystals tend to expel the nanoparticles. The slope of the NIT line on the $T - \phi_c$ plane depends on the value of α_a/n_c.

The smectic A$-$nematic phase transition (ANT) is given by

$$\tau_{AN} = T/T_{NI}^{\circ} = 2.27\gamma\left[1 - (1 + \frac{\omega_1\alpha_a}{n_c\gamma})\phi_c\right]S^2. \tag{24}$$

Since $\gamma > 0$ and $\omega_1 > 0$, the ANT temperature depends on the sign of α_a. For larger negative values of α_a, the ANT temperature increases with increasing ϕ_c. It can also be obtained the direct phase transition from an isotropic to a smectic A phase(Matsuyama, 2009).

We also have the isotropic fluid-crystal phase transition (ICT). The ICT temperature is given by

$$\tau_{IC} = T/T_{NI}^{\circ} = \frac{0.58\alpha_c n_c\phi_c}{n_r(1 - \phi_c/\phi_c^*)}, \tag{25}$$

where

$$\alpha_c \equiv \beta|e_0|/v, \tag{26}$$

shows the strength of the attractive interaction between nanoparticles compared to the nematic interaction parameter v. When $\tau < \tau_{IC}$ the crystalline phase is stable. The ICT temperature increases with increasing ϕ_c and diverges at ϕ_c^*. This corresponds to the entropically driven-liquid-solid transition for hard spherical particles due to the excluded volume interactions.(Alder & Wainwright, 1957; Cates & Evans, 2000)

Figure 3(a) shows the first-order phase transition lines for NIT (red dotted-line, Eq. (22)), ANT (blue dotted-line, Eq. (24)), and ICT (black dotted-line, Eq. (25)) on the reduced temperature (T/T_{NI}°)−concentration (ϕ_c) plane. We set $n_c = 3$, $n_r = 2$, α_c=0.1, $\alpha_a = -2.5$, $\alpha_n(\equiv v/\chi)$=5, $\gamma = 0.87$, $\omega_1 = 1$ for a typical example. When $\phi_c = 0$, or pure liquid crystals, the ANT appears at $T/T_{NI}^{\circ} \approx 0.938$, which is consistent with the result of the MacMillan theory. At high temperatures and low concentrations, we have the isotropic (I) liquid phase. On decreasing temperature, the N phase appears, where nanoparticles are in an isotropic liquid state but liquid crystals are in a nematic state. Further decreasing temperature the smectic A phase

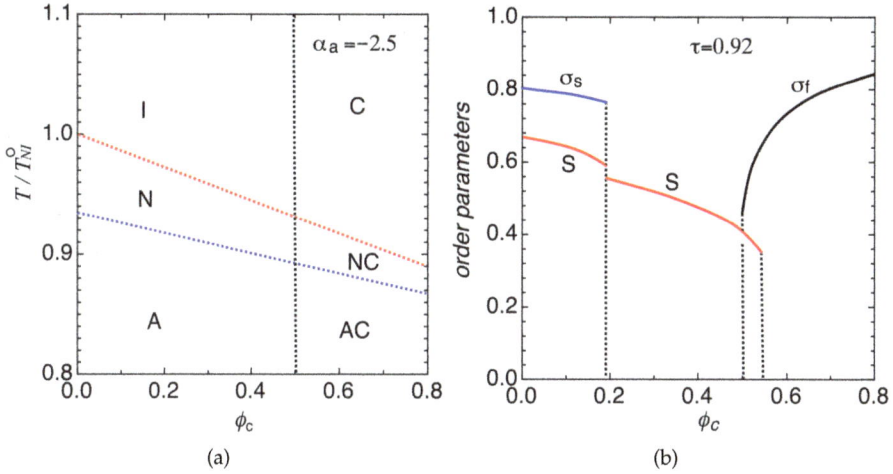

(a)   (b)

Fig. 3. (a) The first-order phase transition lines (the red dotted-line for NIT (Eq. (22)), the blue dotted-line for ANT (Eq. (24)), and the black dotted-line for ICT (Eq. (25))) on the reduced temperature (T/T°_{NI})−concentration (ϕ_c) plane. (b) Order parameters plotted against the volume fraction of colloidal particles at T/T°_{NI}=0.92 in Fig. 3(a).

appears, where nanoparticles are in an isotropic liquid state but liquid crystals are in a smectic A phase. At high temperatures and high concentrations, we have the crystalline (C) phase of colloidal particles. On decreasing temperature, the NC phase appears, where colloidal particles are in a crystalline state and liquid crystals are in a nematic state. Further decreasing temperature the AC phase appears, where colloidal particles are in a crystalline state and liquid crystals are in a smectic A phase. The slope of the transition lines depends on the anchoring energy (α_a) as discussed in Eq. (23). For larger negative values of α_a, the slopes of the NIT and ANT lines become positive on the temperature-concentration plane and the nematic and smectic A phase appear at higher temperatures.

Figure 3(b) shows order parameters plotted against the volume fraction ϕ_c at $T/T^\circ_{NI} = 0.92$ in Fig. 3(a). On increasing ϕ_c, we find the first-order phase transition from the smectic A to nematic (N) phase at $\phi_c \simeq 0.2$, where the order parameters S and σ_s jump. At $\phi_c \simeq 0.5$, the first-order phase transition from the N to NC phase appears. Further increasing ϕ_c the first-order phase transition from the NC to C phase appears at $\phi_c \simeq 0.55$.

2.5 Phase diagrams of nanoparticle/liquid crystal mixtures

In this subsection we show some phase diagrams calculated from the free energy(15). The coexistence curve (binodal) can be obtained by solving the two-phase coexistence conditions: the chemical potentials of each component are equal in each phase. This binodal curve can also be derived by a double tangent method where the equilibrium volume fractions fall on the same tangent line to the free energy curve.

Figure 4 shows the phase diagrams for $n_r = 2$, $n_c = 3$, $\omega_1 = 1$, $\alpha_c = 0.1$, $\alpha_n = 5$, and $\gamma = 0.87$ for an example. The value of α_a is changed: (a) $\alpha_a = 1$; (b) $\alpha_a = -2$; (c) $\alpha_a = -3.5$.

We here discuss the effects of the anchoring strength α_a on the phase behavior. The negative values of α_a mean that the nanoparticles prefer to disperse into liquid crystalline phases. The solid curve shows the binodal curve. The red, blue, and black dotted lines show the NIT, ANT, and ICT line, respectively (see Fig. 3(a)). When $\phi_c = 0$, the smectic A phase appears at $T/T_{NI}^\circ = 0.94$. When $\alpha_a = 1$ [Fig. 4(a)], we have the broad nematic-isotropic $(N + I)$ phase separation between $1 > T/T_{NI}^\circ > 0.94$. Below $T/T_{NI}^\circ < 0.94$, we have the smectic A-isotropic $(A + I)$ phase separation. The nematic and smectic A phase at the lower concentrations consist of almost pure liquid crystals. The triple point $(A + I + C)$ appears at $T/T_{NI}^\circ = 0.89$, where the smectic A, isotropic, and crystalline phases can simultaneously coexist. Below the triple point, we have the two-phase coexistence $(A + C)$ between a smectic A and a crystalline phase. Above the triple point, two-phase coexistence $(I + C)$ between an isotropic and a crystalline phase appears.

On decreasing the anchoring parameter α_a the phase behavior is drastically changed. When $\alpha_a = -2$ [Fig. 4(b)], the NIT (Eq. (22)) and ANT (Eq. (24)) lines shift to higher concentrations and the stable single N and A phases appear at low concentrations of nanoparticles. Two tie lines with arrows show the three-phase coexistence: $A + N + I$ and $A + I + C$. Above the triple point $A + N + I$, we have two-phase coexistence $A + N$ and $N + I$. Below the triple point $A + N + I$, we have A+I phase separation. Below the triple point $A + I + C$, we have the broad A+C phase separation.

Further decreasing α_a, Fig. 4(c), the nematic and smectic A ordering are promoted by adding nanoparticles and the NIT and ANT lines shift to higher temperatures. This increase of the NIT and ANT temperature indicates the attractive interactions between a liquid crystal and a colloidal particle. For example, it has been observed that doping low concentrations of ferroelectric $BaTiO_3$ nanoparticles into liquid crystals increases NIT temperature(Li et. al., 2006a). In this case, ferroelectric nanoparticle with electric dipole moment, which produces an electric field, interacts with orientational order of liquid crystals and stabilizes the nematic phase.(Lopatina & Selinger, 2009) This corresponds to negative anchoring energy in our model. We also have three triple points: $I + C + NC$, $N + NC + AC$, and $N + A + AC$. Above the $I + C + NC$ triple point, we have the $I + C$ and $C + NC$ phase separations. Below the $I + C + NC$ triple point, the $I + NC$ and $NC + AC$ phase separations appear. Below the triple point $N + NC + AC$, we have the $I + N$ and $N + AC$ phase separations. Below the triple point $N + A + AC$ we have $N + A$ and $A + AC$ phase separations. The anchoring energy between liquid crystals and nanoparticles becomes an important parameter to derive a stable N, A, NC, and AC phases in the mixture of nanoparticles and liquid crystals.

Anderson et al. have observed the phase ordering of colloidal (PMMA) particles dispersed in a liquid crystal, 5CB or MBBA.(Anderson et.al., 2001) Particles are covered with chemically grafted short chains, making hairy particles. In a nematic phase, the grafted chains tend to provide a homeotropic (radial) director anchoring. In an isotropic liquid, these particles behave like almost hard spheres and so the $I + C$ phase separation takes place at high concentrations of the colloidal particles. Such $I + C$ phase separation, calculated in Fig. 4, has been observed in colloidal dispersions(Pusey & van Megen, 1986) and protein solutions(Tanaka et. al, 2020). At dilute concentrations of the colloidal particles, Anderson et al. observed a decrease in the NIT temperature T_{NI} as a function of ϕ_c, which follows a linear law. This is consistent with Eq. (22). The N+I and N+C phase separations have also been reported in Latex polyballs suspended in an isotropic micellar solution which exhibits

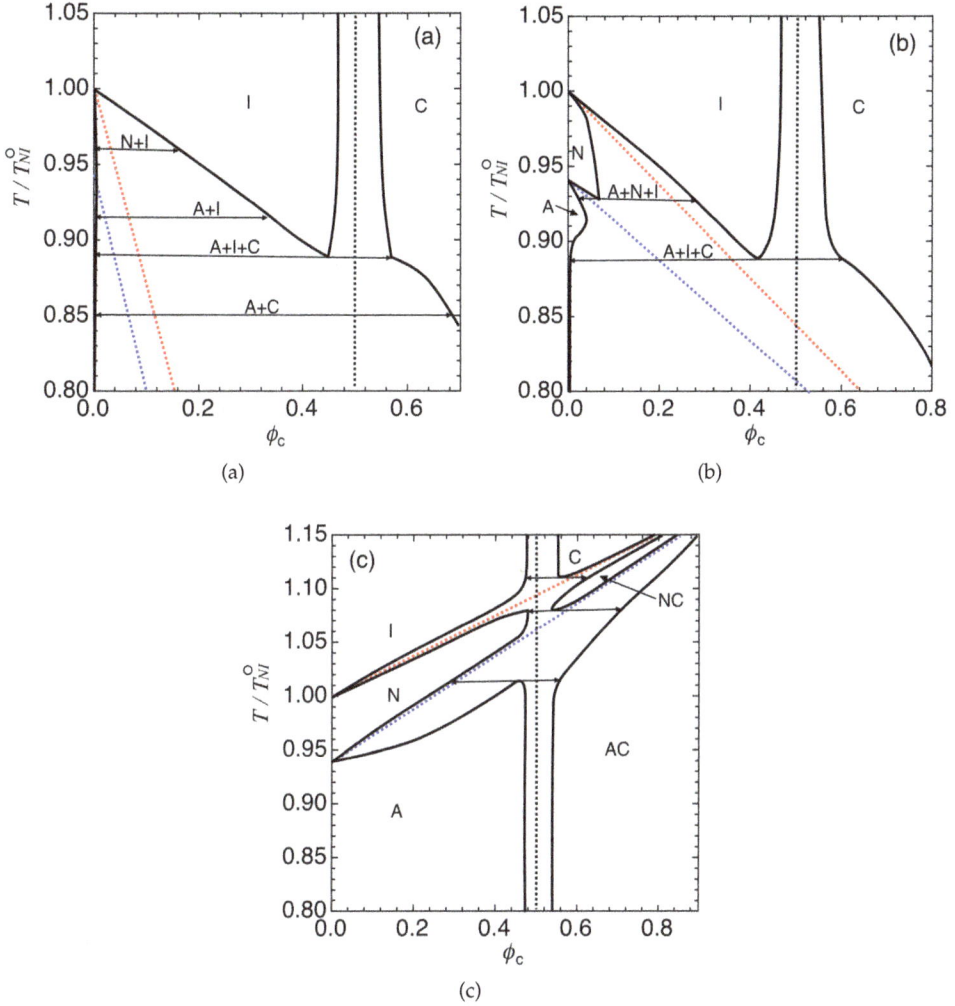

Fig. 4. Phase diagrams for $\alpha_c = 0.1$, $\alpha_n = 5$, $\gamma = 0.87$. The value of α_a is changed: (a) $\alpha_a = 1$; (b) $\alpha_a = -2$; (c) $\alpha_a = -3.5$.

a nematic phase at low temperature.(Pouling et. al., 1994) The observed phase diagram are qualitatively consistent with Fig. 4(a).

The binodal lines calculated at high concentrations of nanoparticles may not be experimentally observed because of high viscosity, however, it is important to understand the phase ordering kinetics(Matsuyama et .al., 2000; Matsuyama, 2008b). The cooperative phenomena between liquid crystalline ordering and crystalline ordering induce a variety of phase separations.

3. Nanotubes dispersed in liquid crystals

Since the discovery of carbon nanotubes (CNTs)(Iijima, 1991), extensive studies of physical and chemical properties of CNTs have been received great attention for many practical applications such as nano-sensors and devices. Windle's group first reported the nematic liquid crystalline behavior of an aqueous suspension of CNTs above a certain concentration and the isotropic-nematic phase separations.(Shaffer & Windle, 1999; Song et. al., 2003). Long nanotubes segregate preferentially to the liquid crystalline phase, whereas shorter nanotubes segregate preferentially to the isotropic phase(Zhang et.al., 2006). Recently such nanotubes as liquid crystalline materials become an important to be used in biological applications such as biosensors, biology imaging, artificial muscles, gene delivery, etc(Woltman et al., 2007).

In order to prepare CNT dispersions, strong van der Waals attractions between nanotubes must be screen out. To do this, the surface of nanotubes can be modified by acid oxidation, acid protonation, polymer or surfactant wrapping, etc.(Zhang & Kumar, 2008). For example, a water-soluble polymer, biopolymers such as DNA, and surfactant molecules have been used to wrap CNT and to increase the dispersibility in water(Badaire et. al., 2005). The polymer-wrapped nanotubes can be dispersed in a solvent with a considerable concentration. The excluded volume and electro-static repulsion between polymers can overcome the intermolecular van der Waals attractions and therefore the polymer-wrapped CNT can be dispersed in water. Thus it is possible to change the strength of the intermolecular interaction between nanotubes by using polymer-wrapping and negative or positive charging of nanotube surface, etc.

Alignment of such CNTs, or rigid-rodlike polymers (rods), with the aid of low molecular-weight-liquid crystalline molecules is an alternative approach. Indeed thermotropic(Basu & Iannacchione, 2008; Dierking et.al., 2005; Jayalakshmi & Prasad, 2009; Lynch & Patrick, 2002; Russell et.al., 2006) as well as lyotropic nematic liquid crystals(Courty et.al., 2003; Lagerwall et. al., 2007; Schymura et. al., 2009; Weiss et. al., 2006) have been applied as nematic solvents for the alignment of nanotubes. Anisotropic interactions between the nanotube and liquid crystal drastically change the alignments and physical properties of the mixtures. Duran et.al. have observed the nematic-isotopic phase transition temperature (T_{NI}) is enhanced by the incorporation of a multi-wall CNT within a small composition gap(Duran et.al., 2005).

In this section, we discuss phase separations in binary mixtures of a low molecular-weight-liquid crystal and a nanotube, such as CNT. We discuss uniaxial and biaxial nematic phases.

3.1 Nematic phases in mixtures of a nanotubes and a liquid crystal

We here consider the effect of the anisotropic interaction between a nanotube and a liquid crystal and that between rods.(Matsuyama, 2010) Depending on the interaction between a nanotube and a liquid crystal, we can expect various nematic phases. Figure 5 schematically shows the four nematic phases, defined by using the orientational order parameter (S_1) of a liquid crystal and that (S_2) of a nanotube. When the orientational order parameter of one component is positive, determining a nematic director, and the orientational order parameter of the second component is negative, we have planer nematic phase, where the second

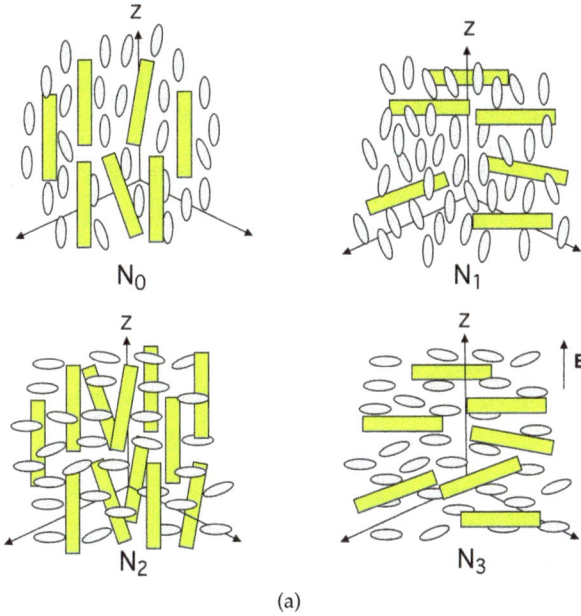

(a)

Fig. 5. Schematically illustrated four possible nematic phases. Four nematic phases are defined using the orientational order parameter S_1 of the liquid crystal and that S_2 of the nanotube: the nematic N_0 phase with $S_1 > 0$ and $S_2 > 0$, the nematic N_1 phase with $S_1 > 0$ and $S_2 < 0$, and the nematic N_2 phase with $S_1 < 0$ and $S_2 > 0$. When an external field (**E**) is applied along to the z axis for the particles of the dielectric anisotropy $\Delta\epsilon_1 < 0$ and $\Delta\epsilon_2 < 0$, the N_3 phase with $S_1 < 0$ and $S_2 < 0$ can appear: the nanotubes and liquid crystals are randomly oriented on the plane perpendicular to the direction of the external field.

component is randomly oriented within in the perpendicular plane to the nematic director. The nematic N_0 phase shows the nanotube and the liquid crystal are parallel to each other: $S_1 > 0$ and $S_2 > 0$. The nematic N_1 phase is defined as that the nanotube and the liquid crystal are perpendicular with each other: $S_1 > 0$ and $S_2 < 0$. In this phase, the nematic director (z axis) can be defined by the orientational direction of the liquid crystals. These perpendicular alignments can be obtained by modifying the surface of a nanotube, or CNT, with polymers or surfactants. The nematic N_2 phase is defined as the nanotube and the LC are perpendicular each other with $S_1 < 0$ and $S_2 > 0$. In this phase, the nematic director (z axis) can be defined by the orientational direction of the nanotube. Biaxial nematic phases are discussed in Section 3.3. When an external field (**E**) is applied along to the z axis for the particles of the dielectric anisotropy $\Delta\epsilon_1 < 0$ and $\Delta\epsilon_2 < 0$, the N_3 phase with $S_1 < 0$ and $S_2 < 0$ may appear, where the nanotubes and liquid crystals are randomly oriented on the plane perpendicular to the direction (z axis) of the external field.

3.2 Free energy of nanotube/liquid crystal mixtures

We consider a binary mixture of a liquid crystal of the length L_1 and the diameter D_1 and and a nanotube of the length L_2 and the diameter D_2: $L_1 < L_2$. The volume of the liquid crystal and that of the nanotube is given by $v_1 = (\pi/4)D_1^2 L_1$ and $v_2 = (\pi/4)D_2^2 L_2$, respectively. We here assume $D \equiv (D_1 = D_2)$. Let $\rho_1(\mathbf{r}, \mathbf{u})$ and $\rho_2(\mathbf{r}, \mathbf{u})$ be the number density of the liquid crystals and the nanotubes with an orientation $\mathbf{u}$ (or its solid angle Ω) at a position $\mathbf{r}$, respectively. The free energy F of the dispersion at the level of second virial approximation is given by Eq. (2). The volume fraction of liquid crystals is given by $\phi_1 = v_1 \rho_1$ and tat of nanotubes $\phi_2 = v_2 \rho_2$. As discussed in Eq. (14), we here consider the incompressible fluids: $\phi_1 + \phi_2 = 1$.

Consider a uniaxial nematic phase, which is spatially uniform but nonuniform for orientation. Let $f_i(\mathbf{u})$ be the distribution function of the particle $i(= 1, 2)$ and then the density can be expressed as

$$\rho_i(\mathbf{r}, \mathbf{u}) = c_i f_i(\mathbf{u}), \tag{27}$$

where $c_i \equiv N_i/V$ is the average number density of the particle i . The total number N_1 of the liquid crystals and N_2 of the nanoparticles must be conserved and then we have the normalization conditions:

$$\int \rho_i(\mathbf{r}, \mathbf{u}) d\mathbf{r} d\Omega = N_i/V, \tag{28}$$

where $d\Omega = 2\pi \sin\theta d\theta$ for a uniaxial nematic phase.

For the interaction between liquid crystals in Eq. (2), we take the attractive (Maier-Saupe) interaction:

$$\beta_{11} = -v_1 v_1 P_2(\cos\theta) P_2(\cos\theta'), \tag{29}$$

where $v_1(\equiv U_1/k_B T > 0)$ and U_0 is the anisotropic attractive (Maier-Saupe) interaction between liquid crystals(Brochard et.al., 1984; Maier & Saupe, 1958). (The subscript symbols c and r in Eq. (2) are changed to 1 and 2, respectively.) For the interaction between nanotubes, we here take into account both the attractive interaction and excluded volume one:(Matsuyama & Kato, 1996)

$$\beta_{22} = 2L^2 D|\mathbf{u} \times \mathbf{u}'| - v_2 v_2 P_2(\cos\theta) P_2(\cos\theta'), \tag{30}$$

where the first term is the excluded volume interaction between nanotubes, or rods,(Onsager, 1949) and the $v_2(\equiv U_2/k_B T > 0)$ is the attractive (Maier-Saupe) interaction between nanotubes. The interaction between a liquid crystal and a nanoparticle is given by

$$\beta_{12} = -v_{12} v_{12} P_2(\cos\theta) P_2(\cos\theta'), \tag{31}$$

where the anisotropic interaction $v_{12}(\equiv U_{12}/k_B T)$ between a liquid crystal and a rod can be positive or negative value. We here assume that the excluded volume interaction of a liquid crystal can be negligible because the length of liquid crystal is short. The volume $v_{12} = (\pi/4)L_1 L_2 D$ is the average excluded volume between a rod and a liquid crystal in an isotropic phase, Using Eqs. (29), (30), and (31), we can obtain the mixing free energy (15) for nanotube/liquid crystal mixtures. We here define an interaction parameter between a nanotube and a liquid crystal:

$$c_{12} = v_{12}/v_1, \tag{32}$$

which becomes an important parameter in the phase behavior.

3.3 Phase diagrams of nanotube/liquid crystal mixtures

In this subsection, we show some phase diagrams calculated from the free energy(Matsuyama, 2010).

3.3.1 Uniaxial nematic N_0 phase

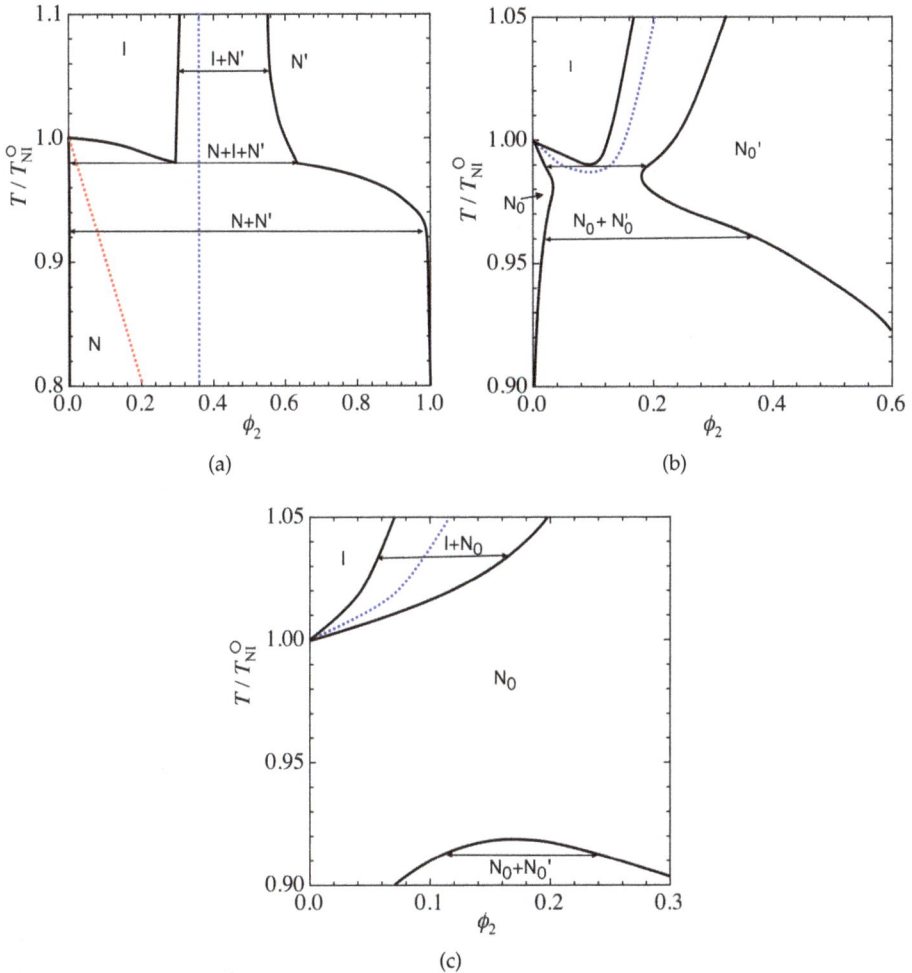

Fig. 6. Phase diagrams for $c_{12}=0$ (a), $c_{12}=0.3$ (b), and $c_{12}=0.4$(c) with $n_1 = 2$ and $n_2 = 10$.

We first show the phase diagram for $c_{12} = 0$ (see Fig. 6(a)), where the excluded volume interaction between nanotubes only prevails. The solid curve shows the binodal. The red and blue dotted lines show the first-order NIT line of a liquid crystal and that of a nanotube, respectively. Above $T/T^\circ_{NI} = 1$, the NIT of nanotubes takes place with increasing the concentration of the nanotube due to the excluded volume interaction between nanotubes, and we have the isotropic (I)-nematic (N) phase separation (I+N'), which has been obtained by Onsager theory(Onsager, 1949) and Flory's lattice theory(Flory, 1956; 1979). At the low temperatures of the NIT line (red line) of liquid crystals, we have a nematic (N) phase, where liquid crystals are in a nematic state but nanotubes are in an isotropic state. We predict the chimney type phase diagram with a triple point (N+I+N'). Below the triple point, we have the broad nematic-nematic (N+N') phase separation. The nematic N phase at lower concentrations consists of almost pure liquid crystals and the N' phase are formed by the orientational ordering of rods. Near $T/T_{NI}^{\circ} < 1$, we have the N+I phase separation.

Figure 6(b) shows the phase diagram for $c_{12} = 0.3$. On increasing the coupling constant c_{12}, the NIT lines shift to higher temperatures and lower concentrations and two NIT curves appeared in Fig. 6(a) merge. Below the NIT line (blue dotted line), we have a nematic N_0 phase, where the rods and the liquid crystals are oriented to be parallel to each other ($S_1 > 0$ and $S_2 > 0$). The width of the biphasic region I+N_0' decreases with decreasing temperature. We find the triple point (N_0+I+N_0'), where the nematic N_0, isotropic(I), and nematic N_0' phases simultaneously coexist. The binodal line of the N_0 phase shifts to higher concentrations and that of the N_0' phase shifts to lower concentrations with increasing c_{12}. Below the triple point we have the phase separation N_0+N_0', where the two nematic N_0 phases with the different concentrations can coexist.

Figure 6(c) shows the phase diagram for $c_{12} = 0.4$. The binodal curve sprits into two parts: one is the phase separation I+N_0 with the lower critical solution temperature (LCST) at $T/T_{NI}^{\circ} = 1$ and the other is the phase separation N_0+N_0' with the upper critical solution temperature (UCST). We have the stable nematic N_0 phase between the LCST and UCST. The length of a nanotube is also important to understand the phase diagrams. On increasing the length of the nanotube, the biphasic regions are broadened. Such LCST type phase diagram has been observed in mixtures of a main-chain nematic polyesters (poly[oxy(chloro-1,4-phenylene)oxycarbonyl][(trifluoromethyl)-1,4-phenylene]carbonyl)(PTFC) with a nematic liquid crystal (p-azoxyanisole)(PAAd14)(Ratto et.al., 1991). The theory can qualitatively describe the observed phase diagram.

3.3.2 Uniaxial nematic N_1 and N_2 phases

When the coupling parameter c_{12} is negative, we can expect that the nanotubes and liquid crystals are oriented to be perpendicular with each other.

Figure 7 shows the phase diagram for $c_{12} = -0.2(a)$ and $c_{12} = -0.5(b)$. The binodal line (solid line) is similar to Fig. 6(a), however, the structure of the nematic phases is different. In Fig. 7(a), the red dotted line at low concentrations shows the first-order nematic (N_1)-isotropic phase transition (1st-N_1IT) and the red dotted line at high concentrations shows the first-order nematic N_1-N_2 phase transition (N_1N_2T). The blue broken line corresponds to the second-order N_1-I phase transition (2nd-N_1IT), where the orientational order parameters continuously change. We also find the tricritical point (TCP) at which the 1st-N_1IT meets the 2nd-N_1IT. The phase diagram shows the three phase coexistence between N_1, I and N_2 phases at $T/T_{NI}^{\circ} \approx 0.92$. Above the triple point, the I+N_2 and N_1+I phase separations appear. Below the triple point, we have the N_1+N_2 phase separation.

On decreasing $c_{12}(< 0)$, the system favors more perpendicular alignment. Figure 7(b) shows the phase diagram for $c_{12} = -0.5$. (Note that the phase diagram is only shown for low concentrations.) The two nematic-isotropic phase transitions: 1st-N_1IT and 2nd-N_1IT, shift to higher temperatures and pass the binodal line of the isotropic phase in Fig. 7(a). The 2nd-N_1IT (blue broken line) appears at lower concentrations than the binodal line and we have the homogeneous N_1 phase. Near $T/T_{NI}^\circ = 1$, the narrow biphasic region N_1+I appears. Inside the binodal region, we have the 1st-N_1IT line (red dotted line). This N_1+I phase separation disappears at TCP. At higher concentrations, the N_1+N_2 phase separation appears. The binodal curve of the coexisting N_2 phase exists at $\phi_2 \approx 0.7$, although it is not depicted in this figure. Further decreasing $c_{12}(< 0)$, the 1st-N_1IT disappears and we have the 2nd-N_1IT and the 2nd-N_1IT temperature increases with increasing ϕ_2.

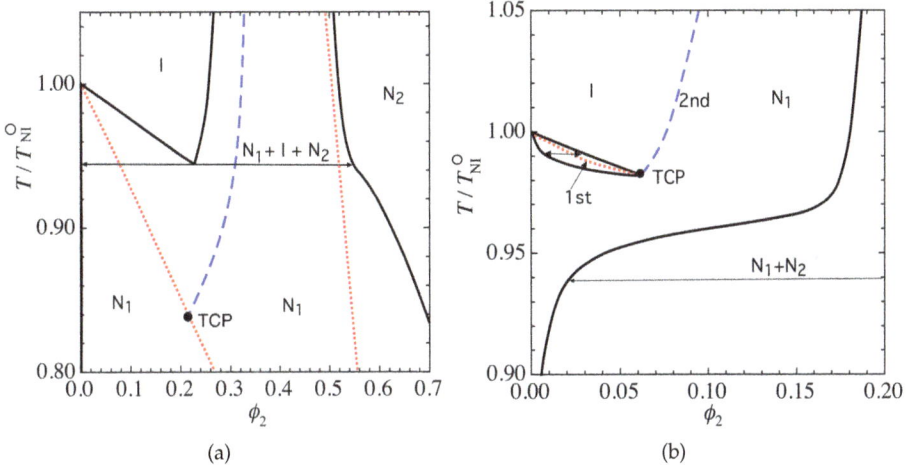

Fig. 7. Phase diagrams for c_{12}=-0.2 (a) and c_{12}=-0.5 (b).

Recent experimental studies of multi-wall carbon nanotube(CNT)/nematic liquid crystal mixtures(Duran et.al., 2005) have observed the NIT temperature of the liquid crystal is enhanced by the incorporation of CNT within a small composition gap and suggested that this enhanced NIT temperature phenomenon is attributed to anisotropic alignment of liquid crystals along the CNT bundles. Our model predicts two kind of phase behavior. When the CNTs and liquid crystals are parallel, the system shows the first-order isotropic-nematic (N_0) phase transition. On the other hand, if the CNTs and liquid crystals favor to be perpendicular each other, we have the 1st- and 2nd-N_1IT. The appearance of these phase transitions is strongly effected by the orientational order of nanotubes and liquid crystals.

3.3.3 Effect of external fields

To form a nematic N_3 phase, external forces such as electric or magnetic fields will be important. When the external magnetic or electric field **E** is applied to the nanotubes and liquid crystals having a dielectric anisotropy $\Delta\epsilon_i \equiv \epsilon_{\parallel,i} - \epsilon_{\perp,i}$ ($i = 1, 2$), the free energy changes due to the external field is given by(de Gennes & Prost, 1993)

$$a^3\beta F_{ext}/V = -\phi_1\beta\Delta\epsilon_1 \int (\mathbf{n} \cdot \mathbf{E})^2 f_1(\mathbf{u})d\Omega - \phi_2\beta\Delta\epsilon_2 \int (1 \cdot \mathbf{E})^2 f_2(\mathbf{u})d\Omega \qquad (33)$$

where **n** and **l** are the unit orientation vector of the liquid crystal and the nanotube, respectively.

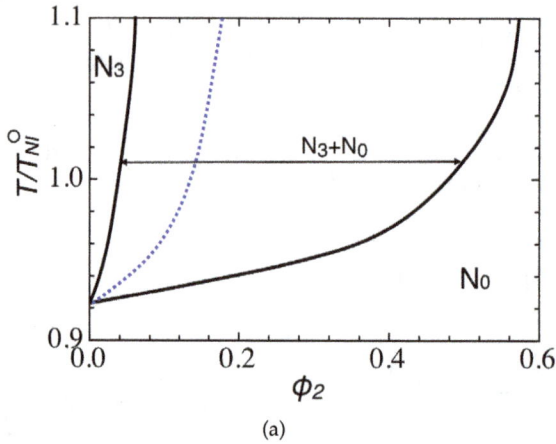

(a)

Fig. 8. Phase diagram under an external field for $\Delta\epsilon_1 = -1$ and $\Delta\epsilon_2 = 1$, where the liquid crystals tend to align perpendicular to the electric field **E**, while the nanotubes tend to parallel to **E**.

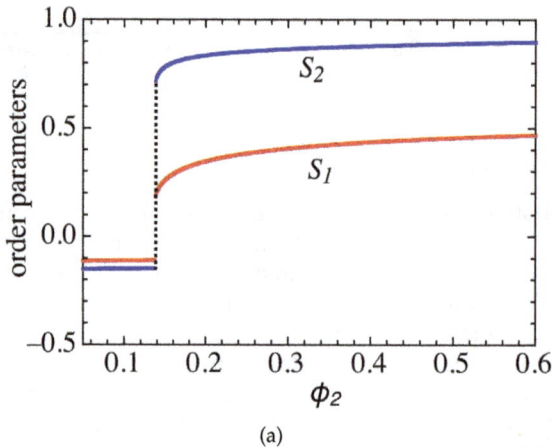

(a)

Fig. 9. Order parameters S_1 and S_2 plotted against ϕ_2 at $T/T_{NI}^\circ = 0.98$ in Fig. 8. We find the phase transition from the N_3 phase with $S_1 < 0$ and $S_2 < 0$ to the N_0 phase with $S_1 > 0$ and $S_2 > 0$.

We here consider the case of $\Delta\epsilon_1 < 0$ and $\Delta\epsilon_2 > 0$: the liquid crystals tend to align perpendicular to the electric field **E**, while the nanotubes tend to parallel to **E**. We apply the external field on Fig. 6(c), where the coupling $c_{12}(= 0.4)$ between the liquid crystal and nanotube is strong. Figure 8 shows the phase diagram under an external field for $\Delta\epsilon_1 = -1$ and $\Delta\epsilon_2 = 1$. The binodal line is broadened, compared with Fig. 6(c). We find the N_3 phase at low concentrations of nanotubes, where most liquid crystals tend to perpendicular to the eternal field and nanotubes favor to be parallel to liquid crystals because of the strong coupling c_{12} even $\Delta\epsilon_2 = 1$. The blue dotted line shows the 1st-order N_3-N_0 phase transition. Figure 9 shows the order parameters plotted against ϕ_2 at $T/T_{NI}^\circ = 0.98$. We also find the phase separation between N_3 and N_0 phases. We emphasize that we can control the four nematic phases by applying external fields.

3.4 Biaxial nematic ordering in nanotube/liquid crystal mixtures

Biaxial nematic phase has been first theoretically predicted by Freiser(Freiser, 1970). Since then, it has been the subject of much experimental(Galerne & Marcerou, 1983; Madsen et. al., 2004; Yu & Saupe, 1980), computational(Biscarini et. al., 1995; Hudson & Larson, 1993), and theoretical(Alben, 1973; Palffy-Muhoray et. al., 1984; Sharma et. al., 1985; Straley, 1974) work (see a recent review(Tschierske & Photinos, 2010)). Biaxiality occurs if anisotropic particles orient along a second axis perpendicular to a main director of the particles(Singh, 2000). Recently it has been experimentally observed a biaxial phase in colloidal dispersions of boardlike particles(van den Pol et. al., 2009). Such biaxiality is expected significant advantages in display applications with a fast response.(Luckhurst, 2001)

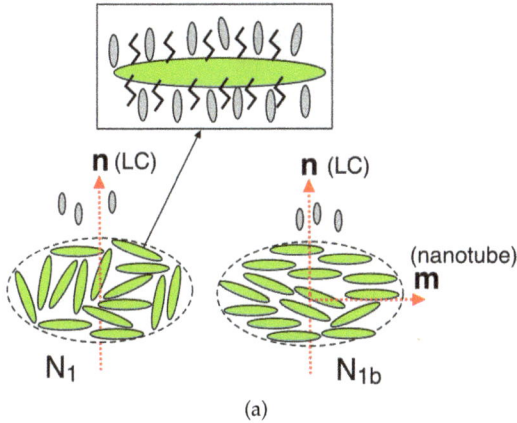

(a)

Fig. 10. Uniaxial planar nematic phase (N_1) and biaxial nematic phase (N_{1b}) in mixtures of a long nanotube and a short liquid crystal, which favor perpendicular orientations with each other. Nanotubes on an easy plane induce the additional ordering of nanotubes in the direction **m** perpendicular to the director **n** and yield a biaxial nematic phase N_{1b}.

As discussed in Section 3.1, when the order parameter of one component is positive, determining the nematic director, and the order parameter of the second component is negative, we have planar nematic phases (N_1 and N_2), where the second component is randomly distributed within the perpendicular plane to the director. In these nematic phases (N_1, N_2), we can expect either a uniaxial or a biaxial nematic phase.

Figure 10 schematically shows a novel biaxial nematic in nanotube/liquid crystal mixtures, where the two components favor a mutually perpendicular orientation.(Matsuyama, 2011) The mutually perpendicular alignments of nanotubes and liquid crystals can be achieved by wrapping polymers or surfactants on nanotube's surface(Badaire et. al., 2005; Zhang &

Kumar, 2008). To form such mutually perpendicular alignments, the anisotropic interaction (enthalpy) between a nanotube and a liquid crystal is needed. Moreover, in the planar nematic N_1 phase, on increasing concentration of nanotubes, we can expect that the excluded volume interaction (entropy) between nanotubes on an easy plane induces the additional ordering of nanotubes in the direction $\mathbf{m}$ (the second "minor" director) perpendicular to the director $\mathbf{n}$ (the first "major" director) of liquid crystals and yields a biaxial nematic phase (N_{1b}). In the N_2 phase, we may have a biaxial nematic phase (N_{2b}), where the additional ordering of liquid crystals appears in the direction $\mathbf{m}$ (minor director) perpendicular to the alignment $\mathbf{n}$ (major director) of nanotubes. Such a biaxiality in mixtures of two types of rodlike molecules has been first suggested by Alben(Alben, 1973). In this subsection, we introduce phase diagrams including such biaxial nematic phases. The phase diagrams appeared in Fig. 7 are drastically changed.

Using the distribution function $f_i(\theta, \varphi)$ of the component $i(= 1, 2)$, defined by a polar angle θ and an azimuthal angle φ, the biaxial order parameter is given by

$$\Delta_i = \int D(\theta, \varphi) f_i(\theta, \varphi) d\Omega, \tag{34}$$

where $D(\theta, \varphi) \equiv (\sqrt{3}/2) \sin^2 \theta \cos(2\varphi)$. Using the tensor order parameter

$$S_{i,\alpha\beta} = (3/2) S_i (n_\alpha n_\beta - \delta_{\alpha\beta}/3), \tag{35}$$

($\alpha, \beta = x, y, z$), we have $\Delta_i = S_{i,yy} - S_{i,xx}$ and $S_i = S_{i,zz}$(Singh, 2000). Here $S_{i,zz}$ describes alignment of molecules along the z axis (major director), whereas the nonzero value of Δ_i describes ordering along the x or y axis. Using the order parameters, we can define an isotropic (I) phase with $S_i = \Delta_i = 0$, a uniaxial N_1 phase: $S_1 > 0, S_2 < 0, \Delta_i = 0$, a uniaxial N_2 phase: $S_1 < 0, S_2 > 0, \Delta_i = 0$, a biaxial N_{1b} phase: $S_1 > 0, S_2 < 0, \Delta_i \neq 0$, and a biaxial N_{2b} phase: $S_1 < 0, S_2 > 0, \Delta_i \neq 0$. Using the additional theorem of a spherical harmonics in Eqs. (29), (30), and (31), we have $P_2(\cos \gamma) = P_2(\cos \theta) P_2(\cos \theta') + D(\theta, \varphi) D(\theta', \varphi')$ and can calculate phase separations(Matsuyama, 2011).

Figure 11 shows the phase diagrams numerically calculated for $c_{12} = -0.5$ (a) and -0.8 (b). Black lines show the binodal line. The red (blue) lines show a first (second)-order phase transition, where the order parameters discontinuously (continuously) change. The biaxial nematic phase N_{1b}, which includes an unstable biaxial phase, a metastable biaxial, and a stable biaxial phase, is indicated by the yellow area. In Fig. 11(a), at high temperatures, we have the phase separation (I+N_2) between an isotropic (I) phase at $\phi_2 \simeq 0.14$ and a uniaxial N_2 phase at $\phi_2 \simeq 0.63$. Such a chimney type's phase diagram with a coexistence between I and N phases is induced by the excluded volumes between long rods(Flory, 1956; 1979; Matsuyama & Kato, 1996; Onsager, 1949). Inside the binodal lines, we find the first-order isotropic-biaxial N_{1b} phase transition at $\phi_2 \simeq 0.22$ and the first-order biaxial N_{1b}-uniaxial N_2 phase transition at $\phi_2 \simeq 0.5$. Above $\phi_2 \simeq 0.6$, we have a stable uniaxial N_2 phase. We also find the three phase coexistence, or triple point (TP), between N_1+I+N_2 at $\tau(\equiv T/T_{NI}^\circ) \simeq 0.98$. Below the TP, we have the N_1+N_2 phase separation. At low concentrations, the N_1+I phase separation appears. The biaxial nematic phase is hidden inside the binodal lines.

Further increasing c_{12} (Fig. 11(b)), the coupling between a liquid crystal and a nanotube drastically changes the phase diagram. The biaxial regions shift to lower concentrations and

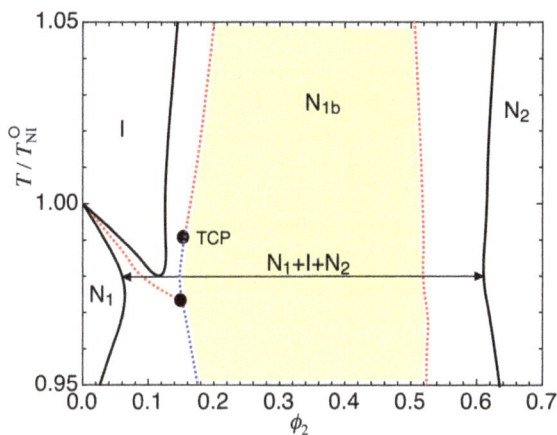

(a)

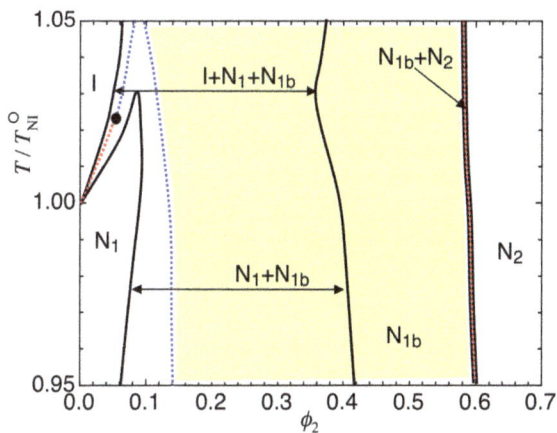

(b)

Fig. 11. Phase diagrams on the temperature ($\tau \equiv T/T^\circ_{NI}$)-volume fraction ($\phi_2$) plane for $c_{12} = -0.5$ (a) and -0.8 (b). The black lines indicate the binodal. The red (blue) lines show a first (second)-order phase transition, where the order parameters discontinuously (continuously) change. The biaxial nematic phase N_{1b}, which includes an unstable biaxial phase, a metastable biaxial, and a stable biaxial phase, is indicated by the yellow area. The stable biaxial phase N_{1b} appears on (b).

the thermodynamically stable biaxial N_{1b} phase appears between $\phi_2 \sim 0.4$ and $\phi_2 \sim 0.6$. We find the phase separations: $I+N_{1b}$, N_1+N_{1b}, $I+N_1$, $N_{1b}+N_2$, and the three phase coexistence $I+N_1+N_{1b}$ at $\tau \simeq 1.03$. Note that the coexistence region $(N_{1b}+N_2)$ at $\phi_2 \simeq 0.6$ is very narrow. At low concentrations, the I-N_1 phase transition temperature increases with increasing ϕ_2 and the TP shifts to higher temperatures. Note that the stable biaxial phase N_{1b} appears on (b).

Duran et.al have observed in multiwall CNT/liquid crystal mixtures that the NIT temperature of the liquid crystal is enhanced by the incorporation of CNT(Duran et.al., 2005). Our theory demonstrates that this enhanced NIT temperature phenomena is attributed to anisotropic coupling between CNTs and liquid crystals. A mutually perpendicular orientation between rods and LCs can be achieved by wrapping surfactants on nanotube's surface, like a Langmuir-Blodgett film with liquid crystals(Barbero & Durand, 1996), where liquid crystals in contact with the surfactants are oriented by steric interaction with the molecules on rods. These modifications can change the strength of the interaction parameter ν_{12} in our model and give a possibility of a novel biaxial phase in this mixture. The biaxial N_{2b} phase does not appear on the phase diagrams because the length of liquid crystal is too short to form the N_{2b} phase.

4. Summary

In this chapter we have reviewed the possible phase separations in mixtures of a nanoparticle and a liquid crystal, based on the mean field theory. In Section 2, we have introduced mixtures of a spherical nanoparticle and a liquid crystal. Ferroelectric spherical nanoparticles dispersed in liquid crystal have a possibility of various phase separations, discussed in this chapter. In Section 3, we have introduced phase diagrams in mixtures of a nanotube and a liquid crystal. Novel uniaxial and biaxial nematic phases are theoretically predicted. We also discuss the effect of external fields in nanotube/liquid crystal mixtures. Phase diagrams introduced in this chapter have not been experimentally observed yet, however, it will be a challenging subject from both an experimental and theoretical point of view.

5. Acknowledgment

These studies were supported by Grant-in Aid for Scientific Research (C) (Grant No. 23540477) and that on Priority Area "Soft Matter Physics" from the Ministry of Education, Culture, Sports, Science and Technology of Japan (Grant No. 21015025).

6. References

Alben, R. (1973).Liquid crystal phase transitions in mixtures of rodlike and platelike molecules,*J. Chem. Phys.*, Vol. 59, No. 3, (Oct. 1973) pp.4299-4304. DOI:10.1063/1.1680625.

Alder, B. J.; Wainwright, T. E. (1957). Phase Transition for a Hard Sphere System,*J. Chem. Phys.*, Vol. 27, No. 5, (Nov. 1957) pp.1208-1209. DOI:10.1063/1.1743957.

Anderson, V. J.; Terentjev, E. M.; Meeker,S.P.; Crain, J.; Poon, W. C. K. (2001). Cellular solid behaviour of liquid crystal colloids 1. Phase separation and morphology.*Eur. Phys. J. E*, Vol. 4, No. 1, (Jan. 2001) pp.11-20.

Anderson, V. J.;Terentjev, E. M. (2001). Cellular solid behaviour of liquid crystal colloids 2. Mechanical properties.*Eur. Phys. J. E*, Vol. 4, No. 1, (Jan. 2001) pp.21-28.

Araki, T.; Tanaka, H. (2004). Nematohydrodynamic Effects on the Phase Separation of a Symmetric Mixture of an Isotropic Liquid and a Liquid Crystal,*Phys. Rev. Lett.*, Vol. 93, No. 1, (Jun. 2004) pp.015702-1-015702-4. DOI:10.1103/PhysRevLett.93.015702.

Badaire, S.; Zakri, C.; Maugey, M.; Derre, A.; Barisci, J. N.; Wallace, G.; Poulin, P. (2005) Liquid Crystals of DNA-Stabilized Carbon Nanotubes,*Adv. Mater.*, Vol. 17, No. 13, (Jul. 2005) pp.1673-1676. DOI:10.1002/adma.200401741.

Barbero, G. & Durand,G. (1996). Surface Anchoring of Nematic liquid Crystals, In: *Liquid Crystals in Complex Geometries*, Crawford, G.P.; Zumer, S., (Ed.), pp.21-52, Taylor & Francis, ISBN:0-7484-0464-3, London

Basu, R.; Iannacchione, G. S. (2008) Carbon nanotube dispersed liquid crystal: A nano electromechanical system,*Appl. Phys. Lett.*, Vol. 93, No. 18, (Nov. 2008) pp.183105-1-183105-3. DOI: 10.1063/1.3005590.

Biscarini, F.; Chiccoli, C.; Pasini, P.; Semeria, F.; Zannoni, C. (1995) Phase Diagram and Orientational Order in a Biaxial Lattice Model: A Monte Carlo Study,*Phys. Rev. Lett.*, Vol. 75, No. 9, (Aug. 1995) pp.1803-1806. DOI: 10.1103/PhysRevLett.75.1803.

Brochard, F.; Jouffroy, J.; Levinson, P. (1984). Phase diagrams of mesomorphic mixtures, *J. de Phys.*, Vol. 45, No. 7, (Jul. 1984) pp.1125-1136. DOI: 10.1051/jphys:019840045070112500.

Cates, M. E.; Evans, M. R. (2000). *Soft and Fragile Matter*, Institute of Physics Publishing, ISBN: 0-7503-0724-2, Bristol.

Caggioni, M.; Giacometti, A.; Bellini, T.; Clark, N. A.; Mantegazza, F.; Maritan, A. (2005). Pretransitional behavior of a water in liquid crystal microemulsion close to the demixing transition: Evidence for intermicellar attraction mediated by paranematic fluctuations.*J. Chem. Phys.*, Vol. 122, No. 21, (Jun. 2005) pp.214721-1-214721-19. DOI:10.1063/1.1913444.

Chiu, H. W.; Kyu, T. (1999). Spatio-temporal growth of nematic domains in liquid crystal polymer mixtures.*J. Chem. Phys.*, Vol. 110, No. 12, (Mar. 1999) pp.5998-6006. DOI:10.1063/1.478502.

Copic, M.; Mertelj, A.; Buchev, O.;Reznikov, Y. (2007). Coupled director and polarization fluctuations in suspensions of ferroelectric nanoparticles in nematic liquid crystals, *Phys. Rev. E*, Vol. 76, No. 1, (Jul. 2007) pp.011702-1-011702-5. DOI:10.1103/PhysRevE.76.011702.

Courty, S.; Mine, J.; Tajbakhsh, A. R.; Terentjev, E. M. (2003). Nematic elastomers with aligned carbon nanotubes: New electromechanical actuators, *Europhys. Lett.*, Vol. 64, No. 5, (Dec. 2003) pp.654-660, DOI:10.1209/epl/i2003-00277-9.

Das, S. K.; Ray, A. D., (2005). Colloidal crystal formation via polymer-liquid-crystal demixing. *Europhys. Lett.*, Vol. 70, No. 5, (May 2005) pp.621-627. DOI: 10.1209/epl/i2005-10034-2.

de Gennes, P. G. & Prost, J. (1993). *The Physics of Liquid Crystals*, Oxford University Press, ISBN: 019-851785-8, New York.

Dierking, I.; Scalia, G.; Morales, P., (2005). Liquid crystal?carbon nanotube dispersions, *J. Appl. Phys.*, Vol. 97, No. 4, (Jan. 2005) pp.044309-1-044309-5. DOI: 10.1063/1.1850606.

Dubaut, A.; Casagrande, C.; Veyssie, M., Deloche, B., (1980). Pseudo Clearing Temperature in Binary Polymer-Nematic Solutions. *Phys. Rev. Lett.*, Vol. 45, No. 20, (Nov. 1980) pp.1645-1648. DOI:10.1103/PhysRevLett.45.1645.

Duran, H.; Gazdecki,; Yamashita, A.; Kyu, T. (2005). Effect of carbon nanotubes on phase transitions of nematic liquid crystals, *Liq. Cryst.*, Vol. 32, No. 7, (Jul. 2005) pp.815-821. DOI:10.1080/02678290500191204.

Flory, P. J. (1953). *Principles of Polymer Chemistry*, Cornell University, ISBN 0-8014-0134-8, Ithaca, New York.

Flory, P. J. (1956). Statistical Thermodynamics of Semi-Flexible Chain Molecules, *Proc. R. Soc. London Ser. A*, Vol. 234, No. 1196, (Jan. 1956) pp.60-73. DOI:10.1098/rspa.1956.0015.

Flory, P. J.; Ronca, G. (1979). Theory of Systems of Rodlike Particles: I. Athermal systems, *Mol. Cryst. Liq. Cryst.*, Vol. 54, No. 3, (Mar. 1979) pp.289-309. DOI:10.1080/00268947908084861.

Freiser, M. J. (1970). Ordered States of a Nematic Liquid, *Phys. Rev. Lett.*, Vol. 24, No. 19, (May 1970) pp.1041-1043. DOI:10.1103/PhysRevLett.24.1041.

Fukuda, J.; Yokoyama, H., (2005). Separation-Independent Attractive Force between Like Particles Mediated by Nematic-Liquid-Crystal Distortions, *Phys. Rev. Lett.*, Vol. 94, No. 14, (Apr. 2005) pp.148301-1-148301-4. DOI:10.1103/PhysRevLett.94.148301.

Fukuda, J. (2009). Liquid Crystal Colloids: A Novel Composite Material Based on Liquid Crystals, *J. Phys. Soc. Jpn.*, Vol. 78, No. 4, (Apr. 2009) pp.041003-1-041003-9. DOI: 10.1143/JPSJ.78.041003.

Galerne, Y.; Marcerou, J.P. (1983). Temperature Behavior of the Order-Parameter Invariants in the Uniaxial and Biaxial Nematic Phases of a Lyotropic Liquid Crystal, *Phys. Rev. Lett.*, Vol. 51, No. 23, (Dec. 1983) pp.2109-2112. DOI:10.1103/PhysRevLett.51.2109.

Hudson, S.D.; Larson, R.G. (1993). Monte Carlo simulation of a disclination core in nematic solutions of rodlike molecules, *Phys. Rev. Lett.*, Vol. 70, No. 19, (May 1993) pp.2916-2919. DOI:10.1103/PhysRevLett.70.2916.

Iijima, S. (1991). Helical microtubules of graphitic carbon, *Nature*, Vol. 354, No. 6348, (Nov. 1991) pp.56-58. DOI:10.1038/354056a0.

Jayalakshmi, V.; Prasad, S. K. (2009). Understanding the observation of large electrical conductivity in liquid crystal-carbon nanotube composites, *Appl. Phys. Lett.*, Vol. 94, No. 20, (May 2009) pp.202106-1-202106-3. DOI: 10.1063/1.3133352.

Kirkwood, G.; Monroe, E. (1941). Statistical Mechanics of Fusion, *J. Chem. Phys.*, Vol. 9, No. 7, (Jul. 1941) pp.514-526. DOI:10.1063/1.1750949.

Kralj, S.; Bradac, Z.; Popa-Nita, V. (2008). The influence of nanoparticles on the phase and structural ordering for nematic liquid crystals, *J. Phys.: Condens. Matter*, Vol. 20, No. 24, (Jun. 2008) pp.244112. DOI:10.1088/0953-8984/20/24/244112.

Kuksenok, O. V.; Ruhwandl, R. W.; Shiyanovskii, S. V.;, Terentjev, E. M., (1996). Director structure around a colloid particle suspended in a nematic liquid crystal, *Phys. Rev. E*, Vol. 54, No. 5, (Nov. 1996) pp.5198-5203. DOI:10.1103/PhysRevE.54.5198.

Kventsel, G. F.; Luckhurst, G. R.; Zewdie, H. B. (1985). A molecular field theory of smectic A liquid crystals, *Molc. Phys.*, Vol. 56, No. 3, (Mar. 1985) pp.589-610. DOI: 10.1080/00268978500102541.

Lagerwall, J. P. F. ; Scalia, G.; Haluska, M.; Dettlaff-Weglikowska, U.; Roth, S.; Giesselmann, F. (2007). Nanotube Alignment Using Lyotropic Liquid Crystals, *Adv. Mater.*, Vol. 19, No.3, (Feb. 2007) pp.359-364, DOI:10.1002/adma.200600889.

Li, F.; Buchnev, O.; Cheon, C.; Glushchenko, A.; Reshetnyak, V.;Reznikov, Y.; Sluckin, T. J.; West, J. L. (2006). Orientational Coupling Amplification in Ferroelectric Nematic Colloids, *Phys. Rev. Lett.*, Vol. 97, No.14, (Oct. 2006) pp.147801-1-147801-4, DOI: 10.1103/PhysRevLett.97.147801.

Li, F.; West, J.; Glushchenko, A.; Cheon, C. I.; Reznikov,Y. (2006). Ferroelectric nanoparticle/liquid-crystal colloids for display applications, *J. Soc. Info. Disp.*, Vol. 14, No. 6, (Jun. 2006), pp.523-527, DOI: 10.1889/1.2210802.

Lopatina, L. M.; Selinger, J. V. (2009). Theory of Ferroelectric Nanoparticles in Nematic Liquid Crystals. *Phys. Rev. Lett.*, Vol. 102, No.19, (May 2009) 197802-1-197802-4, DOI: 10.1103/PhysRevLett.102.197802.

Loudet, J. C.; Barois, P.; Auroy, P.; Keller, P.; Richard, H.; Pouling, P. (2004). Colloidal Structures from Bulk Demixing in Liquid Crystals, *Langmuir*, Vol. 20, No.26, (Sept. 2004) pp.11336-11347, DOI: 10.1021/la048737f.

Lubensky, T. C.; Pettey, D.; Currier, N.;Stark, H. (1998). Topological defects and interactions in nematic emulsions, *Phys. Rev. E*, Vol. 57, No.1, (Jan. 1998) pp.610-625, DOI: 10.1103/PhysRevE.57.610.

Luckhurst, G. R. (2001). Biaxial nematic liquid crystals: fact or fiction?, *Thin Solid Films*, Vol. 393, No.1, (Aug. 2001) pp.40-52, DOI: 10.1016/S0040-6090(01)01091-4.

Lynch, M. D.; Patrick, D. L. (2002). Organizing Carbon Nanotubes with Liquid Crystals, *Nano Lett.*, Vol. 2, No.11, (Nov. 2002) pp.1197-625, DOI: 10.1021/nl025694j.

Madsen, L.A.; Dingemans, T.J.; Nakata, M.; Samulski, E.T. (2004). Thermotropic Biaxial Nematic Liquid Crystals, *Phys. Rev. Lett.*, Vol. 92, No.14, (Apr. 2004) pp.145505-1-145505-4, 10.1103/PhysRevLett.92.145505.

Maier, W.; Saupe, A. (1958). A simple molecular theory of the nematic liquid-crystalline state, *Z. Naturforsch*, Vol. 13a, (Mar. 1958) pp.564-566.

Matsuyama, A. & Kato, T. (1996). Theory of binary mixtures of a flexible polymer and a liquid crystal. *J. Chem. Phys.*, Vol. 105, No.4, (Jul. 1996) pp. 1654-1660. DOI:10.1063/1.472024

Matsuyama, A.; Kato, T. (1998). Phase diagrams of polymer dispersed liquid crystals, *J. Chem. Phys.*, Vol. 108, No.5, (Feb. 1998), pp. 2067-2072, DOI: 10.1063/1.475585.

Matsuyama, A.; Evans, R. M. L.; Cates, M. E. (2000). Orientational fluctuation-induced spinodal decomposition in polymer-liquid crystal mixtures. *Phys. Rev. E*, Vol. 61, No.3, (Mar. 2000), pp. 2977-2986, DOI: 10.1103/PhysRevE.61.2977.

Matsuyama, A. (2006a). Mean Field Theory of Crystalline Ordering in Colloidal Solutions, *J. Phys. Soc. Jpn.*, Vol. 75, No.3, (Mar. 2006), pp. 034604-1-034604-9, DOI: 10.1143/JPSJ.75.034604.

Matsuyama, A. (2006b). Spinodal in a Liquid-Face Centered Cubic Phase Separation, *J. Phys. Soc. Jpn.*, Vol. 75, No.8, (Aug. 2006), pp. 084602-1-084602-10, DOI: 10.1143/JPSJ.75.084602.

Matsuyama, A. & Hirashima, R. (2008a). Phase separations in liquid crystal-colloid mixtures. *J. Chem. Phys.*, Vol. 128, No.4, (Jan. 2008) 044907-1-044907-11, DOI: 10.1063/1.2823737.

Matsuyama, A. (2008b).Morphology of spinodal decompositions in liquid crystal-colloid mixtures. *J. Chem. Phys.*, Vol. 128, No.22, (Jun. 2008) pp. (224907-1)-(224907-8), DOI: 10.1063/1.2936831.

Matsuyama, A. (2009). Phase separations in mixtures of a liquid crystal and a nanocolloidal particle. *J. Chem. Phys.*, Vol. 131, No.20, (Nov. 2009) 204904-1-204904-12, DOI: 10.1063/1.3266509.

Matsuyama, A. (2010). Theory of binary mixtures of a rodlike polymer and a liquid crystal. *J. Chem. Phys.*, Vol. 132, No.21, (Jun. 2010) 214902-1-214902-10, DOI: 10.1063/1.3447892.

Matsuyama, A. (2010a). Thermodynamics of flexible and rigid rod polymer blends, In: *Encyclopedia of polymer blends, Vol. 1: Fundamentals*, Isayev, A. I., (Ed.), Chap.2, pp.45-100, WILEY-VCH Verlag GmbH & Co. KGaA, ISBN: 978-3-527-31929-9, Weinheim.

Matsuyama, A. (2011). Biaxial nematic phases in rod/liquid crystal mixtures. *Liq. Cryst.*, Vol. 38, No.6, (Jun. 2011), pp. 729-736, DOI: 10.1080/02678292.2011.570795.

Meeker,S. P.; Poon, W. C. K.; Crain, J.; Terentjev, E. M. (2000). Colloid?liquid-crystal composites: An unusual soft solid. *Phys. Rev. E*, Vol. 61, No.6, (Jun. 2000), pp.R6083-R6086, DOI: 10.1103/PhysRevE.61.R6083.

McMillan, W. L. (1971). Simple Molecular Model for the Smectic A Phase of Liquid Crystals, *Phys. Rev. A*, Vol. 4, No.3, (Sep. 1971), pp.1238-1246, DOI: 10.1103/PhysRevA.4.1238.

Musevic, I.; Skarabot, M.; Tkalec, U.; Ravnik, M.; Zummer, S. (2004). Two-Dimensional Nematic Colloidal Crystals Self-Assembled by Topological Defects. *Science*, Vol. 313, No.5789, (Aug. 2006), pp.954-958, DOI: 10.1126/science.1129660.

Nazarenko, V. G.; NychA. B.; Lev, B. I. (2001). Crystal Structure in Nematic Emulsion. *Phys. Rev. Lett.*, Vol. 87, No.7, (Jul. 2001), pp.075504-1-075504-4, DOI: 10.1103/PhysRevLett.87.075504.

Onsager, L. (1949). The Effects of Shape on The Interaction of Colloidal Particles, *Ann. N. Y. Acad. Sci.*, Vol. 51, (May 1949), pp.627-659, DOI:10.1111/j.1749-6632.1949.tb27296.x.

Palffy-Muhoray, P. ; Berlinsky, A. J.; De Bruyn, J. R.; Dunmur, D. A. (1984). Coexisting nematic phases in binary mixtures of liquid crystals, *Phys. Lett. A*, Vol. 104, No.3, (Aug. 1984), pp.159-162, DOI:10.1016/0375-9601(84)90367-0.

Popa-Nita, V.; van-der Schoot,P.; Kralj, S. (2006). Crystal Structure in Nematic Emulsion. *Eur. Phys. J. E*, Vol. 21, No.3, (Nov. 2006), pp.189-198, DOI:10.1140/epje/i2006-10059-3.

Pouling, P.; Stark, H.; Lubenski, T. C.; Weitz, D. A. (1997). Novel Colloidal Interactions in Anisotropic Fluids, *Science*, Vol. 275, No. 5307, (Mar. 1997), pp.1770-1773, DOI: 10.1126/science.275.5307.1770.

Pouling, P.; Raghunathan, V. A.; Richetti, P.; Roux, D. J. (1994). On the dispersion of latex particles in a nematic solution. I. Experimental evidence and a simple model, *J. Phys. II* , Vol. 4, No. 9, (Sept. 1994), pp.1557-1569, DOI: 10.1051/jp2:1994217.

Pusey, P. N.; van Megen, W. (1986). Phase behaviour of concentrated suspensions of nearly hard colloidal spheres, *Nature* , Vol. 320, No. 6060, (Mar. 1986), pp.340-1569, DOI: 10.1038/320340a0.

Ratto, J. A.; Volino, F.; Blumstein, R. B. (2006). Phase behavior and order in mixtures of main-chain nematic polyesters with small molecules: a combined proton and deuterium NMR study, *Macromolecules*, Vol. 24, No. 10, (May 1991), pp.2862-2867, DOI: 10.1021/ma00010a035.

Russell, M.; Oh, S.; Larue, I.; Zhou, O.; Samulski. E. T. (2006). Alignment of nematic liquid crystals using carbon nanotube films, *Thin Solid Films*, Vol. 509, No. 1-2, (Jun. 2006), pp.53-57, DOI: 10.1016/j.tsf.2005.09.099.

Skarabot, M.; Ravnik, M.; Zumer, S.; Tkalec, U.; Poberaj, I.; Babic, D.; Musevic, I. (2008). Hierarchical self-assembly of nematic colloidal superstructures, *Phye. Rev. E*, Vol. 77, No. 6, (Jun. 2008), pp.061706-1-061706-4, DOI: 10.1103/PhysRevE.77.061706.

Shaffer, M. S. P.; Windle, A. H. (1999). Analogies between Polymer Solutions and Carbon Nanotube Dispersions, *Macromolecules*, Vol. 32, No. 20, (Oct. 1999), pp.6864-6866, DOI:10.1021/ma990095t.

Sharma, S.R.; Palffy-Muhoray, P.; Bergersen, B.; Dunmur, D.A. (1985). Stability of a biaxial phase in a binary mixture of nematic liquid crystals, *Phys. Rev. A*, Vol. 32, No. 6, (Dec. 1985), pp.3752-3755, DOI:10.1103/PhysRevA.32.3752.

Shen, C.; Kyu, T., (1995). Spinodals in a polymer dispersed liquid crystal, *J. Chem. Phys.*, Vol. 102, No. 1, (Jan. 1995), pp.556-562, DOI:10.1063/1.469435.

Singh, S., (2000). Phase transitions in liquid crystals, *Phys. Rep.*, Vol. 324, No. 2, (Feb. 2000), pp.107-269, DOI:10.1016/S0370-1573(99)00049-6.

Stark, H., (1999). Director field configurations around a spherical particle in a nematic liquid crystal, *Eur. Phys. J. B*, Vol. 10, No. 2, (Jul. 1999), pp.311-321, DOI:10.1007/s100510050860.

Stark, H., (2001). Physics of colloidal dispersions in nematic liquid crystals, *Phys. Rep.*, Vol. 351, No. 6, (Oct. 2001), pp.387-474, DOI:10.1016/S0370-1573(00)00144-7.

Straley, J.P., (1974). Ordered phases of a liquid of biaxial particles, *Phys. Rev. A*, Vol. 10, No. 5, (Nov. 1974), pp.1881-1887, DOI:10.1103/PhysRevA.10.1881.

Song, W.; Kinloch, I. A.; Windle, A. H. (2003). Nematic Liquid Crystallinity of Multiwall Carbon Nanotubes, *Science*, Vol. 302, No. 5649, (Nov. 2003), pp.1363, DOI:10.1126/science.1089764.

Schymura, S.; Enz, E.; Roth, S.; Scalia, G.; Lagerwall, J. P. F. (2009). Macroscopic-scale carbon nanotube alignment via self-assembly in lyotropic liquid crystals, *Synth. Mater.*, Vol. 159, No. 21, (Nov. 2009), pp.2177-2179, DOI:10.1016/j.synthmet.2009.08.021.

Tanaka, S.; Ataka, M.; Ito, K. (2002). Pattern formation and coarsening during metastable phase separation in lysozyme solutions, *Phys. Rev. E*, Vol. 65, No. 5, (May 2020), pp.051804-1-051804-6, DOI:10.1103/PhysRevE.65.051804.

Tschierske, C.; Photinos, D.J. (2010). Biaxial nematic phases, *J. Mater. Chem.*, Vol. 20, No. 21, (Apr. 2010), pp.4263-4294, DOI:10.1039/B924810B.

van den Pol, E.; Petukhov, A.V.; Thies-Weesie, D.M.E.; Byelov, D.V.; Vroege, G.J. (2009). Experimental Realization of Biaxial Liquid Crystal Phases in Colloidal Dispersions of Boardlike Particles. *Phys. Rev. Lett.*, Vol. 103, No.25, (Dec. 2009), pp.258301-1-258301-4, DOI:10.1103/PhysRevLett.103.258301.

Weiss, V.; Thiruvengadathan, R.; Regev, O. (2006). Preparation and Characterization of a Carbon Nanotube¿Lyotropic Liquid Crystal Composite, *Langmure*, Vol. 22, No. 3, (Jan. 2006), pp.854-856, 10.1021/la052746m.

Woltman, S. J.; Jay, G. D.; Crawdord, G. P. (Eds.) (2007). *Liquid Crystals*, World Scientific Publishing, ISBN: 10-981-270-545-7, Singapore.

Yada, M.; Yamamoto, J.; Yokoyama, H., (2004). Direct Observation of Anisotropic Interparticle Forces in Nematic Colloids with Optical Tweezers, *Phys. Rev. Lett.*, Vol. 92, No. 18, (May 2004), pp.185501-1-185501-4, DOI:10.1103/PhysRevLett.92.185501.

Yamamoto, J.; Tanaka, H., (2001). Transparent nematic phase in a liquid-crystal-based microemulsion, *Nature*, Vol. 409, (Jan. 2001), pp.321-325, DOI:10.1038/35053035.

Yamamoto, R. (2001). Simulating Particle Dispersions in Nematic Liquid-Crystal Solvents, *Phys. Rev. Lett.*, Vol. 87, No. 7, (Jul. 2001), pp.075502-1-075502-4, DOI:10.1103/PhysRevLett.87.075502.

Yu, L. J.; Saupe, A. (1980). Observation of a Biaxial Nematic Phase in Potassium
    Laurate-1-Decanol-Water Mixtures, *Phys. Rev. Lett.*, Vol. 45, No. 12, (Sept. 1980),
    pp.1000–1003, DOI:10.1103/PhysRevLett.45.1000.
Zapotocky, M.; Ramos, L.; Pouling, P.; Lubensky, T. C.; Weitz, D. A., (1999). Particle-Stabilized
    Defect Gel in Cholesteric Liquid Crystals, *Science*, Vol. 283, No. 5399, (Jan. 1999),
    pp.209-212, DOI: 10.1126/science.283.5399.209.
Zhang, S.; Kinloch, I. A.; Windle, A. H. (2006). Mesogenicity Drives Fractionation in Lyotropic
    Aqueous Suspensions of Multiwall Carbon Nanotubes, *Nano Lett.*, Vol. 6, No. 3, (Mar.
    2006), pp.568-572, DOI: 10.1021/nl0521322.
Zhang, S.; Kumar, S. (2008). Carbon Nanotubes as Liquid Crystals, *Small*, Vol. 4, No. 9, (Sep.
    2008), pp.1270-1283, 10.1002/smll.200700082.

12

Polymeric Nanoparticles Stabilized by Surfactants: Controlled Phase Separation Approach

Sergey K. Filippov, Jiri Panek and Petr Stepanek
Institute of Macromolecular Chemistry, Academy of Sciences of the Czech Republic,
Czech Republic

1. Introduction

It is accepted nowadays that the self-assembly or self-organization occurs in a system when two types of interactions exist simultaneously between various elements of a system – a short-range attraction and a long-range repulsion. If a combination of such interactions is manifested in a system, equilibrium nanostructures/nanoparticles could occur. This general principle applies for many different systems – *e.g.*, liquid crystals, ferrofluids, lyotropic systems, surfactants and polymers. Polymers and copolymers in good solvent are widely used for creation of self-assembled nanoparticles in solution since they offer an extremely wide range of different monomers and compositions, the possibility to vary the polymer chain length and use tailor-made polymers for producing materials with specific properties and functionalities. For such polymers, no additives are required to form equilibrium nanoparticles.

This chapter reviews another technique of creating self-assembled and self-organized polymeric nanoparticles - *controlled phase separation approach*. Such approach exploits mutual interactions of a polymers and surface active molecules (surfactants or amphiphilic block copolymers) in a common solvent. We shall explore particularly dilute systems where various types of nanoparticles will be investigated. The nanoparticles will be studied keeping in mind their possible applications, especially for biological purposes - encapsulation and delivery of active substances in the case of particles and immobilization.

2. Background

The common approach applied to all types of physical systems described below is based on controlling the extent of *macrophase* separation that occurs in a mixture of two compounds (solvent and polymer) that became immiscible or incompatible as a result of a change of an external variable. This parameter can be temperature, pH or addition of a another solvent, in principle it could also be a change in pressure but the latter is not very practical since usually large pressure changes are needed to achieve relatively small changes in phase diagrams.

A phase diagram for a polymer A/polymer B or polymer/solvent system is schematically represented in Fig. 1. In the classical case the energy of the system is given by enthalpic and entropic contributions and the interaction parameter χ is given by the Flory-Huggins relation $\chi = a / T + b$ where T is absolute temperature and a and b are specific for the polymer/polymer or polymer/solvent pair. For more complex systems, a third term χ_s has to be included:

$$\chi = a / T + b + \chi_s \tag{1}$$

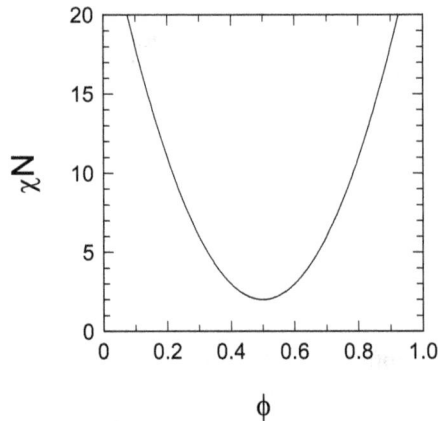

Fig. 1. Schematic phase diagram for a polymer A/polymer B or polymer/solvent. ϕ is the volume fraction of the first component of the system, N the number of monomers in the polymer chain and χ the interaction parameter describing the strength of interaction between polymer A and B or between the polymer and solvent.

In this simple representation the third term includes all additional interactions in the system, in particular the effect of different temperature expansions of the system components and that of various specific interactions in the system (hydrogen bonds, ionic interactions, ...) that may be dominant compared to the enthalpic/entropic terms a, b.

Once a macrophase separation has been initiated, the spatial extent of inhomogeneities produced by nucleation or spinodal decomposition is controlled by addition of amphiphilic molecules chosen in such a way that one part of this amphiphile interacts attractively with the nucleated material while the other part interacts attractively with the surrounding solvent. The surface of the nucleated material then becomes covered with the amphiphilic molecules which effectively terminate the phase separation and stabilizes the system in a dispersed state. The formation of nanoparticles is always driven by thermodynamics (increase of χ-parameter in Eq. (1)) but controlled by specific factors of interaction with the amphiphilic molecules including kinetic and hydrodynamic effects. Assessment of these effects and their importance for nanoparticles preparation is a primary goal in this manuscript. In the following chapter we will describe several systems where this approach can be realized:

- Controlled phase separation induced by a change in temperature.

- Controlled phase separation induced by a change of solvent
- Controlled phase separation induced by a change in pH

3. Experimental methods

To observe and to prove the formation of well defined nanoparticles we have exploited a variety of methods such as static and dynamic light scattering (SLS/DLS), small-angle neutron and X-ray scattering (SANS/SAXS) and Cryo-TEM methods.

Since kinetic factors are involved into formation of nanoparticles created by controlled phase separation method we have examined the nucleation and growth of polymeric nanoparticles using the stopped-flow technique combined with SAXS on the time scale of milliseconds.

3.1 Small-Angle Neutron Scattering (SANS)

SANS experiments were performed at CEA-Saclay on the spectrometer PAXY of the Laboratoire Leon-Brillouin. Measurements were run on a 128×128 multidetector (pixel size 0.5×0.5 cm) using a non-polarized, monochromatic (wavelength λ set by a velocity selector) incident neutron beam collimated with circular apertures for two sample-to-detector distances, namely, 1 m (with λ =0.6 nm) and 7 m (with λ = 0.8 nm). With such a setup, the investigated range of scattering wave vector modulus was 5.9×10^{-2} to 4.3 nm^{-1}. In all the cases reported in this paper, the two-dimensional scattering patterns were isotropic so that they were azimuthally averaged to yield the dependence of the scattered intensity $I_s(q)$ on the scattering vector q. Data were corrected for background scattering and detector efficiency. Intensities of neutron scattering are given in arbitrary units.

3.2 Dynamic and Static Light Scattering (DLS/SLS)

Static light scattering measurements were carried out on an ALV-6010 instrument equipped with a 22 mW He–Ne laser in the angular range 30–150°. Dynamic light scattering measurements were carried out at 90° angle. The obtained correlation functions were analyzed by REPES (Jakes, 1995) analytical software providing a distribution function, $G(R_h)$ of hydrodynamic radii R_h. To account for the logarithmic scale on the R_h axis, all DLS distribution diagrams are shown in the equal area representation, $R_hG(R_h)$. The static light scattering data were analyzed by a Zimm plot:

$$\frac{Kc}{R(q)} = \frac{1}{M_w} + \frac{R_G^2 q^2}{3M_w} \qquad (2)$$

where $R(q)$ is the Rayleigh ratio of the scattering intensity, $q=(4\pi n/\lambda)\sin\Theta/2$ is the scattering vector, $\lambda-$wave length in the medium, $\theta-$ scattering angle between the incident and the scattered beam, K is a contrast factor containing the optical parameters, c is a particle concentration, M_w is the weight average of the molar mass of the particles, and R_G is their radius of gyration. The concentration dependence was neglected which was acceptable because of the low concentrations of the solutions.

3.3 Cryo-transmission electron microscopy - Cryo-TEM

To carry out a Cryo-TEM experiment, a drop of the solution under study was placed on a pretreated copper grid which was coated with a perforated polymer film. Excess solution was removed by blotting with a filter paper. The preparation of the sample film was done under controlled environment conditions, *i.e.*, in a chamber at a constant temperature of 25°C and with a relative humidity of 98-99% to avoid evaporation of the liquid. Rapid vitrification of the thin film was achieved by plunging the grid into liquid ethane held just above its freezing point. The sample was then transferred to the electron microscope, a Zeiss 902A instrument (Carl Zeiss NTS, Oberkochen, Germany), operating at an accelerating voltage of 80 kV and in zero-loss bright-field mode. The temperature was kept below -165 °C and the specimen was protected against atmospheric conditions during the entire procedure to prevent sample perturbation and formation of ice crystals. The resolution in this method was 3-5 nm. Digital images were acquired with a BioVision Pro-SM Slow Scan CCD camera (Proscan electronische systeme, GmbH, Germany). iTEM software (Olympus Soft Imaging Solutions, GmbH, Germany) was used for image processing. The polymer concentration used was $5 \cdot 10^{-3}$ g mL^{-1}.

3.4 Small angle X-ray scattering – SAXS

All time-resolved SAXS (TR-SAXS) experiments were performed on the high brilliance beam line ID02 at the ESRF (Grenoble, France). The SAXS setup is based on a pinhole camera with a beam stop placed in front of a two-dimensional detector (X-ray image intensifier coupled to a CCD camera). The X-ray scattering patterns were recorded on the detector that was located 2 m from the sample, using a monochromatic incident X-ray beam (λ=0.1 nm). The available wave vector range was 0.04 - 2.71 nm^{-1}. Data acquisition and counting of the time t was hardware-triggered within 1 ms before the final mixing process was initiated. SAXS data were acquired with an exposure time of 50 ms per frame.

The fast mixing experiments were performed using a stopped-flow device (SFM-3, Bio-Logic) that has been specifically adapted for SAXS experiments. The device was thermostated at 25.0±0.5 °C

4. Controlled phase separation induced by a change in temperature

To fulfill this task we have exploited the phase separation of thermally sensitive PNIPAM polymer on heating above the *lower critical solubility temperature* (LCST). PNIPAM is a typical temperature-sensitive polymer that has LCST around of 32°C (detailed information on pure PNIPAM, can be found in the review (Aseyev et al., 2010) and references therein). Heating of aqueous solution of PNIPAM above LCST will initialize coil-to-globule transformation with following precipitation of the polymer. Such macrophase separation could be terminated if surface active molecules are presented in solution. Earlier we have demonstrated that well defined nanoparticles of PNIPAM could be prepared in presence of ionic and non-ionic surfactants (SDS, CTAB, Brij98, Brij97) (Konak et al., 2007). The effect of PNIPAM and surfactant concentration, and molecular weight of PNIPAM on nanoparticle parameters and on the phase transition temperature of PNIPAM solutions was investigated. It was proposed that the structure of particles is supposed to be similar to block copolymer micelles. Hydrophobic PNIPAM molecules form the insoluble core of particles and their hydrophilic shell consists of hydrophilic parts of surfactants. An intermediate shell at the core–shell interface contains both the hydrophobic parts of surfactants and PNIPAM chains. The feature of that research is that in contrast to previous studies, where surfactants were used in excess, lower concentrations of surfactants were used.

To validate the proposed model, for PNIPAM and SDS system, a contrast variation study was performed by SANS. It is important to note here that the system studied in our research is different from a so-called mesoglobule state that was also observed (Siu et al., 2003; (Aseyev et al., 2005; Kujawa et al., 2006a; Kujawa et al., 2006b) for PNIPAM. It was established in variety of papers that PNIPAM macromolecules of high molar masses on very diluted solutions might undergo through intermediate mesoglobule state with increasing of temperature above LCST. These mesoglobules are aggregates of PNIPAM molecules that consist of one or more macromolecules. They are metastable particles that are stabilized either by electrostatic or steric interactions (Kujawa et al., 2006b). No surfactant is required. Nevertheless when PNIPAM concentration in solution is rather high macroscopic precipitation occurs. In this case, surface active molecules are needed to create stable polymeric nanoparticles.

Three types of nanoparticles were tested: (i) deuterated d7-PNIPAM + protonated SDS in a 72%D_2O/28%H_2O volume mixture where the coherent scattering originates only from the surfactant. (ii) protonated PNIPAM + protonated SDS in pure D_2O where the scattering comes from the polymer and the surfactant. The whole nanoparticle should be visible. (iii) protonated PNIPAM + deuterated d25-SDS in D_2O. The scattering length density of the deuterated surfactant is almost matched by D_2O. In this case most of the scattering is produced by the polymer. The experiments have been conducted at T=42 $°C$.

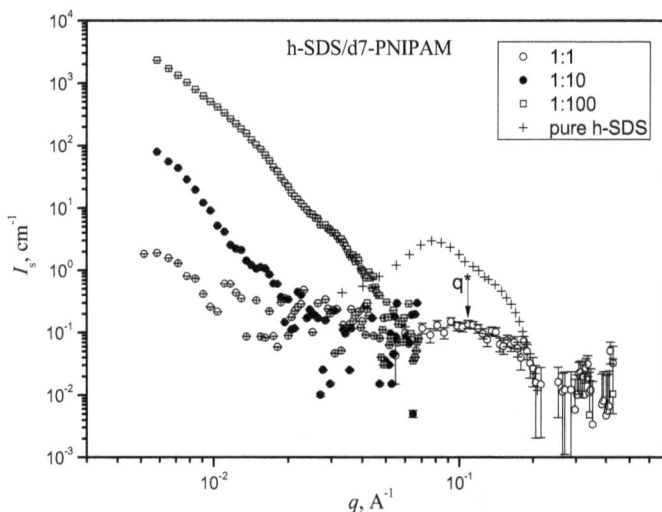

Fig. 2. Scattered intensity I_s as a function of the scattering vector q for systems d7-PNIPAM/h-SDS sample, $c_{d7-PNIPAM}$=5 g/L; (○) c_{SDS}/c_{PNIPAM} =1:1; (•) c_{SDS}/c_{PNIPAM} =1:10; (□) c_{SDS}/c_{PNIPAM} =1:100; (+) pure SDS. Data are taken from the reference (Lee & Cabane, 1997).

Earlier Cabane and Lee in their pioneer work have investigated similar the PNIPAM-SDS system by SANS (Lee & Cabane, 1997) . The polymer molar mass that have been used in their study was $1 \cdot 10^6$ g/mole and concentration of solution was mainly 30 g/L. To avoid a mesoglobule state we have selected the h-PNIPAM with $M_w = 1.88 \cdot 10^5$ g/mole and d7-PNIPAM with $M_w = 3.6 \cdot 10^5$ g/mole that is somewhat smaller than the one used by Cabane *et.al*. For the same reason, concentration of PNIPAM in all solution was kept of 5 g/L. Our work is thus a research on a similar system with different conditions.

Fig. 2-4 represents the data for different surfactant-to-polymer ratios. For all systems, the scattered intensity extrapolated to zero q is increasing with decrease of the ratio. In other words, the growth of colloidal nanoparticles is observed with decrease of surfactant-to-polymer ratio. One can see continuous evolution of the characteristic features of colloids.

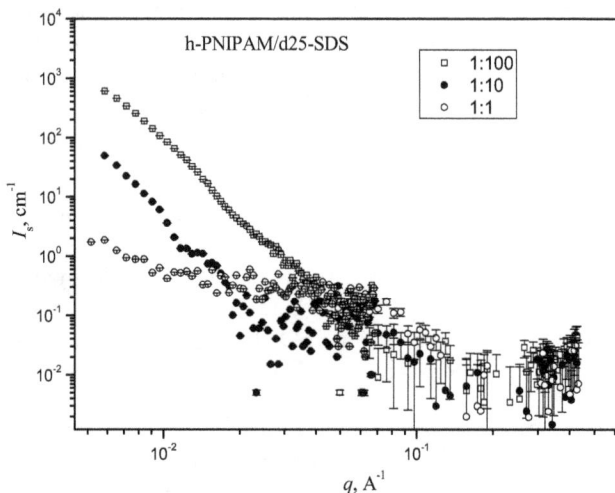

Fig. 3. Scattered intensity I_s as a function of the scattering vector q for systems h-PNIPAM/d25-SDS sample, c-PNIPAM=5 g/L; (○)c_{SDS}/c_{PNIPAM} =1:1; (●)c_{SDS}/c_{PNIPAM} =1:10; (□) c_{SDS}/c_{PNIPAM} =1:100

4.1 h-SDS/d7-PNIPAM

When a surfactant is protonated, coherent scattering comes only from the surfactant in a 72%D_2O/28%H_2O volume mixture. At low q, a q^{-4} decay is visible at c_{SDS}/c_{PNIPAM} =1:100 (Fig. 2). At high q a signal is too low. For comparison reason, the scattering of pure SDS micelles is presented on Fig. 2. No peaks that correspond to the distance between consecutive SDS micelles at high q range observed by Cabanne and Lee[8] (located at q of about 0.1 Å^{-1}) appear on the graph. We conclude that all surfactant molecules are uniformly incorporated inside of a colloidal particle or on its surface. It is worth to note that such strong q dependence indicates that the surfactant forms big structures. At c_{SDS}/c_{PNIPAM} =1:1, one can see that the scattering at low q is week and simultaneously a peak at high q range appears ($q^* = 0.11$ Å^{-1}). Obviously, colloidal particles are completely dissolved now; a pearl-necklace complex exists with SDS micelles bound to a polymer chain with the distance of 58Å ($d = 2\pi/q^*$). That finding is in good agreement with results of Cabanne where such distance was about 63 Å.

4.2 h-PNIPAM/d25-SDS

Similar features are observed for the system where the PNIPAM is only visible. Again, the formation of nanoparticles could be monitored by the growth of the scattering intensity with decrease of the composition ratio. No peaks at high q range observed are visible in this case. We conclude that PNIPAM is also uniformly distributed inside a nanoparticle.

4.3 h-PNIPAM/h-SDS

In this case both the polymer and the surfactant are visible in SANS.

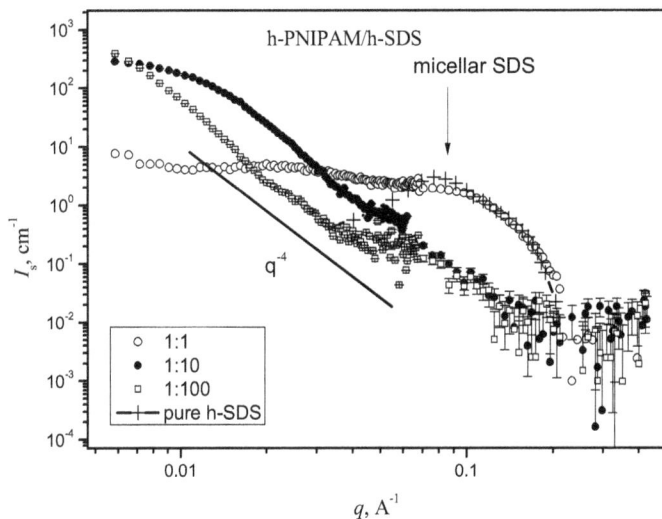

Fig. 4. Scattered intensity I_s as a function of the scattering vector q for systems h-PNIPAM/h-SDS sample, c_{PNIPAM}=5 g/L; ($\circ$)c_{SDS}/c_{PNIPAM}=1:1; ($\bullet$) c_{SDS}/c_{PNIPAM}=1:10; ($\square$) c_{SDS}/c_{PNIPAM}=1:100; (+) pure SDS. Data are taken from reference (Lee & Cabane, 1997).

The scattering curve at c_{SDS}/c_{PNIPAM} =1:100 begins at low q at high intensity; then it curves downward and continuous with q^{-4} decay (Fig. 4). This part of the scattering curve corresponds the scattering from colloidal particles. Fitting the scattering curves by form-factor of a hard sphere with Schultz-Zimm distribution provides R_g values of nanoparticles. Obtained values nanoparticles are 216, and 96 Å for ratios 1:100 and 1:10, respectively, giving corresponding outer radii 279, and 124 Å. Polydispersity value obtained from the fitting routine was 0.37 and 0.44, respectively. Cabanne et. al. reported the similar value of about 0.5. Such high numbers imply strong polydispersity in size for nanoparticles in solution.

At c_{SDS}/c_{PNIPAM} =1:1 the scattering is flat at low q (Fig. 4, 5a,b) as it could be visible from a comparison with the spectra of samples made at lower ratios (1:10 and 1:100). At high q, one can see a plateau and, beyond q=0.1 Å^{-1}, a steeper decay. This spectrum is identical to the scattering from a micellar solution of SDS at the same concentration in the absence of polymer (Fig. 4, 5a). In particular, the peak position matches the average intermicellar distance in pure SDS solutions.

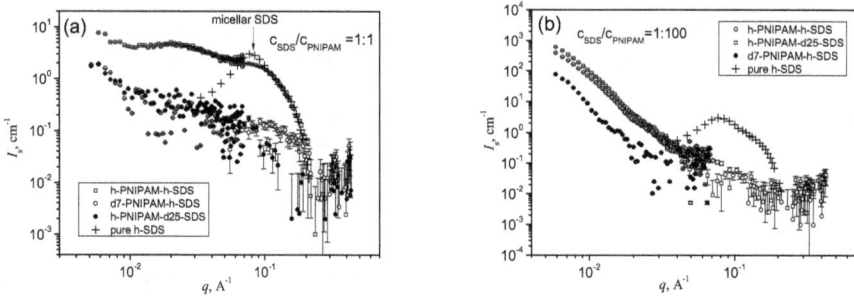

Fig. 5a. Scattered intensity I_s as a function of the scattering vector q for systems PNIPAM/SDS sample c_{PNIPAM}/c_{SDS}=1:1, c_{PNIPAM}=5 g/L; (○) d7-PNIPAM/h-SDS; (●) h-PNIPAM/d25-SDS; (□) h-PNIPAM/h-SDS; (+) pure SDS. Data are taken from reference (Lee & Cabane, 1997).

Fig. 5b. Scattered intensity I_s as a function of the scattering vector q for systems PNIPAM/SDS sample c_{PNIPAM}/c_{SDS}=1:00, c_{PNIPAM}=5 g/L; (○) h-PNIPAM/h-SDS; (●) d7-PNIPAM/h-SDS; (□) h-PNIPAM/d25-SDS; (+) pure SDS. Data are taken from reference (Lee & Cabane, 1997).

In order to determine the shape and geometric size of the particles we performed Cryo-TEM measurements for the samples at composition 1:100. The Cryo-TEM images are shown in Fig. 6.

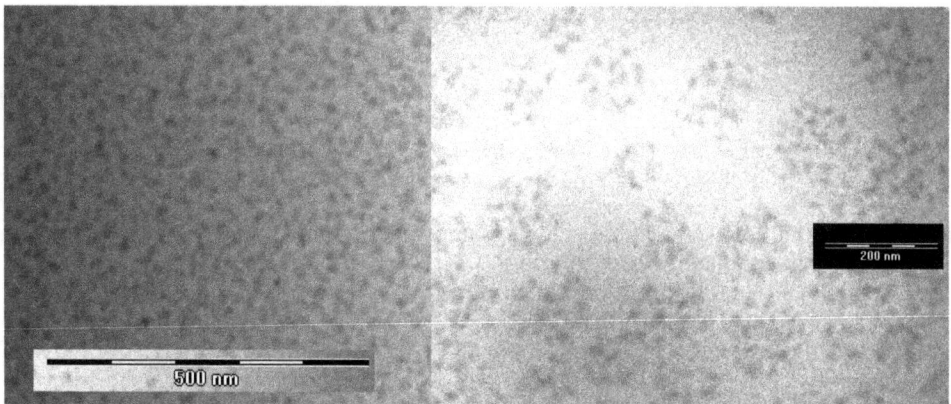

Fig. 6. Cryo-TEM micrographs. h-PNIPAM/h-SDS sample c_{PNIPAM}=5·10⁻³g mL⁻¹; c_{SDS}=5·10⁻⁵g mL⁻¹. c_{SDS}/c_{PNIPAM} =1:100

Fig. 6 shows objects thate are rather polydisperse in size. The average size of moieties is in agreement with SANS data, giving value roughly 20-30 nm in radius. The TEM images are 2D projection of the particles, observed under different angles. Therefore, we believe that the structures we see in Fig. 6 are of more or less spherical shape. Individual micelles seen in SANS experiments were not imaged in these samples, because of their small size.

5. Controlled phase separation induced by a change of solvent

5.1 Density

For a specific polymer/solvent system, phase separation could be induced by a change of the solvent. This particular case of spontaneous macrophase separation leading to formation of nano-sized droplets is frequently referred to as Ouzo effect (Ganachaud & Katz , 2005), although it has a variety of names. Some authors call this process solvent shifting (Brick et al., 2003; Van Keuren, 2004), solvent displacement(Potineni et al., 2003; Trimaille et al., 2003; Lince et al., 2008; Chu et al., 2008; Vega et al., 2008; Nguyen et al., 2008; Beck-Broichsitter et al., 2009) spontaneous emulsification (Gallardo et al., 1993; QuintanarGuerrero et al., 1997; Baimark et al., 2007; Tan et al., 2008; Katas et al., 2009) or micro/nano precipitation (Leroueil-Le Verger et al., 1998; Peracchia et al., 1999; Bilati et al., 2005; Leo et al., 2006; Legrand et al., 2006). In our previous papers (Panek et al., 2011a; Panek et al., 2011b) we have successfully tested this procedure: polymeric nanoparticles were prepared by mixing a polymer - poly(methyl methacrylate) or polystyrene - dispersed in an organic solvent with an aqueous solution of a surfactant (SDS). Since water is a bad solvent for either of the polymers they start to precipitate but the presence of a surfactant terminates the phase separation and nearly monodisperse nanoparticles appear with a typical size in the range of 50 to 300 nm. Finally the organic solvent is evaporated. We demonstrated that a variety of parameters such as polymer molar mass, surfactant hydrophobicity, solution temperature and composition influence the physico-chemical properties of nanoparticles formed in solution (Panek et al., 2011a; Panek et al., 2011b). Nevertheless, detailed information on nanoparticles structure is still missing.

Here we describe new experiments with static and dynamic light scattering that were conducted to get further insight on the internal structure of the nanoparticles. Using a combination of both methods, for the first time, we calculated such parameters as a structure factor R_g/R_h and density of nanoparticles (ρ). R_g values were measured by static light scattering. In contrast with R_h (Panek et al., 2011a) there is no detectable difference in the factor $\rho=R_g/R_h$ of nanoparticles made from ionic and non-ionic surfactants (Fig. 7).

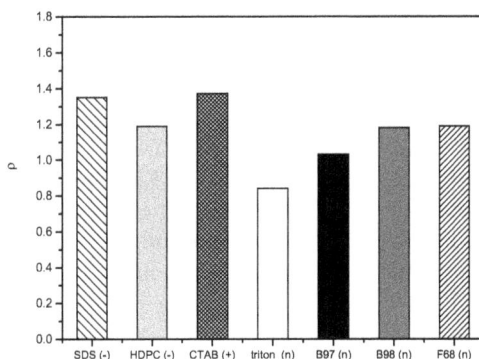

Fig. 7. Histogram of structure factor $\rho=R_g/R_h$ obtained from SLS/DLS data for PMMA(1) $c_P = 2 \cdot 10^{-4}$ g mL^{-1} for various surfactants.

According to Burchard (Burchard, 1999) this generalized ratio ρ is of special interest for establishing the particle architecture. It is varying in the range 0.8-1.4. The lowest value is for a Triton surfactant, where the particle behaves as a hard sphere (ρ is close to 0.778). The value of ρ for CTAB, SDS, Brij 97 and 98 is about 1.2-1.4. This is characteristic for several models, in particular for branched polymers, soft spheres and dendrimers.

In contrast, the density of nanoparticles is very sensitive to the nature of a surfactant (Fig. 8). The density of nanoparticles was calculated by dividing their mass obtained from static light scattering by their volume based on the R_g. Nanoparticles composed of low molecular ionic surfactants have almost two-fold higher density then the ones with polymeric non-ionic surfactants. Since all polymeric surfactants are diblock copolymers, we can assume that polymers can't adopt maximum packing structure due to steric factors.

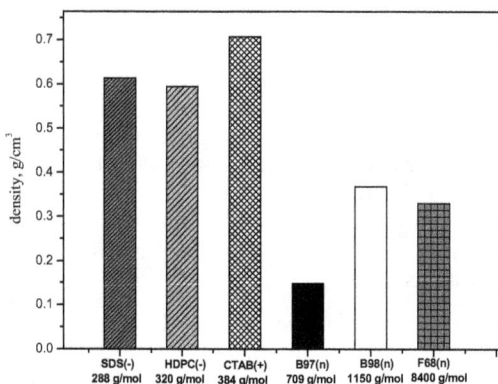

Fig. 8. Histogram of density obtained from SLS/DLS data for PMMA(1) c_{PMMA} = $2 \cdot 10^{-4}$ g mL^{-1} for various surfactants.

Obviously some voids should be inside. Such conclusion is in agreement with previous SANS and Cryo-TEM (Panek et al., 2011a). Analysis of the SANS curves supports neither a core-shell structure model of the nanoparticles nor a polymeric sphere with surfactant inclusions. Nevertheless a closer inspection of some micrographs reveals the presence of thin white hallo around a nanoparticle. Possible distribution of surfactant inside of a nanoparticle is presented on Fig. 9. The permanent entrapment of a surfactant inside nanoparticle may occur because the polymer (PMMA or PS) is in the glassy state. Plausibility of such scenario has been proven by J. Kriz *et al.* (Kriz et al., 1996) who demonstrated that the mobility of PS moieties in the core of polystyrene-block-poly(methacrylic acid) (PS-PMAc) micelles is significantly decreased, which indicates that the polymer including the surfactant inside a micelle is vitrified.

5.2 Influence of mixing rate

The effect of mixing rate (i.e. the rate at which the water solution of a surfactant is delivered into the organic solution of a polymer) on the self-assembled nanoparticles formed in the PS/SDS mixed solutions was also investigated (Fig. 10). The molecular weight of PS was varied in the range 0.9 - 30 $\cdot 10^6$ g mol^{-1}. At low molecular weights of PS, the mixing rate has

Fig. 9. Possible distribution of a surfactant (red color) and polymer (black color) inside of a nanoparticle by SANS data.

an important influence on the size of nanoparticles. Changing the mixing rate from 0.5 mL/min up to 2 mL/min makes two times smaller particles. At higher polymer molecular weights the influence of mixing rate is smaller. We conclude that lower mixing rate reduces the number of surfactant molecules in the neighborhood of polymeric nuclei formed after solvent shifting. Smaller number of surfactant molecules slow down stabilization of polymeric nuclei thus leading to forming bigger nanoparticles.

The difference in composition ratio is responsible for molecular weight dependence of nanoparticle dimensions at constant mixing rate and polymer weight concentration (Fig. 10). The bigger molecular weight of a polymer the smaller its molar concentration in mixed solution that leads to a decrease in composition ratio which governs the nanoparticle dimensions. At higher molecular weights of the polymer, the tendency is reverse, showing the growth of sizes (Fig. 10). One of the possible explanations is that macromolecules with

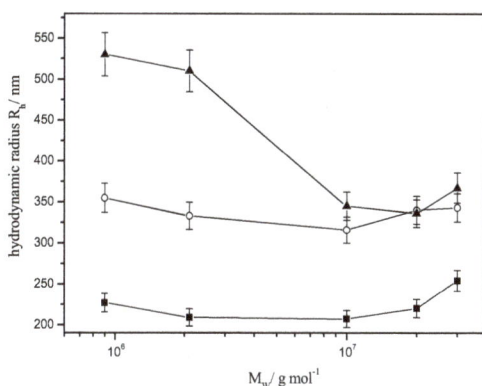

Fig. 10. Dependence of hydrodynamic radius of nanoparticles on molecular weight of polymer for PS $c_{PS} = 2 \cdot 10^{-4}$ g mL^{-1} and SDS $c_{SDS} = 5 \cdot 10^{-3}$ g mL^{-1} system at different mixing rate. ($\blacksquare$) 2 mL/min; ($\circ$) 1 mL/min; ($\blacktriangle$) 0.5 mL/min

extra large molecular weight have very low diffusion which limits the probability of surfactant molecules to find enough polymer molecules in surroundings. Fast diffusion of water molecules into polymer interface during mixing forms a surfactant-abandoned layer. In these conditions, it's energetically more favorable for a polymer chain to merge with other polymer molecules and form bigger nanoparticles in comparison with solutions of the same surfactant concentration and mixing rate.

The results presented so far show that the self-assembly in the mixed polymer/surfactant systems is rather complex. Mixing of the surfactant solutions with the polymer solutions in organic solvent results in the formation of nanoparticles, whose size can be tuned by changing the relative amounts of surfactant and polymer, as well as mixing rate.

6. Controlled phase separation induced by a change in pH

Changing of pH for a pH sensitive polymer is another way to construct polymeric nanoparticles. The pH of the solution is gradually changed so that the polymer which is in the beginning in an environment where it is molecularly soluble (i.e. in the mostly ionized form), starts to precipitate. Adding a surfactant terminates the phase separation in a controlled way leading to formation of well-defined nanoparticles with low polydispersity. We have demonstrated this procedure using a pH-sensitive hydrophobic polymer – *i.e.* poly(N-methacryloyl-L-valine) (pNMV), the extent of macrophase separation was controlled by the amphiphilic molecule Brij98. We have shown previously (Filippov et al., 2008; Filippov et al., 2010) that in a certain range of concentration and composition of the polymer/amphiphile system very monodisperse particles with size ca. 50 nm could be reproducibly prepared after a change of pH from 7 to 3.5. This change is reversible and the nanoparticles can be repeatedly created and dissolved by variation in pH. This type of particles can be very useful, since they may be able to solubilize hydrophobic drugs in large amounts and release them after a change of pH. For example the pH of stomach is 1 to 3 (nanoparticles associated), while the pH of duodenum is 7 to 8 (particles dissolved, drug released).

The nucleation of these nanoparticles has not yet been investigated. The early stages of nucleation in such systems determine the nanoscopic structure of the particles that is so far unknown, but important for their envisaged applications. Recently, new technical possibilities to study the kinetics of self-assembly were developed. Primary, it concerns so-called stopped-flow experiments combined with small-angle scattering equipments. (Narayanan et al., 2001; Grillo et al., 2003; Panine et al., 2006). A variety of nanostructures were tested by time-resolved light scattering, SAXS, and SANS. The kinetics of micelles-to-vesicles (Schmolzer et al., 2002; Weiss et al., 2005; Weiss et al., 2008; Shen et al., 1989) and lamellar-to-microemulsion (Deen et al., 2009; Tabor et al., 2009) phase transition was studied in details. Another challenging areas for time-resolved experiments are the life time of micelles (Lund et al., 2009) monomers-micelles exchange rate (Eastoe et al., 1998; Tucker et al., 2009), and nucleation of gold (Abecassis et al., 2007; Abecassis et al., 2008) and mineral nanoparticles. (Pontoni et al., 2002; Ne et al., 2003; Bolze et al., 2004). For further details on this topic, the reader is referred to reviews on the application of stopped-flow technique in SANS and SAXS. (Grillo et al., 2009; Gradzielski et al., 2003; Gradzielski et al., 2004)

We have exploited stopped-flow technique combined with SAXS to monitor early stages of nucleation on the time scale of seconds (Fig. 11). The main difference from solvent-shifting experiments described above is absence of macroscopic fluxes and solution inhomogenuities caused by mixing. In stopped-flow experiments very fast mixing setup provides solution with uniform density where nanoparticles are growing in time.

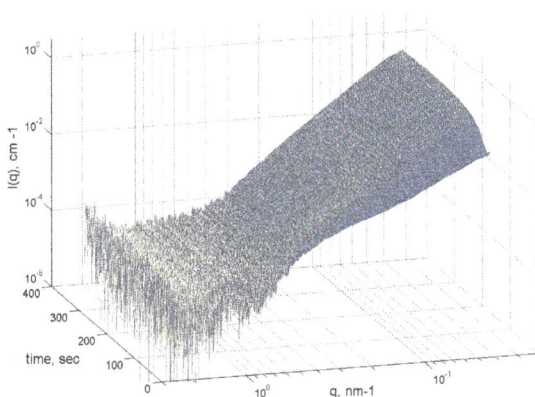

Fig. 11. The temporal evolution of the SAXS intensity for pNMGL (c_p/c_{surf}=2.0) system.

The aim of the experiment was to measure the kinetics of self-assembly of pH-sensitive polymeric nanoparticles stabilized by surfactants. The four types of pH-sensitive hydrophobic polymers that have been used in our research for the growth of nanoparticles were: (a) poly(N-methacryloyl-L-valine) (pNMV); (b) poly(N-methacryloyl glycyl-L-phenylalanyl-L-leucinyl-glycine) (pNMGPLG), and (c) poly(N-methacryloyl glycyl-L-leucine) (pNMGL). The extent of macrophase separation was controlled by the surfactants Brij 97, and Brij 98. The surfactants were different in the length of hydrophilic PEO chain.

6.1 The self-assembly of nanoparticles

Fig. 11 displays the intensity of scattered X-rays from the mixture of aqueous solution of Brij 98 surfactant and pNMGL (c_p/c_{surf}=2.0) solution as a function of time. The SAXS technique is commonly used to extract information on molecular architecture and size of nano-objects in solution that can be performed by the analysis of Kratky or Guinier plots.(Glatter & Kratky , 1982) The scattering from nanostructures reveals three regions in the dependence of scattering intensity on scattering vector, $I \sim q^\alpha$ with different behaviors characteristic for the various length scales. At low q-range, the "Guinier" regime ($qR_g < 1$) is usually attained. Middle q-range is usually sensitive to the shape of the scattering object; α = -4 stands for hard spheres, -2 stands for planar objects, and α = -1 stands for rod-like structures. It was proved experimentally that in some cases, the α value in middle q-range is not integral but rather fractional. This situation corresponds to the so-called fractal structure. In the high q-range, local stiffness of macromolecules (due to shorter length scales probed) can be revealed with $I(q) \sim q^{-1}$. Nanoparticles with sharp interface and smooth surface obey a q^{-4} law that is usually referred to as „Porod" behavior.

Several things should be noted. $I(q)$ value at the lowest experimental q grows with time, which is clearly an indication of particle growth (Fig. 11). For the highest composition ratio c_{Brij98}/c_{pNMGL} =2.0, in the middle q-range, the exponent value α is growing from -2.2 at the beginning up to -3.4 for the longer time (Fig. 12). The $q^{-2.2}$ dependence of $I(q)$ observed on early stages of self-assembly is attributed to the scattering from a loose, fractal structure. In contrast, α value of -3.4 suggests large compact objects. Thus using TRSAXS we can monitor the self-assembly of nanoparticles when particles transform through fractal structure with loose surface into hard spheres with sharp interface.

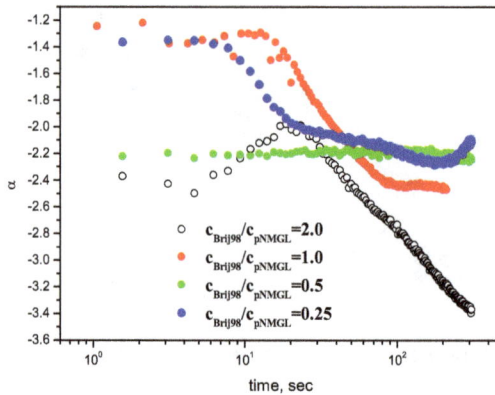

Fig. 12. The temporal evolution of the exponent value α for a pNMGL–Brij 98 system at different composition ratio c_{Brij98}/c_{pNMGL}.

For the lowest composition ratio c_{Brij98}/c_{pNMGL} =0.25, the behavior changes greatly. $I(q)$ value at the lowest q as well as the α exponent do not evolve with time (Fig. 12). Obviously, nanoparticles have been already formed prior to the first measurement. Those nanoparticles do not have sharp boundaries and have fractal structure that is a characteristic for loose entities. Surfactant molecules are not enough to cover the whole nanoparticles.

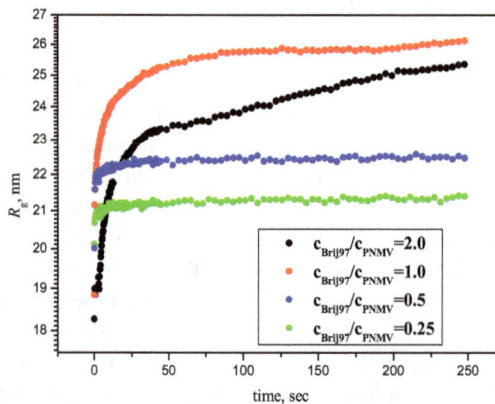

Fig. 13. The temporal evolution of the exponent value α for a pNMV –Brij 97 system at different composition ratio c_{Brij97}/c_{pNMV}.

To extract further information on the kinetics of nanoparticle formation, the radius of gyration was calculated and compared for different polymers and surfactants. The results are shown in Fig. 13-15. We observe that the radius strongly depends on the composition ratio. Moreover, two distinct regimes separated in time can be observed. During the first seconds, there is a rapid increase in the R_g value. This behavior could be explained as a nucleation regime when preliminary nuclei are formed. After a short period of time that depends also on composition ratio, the R_g value of nanoparticles increases by consumption of the remaining surfactant molecules in solution, thus defining the growth regime. The higher the composition ratio, the growth regime is more expressive (Fig. 13, 14). Nevertheless, sometimes a decrease in R_g is observed at the first seconds. We can assume that such scenario could be realized when several aggregates of pearl-necklace micelles disassembling prior to formation of original nuclei.

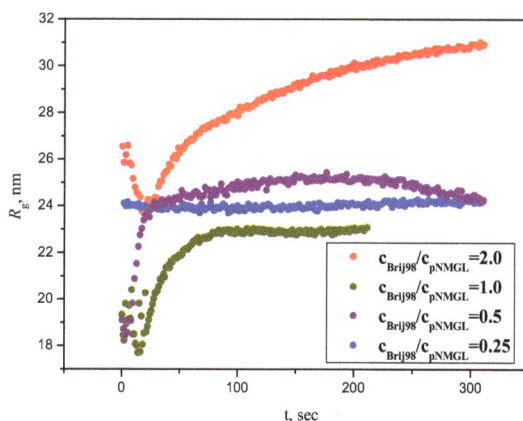

Fig. 14. The temporal evolution of the exponent value α for a pNMV–Brij 98 system at different composition ratio c_{Brij98}/c_{pNMGL}.

The conclusion that the growth regime is governed by a surfactant only is further supported from a comparison of kinetic curves of different polymers but the same composition ratio (Fig. 15). The polymers of different nature but the same surfactant (Brij 98) and composition ratio (1.0) could be arranged into a master-curve (inset of Fig. 15).

In contrast, the system with Brij 97 (red circles) is undoubtedly couldn't be superimposed into the master curve. We conclude that the number and hydrophobicity of monomeric units of a polymer determine the final size of a nanoparticle whereas the growth nucleation rates are controlled by the nature and amount of a surfactant.

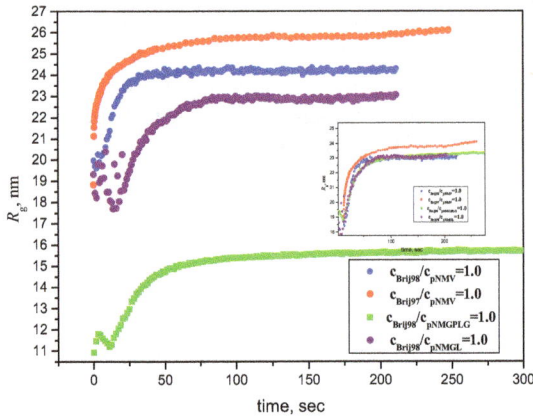

Fig. 15. The temporal evolution of the exponent value α for a pNMV, pNMGL, and pNMGPLG +Brij 98, 97 system at composition ratio 1.0. Inset: master curve.

7. General conclusions

We have systematically investigated nanoparticles prepared by *controlled phase separation approach*. On the basis of our research we have established that the most important parameter for steady-state nanoparticles dimensions is the composition ratio c(surfactant)/c(polymer). Our study demonstrates that full grown nanoparticles have a spherical shape. For the first time we have investigated the architecture of nanoparticles prepared by the solvent shifting method. The density, and factor $\rho=R_g/R_h$ measurement together with SANS experiment shows that nanoparticles are entities with uniform density and without internal structure. Polymeric and surfactant molecules are evenly distributed within a nanoparticle.

When macroscopic non-equilibrium hydrodynamic forces are involved into nanoparticle formation, the nature of the surfactant, its hydrophobicity and charge, insignificantly influences the nanoparticles sizes. A mixing rate is of primary importance for that case.

When hydrodynamic fluxes are eliminated by fast mixing again, the surfactant/polymer composition ratio is of primary importance in nanoparticle formation, thus confirming previous results. Excess of a surfactant results in much faster kinetics in comparison with the solution where a polymer is in excess. Our results suggest that the formation of the nanoparticles is a two stage process. In the beginning a nucleation stage occurs followed by a growth regime. The hydrophilicity/hydrophobicity of surfactants plays an important role in the formation of nanoparticles.

8. Acknowledgements

We gratefully acknowledge the European Synchrotron Radiation Facility (Grenoble, France) for the provision of synchrotron beam time (SC2883 and SC3113). This work was supported by the Grant Agency of the Czech Republic (202/09/2078) and also by Grant No. IAA400500805 of the Grant Agency of the Academy of Sciences of the Czech Republic. Also, we would like to thank Prof. Katarina Edwards and Dr. Goran Karlsson, Uppsala University, Department of Physical and Analytical Chemistry for help with Cryo-TEM experiments.

9. References

Abecassis, B., Testard, F., Spalla, O. & Barboux, P. (2007). Probing in situ the nucleation and growth of gold nanoparticles by small-angle x-ray scattering. *Nano Letters* Vol. 7, pp. 1723-1727.

Abecassis, B., Testard, F., & Spalla, O. (2008). Gold nanoparticle superlattice crystallization probed in situ. *Physical Review Letters*, Vol. 100, pp. 115504.

Aseyev, V., Hietala, S., Laukkanen, A., Nuopponen, M., Confortini, O., Du Prez, F.E. & Tenhu, H. (2005). *Polymer*, Vol. 46, pp. 7118-7131.

Aseyev, V., Tenhu, H. & Winnik, F. (2010). Non-ionic Thermoresponsive Polymers in Water, Self Organized Nanostrcutures of Amphiphilic Block Copolymers. Springer Berlin Heidelberg, pp. 1-61. (Advances in Polymer Science).

Baimark, Y., Srisaard, M., Threeprom J. & Narkkong, N. A. (2007). *Colloid and Polymer Science*, Vol. 285, pp. 1521-1525.

Beck-Broichsitter, M., Gauss, J., Packhaeuser, C.B. , Lahnstein, K. , Schmehl, T. , W. Seeger, T. Kissel & T. Gessler, (2009). *International Journal of Pharmaceutics*, Vol. 367, pp. 169-178.

Bilati, U., Allemann, E. & Doelker, E. (2005). *European Journal of Pharmaceutical Sciences*, Vol. 24, pp. 67-75.

Bolze, J., Pontoni, D., Ballauff, M., Narayanan, T., & Colfen, H. (2004). Time-resolved SAXS study of the effect of a double hydrophilic block-copolymer on the formation of CaCO$_3$ from a supersaturated salt solution. *Journal of Colloid and Interface Science*, Vol. 277, pp. 84-94.

Brick, M. C., Palmer, H. J. & Whitesides T. H. (2003). *Langmuir*, Vol. 19, pp. 6367-6380.

Burchard, W. (1999) *Adv Polym Sci*, Vol. 143, pp. 113-194.

Chu, B. S. , Ichikawa, S. , Kanafusa S. & Nakajima, M. (2008). *Journal of the Science of Food and Agriculture*, Vol. 88, pp. 1764-1769.

Deen, G.R., Oliveira, C.L.P. & Pedersen, J.S. (2009). Phase Behavior and Kinetics of Phase Separation of a Nonionic Microemulsion of C12E5/Water/1-Chlorotetradecane upon a Temperature Quench. *Journal of Physical Chemistry B*, Vol. 113, pp. 7138-7146.

Eastoe, J., Dalton, J.S., Downer, A., Jones, G. & Clarke, D. (1998). Breakdown kinetics of fluorocarbon micelles studied by stopped-flow small-angle X-ray scattering. *Langmuir*, Vol. 14, pp. 1937-1939.

Filippov, S., Hruby, M., Konak, C., Mackova, H., Spirkova, M. Stepanek, P. (2008). *Langmuir*, Vol. 24, pp. 9295- 9301.

Filippov, S., Starovoytova, L., Koňák, C., Hrubý, M., Macková, H., Karlsson, G. & Štěpánek, P. (2010). *Langmuir*, vol. 26, pp. 14450–14457

Gallardo, M. , Couarraze, G. , Denizot, B., Treupel, L. , Couvreur P., and Puisieux, F. (1993). *International Journal of Pharmaceutics*, Vol. 100, pp. 55-64.

Ganachaud, F. & Katz, J.L. (2005). *Chemphyschem*, Vol. 6, pp. 209-216.

Glatter, O. & Kratky, O. (1982) Small-Angle X-ray Scattering. London: Academic Press

Gradzielski, M. (2003). Kinetics of morphological changes in surfactant systems. *Current Opinion in Colloid & Interface Science*, Vol. 8:, pp. 337-345.

Gradzielski, M. (2004). Investigations of the dynamics of morphological transitions in amphiphilic systems. *Current Opinion in Colloid & Interface Science*, Vol. 9, pp. 256-263.

Grillo, I., Kats, E.I. & Muratov, A.R. (2003). Formation and growth of anionic vesicles followed by small-angle neutron scattering. *Langmuir*, Vol.19, pp. 4573-4581.

Grillo, I. (2009). Applications of stopped-flow in SAXS and SANS. *Current Opinion in Colloid & Interface Science,*Vol 14, pp. 402-408.

Jakeš, J. (1995*). Collect. Czech. Chem. C*, Vol. 60, pp. 1781-1797.

Katas, H. , Cevher E. & Alpara H. O. (2009). *International Journal of Pharmaceutics*, Vol. 369, pp. 144-154

Koňák, C., Pánek, J. & Hrubý, M. (2007). *Colloid Polym Sci.* Vol. 285, pp. 1433–1439.

Kriz, J., Masar, B., Pospisil, H., Plestil, J., Tuzar Z. & Kiselev, M.A. (1996). *Macromolecules*, Vol. 29, pp. 7853-7858

Kujawa, P., Tanaka, F. & Winnik, FM. (2006). *Macromolecules*, Vol. 39, pp. 3048-3055

Kujawa, P., Aseyev, V., Tenhu, H. & Winnik, FM. (2006). *Macromolecules* , Vol. 39, pp. 7686-7693

Lee, L.T. & Cabane, B. (1997). *Macromolecules*, Vol. 30, pp. 6559-6566

Lince, F., Marchisio D. L. & Barresi A. A. (2008). *Journal of Colloid and Interface Science*, Vol. 322, pp. 505-515.

Lund, R., Willner, L., Monkenbusch, M., Panine, P., Narayanan, T., Colmenero, J. & Richter, D. (2009). Structural Observation and Kinetic Pathway in the Formation of Polymeric Micelles. *Physical Review Letters*, Vol. 102, pp.188301

Leroueil-Le Verger, M., Fluckiger, L., Kim, Y. I. , Hoffman M. & Maincent, P. (1998*). European Journal of Pharmaceutics and Biopharmaceutics*, Vol. 46, pp. 137-143.

Leo E., Scatturin, A., Vighi E., & Dalpia A. (2006). *Journal of Nanoscience and Nanotechnology*, Vol. 6, pp. 3070-3079.

Legrand, P., Lesieur, S., Bochot, A., Gref, R., Raatjes, W., Barratt G. & Vauthier C. (2006). Conference on New Trends in Drug Delivery Systems, Paris, FRANCE.

Narayanan, T., Diat, O. & Bosecke, P. (2001). SAXS and USAXS on the high brilliance beamline at the ESRF. *Nuclear Instruments & Methods in Physics Research Section a-*

Accelerators Spectrometers Detectors and Associated Equipment, Vol. 467, pp. 1005-1009

Ne, F., Testard, F., Zemb T, & Grillo, I. (2003). How does ZrO2/surfactant mesophase nucleate? Formation mechanism. *Langmuir*, Vol.19, pp. 8503-8510.

Nguyen, J. , Steele, T. W. J., Merkel, O., Reul R. & Kissel, T. (2008). *Journal of Controlled Release*, Vol. 132, pp. 243-251.

Pánek, J., Filippov, S.K. , Koňák, Č.,Nallet, F., Noirez, L., Karlsson, G. & Štěpánek, P. (2011). *Journal of Dispersion Science and Technology*, Vol. 32, N. 6, pp. 888-897

Pánek, J., Filippov, S.K., Koňák, C., Steinhart, M. & Štěpánek, P. (2011). *Journal of Dispersion Science and Technology*, Vol. 32, N. 8, pp. 1105-1110

Panine, P., Finet, S., Weiss, T.M. & Narayanan, T. (2006). Probing fast kinetics in complex fluids by combined rapid mixing and small-angle X-ray scattering. *Advances in Colloid and Interface Science*, Vol. 127, pp. 9-18.

Peracchia, M. T., Fattal, E., Desmaele, D. , Besnard, M., Noel, J. P. , Gomis, J. M. , Appel, M. , d'Angelo J. & Couvreur P. (1999). *Journal of Controlled Release*, Vol. 60, pp. 121-128.

Potineni, A., Lynn, D. M., Langer R. & Amiji, M. M. (2003). *Journal of Controlled Release*, Vol. 86, pp. 223-234.

Pontoni, D., Narayanan, T., & Rennie AR (2002). Time-resolved SAXS study of nucleation and growth of silica colloids. *Langmuir*, Vol. 18, pp. 56-59.

QuintanarGuerrero D., Allemann, E., Doelker E. & Fessi, H. (1997). Colloid and Polymer Science, Vol. 275, pp. 640-647.

Schmolzer, S., Grabner, D., Gradzielski, M. & Narayanan T (2002). Millisecond-range time-resolved small-angle x-ray scattering studies of micellar transformations. *Physical Review Letters*, Vol. 88, pp. 258301

Shen, L., Du, J.Z., Armes, S.P. & Liu, S.Y. (1989). Kinetics of pH-induced formation and dissociation of polymeric vesicles assembled from a water-soluble zwitterionic diblock copolymer. *Langmuir*, Vol. 24, pp.10019-10025.

Siu, M.H., He, C. &Wu C. (2003). *Macromolecules,* Vol. 36, pp. 6588-6592

Tabor, R.F., Eastoe, J. & Grillo, I. (2009) Time-resolved small-angle neutron scattering as a lamellar phase evolves into a microemulsion. *Soft Matter*, Vol. 5, pp. 2125-2129.

Trimaille, T. , Chaix, C., Pichot C. & Delair, T. (2003). *Journal of Colloid and Interface Science*, Vol. 258, pp. 135-145.

Tucker, I., Penfold, J., Thomas, R.K. & Grillo, L. (2009). Monomer-Aggregate Exchange Rates in Dialkyl Chain Cationic-Nonionic Surfactant Mixtures. *Langmuir*, Vol. 25, pp. 2661-2666.

Van Keuren, E. R. (2004). *Journal of Dispersion Science and Technology*, Vol. 25, pp. 547-553.

Vega, E. , Gamisans, F., Garcia, M. L, Chauvet, A., Lacoulonche F. & Egea, M. A. (2008). *Journal of Pharmaceutical Sciences*, Vol. 97, pp. 5306-5317

Weiss, T.M., Narayanan, T., Wolf, C., Gradzielski, M., Panine, P., Finet, S. & Helsby, W.I. (2005). Dynamics of the self-assembly of unilamellar vesicles. *Physical Review Letters*, Vol. 94, pp. 038303.

Weiss, T.M., Narayanan, T. & Gradzielski, M. (2008). Dynamics of spontaneous vesicle formation in fluorocarbon and hydrocarbon surfactant mixtures. *Langmuir*, Vol. 24, pp. 3759-3766.

Yan, C. H. , Yuan, X. B., Kang, C. S., Zhao, Y. H. , Liu, J., Guo, Y. S. , Lu, J. , Pu P. Y. & Sheng J. (2008). *Journal of Applied Polymer Science*, Vol. 110, pp. 2446-2452.

13

On the Optical Response of Nanoparticles: Directionality Effects and Optical Forces

Braulio García-Cámara[1], Francisco González[1], Fernando Moreno[1],
Raquel Gómez-Medina[2], Juan José Sáenz[2] and Manuel Nieto-Vesperinas[3]
[1]*Grupo de Óptica, Departamento de Física Aplicada, Universidad de Cantabria*
[2]*Departamento de Física de la Materia Condensada, Universidad Autónoma de Madrid*
[3]*Instituto de Ciencias de los Materiales de Madrid, CSIC*
Spain

1. Introduction

Nowadays, miniaturization is a general challenge for technology. Researchers in science and technology claim to study ever smaller systems and develop ever smaller devices. The nanometric range is, at present, an important focus of attention of scientists and engineers following the famous prediction by Prof. Feynman: "There's plenty of room at the bottom". Reduction of dimensions, at this level, involves that more specific and more complex tools are needed.

Light has appeared as a convenient solution for these tasks because of its wavelength (hundreds of nanometers) and the large amount of information it contains about systems with which it interacts (Prasad, 2004). The interaction of light with small systems, either particles or structures, gives rise to several scattering phenomena which are strongly dependent on both the characteristics of the incident radiation (frequency, polarization) and those of the object (size, shape, optical properties). These interactions can be used either to obtain information about the interacting object (e.g. particle sizing) (Zhu et al., 2010) or to produce light scattering phenomena "à la carte" by means of suitable nanoobjects.

At the nanoscopic level, the interaction between an incident beam and a metallic system produces an interesting physical phenomenon which is the base of many technological applications in diverse fields like medicine, biology, communications, information storing, energy transformation, photonics, etc (Anker et al., 2008; Maier et al., 2003). This is the excitation of *localized surface plasmon resonances* (*LSPR*) (Prasad, 2004). For these, the electromagnetic field experiences a high localization in the scatterer and a strong enhancement out of the scatterer.

These advances have stimulated new research devoted to obtain a greater control over how light is scattered by these systems. Researchers have analyzed emerging structures (nanoholes (Gao et al., 2010), nanocups (Mirin & Halas, 2009), etc). But, what it is more interesting, new engineered materials, called *metamaterials* and whose optical properties can be manipulated, have been developed (Boltasseva & Atwater, 2011). The possibility to obtain structures with optical properties "à la carte" allows getting scattering phenomena never observed before in natural media, for instance negative refraction (Shalaev, 2008). The main consequences of negative refraction are the two interesting potential applications: cloacking (Pendry et al., 2006) and perfect lens (Pendry, 2000; Nieto-Vesperinas & Garcia, 2003).

The control over the values of both the electric permittivity and the magnetic permeability of an object gives us a control over the way it scatters light, and in particular, the angular distribution of the scattered radiation. This control could involve a dramatic evolution on the field of nanodevices. For this reason, the objective of this chapter is to analyze directional effects on both light scattering and optical forces of a nanoparticle with convenient optical constants. The structure of the chapter is as follows: while sections 2, 3 and 4 are devoted to the directional features on light scattering by nanoparticles, section 5 summarizes the main results on optical forces. Finally, the most important conclusions about these results are recapitulated in section 6.

2. Light scattering by nanoparticles

2.1 Mie theory

The problem of the electromagnetic scattering from an isolated and spherical particle was firstly solved in 1908 by Gustav Mie (Mie, 1908). However, this simple system still involves interesting physical behaviors that are worthy of further study.

Mie theory considers a spherical particle of radius a and optical constants given by an electric permittivity, ε_p, and a magnetic permeability, μ_p, immersed in a homogeneous and isotropic medium. This is illuminated by a linear polarized plane wave, as in Figure 1. Without loss of generality, we assume that the surrounding medium is vacuum ($\varepsilon_s=\mu_s=1$). The scattered electromagnetic field ($\mathbf{E}_s$, $\mathbf{H}_s$) can be expressed as a multipole expansion of *Vector Spherical Harmonics (VSH)*, called Mie expansion, as follows

$$\mathbf{E}_s = \sum_{n=1}^{\infty} E_n \left(i a_n \mathbf{N}_{e1n}^{(3)} - b_n \mathbf{M}_{o1n}^{(3)} \right) \tag{1}$$

$$\mathbf{H}_s = \frac{k}{\omega \mu_p} \sum_{n=1}^{\infty} E_n \left(i b_n \mathbf{N}_{o1n}^{(3)} + a_n \mathbf{M}_{e1n}^{(3)} \right) \tag{2}$$

where $k = m\omega / c = m 2\pi / \lambda$, λ being the incident wavelength in vacuum, $m = \sqrt{\varepsilon_p \mu_p}$ the refractive index of the particle, c the speed of light in vacuum and ω the angular frequency of the incident wave. E_n is defined as $E_n = E_0 i^n \dfrac{2n+1}{n(n+1)}$, E_0 being the amplitude of the incident plane wave. The series are characterized by the a_n and b_n Mie coefficients which are defined as (Bohren & Huffman, 1983)

$$a_n = \frac{\mu_s m^2 j_n(mx) \left[x j_n(x) \right]' - \mu_p j_n(x) \left[m x j_n(mx) \right]'}{\mu_s m^2 j_n(mx) \left[x h_n^{(1)}(x) \right]' - \mu_p h_n^{(1)}(x) \left[m x j_n(mx) \right]'} \tag{3}$$

$$b_n = \frac{\mu_p m^2 j_n(mx)[xj_n(x)]' - \mu_s j_n(x)[mxj_n(mx)]'}{\mu_p m^2 j_n(mx)[xh_n^{(1)}(x)]' - \mu_s h_n^{(1)}(x)[mxj_n(mx)]'} \tag{4}$$

x being the size parameter, that is defined as

$$x = ka = \frac{2\pi a}{\lambda}, \tag{5}$$

In addition, j_n are the *spherical Bessel functions* and $h_n^{(1)}$ the *spherical Bessel functions of third kind* or *Hankel functions*. As the electric and magnetic dipolar contributions are weighted by coefficients a_1 and b_1, respectively, the quadrupolar ones by a_2 and b_2 and so on, Mie coefficients a_n are associated to the electric part of the scattered electromagnetic radiation, while b_n are associated to the magnetic one.

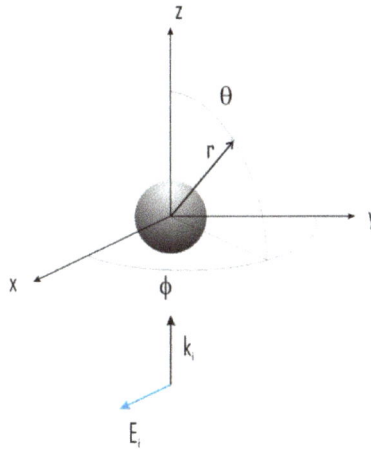

Fig. 1. Scheme of the geometry of the problem.

These coefficients contain the relevant information about essential scattering parameters as the extinction, C_{ext}, and scattering, C_{sca}, cross sections. These can be written as

$$C_{sca} = \frac{2\pi}{k^2} \sum_{n=1}^{\infty} (2n+1)(|a_n|^2 + |b_n|^2) \tag{6}$$

$$C_{ext} = \frac{2\pi}{k^2} \sum_{n=1}^{\infty} (2n+1)\text{Re}(a_n + b_n) \tag{7}$$

2.2 Details of Mie theory at the nanoscopic level

If particle size is very small compared with the incident wavelength, that is $a/\lambda \ll 1$, dipolar contributions ($n=1$ in Eqs. (1)- (2)) clearly dominate and Mie coefficients of order higher than 1 can be neglected. Thus, the Mie expansion can be simplified and the previous parameters have simple expressions

$$C_{sca} = \frac{2\pi}{k^2}[3(|a_1|^2 + |b_1|^2)]$$ (8)

$$C_{ext} = \frac{2\pi}{k^2}[3\,\mathrm{Re}(a_1 + b_1)]$$ (9)

This is the case of a nanoparticle ($a < 50\ nm$) when it is illuminated by an incident wave in the visible or near infrared (NIR) part of the spectrum ($\lambda > 500\ nm$).

The predominant dipolar conduct, either electric or magnetic, of nanoparticles is usually described by the electric and/or magnetic complex polarizabilities, α_e and α_m, respectively. Both can also be expressed as a function of the two first Mie coefficients

$$\alpha_e = \frac{\alpha_e^{(0)}}{1 - i\frac{2}{3}k^3\alpha_e^{(0)}} = \frac{3i}{2k^3}a_1$$ (10)

$$\alpha_m = \frac{\alpha_m^{(0)}}{1 - i\frac{2}{3}k^3\alpha_m^{(0)}} = \frac{3i}{2k^3}b_1$$ (11)

where

$$\alpha_e^{(0)} = 4\pi a^3 \frac{\varepsilon_p - 1}{\varepsilon_p + 2}$$ (12)

$$\alpha_m^{(0)} = 4\pi a^3 \frac{\mu_p - 1}{\mu_p + 2}$$ (13)

are the static polarizabilities, defined in the limit $ka \to 0$.

3. Directional effects on light scattering by nanoparticles with arbitrary values of ε and μ.

3.1 Kerker's theory

In the early eighties, M. Kerker and co-authors (Kerker et al., 1983) presented an interesting study about the scattering properties, in the far field, of a spherical particle much smaller than the incident wavelength, illuminated by a plane wave and without any restriction for the values of its relative optical constants (ε and μ). Some interesting electromagnetic scattering effects were described in this work such as the zero-backward and the zero-forward scattering. Although the idea of a magnetic permeability different from 1 in the visible range was hypothetical and the described effects were thought to be impossible to be observed when the work was presented, the engineered metamaterials have currently revitalized these electromagnetic studies (Zhedulev, 2010).

In this section, the main theoretical aspects described by M. Kerker et al. are briefly reviewed.

3.1.1 Zero-backward scattering: First Kerker's condition

When we consider a system, like that of Figure 1, the scattered intensity in the scattering plane can be described by means of two polarized components: I_{TE} and I_{TM}. While I_{TE} corresponds to an incident electric field parallel to the scattering plane, I_{TM} corresponds to a perpendicular one. These components can be written as (Bohren & Huffman, 1983)

$$I_{TE} = \frac{\lambda^2}{4\pi r^2} \left| \sum_n \frac{2n+1}{n(n+1)} (a_n \pi_n + b_n \tau_n) \right|^2 \tag{14}$$

$$I_{TM} = \frac{\lambda^2}{4\pi r^2} \left| \sum_n \frac{2n+1}{n(n+1)} (a_n \tau_n + b_n \pi_n) \right|^2 \tag{15}$$

where r is the distance from the particle to the observer ($2\pi r/\lambda >> 1$) and π_n and τ_n are angular functions defined in (Bohren & Huffman, 1983). As we are considering a very small or dipole-like particle ($a \rightarrow 0$), only the two first Mie coefficients (a_1 and b_1) are introduced in the expressions. In addition some approximations can be applied to these coefficients in such a way that the scattered intensity components can be approximated by

$$I_{TE} = \frac{\lambda^2 x^6}{4\pi r^2} \left| (a_1 + b_1 \cos\theta) \right|^2 = \frac{\lambda^2 x^6}{4\pi r^2} \left| \left((\frac{\varepsilon-1}{\varepsilon+2}) + (\frac{\mu-1}{\mu+2}) \cos\theta \right) \right|^2 \tag{16}$$

$$I_{TM} = \frac{\lambda^2 x^6}{4\pi r^2} \left| (a_1 \cos\theta + b_1) \right|^2 = \frac{\lambda^2 x^6}{4\pi r^2} \left| \left((\frac{\varepsilon-1}{\varepsilon+2}) \cos\theta + (\frac{\mu-1}{\mu+2}) \right) \right|^2 \tag{17}$$

θ being the scattering angle, defined as the angle between the incident and the scattered directions (see Figure 1).

For the backward scattering direction ($\theta = 180°$) the previous expressions adopt the following forms

$$I_{TE}(180°) = \frac{\lambda^2}{4\pi r^2} x^6 \left| \left((\frac{\varepsilon-1}{\varepsilon+2}) - (\frac{\mu-1}{\mu+2}) \right) \right|^2 \tag{18}$$

$$I_{TM}(180°) = \frac{\lambda^2}{4\pi r^2} x^6 \left| \left(-(\frac{\varepsilon-1}{\varepsilon+2}) + (\frac{\mu-1}{\mu+2}) \right) \right|^2 \tag{19}$$

It easy to observe that when $\varepsilon = \mu$, or equivalently when $\alpha_e = \alpha_m$, the scattered intensity in the backward direction is zero for both incident polarizations. This is the *zero-backward scattering condition* and we shall call in the following the *first Kerker's condition*. In Figure 2 the scattering pattern of a dipole-like particle with relative optical properties, $\varepsilon = \mu = 3$ is shown. Only a TM polarization is considered because, from Eqs. (18) and (19), the scattered intensity is equal for both polarizations under this condition.

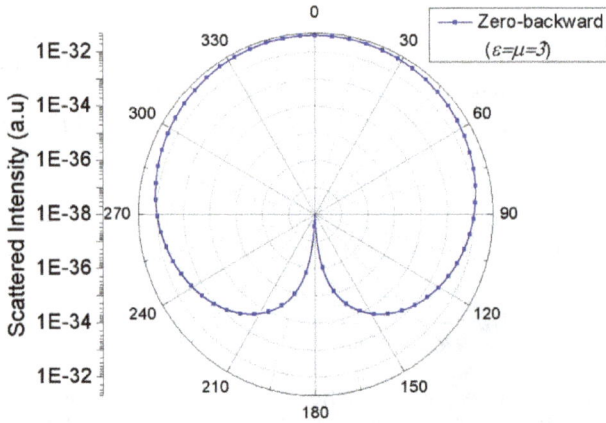

Fig. 2. Scattering diagram of a dipole-like particle ($a = 10^{-6}\lambda$) with relative optical properties fulfilling the zero-backward condition and for a TM incident polarization

3.1.2 Zero-forward scattering: Second Kerker's condition

For $\theta = 0°$ (forward scattering direction), Eqs. (16)-(17) become

$$I_{TE}(0°) = \frac{\lambda^2}{4\pi r^2} x^6 \left| \left(\frac{\varepsilon-1}{\varepsilon+2} \right) + \left(\frac{\mu-1}{\mu+2} \right) \right|^2 \qquad (20)$$

$$I_{TM}(0°) = \frac{\lambda^2}{4\pi r^2} x^6 \left| \left(\frac{\varepsilon-1}{\varepsilon+2} \right) + \left(\frac{\mu-1}{\mu+2} \right) \right|^2 \qquad (21)$$

In this case, the $\varepsilon-\mu$ relation which cancel $I_{TE}(0°)$ and $I_{TM}(0°)$ is not as evident as before. However, Kerker et al. (Kerker et al., 1983) demonstrated that this happens when

$$\varepsilon = \frac{4-\mu}{2\mu+1} \qquad (22)$$

which is equivalent to $Re(\alpha_e) = -Re(\alpha_e)$ and $Im(\alpha_e) = Im(\alpha_e)$. This is the *zero-forward scattering condition*, that we shall call the *second Kerker's condition*.

It is interesting to highlight that this condition is symmetric. This means that it remains invariant by interchanging ε and μ. An example of the angular distribution of the scattered intensity of a very-small particle satisfying this condition is shown in Figure 3 for a TM polarized incident beam (TE polarization produces a similar result).

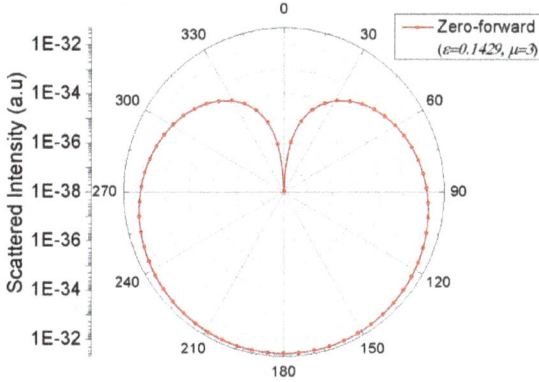

Fig. 3. Scattering diagram of a dipole-like particle ($a = 10^{-6}\lambda$) with relative optical properties fulfilling the zero-forward condition, $(\varepsilon; \mu) = (0.1429; 3)$, and TM incident polarization.

3.2 An analysis of Kerker's conditions

Kerker's theory was developed under the far-field approximation and for the very particular case of dipole-like particles for which only the two first Mie coefficients (a_1 and b_1) are non negligible. However, as particle size increases or the observer approaches, other multipolar terms become important and the directional features, described previously, can be modified.

3.2.1 Size effects on the directionality conditions

One of the responsibles for the appearance of multipolar contributions is the size of the particle, a. When this deviates from the condition $a/\lambda \ll 1$, orders in the Mie expansion greater than 1 start to be non negligible. The purpose of this section is to analyze size effects on the two Kerker's conditions (García-Cámara et al., 2010a).

Zero-backward scattering condition can be extended even for large particle sizes. This is possible because the ε–μ symmetry of Mie coefficients (Eqs. (3)-(4)) ensures that all the electric and magnetic multipolar contributions are equal and with opposite sign at backward direction. This produces a destructive interferential effect between both contributions for every multipolar order and for a given particle size, a. Figure 4 shows the scattering diagrams for several particles with different size (a) and optical properties satisfying the zero-backward scattering condition ($\varepsilon = \mu$).

On the contrary, the zero-forward scattering condition is much more sensitive to size effects. In fact, as a increases and multipolar terms, other than the dipolar ones, become important, the electric and magnetic contributions in the forward direction do not interfere destructively anymore, and the zero-forward-scattering tends to disappear. In spite of this, it is possible to find pairs $(\varepsilon; \mu)$ which minimizes the scattered intensity in the forward direction. In Figure 5, the distribution of the scattered intensity for spherical particles of different sizes is plotted. The values of ε and μ, which are included in the figure caption, were chosen such that a minimum of the scattered intensity in the forward direction appears. For the smallest value of a, the scattered intensity in the forward direction is considerably lower compared to other angles. However, as a increases, this minimum becomes less pronounced due to the influence of quadrupolar terms.

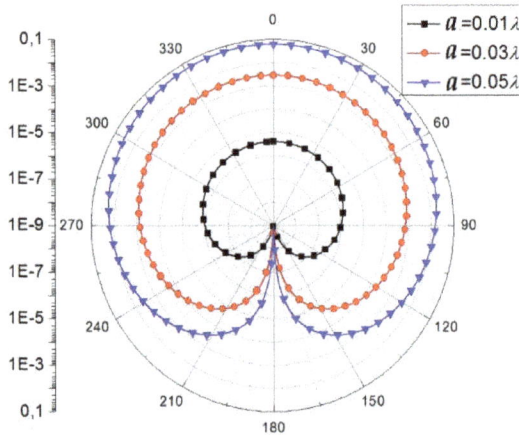

Fig. 4. Scattering diagrams, in logarithmic scale, for a spherical particle with relative optical properties $(\varepsilon; \mu)$ = (-3;-3) and illuminated by a TE-polarized incident light.
Several particlesizes have been considered.

In a recent research (García-Cámara et al., 2010a), it was found that these optical constants which minimize forward scattering don't follow Kerker's conditions but can be fitted to a formally similar expression where fitting coefficients are dependent on particle size.

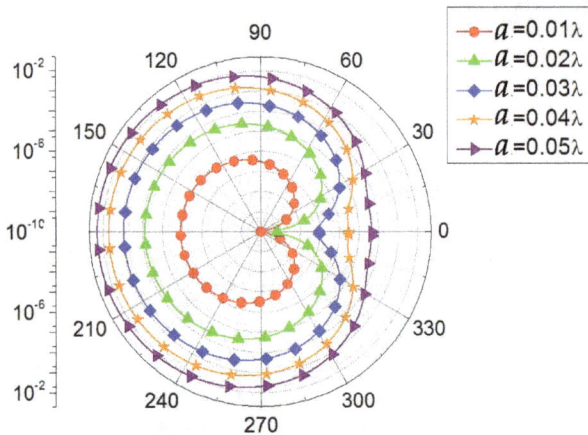

Fig. 5. Scattering diagrams, in logarithmic scale, for a spherical particle illuminated with a TE linearly polarized incident beam. For each particle size, optical properties, in the negative-negative range, are such that the scattered intensity is minimum in the forward direction. In particular, μ=-4.55 for every particle size and ε=-1.06 (a=0.01λ), ε=-1.07 (a=0.02λ), ε=-1.09 (a=0.03λ), ε=-1.11 (a=0.04λ) and ε=-1.13 (a=0.05λ).

3.2.2 Distance effects on the directionality conditions: From far to near-field

Kerker's conditions, as have been remarked above, were deduced under the far-field approximation, that is $(2\pi r/\lambda >> 1)$. If the observer tends to approach $(r/\lambda \leq 1)$, directional effects on light scattering are affected. In a recent work (García-Cámara et al. 2010b), it has been shown that directional effects on light scattering of nanoparticles with optical properties under Kerker's conditions tends to disappear as r decreases. Figure 6 shows the scattered intensity measured on a line crossing a nanoparticle $(a \sim 0.01\lambda)$ from the backward to the forward direction (Z-axis). Figure 6(a) is devoted to a particle satisfying the first Kerker's condition, while Figure 6(b) shows the same result when its relative optical constants fulfill the second Kerker's condition (eq. (22)). In both cases incident light is P-polarized (an orthogonal polarization produces similar results) and the case of a particle with the same value of ε and $\mu=1$ (non-directional case) is also plotted, for comparison purposes. For observation distances, $r>0.16\lambda$, the directionality effects appears through a remarkable drop of the scattered intensity in either the backward (Figure 6a) or the forward direction (Figure 6b). However, as the observer approaches $(r \leq 0.16\lambda)$, the evolution with the observation distance of the scattered intensity of a nanoparticle with directional features tends to that of a nanoparticle which optical constants do not satisfied any Kerker's condition.

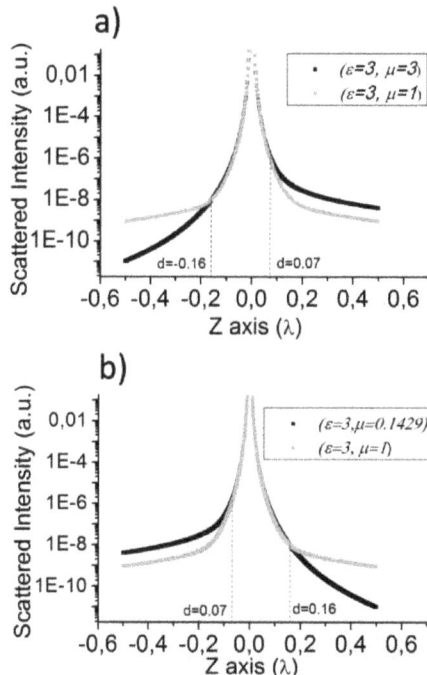

Fig. 6. Scattered intensity by a nanoparticle of radius $a=0.01\lambda$ and relative optical constants satisfying a) the first Kerker's condition $(\varepsilon=\mu=3)$ or b) the second Kerker's condition $(\varepsilon=3; \mu=0.1429)$ as a function of the distance from the particle surface in a direction parallel to the incident direction. For comparison, we have also included the case of a particle with $(\varepsilon=3; \mu=1)$. In both cases the incident beam is polarized with the electric field parallel to the scattering plane.

3.3 A generalization of the Kerker's conditions

3.3.1 The zero-forward scattering condition and the optical theorem

In a recent research, Alù et al. (Alù & Engheta, 2011) stated that the *zero-forward scattering condition* (Eq. 22) is incongruent with the Optical Theorem. This relates the extinction efficiency (Q_{ext}) and the scattering amplitude in the forward direction $[S(0^o)]$ as follows (Bohren & Huffman, 1983)

$$Q_{ext} = \frac{4}{x^2} \text{Re}\{S(0^\circ)\}$$ (23)

When the *zero-forward scattering condition* holds, $S(0^o)=0$ and then $Q_{ext}=0$. This would imply that the particle would not scatter neither absorb electromagnetic radiation. However, in the examples shown in Figures 3 and 5, while the absorption is null because the optical constants are real, light scattering, and then the extinction efficiency, is non-zero at scattering angles other than $\theta=0^o$.

A first attempt to solve this apparent paradox is found in (Chylek & Pinnick, 1979) where they conclude that the dipolar approximation used by Kerker and co-workers is a non-unitary approximation because $Re(a_n) \geq |a_n|^2$, $Re(b_n) \geq |b_n|^2$ are not satisfied, and therefore the Optical Theorem cannot be applied. However, other more specific solutions to this paradox have been proposed recently. Alù et al (Alù & Engheta, 2011) established that, for a correct estimation of Q_{ext} it is crucial to include the radiative correction (Draine & Flatau, 1994) into the two first Mie coefficients (a_1 and b_1). From these considerations, energy conservation is warranted and, although the forward scattering is not zero, it is minimum with respect to other scattering angles. In addition, if the radiative correction is also included in the deduction of the *zero-forward scattering condition* (García-Cámara et al., 2011), a new condition can be found where both the Optical Theorem and the zero scattering at $\theta=0^o$ hold. This condition follows the equation

$$\varepsilon = \frac{\pi(4-\mu)-iVk^3(\mu-1)}{\pi(2\mu+1)-iVk^3(\mu-1)}$$ (24)

where V is the volume of the particle.

3.3.2 Directional effects at scattering angles other than forward and backward directions

Previous analysis on the distribution of the scattered intensity by a nanoparticle at both the forward and the backward direction can also be extended to other scattering angles. In a previous work (García-Cámara, 2010a), it is shown that by choosing a certain scattering angle different from 0° and 180°, there are pairs $(\varepsilon; \mu)$, which produce minimum scattered intensity within the scattering plane.

In Figure 7, we plot the scattering diagrams of a nanoparticle ($a = 0.01\lambda$) illuminated by a TE polarized incident beam. The optical constants are such that the scattered intensity is minimum at representative angles like 30°, 60°, 120° and 150°. Each diagram shows a double-lobe structure with the position of the minimum depending on the particular values of the relative electric permittivity (ε) and the relative magnetic permeability (μ). Therefore, a suitable tuning of the material optical constants serves to control the angular position of the minimum of the scattered intensity.

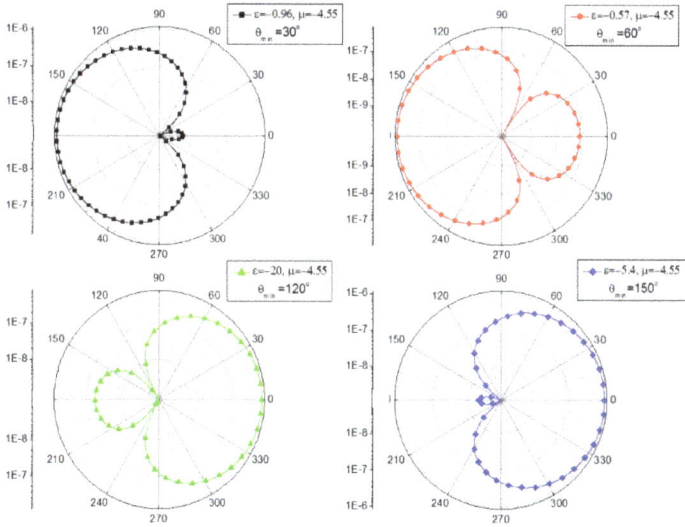

Fig. 7. Scattering diagrams of a spherical particle with $a = 0.01\lambda$ and relative optical constants in the negative-negative range (labeled in the figure) which produce a minimum scattering at certain scattering angles. The particle is illuminated with a linearly polarized incident plane wave with the electric field perpendicular to the scattering plane (TE polarization).

4. Directional effects on light scattering by dielectric particles

Previous analysis presented in this chapter about directional effects of light scattering have been done for nanoparticles with arbitrary values of both the relative electric permittivity (ε) and the relative magnetic permeability (μ) which do not correspond to any real material. In general, conventional materials do not show any stimulus to the magnetic field of electromagnetic radiation in the visible or the near-infrared region of the electromagnetic spectrum. For this reason, previous analyses have been considered an entelechy, as V. Veselago did in his work to generate and enrich scientific knowledge (Veselago, 1968). Very recently and looking for real situations, it has been shown that submicrometer particles made of Silicon (Evlyukhin et al, 2010 ; García-Extarri et al., 2011) or Germanium (Gómez-Medina et al, 2011b) present both effective electric and magnetic responses, corresponding to the dipolar contributions characterized by their first-order Mie coefficients, in the near-infrared range. Either of them can be selected by changing the illumination wavelength.

For this kind of nanoparticles, the spectral proximity of both dipolar electric and magnetic responses allows the appearance of coherent effects between dipolar modes. Consequently, under certain conditions, these scatterers are able to satisfy Kerker's conditions. Following the work made by Gómez-Medina et al. (Gómez-Medina et al, 2011b), in Figure 8, the electric (α_e) and the magnetic (α_m) polarizabilities of a Ge nanoparticle of radius $a=240nm$ are plotted as a function of the wavelength (λ) of the incident radiation. In the considered spectral range, Germanium has a refractive index which can be well approximated by a real constant $m=4$ (Palik, 1985). Also the spectral evolution of the extinction efficiency (Q_{ext}) has been included in order to show the resonant behaviors that appear in a Ge nanoparticle. A dipolar electric (DE) mode arises at $\lambda=1823\ nm$, while a dipolar magnetic (DM) resonance is located at $\lambda=2193\ nm$. The vertical lines point the wavelengths at which either the first ($\alpha_e=\alpha_m$) or the second ($Re(\alpha_e)=-Re(\alpha_e)$ and $Im(\alpha_e)=Im(\alpha_e)$) Kerker's condition are fulfilled.

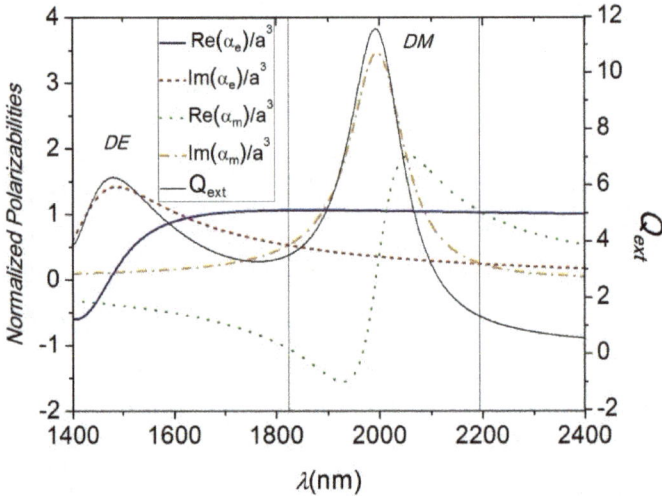

Fig. 8. Real and imaginary parts of the electric (α_e) and the magnetic (α_m) polarizabilities for a Ge nanoparticle ($a=240nm$). The refractive index of Germanium, in the considered range, can be considered as real and constant, $m\cong4+0i$. The wavelengths at which the first and second Kerker's conditions ($\lambda=2193nm$ and $\lambda=1823\ nm$, respectively) are satisfied, are identified with vertical lines. Also, for comparison purposes, the extinction efficiency is plotted identifying the dipolar electric (DE) and the dipolar magnetic (DM) resonances.

The fact that a dielectric and non-magnetic particle ($\varepsilon>0$ and $\mu=1$) presents both dipolar electric and also dipolar magnetic modes is quite interesting and could be useful for potential applications. For instance, this kind of resonances has been currently used for several tasks in a wide range of fields, ranging from the design of nanodevices (Maier et al, 2003; Anker et al., 2008) to biomedical treatments (Zemp, 2009). Unfortunately, they were observed only in metallic materials which present strong absorption losses. One of the advantages of dielectric materials, like Germanium or Silicon, is that they show negligible absorption in the considered range (Palik, 1985) and then losses are almost absent.

The position and shape of the dipolar resonances shown in Fig. 8 for Ge particles (similarly for Si particles) produces interesting coherent effects between them and consequently a natural way of reproducing Kerker's conditions by means of real materials. In order to verify that these directional features show up, Figure 9 plots the scattering diagrams of a Ge nanosphere (*a=240nm*) when the incident wavelengths are those marked by vertical lines in Figure 8. The *zero-backward scattering condition* is satisfied for *λ=1823 nm,* and there is no scattered intensity in this direction (Figure 9a). However, the *zero-forward scattering condition* is strongly affected by size effects (Figure 9b). As was described in Section 3.2.1, the size of the particle prevents scattered intensity to be completely suppressed in the forward direction. However, its value is very small compared with those at other scattering angles and most part of the scattered intensity is located in the backward hemisphere ($\pi<\theta<2\pi$).

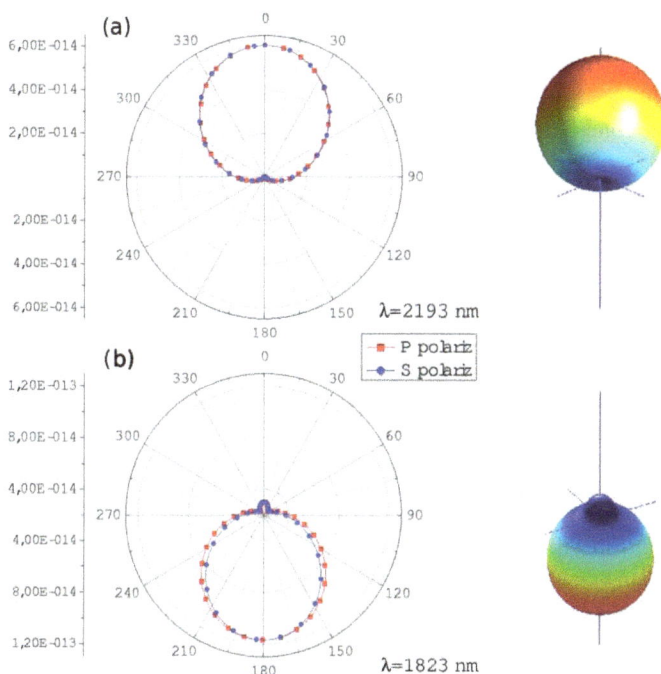

Fig. 9. Scattering diagrams for a Ge nanoparticles (*a=240nm*) illuminated by a linear polarized plane wave. Both polarizations, with the incident electric field parallel (TM or P polarization) or normal (TE or S polarization) to the scattering plane are considered. The incident wavelength is labeled in the figure. From (Gómez-Medina et al., 2011b).

Previous results for Germanium can also be extended to Silicon nanoparticles. These behaviors in Silicon could be even more interesting due to the wide range of applications of this material. Silicon is the base of microelectronics due to its semiconductor character and also to its abundance in Earth. For this reason, the industry of Silicon is very well developed. These new scattering features in the nanometric range could be the base for the development of new silicon applications as, for instance, optical nanocircuits.

5. Optical forces

Light carries energy and both linear and angular momenta that can be transferred to atoms, molecules and particles. Demonstration of levitation and trapping of micron-sized particles by radiation pressure dates back to 1970 and the experiments reported by Ashkin and co-workers (Ashkin, 1970). Light forces on small particles are usually described as the sum of two terms: the dipole or gradient force and the radiation pressure or scattering force (Askhin et al., 1986; Neuman & Block, 2004; Novotny & Hecht, 2006; Chaumet & Nieto-Vesperinas 2000b; Gómez-Medina et al., 2001; Chaumet & Nieto-Vesperinas, 2002; Nieto-Vesperinas et al., 2004; Gómez-Medina & Saénz, 2004). There is an additional non-conservative curl force arising in a light field of non-uniform ellipticity that is proportional to the curl of the spin angular momentum of the light field (Albaladejo et al., 2009a; Nieto-Vesperinas et al., 2010). In analogy with electrostatics, small particles develop an electric (magnetic) dipole moment in response to the light electric (magnetic) field. The induced dipole is then drawn by field intensity gradients which compete with radiation pressure due to momentum transferred from the photons in the beam. By fashioning proper optical field gradients it is possible to trap and manipulate small dielectric particles with optical tweezers (Askhin et al, 1986; Neuman & Block, 2004) or create atomic arrays in optical lattices (Verkerk et al., 1992; Hemmerich & H'ansch, 1993). Intense optical fields can also induce significant forces between particles (Burns et al., 1989; Burns et al., 1990; Tartakova et al., 2002; Chaumet & Nieto-Vesperinas, 2001; Gómez-Medina & Saénz, 2004). Some previous work focused on optical forces on macroscopic media, either with electric (Mansuripur, 2004) or magnetic response (Kemp et al., 2005; Mansuripur, 2007), or particles with electric response (Kemp et al., 2006a). Radiation pressure forces on dielectric and magnetic particles under plane wave incidence have been computed for both small cylinders (Kemp et al., 2006b) and spheres (Lakhtakia & Mulholland, 1993; Lakhtakia, 2008). The total force on an electric and magnetic dipolar particle has been shown (Chaumet & Rahmani, 2009; Nieto-Vesperinas et al., 2010; Nieto-Vesperinas et al., 2011; Gómez-Medina et al., 2011a; Gómez-Medina et al., 2011b) to have a similarity with that previously obtained for electric dipoles. Moreover, in the presence of both electric and magnetic responses, the force presents an additional term proportional to the cross product of the electric and magnetic dipoles (Chaumet & Rahmani, 2009; Nieto-Vesperinas et al., 2010; Nieto-Vesperinas et al., 2011; Gómez-Medina et al., 2011a; Gómez-Medina et al., 2011b). The relevance and physical origin of this electric-magnetic dipolar interaction term for a single particle has been recently discussed (Nieto-Vesperinas et al., 2010; Nieto-Vesperinas et al., 2011; Gómez-Medina et al., 2011a; Gómez-Medina et al., 2011b)

5.1 Force on a small particle with electric and magnetic response to an electromagnetic wave

We consider a dipolar particle embedded in a non-dissipative medium with relative dielectric permittivity ε and magnetic permeability μ, subjected to an incident electromagnetic field whose electric and magnetic vectors are $\mathbf{E}^{(i)}$ and $\mathbf{B}^{(i)}$, respectively. The total time-averaged electromagnetic force acting on the particle is (Chaumet & Nieto-Vesperinas, 2000; Jackson, 1998; Nieto-Vesperinas et al., 2010):

$$\langle \mathbf{F} \rangle = \frac{1}{8\pi} \Re \left\{ \int_S \left[\varepsilon(\mathbf{E} \cdot \mathbf{s})\mathbf{E}^* + \mu^{-1}(\mathbf{B} \cdot \mathbf{s})\mathbf{B}^* - \frac{1}{2}\left(\varepsilon|\mathbf{E}|^2 + \mu^{-1}|\mathbf{B}|^{-1} \right)\mathbf{s} \right] dS \right\}, \qquad (25)$$

where $\Re$ stands for real part, dS denotes the element of any surface S that encloses the particle.

The fields in Eq. (25) are total fields, namely the sum of the incident and scattered (re-radiated) fields: $E^{(i)}+E^{(r)}$, $B^{(i)}+B^{(r)}$. **s** is its local outward unit normal. A time dependence $e^{(-i\omega t)}$ is assumed throughout. For a small particle, within the range of validity of the dipolar approximation, the scattered field corresponds to that radiated by the induced electric and magnetic dipole moments, **p** and **m**, respectively. In this case, Eq. (25) leads to the expression

$$\langle \mathbf{F} \rangle = \frac{1}{2} \Re \left\{ \mathbf{p} \left(\nabla \otimes \mathbf{E}^{(i)*} \right) + \mathbf{m} \left(\nabla \otimes \mathbf{B}^{(i)*} \right) - \frac{2k^4}{3} \sqrt{\frac{\mu}{\varepsilon}} \left(\mathbf{p} \times \mathbf{m}^* \right) \right\} \tag{26}$$

Equation (26) represents the generalization of the result of (Chaumet & Rahmani, 2009) for the time-averaged force on a particle immersed in an arbitrary medium with refractive index: $m = \sqrt{\varepsilon\mu}$. The wavenumber is $k = m\omega/c$, ω being the frequency. The symbol $\otimes$ represents the dyadic product so that the matrix operation: $\mathbf{W}(\nabla \otimes \mathbf{V})$ has elements $W_j \partial_j V_j$ for $i, j = 1, 2\ 3$. All variables in Eq. (26) are evaluated at a point $r = r_0$ in the particle. The first term of Eq. (26) is the force $< F_e >$ exerted by the incident field on the induced electric dipole, the second and third terms $< F_m >$ and $< F_{em} >$ are the force on the induced magnetic dipole and the force due to the interaction between both dipoles (Chaumet & Rahmani, 2009; Nieto-Vesperinas et al., 2010).

5.2 Optical theorem and forces on an electric and magnetic dipolar particle

The question of energy conservation has been recurrently addressed and debated as regards small particles (Chýlek & Pinnick, 1979; Lock et al., 1995), especially in connection with magnetic particles that produce zero-forward scattering intensity (Alù & Engheta, 2011; Nieto-Vesperinas et al., 2011; García-Cámara et al., 2011; Gómez-Medina et al., 2011b). It is thus relevant to explore the formal analogy between the force as momentum "absorption" rate and the optical theorem expressing the conservation of electromagnetic energy. From the Poynting's theorem (Bohren & Huffman, 1983; Jackson, 1998), the rate $-W^{(a)}$ at which energy is being absorbed by the particle is given by

$$-W^{(a)} = \int_S \left\{ \langle \mathbf{S} \rangle - \langle \mathbf{S}^{(i)} \rangle \right\} \mathbf{s} dS \tag{27}$$

$$= \frac{c}{8\pi m} \Re \left\{ \int_S \varepsilon (\mathbf{E} \cdot \mathbf{s}) \mathbf{E}^* + \mu^{-1} (\mathbf{B} \cdot \mathbf{s}) \mathbf{B}^* - \frac{1}{2} \left(\varepsilon |\mathbf{E}|^2 + \mu^{-1} |\mathbf{B}|^{-1} \right) dS \right\} \tag{28}$$

By introducing the incident field as a decomposition of plane wave components and taking the sphere S in Eq. (27) so large that $k|r - r_0| \to \infty$, and using Jones' lemma based on the principle of the stationary phase, (see Appendix XII of Bohren & Huffman, 1983), and the source-free condition, we get the optical theorem for an arbitrary field (Nieto-Vesperinas et al., 2010):

$$-W^{(a)} = -\frac{\omega}{2} \Im \left\{ \mathbf{p} \cdot \mathbf{E}^{(i)*}(r_0) \right\} - \frac{\omega}{2} \Im \left\{ \mathbf{m} \cdot \mathbf{B}^{(i)}(r_0) \right\} + \frac{c}{m} \frac{k^4}{3} \left(\varepsilon^{-1} |\mathbf{p}|^2 + \mu |\mathbf{m}|^2 \right). \tag{29}$$

The first two terms of Eq. (29), coming from the interference between the incident and radiated fields, are the energy analogue of the electric and magnetic dipolar forces given by first two terms in Eq. (26).

The third and fourth terms of Eq. (29) that come from the integral of the third and fourth terms of Eq. (28), now yield the rate $W^{(s)}$ at which the energy is being scattered, which together with the left hand side of this equation contributes to the rate of energy extinction by the particle $W^{(a)}+W^{(s)}$:

$$W^{(a)}+W^{(s)}=\frac{\omega}{2}\Im\left\{\mathbf{p}\cdot\mathbf{E}^{(i)*}(r_0)\right\}+\frac{\omega}{2}\Im\left\{\mathbf{m}\cdot\mathbf{B}^{(i)*}(r_0)\right\}. \tag{30}$$

Analogously as with the rate of scattered energy, the electric-magnetic dipolar interaction term of the force (third term of Eq. (26)) corresponds to the rate at which momentum is being scattered by the particle. We shall explore in some detail this analogy in order to illustrate the physical origin of $< \mathbf{F}_{em} >$. We notice that the power density of the scattered field can be written as the sum of two terms (Nieto-Vesperinas et al., 2010)

$$\left\langle \mathbf{s}^{(r)}\right\rangle dS=\frac{c}{8\pi m}k^4\left(\varepsilon^{-1}\left|\mathbf{p}\times\mathbf{s}\right|^2+\mu\left|\mathbf{m}\times\mathbf{s}\right|^2\right)sd\Omega$$

$$+\frac{c}{4\pi m}k^4\sqrt{\frac{\mu}{\varepsilon}}\Re\left\{(\mathbf{s}\times\mathbf{p})\cdot\mathbf{m}^*\right\}sd\Omega. \tag{31}$$

where the second term of Eq. (31) corresponds to the interference between the electric and magnetic dipolar fields. After integration over the closed surface S, that second term does not contribute to the radiated power, while it is the only contribution to the electric-magnetic dipolar interaction term of the force in Eq. (26). Namely, $< \mathbf{F}_{em} >$ comes from the interference between the fields radiated by $\mathbf{p}$ and $\mathbf{m}$.

5.3 Forces on an electric and magnetic dipolar particle for plane wave incidence

In order to illustrate the relevance of the different terms in the optical forces, we shall next consider the force from a plane wave $E^{(i)}=e^{(i)}e^{iks_0\cdot r}$, $B^{(i)}=b^{(i)}e^{iks_0\cdot r}$ with $\mathbf{e}^{(i)}=\left\{\mathbf{b}^{(i)}\times\mathbf{s}_0\right\}/m$ on a small dielectric and magnetic spherical particle characterized by its electric and magnetic polarizabilities α_e and α_m. When the induced dipole moments are expressed in terms of the incident field, i.e.

$$\mathbf{p}=\alpha_e\mathbf{e}^{(i)}; \quad \mathbf{m}=\alpha_m\mathbf{b}^{(i)}. \tag{32}$$

For plane wave incidence, the total force is given by (Nieto-Vesperinas et al., 2010):

$$\langle\mathbf{F}\rangle = \langle\mathbf{F}_e\rangle+\langle\mathbf{F}_m\rangle+\langle\mathbf{F}_{em}\rangle$$

$$=s_0\frac{k}{2}\Im\left\{\mathbf{p}\cdot\mathbf{e}^{(i)*}+\mathbf{m}\cdot\mathbf{b}^{(i)*}\right\}-\frac{m}{c}\int_S\left\langle\mathbf{s}^{(r)}\right\rangle dS \tag{33}$$

$$= s_0 F_0 \left[\Im\{\varepsilon^{-1}\alpha_e\} + \Im\{\mu\alpha_m\} - \frac{2k^3}{3}\frac{\mu}{\varepsilon}\Re\{\alpha_e\alpha_m^*\} \right], \tag{34}$$

where $F_0 = \varepsilon \left| \mathbf{e}^{(i)} \right|^2 / 2$. The first two terms, $< \mathbf{F}_e >$ and $< \mathbf{F}_m >$, correspond to the forces on to the sum of radiation pressures for a pure electric and a pure magnetic dipole, respectively. The third term, $< \mathbf{F}_{em} >$, is the time-averaged scattered momentum rate, and we shall see below that it also contributes to radiation pressure (Nieto-Vesperinas et al., 2010; Gómez-Medina et al., 2011a) and it is related to the asymmetry in the scattered intensity distribution (Nieto-Vesperinas et al., 2011; Gómez-Medina et al., 2011b).

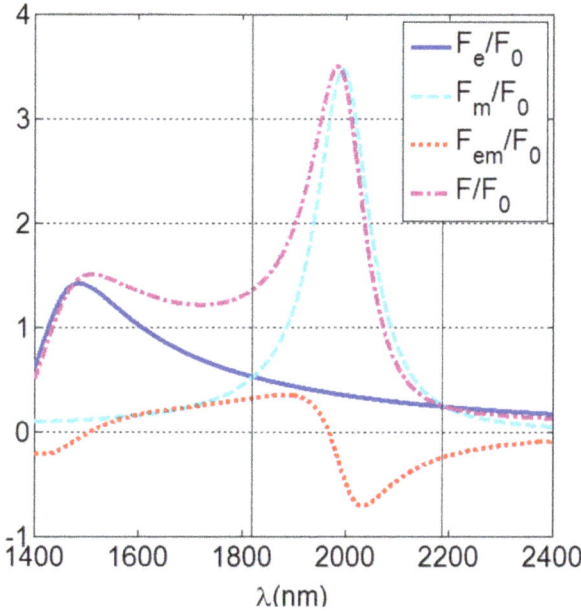

Fig. 10. Different contributions to the total radiation pressure versus the wavelength, for the Ge particle of Figs. 8-9. Normalization is done by $F_0 = \varepsilon \left| \mathbf{e}^{(i)} \right|^2 / 2$. The vertical lines mark the first and second generalized Kerker's conditions. Notice that when the first generalized Kerker's condition is fulfilled, i.e., $\Re\{\alpha_e\} = \Re\{\alpha_m\}$ and $\Im\{\alpha_e\} = \Im\{\alpha_m\}$, $\langle \mathbf{F}\rangle = \langle \mathbf{F}_e\rangle = \langle \mathbf{F}_m\rangle = -\langle \mathbf{F}_{em}\rangle$. From (Gómez-Medina et al. 2011b).

5.4 The generalized Kerker's conditions on optical forces

From Eqs. (6) and (34), one derives for the radiation pressure force (Nieto-Vesperinas et al., 2011):

$$\langle \mathbf{F}\rangle = s_0 F_0 \frac{1}{6k} \left[\frac{dC_{sca}}{d\Omega}(0°) + 3\frac{dC_{sca}}{d\Omega}(180°) - \frac{3}{2\pi}C_{abs} \right]. \tag{35}$$

Equation (35) emphasizes the dominant role of the backward scattering on radiation pressure forces.

At the first generalized Kerker's condition, the interference term of Eqs. (34-35) cancels out the magnetic contribution and we obtain $< \mathbf{F} > = < \mathbf{F}_e >$. At the second generalized Kerker's condition, where the backscattering is enhanced, $< \mathbf{F} > = 3< \mathbf{F}_e >$. Notice that at both generalized Kerker's conditions the scattering cross section is exactly the same; however, the radiation pressures differ by a factor of 3. These properties are illustrated in Figure 10, where we show the different contributions to the total time averaged force on a submicrometer Ge particle.

The strong peak in the radiation pressure force is mainly dominated by the first "magnetic" Mie resonance. This is striking and in contrast with all previous beliefs about optical forces on dipolar dielectric particles, that assumed that these forces would solely be described by the electric polarizability. It is also common to assume that for dielectric particles the real part of the polarizability is much larger than its imaginary part. As a matter of fact, this is behind the development of optical tweezers, in which gradient forces (that are proportional to $\Re(\alpha_e)$), dominate over the radiation pressure or scattering force contribution (which is proportional to $\Im(\alpha_e)$) (Volpe et al., 2006). However, as the size of the particle increases, and for any dielectric particle, there is a crossover from electric to magnetic response as we approach the first Mie resonance, the point at which the response is absolutely dominated by the magnetic dipole. Moreover, just at the resonance, and in absence of absorption, $\Re(\alpha_m) = 0$ and $\Im(\alpha_m) = 3/(\mu 2k^3)$. Then, the radiation pressure contribution of the magnetic term dominates the total force $\langle \mathbf{F} \rangle \cong \langle \mathbf{F}_m \rangle \approx \mathbf{S}_0 \left(\varepsilon \left| \mathbf{e}^{(i)} \right|^2 / 2 \right) \left[3/2k^3 \right]$. Namely, in resonance the radiation pressure force presents a strong peak, the maximum force being independent of both material parameters and particle radius.

6. Conclusion

In this chapter we have analyzed the main aspects of one of the most interesting phenomena of light scattering by nanoparticles: the possibility to control its angular distribution (directionality). As it has been shown, a general magneto-dielectric particle, with suitable values of its relative optical constants (ε, μ), could present directional effects resulting from a coherent effect between real and imaginary parts of both electric and magnetic polarizabilities. The control of this effect could improve the characteristics of many current applications which employ nanoparticles. Also, it can be the base of new potential applications related with light guidance in low dimensions, as for instance, intra- or inter-chip optical communications (García-Cámara; 2011b). In addition, we showed that these scattering effects also affect the radiation pressure on these small particles. Thus, the "non-usual" scattering properties discussed before will strongly affect the dynamics of particle confinement in optical traps and vortex lattices (Albaladejo et al., 2009b; Gómez-Medina et al., 2011a; Albaladejo et al., 2011) governed by both gradient and curl forces.

Finally, we have showed that small dielectric particles made of non magnetic materials present scattering properties similar to those previously reported for hypothetical magneto-dielectric particles. In particular, it has been shown that submicrometer Germanium particles present these directional phenomena in light scattering in the near-infrared range. These studies could serve as a stimulus for new experiments which implement these non-conventional phenomena.

7. Acknowledgment

This work has been supported by the EU NMP3-SL-2008-214107-Nanomagma, the Spanish MICINN Consolider NanoLight (CSD2007-00046), FIS2010-21984, FIS2009-13430-C01-C02, and FIS2007-60158, as well as by the Comunidad de Madrid Microseres-CM (S2009/TIC-1476). B.G.-C. wants to express his gratitude to the University of Cantabria for his postdoctoral fellowship. Work by R.G.-M. was supported by the MICINN "Juan de la Cierva" Fellowship.

8. References

Albaladejo, S.; Marqués, M.I.; Laroche, M. & Sáenz, J.J. (2009). Scattering forces from the curl of the spin angular momentum of a light field. *Physical Review Letters* Vol. 102, No. 11 (March 2009), pp. 113602, ISNN 0031-9007.

Albaladejo, S.; Marqués, M.I. & Sáenz, J.J. (2011). Light control of silver nanoparticle's diffusion. *NanoLetters*, Vol. 9, No. 10 (October 2009), pp. 3527-3531, ISNN 1530-6984.

Albaladejo, S.; Marqués, M.I. & Sáenz, J.J. (2011). Light control of silver nanoparticle's diffusion. *Optics Express*, Vol. 19, No. 12 (June 2011), pp. 11471-11478, ISNN 1094-4087.

Alù, A. & Engheta, N. (2011). How does forward-scattering in magnetodielectric nanoparticles comply with the optical theorem? *Journal of Nanophotonics*, Vol. 4 (May 2010), pp. 041590, ISSN 1934-2608.

Anker, J.N.; Hall, W.P.; Lyandres, O.; Shan, N. C.; Zhao, J. & Van Duyne, R.P. (2008) Biosensing with plasmonic nanosensors. *Nature Materials*, Vol. 7, No. 6, pp. 442-453, ISSN 1476-1122.

Ashkin, A. (1970). Acceleration and Trapping of Particles by Radiation Pressure. *Physical Review Letters*, Vol. 24, No. 4 (January 1970), pp. 156–159, ISSN 1079-7114.

Ashkin, A.; Dziedzic, J. M.; Bjorkholm, J. E. & Chu, S. (1986). Observation of a single-beam gradient force optical trap for dielectric particles. *Optics Letters, Vol.* 11, No. 5 (May 1986), pp. 288–290, ISSN 0146-9592.

Bohren, C.F & Huffman, D.R. (Eds.). (1983). *Absorption and Scattering of Light by Small Particles,* John Wiley& Sons, ISBN 0-471-05772-X, New York.

Burns, M.M.; Fournier, J.M. & Golovchenko, J.A. (1989). Optical Binding. *Physical Review Letters,* Vol. 63, No. 12 (September 1989), pp. 1233–1236, ISSN 1079-7114.

Burns, M.M.; Fournier, J.M. & Golovchenko, J.A. (1990). Optical Matter: Crystallization and Binding in Intense Optical Fields. *Science*, Vol. 249, No. 4970 (August 1990), pp. 749–754, ISSN 0036-8075.

Chaumet, P.C. & Nieto-Vesperinas, M. (2000). Coupled dipole method determination of the electromagnetic force on a particle over a flat dielectric substrate. *Physical Review B*, Vol. 61, No. 20 (May 2000), pp. 14119–14127, ISSN 1098-0121.

Chaumet, P.C. & Nieto-Vesperinas, M. (2000). Electromagnetic force on a metallic particle in the presence of a dielectric surface. *Physical Review B*, Vol. 62, No. 16 (October 2000), pp. 11185–11191, ISSN 1098-0121.

Chaumet, P.C. & Nieto-Vesperinas, M. (2001). Optical binding of particles with or without the presence of a flat dielectric surface. *Physical Review B*, Vol. 64, No. 3 (June 2001), pp. 035422–0354227, ISSN 1098-0121.

Chaumet, P.C.; Rahmani, A. and & Nieto-Vesperinas, M. (2002). Optical trapping and manipulation of nano-objects with an apertureless probe. *Physical Review Letters*, Vol. 88, No. 12 (March 2002), pp. 123601-123604, ISNN 1079-7114.

Chaumet, P.C. & Rahmani, A. (2009). Electromagnetic force and torque on magnetic and negative-index scatterers. *Optics Express*, Vol. 17, No. 4 (February 2009), pp. 2224–2234, ISNN 1094-4087.

Chýlek, P. & Pinnick, R.G. (1979). Nonunitarity of light scattering approximations. *Applied Optics*, Vol. 18, No. 8 (April, 1979), pp.1123-1124, ISSN 1559-128X.

Draine, B.T. & Flatau, P.J. (1994). Discrete-dipole approximation for scattering calculations. *Journal of the Optical Society of America A*, Vol. 11, No. 4 (April 1994), pp. 1491-1499, ISSN 1084-7529.

Evlyyukhin, A.B.; Reinhardt, C.; Seidel, A.; Luk'yanchuk, B.S. & Chichkov, B.N. (2010). Optical response features of Si-nanoparticle arrays. *Physical Review B*, Vol. 82, No. 4 (July 2010), pp. 045404, ISSN 1098-0121.

Gao, H.; Hyun, J.K.; Lee, M.H.; Yang, J.-C.; Lauhon, L.J. & Odom, T.W. (2010). Broadband plasmonic microlenses based on patches of nanoholes. *Nano Letters*, Vol 10. No. 10 (September 2010), pp. 4111-4116, ISSN 1530-6984.

García-Cámara, B.; Saiz, J.M.; González, F. & Moreno, F. (2010). Nanoparticles with unconventional properties: Size effects. *Optics Communications*, Vol. 283, No. 3 (February, 2010), pp. 490-496, ISNN 0030-4018.

García-Cámara B.; Saiz, J.M.; González, F. & Moreno, F. (2010). Distance limit of the directionality conditions for the scattering of nanoparticles. *Metamaterials*, Vol. 4, No. 1 (May 2010), pp. 15-23, ISSN 1873-1988.

García-Cámara, B.; Alcaraz de la Osa, R.; Saiz, J.M.; González, F. & Moreno, F. (2011). Directionality in scattering by nanoparticles: Kerker's null-scattering conditions revisited. *Optics Letters*, Vol. 36, No. 5 (February, 2011), pp. 728-730, ISSN 0146-9592.

García-Cámara, B. (2011), Inta-/Inter-chip optical communications: High speed and low dimensions, In: *Communication architecture for systems-on-chip*, J.L. Ayala (Ed.), pp. 249-322, CRC Press, ISBN 978-1-4398-4170-9, Florida (USA).

García-Etxarri, A.; Gómez-Medina, R.; Froufe-Pérez, L.S.; López, C.; Chantada, L.; Scheffold F.; Aizpurua, J.; Nieto-Vesperinas, M. & Sáenz, J.J. (2011). Strong magnetic response of submicrometer silicon particles in the infrared. *Optics Express*, Vol. 19, No. 6 (February, 2011), pp. 4815-4826, ISSN 1094-4087.

Gómez-Medina, R.; San José, P.; García-Martín, A.; Lester, M.; Nieto-Vesperinas, M. & Sáenz, J.J. (2001). Resonant radiation pressure on neutral particles in a waveguide *Physical Review Letters*, Vol. 86, No. 19 (May 2001), pp. 4275-4277, ISNN 1079-7114.

Gómez-Medina, R. & Saénz J.J. (2004). Unusually Strong Optical Interactions between Particles in Quasi-One-Dimensional Geometries. *Physical Review Letters*, Vol. 93, No. 24 (December 2004), pp. 243602–243605, ISNN 1079-7114.

Gómez-Medina, R.; Nieto-Vesperinas, M. & Saénz J.J. (2011). Nonconservative electric and magnetic optical forces on submicrometer dielectric particles. *Physical Review A*, Vol. 83, No. 3 (March 2011), pp. 033825, ISSN 1050-2947.

Gómez-Medina, R.; García-Cámara, B.; Suárez-Lacalle, I.; González, F.; Moreno, F.; Nieto-Vesperinas, M. & Saénz J.J. (2011). Electric and Magnetic dipolar response of germanium nanospheres: interference effects, scattering anisotropy, and optical forces. *Journal of Nanophotonics* , Vol. 5 (June, 2011), pp. 053512, ISSN 1934-2608.

Hemmerich A. & H'ansch, T.W. (1993). Two-dimesional atomic crystal bound by light. *Physical Review Letters*, Vol. 70, No. 4 (January 1993), pp. 410–413, ISNN 1079-7114.

Jackson, J.D. (1998). *Classical Electrodynamics*, 3rd edition, John Wiley, New York.

Jessen, P.S.; Gerz, C.; Lett, P.D.; Phillips, W.D.; Rolston, S.L.; Spreeuw, R.J.C. & Westbrook, C.I. (1992). Observation of quantized motion of Rb atoms in an optical field. *Physical Review Letters*, Vol. 69, No. 1 (July 1992), pp. 49–52, ISSN 1079-7114.

Kemp, B.A.; Grzegorczyk, T.M. & Kong, J.A. (2005). Ab initio study of the radiation pressure on dielectric and magnetic media. *Optics Express*, Vol. 13, No. 23 (November 2005) pp. 9280-9291, eISNN 1094-4087.

Kemp, B.A.; Grzegorczyk, T.M. & Kong, J.A. (2006). Optical momentum transfer to absorbing Mie particles. *Physical Review Letters*, Vol. 97, No. 13 (September 2006), pp. 133902, ISNN 1079-7114.

Kemp, B.A.; Grzegorczyk, T.M. & Kong, J.A. (2006). Lorentz force on dielectric and magnetic particles. *Journal of Electromagnetics Waves and Applications*, Vol. 20, No. 6 (June 2006), pp. 827–839, ISNN 0920-5071.

Kerker, M.; Wang, D. & Giles G. (1983). Electromagnetic scattering by magnetic spheres. *Journal of the Optical Society of America*, Vol. 73, No. 6 (June, 1983), pp. 756-767, ISSN 0030-3941.

Lakhtakia, A. (2008) Radiation pressure efficiencies of spheres made of isotropic, achiral, passive, homogeneous, negative-phase-velocity materials. *Electromagnetics*, Vol. 28, No. 5 (June 2008), pp. 346–353, ISNN 0272-6343.

Lakhtakia A. & Mulholland, G.W. (1993). On two numerical techniques for light scattering by dielectric agglomerated structures. Journal of Research of the National Institute of Standards and Technology, Vol. 98, No. 6 (December 1993), pp. 699–716, ISNN 1044-677X.

Lock, J. A.; Hodges, J.T. & Gouesbet, G. (1995) Failure of the optical theorem for Gaussian-beam scattering by a spherical particle. *Journal of the Optical Society of America A*, Vol. 12, No. 12 (December 1995), pp. 2708–2715, ISSN 1084-7529.

Maier, S. A.; Kik, P. G.; Atwater, H. A.; Meltzer, S.; Harel, E.; Koel B. E. & Requicha, A. G. (2003). Local detection of electromagnetic energy transport below the diffraction limit in metal nanoparticle Plasmon waveguides. *Nature Materials*, Vol. 2, No. 4 (April 2003), pp. 229-232, ISSN 1476-1122.

Mansuripur, M. (2004). Radiation pressure and the linear momentum of the electromagnetic field. *Optics Express*, Vol.12, No 22 (November 2004). pp. 5375–5401, eISNN 1094-4087.

Mansuripur, M. (2007). Radiation pressure and the linear momentum of the electromagnetic field in magnetic media. *Optics Express*, Vol. 15, No.21 (October 2007), pp. 13502-13518, eISNN 1094-4087.

Mirin, N. A. & Halas, N. J. (2009).Light-bending nanoparticles. *Nano Letters*, Vol 9, No. 3 (February 2009), pp. 1255-1259, ISSN 1530-6984.

Neuman, K. C. & Block, S. M. (2004). Optical trapping. *Review of Scientific Instruments*, Vol. 75, No. 9 (September 2004), pp. 2787–2809, ISSN 0034-6748.

Nieto-Vesperinas, M. & García, N. Comment on "Negative refraction makes a perfect lens". *Physical Review Letters*, Vol. 91, No. 9 (August 2003), pp. 099702, ISSN 1079-7114.

Nieto-Vesperinas, M.; Chaumet, P.C. & Rahmani, A. (2004). Near field photonic forces. *Philosophical Transactions of the Royal Society of London A,* Vol. 362, No. 1817 (February 2004), pp. 719–737, ISSN 1471-2962.

Nieto-Vesperinas, M.; Sáenz, J.J.; Gómez-Medina, R.; & Chantada, L. (2010). Optical forces on small magnetodielectric particles. *Optics Express,* Vol. 18, No. 11 (May 2010), pp. 11428–11443, eISNN 1094-4087.

Nieto-Vesperinas, M.; Gómez-Medina, R. & Sáenz, J.J. (2011). Angle-suppressed scattering and optical forces on submicrometer dielectric particles. *Journal of the Optical Society of America A,* Vol. 28, No. 1 (December 2010), pp. 54–60, ISSN 1084-7529.

Novotny, L. & Hecht, B. (2006). *Principles of Nano-Optics.* Cambridge University Press, Cambridge.

Palik, E.D. (Ed.). (1985), *Handbook of Optical Constants of Solids,* Academis Press, ISBN 0-12-544420-6 , Orlando, Florida

Pendry, J.B.; Schuring D. & Smith D.R. (2006). Controlling electromagnetic fields. *Science,* Vol. 312, N0. 5781 (June 2006), pp. 1780-1782, ISSN 0036-8075.

Pendry, J.B. (2000). Negative refraction makes perfect lens. *Physical Review Letters,* Vol. 85. No. 18 (October 2000), pp. 3966-3969, ISNN 1079-7114.

Prasad, P. N. (2004). *Nanophotonics,* John Wiley& Sons, ISBN 0471649880, New York.

Shalaev, V.M. (2008). Transforming light. *Science,* Vol. 322, No. 5900 (October 2008), pp. 384-386, ISSN 0036-8075.

Tatarkova, S.A.; Carruthers, A.E. & Dholakia, K. (2002). One-Dimensional Optically Bound Arrays of Microscopic Particles. *Physical Review Letters,* Vol. 89, No. 28 (December 2002), pp. 283901–283904, ISNN 1079-7114.

Verkerk, P.; Lounis, B.; Salomon, C.; Cohen-Tannoudji, C.; Courtois, J.-Y. & Grynberg, G. (1992). Dynamics and spatial order of cold cesium atoms in a periodic optical potential. *Physical Review Letters,* Vol. 68, No. 26 (June 1992), pp. 3861–3864, ISNN 1079-7114.

Veselago, V. (1968). The electrodynamics of substances with simultaneously negative values of ε and μ. *Soviet Physics Uspekhi,* Vol. 10, No. 4 (January 1968), pp. 509-514, ISNN 0038-5670.

Volpe, G. ; Quidant, R.; Badenes, G. & Petrov, D. (2006). Surface plasmon radiation forces. *Physical Review Letters,* Vol. 96, No. 23 (June 2006), pp. 238101, ISSN 1079-7114.

Zemp, R. J. (2009) Nanomedicine:Detecting rare cancer cells. *Nature Nanotechnology,* Vol. 4, No. 12 (December 2009), pp. 798-799, ISSN 1748-3387.

Zhedulev, N.I. (2010). The road ahead of metamaterials. *Science,* Vol. 328, No. 5978 (April, 2010), pp. 582-583, ISNN 0036-8075.

Zhu, J.; Ozdemir, S. K.; Xiao, Y.-F.; Li, L.; He, L.; Chen, D.- R. & Yang L. (2010). On-chip single nanoparticle detection and sizing by mode splitting in an ultrahigh-Q microresonator. *Nature Photonics,* Vol. 4, No. 1 (January, 2010), pp. 46-49, ISSN 1749-4885.

14

Deexcitation Dynamics of a Degenerate Two-Level Atom near (Inside) a Body

Gennady Nikolaev
[1]*Institute of Automation and Electrometry, Siberian Branch,
Russian Academy of Sciences, Novosibirsk,*
[2]*Novosibirsk State University, Novosibirsk,*
Russia

1. Introduction

It has long been known that atomic radiation processes near a macroscopic body differ from those in free space substantially (Purcell, 1946). In particular, the lifetime of an excited state of an atom or a molecule near surface (Arnoldus & George, 1988a;b; Barnes, 1998; Chance et al., 1978; Drexhage et al., 1968; Ford et al., 1984; Fort & Grésillon, 2008; Garrett et al., 2004; Hellen & Axelrod, 1987; Kreiter et al., 2002; Lukosz & Kunz, 1977; Macklin et al., 1996; Milonni & Knight, 1973; Snoeks et al., 1995; Steiner et al., 2005; Yeung & Gustafson, 1996) or in the vicinity of (or inside) a nanoparticle (Chew, 1987; 1988; Das & Metiu, 1985; Dung et al., 2000; Gersten & Nitzan, 1981; Klimov, Ducloy & Letokhov, 1996; Klimov et al., 2001; Klimov, Ducloy, Letokhov & Lebedev, 1996; Ruppin, 1982) may be increased or decreased depending on specific conditions. This lifetime change is theoretically calculated in many papers. These calculations made in a variety of ways. Nevertheless all of these papers can be divided into two classes. The first class includes the papers that represent an excited atom as a three-dimensional damped oscillator (Chance et al., 1978; Chew, 1987; 1988; Das & Metiu, 1985; Hellen & Axelrod, 1987; Klimov, Ducloy & Letokhov, 1996; Klimov, Ducloy, Letokhov & Lebedev, 1996; Ruppin, 1982). The second class includes the papers that consider an excited atom by means of quantum mechanics (Agarwal, 1975a;b; Arnoldus & George, 1987; 1988a;b; Barnes, 1998; Dung et al., 2000; Wylie & Sipe, 1984; 1985; Yeung & Gustafson, 1996).

It is shown in the papers that are in the first class that the atomic oscillator rate of damping take a different value in the case of radial and tangential orientation of the oscillating atomic electric dipole. The magnitude of the rate of damping lies between these values in the case of another atomic dipole orientation. However the atomic or molecule decay rate is measured by the fluorescence detection after light pulse excitation of the atom or molecule. So, fluorescence is two-step process, and hence, orientation of the oscillating atomic dipole in general is not the same as exciting light polarization.

In the second class papers the problem of the atomic dipole orientation is either no discussed explicitly or reduced to partitioning of the dipole matrix element on radial and tangential parts as in the case of the classic atomic oscillator. The ratio between these two parts is either no evaluated or assumed to be in the ratio 1:2 as in the case of free space. This approach one cannot consider as correct because of anisotropy of the atomic surroundings. The remark about fluorescence as two-step process mentioned above refers equally to the papers.

To rigorous description of the vector nature of the atomic dipole moment it is necessary to take into account the atomic angular degrees of freedom, that is degeneracy of atomic levels. As far as we know, it was done only in the papers (Arnoldus & George, 1987; 1988a;b). In the papers the steady-state fluorescence of the atom near an axial symmetrical surface was theoretically investigated and influence of the surface was expressed in terms of electric field correlation function.

The purpose of the chapter is to present the correct description of deexcitation dynamics of a degenerate two-level atom in the vicinity of arbitrary body.

We start with a quantum mechanical expression for the atomic deexcitation probability expressed in terms of the normal correlation function of the atomic dipole moment operator and the antinormal correlation function of the electric field strength operator. Then the antinormal correlation function is expressed in terms of the field susceptibility by use of the fluctuation-dissipation theorem. The atomic dipole moment operator as well as the atomic density matrix operator is expressed in terms of irreducible tensor operators. Finally, it is shown that the atomic deexcitation rate at the instant immediately after pulse excitation is proportional to a linear combination of the products of the so-called atomic polarization moments, population and alignment, and anisotropic relaxation matrix.

To find out deexcitation dynamics, a master equation for atomic density matrix is derived from an evolution equation for the total density matrix describing both atom and field. A consistent system of linear first-order ordinary differential equations for the atomic polarization moments is obtained from the master equation. Components of the anisotropic relaxation matrix describing the consistent system are expressed in terms of the field susceptibility tensor. Symmetries of the anisotropic relaxation matrix are found. It is shown that atomic deexcitation in general is multi-exponential. The simple exponential decay of the excited energy level takes place only if its total angular momentum is less then one. Deexcitation dynamics is considered in more detail for the case when the total angular momenta of the upper and lower levels are equal to 1 and 0 respectively. It is shown that in this case deexitation dynamics also may be exponential at certain polarizations of the exciting light.

In conclusion, an intriguing issue that is why the simple model of classical oscillating dipole for description of fluorescence is in good agreement with observational evidence(Amos & Barnes, 1997; Chance et al., 1978; Drexhage et al., 1968; Fort & Grésillon, 2008; Kreiter et al., 2002; Snoeks et al., 1995; Vallée et al., 2001), is clarified.

2. Atomic transition rate of a degenerate two-level atom in the vicinity of a material body

To investigate deexcitation of a degenerate two-level atom in the vicinity of a nanoparticle we consider more general problem of deexcitation of the atom in the vicinity of a material body at first.

Our approach to the problem is based on using correlation functions that appear in linear-response theory. It is about the same as used in number of works (Agarwal, 1975a; Wylie & Sipe, 1984) concerning the quantum electrodynamics and life time of a non-generate atom near an interface. It is most of all close to approach developed in (Klyshko, 2011).

2.1 Transition rate in dipole approximation vs atomic and fluctuating electric field correlation functions

We will assume that both the atom and the electromagnetic field are quantized.

Let the atom and the field be independent at the initial time moment t_0. Therefore at that instant the quantum state of the system $|mk_i\rangle$ is equal to $|m\rangle|k_i\rangle$, where $|m\rangle$, $|k_i\rangle$ are the initial states of the atom and field, respectively. In the first order of the perturbation theory, the amplitude $c_{nk}(t)$ of the transition into some state $|nk\rangle$ is proportional to the matrix element of the interaction operator $\hat{V}$, $\langle nk|\hat{V}|mk_i\rangle$, where $|n\rangle$, $|k\rangle$ are states of the atom and field at the final time moment t, respectively. In the dipole approximation, $\hat{V} = -\hat{\vec{d}}(t)\hat{\vec{E}}(t)$, so,

$$c_{nk} = -\frac{1}{i\hbar} \int_{t_0}^{t} dt' \langle nk|\hat{\vec{d}}(t')\hat{\vec{E}}(t')|mk_i\rangle, \tag{1}$$

where the operators of the atomic dipole moment $\hat{\vec{d}}(t)$ and the electric field strength $\hat{\vec{E}}(t)$ are considered in the interaction picture, i.e., without the account for the perturbation.

In the rotating-wave approximation (Allen & Eberly, 1975), we have

$$-\hat{V}(t) \approx \hat{\vec{d}}^{(-)}(t)\hat{\vec{E}}^{(+)}(t) + \hat{\vec{d}}^{(+)}(t)\hat{\vec{E}}^{(-)}(t), \tag{2}$$

where $\hat{\vec{d}}^{(+)}$ and $\hat{\vec{E}}^{(+)}$ are the positive-frequency parts of the operators, whereas $\hat{\vec{d}}^{(-)}$ and $\hat{\vec{E}}^{(-)}$ are negative-frequency ones. At $t - t_0 \equiv T \gg 1/\bar{\omega}$, fast oscillating (with approximately twice the mean frequency $\bar{\omega}$) products $\hat{\vec{d}}^{(+)}\hat{\vec{E}}^{(+)}$ and $\hat{\vec{d}}^{(-)}\hat{\vec{E}}^{(-)}$ have no contribution into the integral (1).

The initial atomic state $|m\rangle$ has more high energy than the final atomic state $|n\rangle$ for the deexcitation process under consideration. That is why only the second term in (2) gives a nonzero contribution for this process. Hence, the probability of the atomic deexcitation is given by

$$P(nk|mk_i) = \frac{1}{\hbar^2} \int_{t_0}^{t} \int_{t_0}^{t} dt'dt'' \sum_{\alpha\beta} \langle m|\hat{d}_{\alpha}^{(-)}(t')|n\rangle \langle n|\hat{d}_{\beta}^{(+)}(t'')|m\rangle$$
$$\times \langle k_i|\hat{E}_{\alpha}^{(+)}(t')|k\rangle \langle k|\hat{E}_{\beta}^{(-)}(t'')|k_i\rangle, \tag{3}$$

where we have used the equality $\langle r|\hat{A}^{(+)}|s\rangle = \langle s|\hat{A}^{(-)}|r\rangle^*$ for the matrix element of an operator $\hat{A}$ between states $|r\rangle$ and $|s\rangle$. We also have used the Greek letters in subscripts for the notation of the Descartes's components of the vector operators.

One should sum the expression (3) over all possible states $|nk\rangle$ if we are not interested in what specific state the system under consideration has came. These states constitute the complete set and satisfy the completeness condition

$$\sum_{nk} |nk\rangle\langle nk| = \hat{I}. \tag{4}$$

Thus we can represent the total probability of the atomic deexcitation in the following way

$$P = \hbar^{-2} \int_{t_0}^{t} \int_{t_0}^{t} dt' dt'' \sum_{\alpha\beta} f_{\alpha\beta}^{(+)} \left(t', t'' \right) g_{\alpha\beta}^{(-)} \left(t', t'' \right), \tag{5}$$

where

$$f_{\alpha\beta}^{(+)} \left(t', t'' \right) \equiv \left\langle \hat{d}_{\alpha}^{(-)}(t') \hat{d}_{\beta}^{(+)}(t'') \right\rangle,$$

$$\tag{6}$$

$$g_{\alpha\beta}^{(-)} \left(t', t'' \right) \equiv \left\langle \hat{E}_{\alpha}^{(+)}(t') \hat{E}_{\beta}^{(-)}(t'') \right\rangle$$

are normally and anti-normally ordered correlation function (CF) of the atomic dipole moment and the electric field strength in an initial state, respectively. The initial state may be pure as well as mixed, of course.

We suppose that initial unperturbed states of both interacting systems are stationary. In this case correlation functions (6) depend only on the difference of their arguments:

$$f_{\alpha\beta}^{(\pm)} \left(\tau \right) \equiv f_{\alpha\beta}^{(\pm)} \left(t, t+\tau \right) = \left\langle \hat{d}_{\alpha}^{(\mp)}(0) \hat{d}_{\beta}^{(\pm)}(\tau) \right\rangle = \left(f_{\beta\alpha}^{(\pm)} \left(-\tau \right) \right)^{*}, \tag{7}$$

$$g_{\alpha\beta}^{(\pm)} \left(\tau \right) \equiv g_{\alpha\beta}^{(\pm)} \left(t, t+\tau \right) = \left\langle \hat{E}_{\alpha}^{(\mp)}(0) \hat{E}_{\beta}^{(\pm)}(\tau) \right\rangle = \left(g_{\beta\alpha}^{(\pm)} \left(-\tau \right) \right)^{*}. \tag{8}$$

Hence, the total probability of the atomic deexcitation (5) becomes

$$P = \hbar^{-2} \int_{0}^{T} d\tau \, (T-\tau) \sum_{\alpha\beta} \left[f_{\alpha\beta}^{(+)} \left(\tau \right) g_{\alpha\beta}^{(-)} \left(\tau \right) + \left(\tau \to -\tau \right) \right], \tag{9}$$

where $T \equiv t - t_0$ is observation time. When it is much more then the atomic and field correlation time, the total probability of the atomic deexcitation (9) becomes proportional to T. So, atomic transition rate $W \equiv P/T$ independent on time one may introduce

$$W = \hbar^{-2} \int_{-\infty}^{\infty} d\tau \sum_{\alpha\beta} f_{\alpha\beta}^{(+)} \left(\tau \right) g_{\alpha\beta}^{(-)} \left(\tau \right), \tag{10}$$

where limits of integration $\pm T$ are extended to $\pm\infty$. It is convenient rewrite (10) in terms of the Fourier components of the correlation functions in the following way

$$W = \left(1/2\pi\hbar^2 \right) \int_{-\infty}^{\infty} d\omega \sum_{\alpha\beta} f_{\alpha\beta}^{(+)} \left(\omega \right) g_{\alpha\beta}^{(-)} \left(-\omega \right), \tag{11}$$

where the Fourier transform $A\left(\omega\right)$ of a function $A\left(\tau\right)$ is defined by

$$A\left(\omega\right) = \int_{-\infty}^{\infty} d\tau \, e^{i\omega\tau} A\left(\tau\right). \tag{12}$$

2.2 Transition rate in terms of electric field susceptibility

It is known that total correlation function is represented as a sum of normally and anti-normally ordered correlation function in the case of stationary process. Indeed, the total correlation function of the electric field strength may be written as

$$
g_{\alpha\beta}(t, t+\tau) \equiv \left\langle \left[\hat{E}_\alpha^{(+)}(t) + \hat{E}_\alpha^{(-)}(t) \right] \left[\hat{E}_\beta^{(+)}(t+\tau) + \hat{E}_\beta^{(-)}(t+\tau) \right] \right\rangle
$$
$$
= \sum_{s',s=\pm 1} \left\langle \hat{E}_\alpha^{(s')}(t) \hat{E}_\beta^{(s)}(t+\tau) \right\rangle \tag{13}
$$

Expressing $\hat{E}_\alpha^{(s')}(t)$ and $\hat{E}_\beta^{(s)}(t+\tau)$ in terms of Fourier transforms, we obtain

$$
g_{\alpha\beta}(t, t+\tau) \equiv (2\pi)^{-2} \int_{-\infty}^{\infty}\int_{-\infty}^{\infty} d\omega' \, d\omega \, e^{-i\omega\tau} e^{-i(\omega'+\omega)t} \sum_{s',s=\pm 1} \left\langle \hat{E}_\alpha^{(s')}(\omega') \hat{E}_\beta^{(s)}(\omega) \right\rangle. \tag{14}
$$

Note, that

$$
\hat{E}_\alpha^{(s)}(\omega) \equiv \theta(s\omega)\, \hat{E}_\alpha(\omega) \tag{15}
$$

by definition, where $\theta(\omega)$ is step function.

It is clear that (14) is independent on t only when expression in the angle brackets is proportional to Dirac function:

$$
\left\langle \hat{E}_\alpha^{(s')}(\omega') \hat{E}_\beta^{(s)}(\omega) \right\rangle \equiv 2\pi g_{\alpha\beta}^{(s)}(\omega) \, \delta(\omega' + \omega), \tag{16}
$$

where spectral density of the normally ordered correlation function $g_{\alpha\beta}^{(+)}(\omega)$ and anti-normally ordered one $g_{\alpha\beta}^{(-)}(\omega)$ are introduced respectively. In turn, (16) and (15) imply $s' = -s$. Hence, in (14) only two terms are nonzero, and we have

$$
g_{\alpha\beta}(\tau) = g_{\alpha\beta}^{(+)}(\tau) + g_{\alpha\beta}^{(-)}(\tau). \tag{17}
$$

Note that from (14), (16), and (17) it is follows that relationship between $g_{\alpha\beta}^{(\pm)}(\omega)$ and $g_{\alpha\beta}^{(\pm)}(\tau)$ is given by the ordinary formula (12). It is clear also that ordered correlation functions $g_{\alpha\beta}^{(\pm)}(\omega)$ are expressed in terms of the ordinary correlation function $g_{\alpha\beta}(\omega)$ similar to relation (15):

$$
g_{\alpha\beta}^{(\pm)}(\omega) = \theta(\pm\omega)\, g_{\alpha\beta}(\omega) \tag{18}
$$

At thermal equilibrium the correlation function $g_{\alpha\beta}(\tau)$ is simply related with symmetrized correlation function $\{g\}_{\alpha\beta}(\tau)$ defined by

$$
\{g\}_{\alpha\beta}(\tau) \equiv \frac{1}{2}\left\langle \hat{E}_\alpha(0)\hat{E}_\beta(\tau) + \hat{E}_\beta(\tau)\hat{E}_\alpha(0) \right\rangle = \frac{1}{2}\left\{ g_{\alpha\beta}(\tau) + g_{\beta\alpha}(-\tau) \right\}. \tag{19}
$$

There is a simple Kubo-Martin-Schwinger's boundary condition

$$
g_{\beta\alpha}(-\tau) = g_{\alpha\beta}(\tau + i\hbar\xi), \tag{20}
$$

where $\xi \equiv 1/(kT)$, k and T are Boltzmann's constant and temperature respectively. It is easily proofed by using the invariance of the trace under a cyclic permutation of the operators:

$$
\begin{aligned}
g_{\beta\alpha}(-\tau) &= \left\langle \hat{E}_\beta(\tau)\hat{E}_\alpha(0) \right\rangle \equiv \mathrm{tr}\left\{ \hat{\rho}_0 e^{i\hat{H}\tau/\hbar}\hat{E}_\beta e^{-i\hat{H}\tau/\hbar}\hat{E}_\alpha \right\} \\
&= Z^{-1}\mathrm{tr}\left\{ e^{-\xi\hat{H}}e^{i\hat{H}\tau/\hbar}\hat{E}_\beta e^{-i\hat{H}\tau/\hbar}\hat{E}_\alpha \right\} \\
&= Z^{-1}\mathrm{tr}\left\{ \hat{E}_\alpha e^{i(i\xi+\tau/\hbar)\hat{H}}\hat{E}_\beta e^{-i(i\xi+\tau/\hbar)\hat{H}}e^{-\xi\hat{H}} \right\} = g_{\alpha\beta}(\tau+i\hbar\xi),
\end{aligned}
\tag{21}
$$

where $\hat{\rho}_0 = Z^{-1}e^{-\xi\hat{H}}$ is the thermal equilibrium density operator, $Z = \mathrm{tr}\left\{ e^{-\xi\hat{H}} \right\}$, and $\hat{H}$ is unperturbed Hamiltonian of the system.

Using (20), we rewrite relation (19) as follows

$$
\{g\}_{\alpha\beta}(\tau) = \frac{1}{2}\left\{ g_{\alpha\beta}(\tau) + g_{\alpha\beta}(\tau+i\hbar\xi) \right\}.
\tag{22}
$$

In turn, taking the Fourier transform, we obtain

$$
\{g\}_{\alpha\beta}(\omega) = \frac{1}{2}\left\{ 1 + e^{\hbar\omega\xi} \right\} g_{\alpha\beta}(\omega).
\tag{23}
$$

The Fourier transform of symmetrized correlation function $\{g\}_{\alpha\beta}(\vec{r},\vec{r}';\omega)$ is related with dynamical value $G_{\alpha\beta}(\vec{r},\vec{r}';\omega)$, the Fourier transform of the electric field susceptibility $G_{\alpha\beta}(\vec{r},\vec{r}';\tau)$, by the fluctuation-dissipation theorem as follows (Bernard & Callen, 1959; Callen et al., 1952; Callen & Welton, 1951; Landau & Lifshitz, 1980)

$$
\{g\}_{\alpha\beta}(\vec{r},\vec{r}';\omega) = \frac{1}{2}i\hbar \left[G^*_{\beta\alpha}(\vec{r}',\vec{r};\omega) - G_{\alpha\beta}(\vec{r},\vec{r}';\omega) \right] \coth\left(\frac{\hbar\omega\xi}{2} \right),
\tag{24}
$$

where tensor $G_{\alpha\beta}(\vec{r},\vec{r}';\omega)$ relates Fourier transforms of the electric dipole $\hat{d}_\beta(\vec{r}';\omega)$ and induced electric field $\hat{E}_\alpha(\vec{r};\omega)$ as follows

$$
\hat{E}_\alpha(\vec{r};\omega) = \sum_\beta G_{\alpha\beta}(\vec{r},\vec{r}';\omega)\,\hat{d}_\beta(\vec{r}';\omega),
\tag{25}
$$

and the electric field susceptibility tensor $G_{\alpha\beta}(\vec{r},\vec{r}';\tau)$ is defined by

$$
G_{\alpha\beta}(\vec{r},\vec{r}';\tau) \equiv \frac{i}{\hbar}\theta(\tau)\left\langle \left[\hat{E}_\alpha(\tau), \hat{E}_\beta(0) \right] \right\rangle.
\tag{26}
$$

Note that the same tensor $G_{\alpha\beta}(\vec{r},\vec{r}';\omega)$ relates classical, not quantum, values $E_\alpha(\vec{r};\omega)$ and $d_\beta(\vec{r}';\omega)$ by the same way (25). So it can be found from the solution of the classical electrodynamic problem in the same condition.

Using (18), (23), and (24), we obtain

$$
g^{(-)}_{\alpha\beta}(-\omega) = i\hbar\theta(\omega)\frac{1}{2}\left[1 + \coth\left(\frac{\hbar\omega\xi}{2} \right) \right]\left[G^*_{\beta\alpha}(\vec{r}',\vec{r};\omega) - G_{\alpha\beta}(\vec{r},\vec{r}';\omega) \right],
\tag{27}
$$

When there is no external magnetic field, tensor $G_{\alpha\beta}(\vec{r},\vec{r}';\omega)$ is symmetrical one, and its imaginary part is odd in ω. In this case (27) goes over into (Agarwal, 1975a)[1]

$$g_{\alpha\beta}^{(-)}(-\omega) = \hbar\theta(\omega)\left[1+\coth\left(\frac{\hbar\omega\xi}{2}\right)\right]\Im\left[G_{\alpha\beta}(\vec{r}',\vec{r};\omega)\right], \qquad (28)$$

We are interesting in only local field response because of point atom approximation used. Substituting (28) in (11) we find

$$W = (1/2\pi\hbar)\int_0^\infty d\omega \sum_{\alpha\beta} f_{\alpha\beta}^{(+)}(\omega)\left[1+\coth\left(\frac{\hbar\omega\xi}{2}\right)\right]\Im\left[G_{\alpha\beta}(\vec{r}_0,\vec{r}_0;\omega)\right], \qquad (29)$$

where $\vec{r}_0$ is radius vector of the atom.

2.3 Transition rate of a degenerate two-level atom

The explicit form of the atomic CF $f_{\alpha\beta}^{(+)}(\omega)$ depends on the atomic model used. Here we consider a degenerate two-level atom. Its energy levels are degenerate on the total angular momentum projection on any axis. Suppose the excited upper energy level m and lower one n have quantum numbers $J_m M_m$ and $J_n M_n$ respectively, where J_j and M_j label the total angular momentum of the level j and its projection on the z-axis, respectively.

It is convenient describe vector or tensor values in terms of the circular components instead of the Descartes's one. The circular components v_σ of a vector $\vec{v}$, where $\sigma = 0,\pm 1$, are related with the Descartes's one v_i as follows (Varshalovich et al., 1988):

$$v_0 = v_z,$$
$$v_{\pm 1} = \mp (v_x \pm v_y)/\sqrt{2}. \qquad (30)$$

The circular components of the atomic dipole operator can be expressed according to the Wigner-Eckart theorem in terms of the so-called unit irreducible tensor operators $\hat{T}_Q^K(J_m J_n)$ in the following way (Biedenharn & Louck, 1981; Blum, 1996; Fano & Racah, 1959; Varshalovich et al., 1988):

$$\hat{d}_\sigma^{(+)}(t) = \frac{d_{nm}}{\sqrt{3}}\hat{T}_\sigma^1(J_n J_m)\exp(-i\omega_{mn}t),$$
$$\hat{d}_\sigma^{(-)}(t) = \left\{(-1)^\sigma \hat{d}_{-\sigma}^{(+)}(t)\right\}^\dagger, \qquad (31)$$

where d_{mn} and ω_{mn} are reduced matrix element of the atomic dipole moment and resonant frequency of the atomic transition, respectively. The irreducible tensor operator $\hat{T}_Q^K(J_m J_n)$, where K and Q are its rank and component $(-K \leqslant Q \leqslant K)$ correspondingly, is defined as (Biedenharn & Louck, 1981; Blum, 1996; Fano & Racah, 1959; Varshalovich et al., 1988)

$$\hat{T}_Q^K(J_m J_n) = \sum_{M_m,M_n}(-1)^{J_n-M_n}\langle J_m M_m J_n - M_n|KQ\rangle|J_m M_m\rangle\langle J_n M_n|, \qquad (32)$$

where $\langle J_m M_m J_n - M_n|KQ\rangle$ is the vector coupling (Clebsch-Gordan) coefficient. Quantities J_m, J_n, and K of the coefficient obey triangle inequality, so $|J_m - J_n| \leqslant K \leqslant J_m + J_n$.

---

[1] Definition of the ordered correlation functions in this paper differs from ours one by sign of the argument τ and, hence, in sign of ω.

2.3.1 Properties of irreducible tensor operators and density matrix multipole components

The operators $\hat{T}_Q^K(JJ')$ are orthonormal in the following sense

$$\text{tr}\left[\hat{T}_{Q'}^{K'}(J'J)\hat{T}_Q^{K\dagger}(J'J)\right] \equiv \sum_{M'M} \langle J'M'|\hat{T}_{Q'}^{K'}(J'J)|JM\rangle\langle JM|\hat{T}_Q^{K\dagger}(J'J)|J'M'\rangle = \delta_{K'K}\delta_{Q'Q}, \qquad (33)$$

where the Hermitian conjugate operator $\hat{T}_Q^{K\dagger}(J'J)$ is expressed in terms of $\hat{T}_Q^K(JJ')$ as follows

$$\hat{T}_Q^{K\dagger}(J'J) \equiv (-1)^{J'-J-Q}\,\hat{T}_{-Q}^K(JJ'). \qquad (34)$$

Set of the operators $\hat{T}_Q^K(J'J)$ is complete. So, density operator can be decomposed into irreducible parts as follows

$$\hat{\rho} = \sum_{J'JKQ} \rho^{KQ}(J'J)\,\hat{T}_Q^K(J'J). \qquad (35)$$

In turn, coefficients $\rho^{KQ}(J'J)$ known as multipole components are expressed in terms of $\hat{T}_Q^K(J'J)$ and density operator by using (33) and (32) in the following way

$$\rho^{KQ}(J'J) = \text{tr}\left[\hat{\rho}\hat{T}_Q^{K\dagger}(J'J)\right] = \sum_{M'M} (-1)^{J-M}\langle J'M'J-M|KQ\rangle\langle J'M'|\hat{\rho}|JM\rangle. \qquad (36)$$

It is seen that multipole components $\rho^{KQ}(J'J)$ satisfy the following relations similar to relations (34):

$$\left[\rho^{KQ}(J'J)\right]^* = (-1)^{J-J'-Q}\rho^{K-Q}(JJ'), \qquad (37)$$

so multipole components $\rho^{K0}(JJ)$ is real. Note also that $\rho^{KQ}(J'J)$ transform under rotations like $\hat{T}_Q^{K\dagger}(J'J)$, and hence, are contravariant to $\hat{T}_Q^K(J'J)$ because of property (34).

We are interesting only in states of the excited level m, so the relevant density operator $\hat{\rho}(J_m)$ is

$$\hat{\rho}(J_m) = \sum_{KQ}\rho^{KQ}(J_mJ_m)\,\hat{T}_Q^K(J_mJ_m). \qquad (38)$$

In this decomposition the rank K is in the range $0 \leqslant K \leqslant 2J_m$ as was noted after definition (32). All multipole components $\rho^{KQ}(J_mJ_m)$ have clear physical sense (see, for example, (Biedenharn & Louck, 1981; Blum, 1996; Omont, 1977; Varshalovich et al., 1988)). In particular, $\sqrt{2J_m+1}\rho^{00}(J_mJ_m)$ is equal to the total population of the level m, the $\rho^{1Q}(J_mJ_m)$'s are the three standard components of what is generally called "orientation" proportional to the mean magnetic dipole of the state, and the $\rho^{2Q}(J_mJ_m)$'s are the five standard components of the "alignment" proportional to the mean electric quadrupole moment of the state.

2.3.2 Transition rate and material body symmetry

Finally, after some manipulation using the relations (7), (31), and (38), and also properties of irreducible tensor operators, one can represent relation (29) in the form

$$W = \frac{1}{2}\left[1 + \coth\left(\frac{\hbar\omega_{mn}\xi}{2}\right)\right]\sum_{KQ}\gamma_Q^K\rho^{KQ}(J_mJ_m), \qquad (39)$$

where

$$\gamma_Q^K \equiv 2\frac{|d_{mn}|^2}{\hbar}(-1)^{J_m+J_n}\left\{\begin{matrix} 1 & 1 & K \\ J_m & J_m & J_n \end{matrix}\right\}[\mathbf{G}''(\vec{r}_0,\vec{r}_0;\omega_{mn})]_Q^K \tag{40}$$

is irreducible relaxation tensor of the multipole $\rho^{KQ}(J_m J_m)$, $\left\{\begin{matrix} 1 & 1 & K \\ J_m & J_m & J_n \end{matrix}\right\}$ is $6-j$ coefficient, and $[\mathbf{G}''(\vec{r}_0,\vec{r}_0;\omega_{mn})]_Q^K$ is irreducible spherical tensor of the imaginary part of the electric field susceptibility in the ω–representation. Irreducible spherical tensor $G_Q^K(\vec{r}_0,\vec{r}_0;\omega_{mn})$ is related with circular components $G_{\sigma\sigma'}(\vec{r}_0,\vec{r}_0;\omega_{mn})$ as follows

$$G_Q^K(\vec{r}_0,\vec{r}_0;\omega_{mn}) \equiv \sum_{\sigma\sigma'}\langle 1\sigma 1\sigma'|KQ\rangle G_{\sigma\sigma'}(\vec{r}_0,\vec{r}_0;\omega_{mn}). \tag{41}$$

It is follows from properties of the Clebsch-Gordan coefficient $\langle 1\sigma 1\sigma'|KQ\rangle$ that $0\leqslant K\leqslant 2$. Furthermore, symmetry of the tensor $G_{\sigma\sigma'}(\vec{r}_0,\vec{r}_0;\omega_{mn})$ under the interchange $\sigma\leftrightarrows\sigma'$ requires that K have to be even, so $K=0,2$. In other words, deexcitation rate depends on the total population of excited level ($K=0$) and its alignment ($K=2$). Their relative contribution depends according to (39) and (40) on quantum numbers of combining levels m and n, on the excitation type determining the value of $\rho^{KQ}(J_m J_m)$, and on the atom surroundings by $G_Q^K(\vec{r}_0,\vec{r}_0;\omega_{mn})$. Let us consider these factors in more detail.

As was noted after (38), K is in the range of values defining by $0\leqslant K\leqslant 2J_m$. Consequently, if the total momentum J_m of the the excited level is equal to 0, or $1/2$, there is no alignment of the level. So, deexcitation is governed only by γ_0^0 and does not depend on excitation type. In the case of $J_m>1/2$, the ratio of two deexcitation rates corresponding to some two fixed excitation types, differing in initial values of $\rho^{KQ}(J_m J_m)$, is not universal but depends on J_m, J_n.

One can diagonalize symmetrical tensor $G_{\alpha\beta}(\vec{r}_0,\vec{r}_0;\omega_{mn})$. Let us label its principal axes of coordinate by X,Y,Z. In this proper basis only the following irreducible components of the tensor $\mathbf{G}$ are not zero:

$$G_0^0 = -\frac{1}{\sqrt{3}}\mathrm{tr}\,(\mathbf{G}) = -\frac{1}{\sqrt{3}}(G_{XX}+G_{YY}+G_{ZZ}), \tag{42}$$

$$G_0^2 = \sqrt{\frac{2}{3}}\left[G_{ZZ}-\frac{1}{2}(G_{XX}+G_{YY})\right], \tag{43}$$

$$G_{\pm 2}^2 = \frac{1}{2}(G_{XX}-G_{YY}). \tag{44}$$

As is seen from (44), components $G_{\pm 2}^2=0$ if surroundings of the atom is axial symmetric (symmetry axis along Z). In particular, this case is realized when atom is near a half-space boundary or near a spherical particle.

When surroundings of the atom is isotropic, the only nonzero component of the tensor $\mathbf{G}$ is G_0^0 one. It is just the case of an isotropic infinite medium (in particular, vacuum) or when atom is in the center of spherical particle or cavity. In this case $(-1)^{J_m+J_n}\left\{\begin{matrix} 1 & 1 & 0 \\ J_m & J_m & J_n \end{matrix}\right\} = -1/\sqrt{3(2J_m+1)}$ in (40). So, using relations (42), (40) we obtain from (39)

$$W_{is} = \frac{2}{3}\frac{|d_{mn}|^2}{\hbar(2J_m+1)}\Im\left(\sum_{i=X,Y,Z}G_{ii}\right)\sum_{M_m=-J_m}^{J_m}\langle J_m M_m|\hat{\rho}|J_m M_m\rangle. \tag{45}$$

Since we are here interested primarily in atomic transition energies on the order of a Rydberg that implies $\dfrac{\hbar \omega_{mn} \xi}{2} \gg 1$ at room temperature, we have replaced the expression in square brackets in Eq. (39) by 2. The total population of the upper level

$$\sum_{M_m=-J_m}^{J_m} \langle J_m M_m | \hat{\rho} | J_m M_m \rangle = 1$$

because we suppose that atom is excited on level m at the initial time. For free space (Barash, 1988; Lifshitz & Pitaevskii, 1980; Nikolaev, 2006), we have

$$\Im \left(\sum_{i=X,Y,Z} G_{ii} \right) = 2 \left(\frac{\omega_{mn}}{c} \right)^3 . \qquad (46)$$

Substituting these two expressions in Eq. (45) we immediately obtain the well-known expression for the radiative decay rate of the excited state of an isolated atom (see, i.e., (Berestetskii et al., 2008; Sobelman, 1972)) :

$$W_0 = \frac{4}{3} \frac{|d_{mn}|^2}{\hbar (2J_m + 1)} \left(\frac{\omega_{mn}}{c} \right)^3 . \qquad (47)$$

It should be noted that Eq. (39) describes deexcitation rate at the initial time moment just following the excitation. Density matrix multipole components $\rho^{KQ} (J_m J_m)$ will be changed with the passage of time. It is reasonable to suggest that the expression opposite in sign to the right-hand side of Eq. (39) describes the decrease of the upper level population per unit of time. To prove the suggestion let us consider more general problem of the dynamics of the density matrix multipole components caused by interaction of the atom with quantized field.

3. Master equations for the excited density matrix multipole components

3.1 Integro-differential equation for total density matrix operator

Let us consider a large isolated system consisting of an atom, material body and interacting with them quantum electromagnetic field. Atomic surrounding, electromagnetic field and material body that interact among themselves, we will treat as a large subsystem referred to as the thermostat. In the interaction picture representation, the density matrix $\hat{R}$ of the total isolated system obeys the Liouville equation:

$$i\hbar \frac{\mathrm{d} \hat{R}(t)}{\mathrm{d}t} = [\hat{V}(t), \hat{R}(t)] , \qquad (48)$$

where $\hat{V}$ is the atom-field interaction operator that in the rotating-wave approximation is given by Eq. (2). It is known that this equation can be rewritten in the integro-differential form that is suitable for perturbation technique. Indeed, formal integrating this equation in time, we obtain the integral equation:

$$\hat{R}(t) = \hat{R}(0) - (i/\hbar) \int_0^t dt' [\hat{V}(t'), \hat{R}(t')] . \qquad (49)$$

Substituting this expression into Eq. (48), we get the equation for the total density matrix operator in the following form:

$$\frac{d\hat{R}(t)}{dt} = (-i/\hbar)\left[\hat{V}(t), \hat{R}(0)\right] + (-i/\hbar)^2 \int_0^t dt' \left[\hat{V}(t), \left[\hat{V}(t'), \hat{R}(t')\right]\right]. \tag{50}$$

In Eqs. (49) and (50) the lower limit we took 0 since it is assumed that the thermostat and the atom did not interact before this time moment because the atom was unexcited. Consequently, until this moment the thermostat and the atom were uncorrelated, so the total density matrix $\hat{R}$ was equal to the direct product of the density matrices of the system:

$$\hat{R}(0) = \hat{\rho}(0)\hat{\rho}_{th}(0), \tag{51}$$

where $\hat{\rho}$ and $\hat{\rho}_{th}$ are the density matrix operator of the atom and thermostat, respectively.

3.2 Large thermostat approximation

Following the paper (Fano & Racah, 1959) (see also (Blum, 1996)), we will suppose that thermostat is always in the state of the thermal equilibrium because it has a large number of degrees of freedom and, hence, atom almost do not changes its state. The supposition implies that the total density matrix is always equal to the direct product of the density matrices of the system:

$$\hat{R}(t) = \hat{\rho}(t)\hat{\rho}_{th}(0) \tag{52}$$

This relation is referred to as the main condition of the irreversibility.

Substituting (52) in (50) and taking trace over thermostat variables, we get the equation for the reduced atomic density matrix operator, $\hat{\rho}(t) \equiv \text{tr}_{th}\hat{R}(t)$,

$$\frac{d\hat{\rho}(t)}{dt} = -(i/\hbar)\text{tr}_{th}\left[\hat{V}(t), \hat{\rho}(0)\hat{\rho}_{th}(0)\right] - (1/\hbar)^2 \int_0^t dt'\text{tr}_{th}\left[\hat{V}(t), \left[\hat{V}(t'), \hat{\rho}(t')\hat{\rho}_{th}(0)\right]\right]. \tag{53}$$

3.3 Integro-differential equation for atomic multipole components

To obtain dynamics equation for atomic multipole components, we make use of relation (36). Precisely, let us multiply both sides of (53) by $\hat{T}_Q^{K\dagger}(J_m J_m)$ and take trace over atomic variable. So, we get

$$\frac{d\rho^{KQ}(J_m J_m)(t)}{dt} = -(i/\hbar)\text{tr}_{all}\left\{\hat{T}_Q^{K\dagger}(J_m J_m)\left[\hat{V}(t), \hat{\rho}(0)\hat{\rho}_{th}(0)\right]\right\}$$

$$-(1/\hbar)^2 \int_0^t dt'\text{tr}_{all}\left\{\hat{T}_Q^{K\dagger}(J_m J_m)\left[\hat{V}(t), \left[\hat{V}(t'), \hat{\rho}(t')\hat{\rho}_{th}(0)\right]\right]\right\}, \tag{54}$$

where tr_{all} stands for the trace over all isolated system variables including atomic and thermostat one.

We will now transform this equation in such a way that terms include the trace of the product of $\hat{\rho}(t')\hat{\rho}_{th}(0)$ by an operator.

To do this, we make use of the identity (Il'inskii & Keldysh, 1994)

$$\text{tr}\left\{\hat{A}\left[\hat{A}_1,\left[\hat{A}_2,\cdots\left[\hat{A}_k,\hat{B}\right]\cdots\right]\right]\right\}=\text{tr}\left\{\left[\cdots\left[\left[\hat{A},\hat{A}_1\right],\hat{A}_2\right]\cdots\hat{A}_k\right]\hat{B}\right\} \tag{55}$$

which holds for arbitrary operators $\hat{A},\hat{A}_1,\hat{A}_2,\cdots,\hat{B}$.

Using identity (55) and the atomic density matrix decomposition (35), we can rewrite (54) as

$$\frac{d\rho^{KQ}(J_mJ_m;t)}{dt}=-\frac{i}{\hbar}\sum_{J'JK'Q'}\rho^{K'Q'}(J'J;0)\,\text{tr}_{all}\left\{\left[\hat{T}_Q^{K\dagger}(J_mJ_m),\hat{V}(t)\right]\hat{T}_{Q'}^{K'}(J'J)\hat{\rho}_{th}(0)\right\}$$

$$-\frac{1}{\hbar^2}\sum_{J'JK'Q'}\int_0^t dt'\rho^{K'Q'}(J'J;t')\,\text{tr}_{all}\left\{\left[\left[\hat{T}_Q^{K\dagger}(J_mJ_m),\hat{V}(t)\right],\hat{V}(t')\right]\hat{T}_{Q'}^{K'}(J'J)\hat{\rho}_{th}(0)\right\}. \tag{56}$$

Substituting in (56) the interaction Hamiltonian (2), using (31), and also taking into account that scalar product $\vec{d}\hat{E}$ in the circular basis (30) has the form $\sum_\sigma(-1)^\sigma d_\sigma E_{-\sigma}$, we obtain

$$\frac{d\rho^{KQ}(J_mJ_m;t)}{dt}=-\frac{d_{mn}d_{nm}}{3\hbar^2}\sum_{J'JK'Q'}\int_0^t dt'\rho^{K'Q'}(J'J;t')\sum_{\alpha\beta}\{e^{i\omega_{mn}(t-t')}\left[g_{\alpha\beta}^{(-)}(t'-t)\,A_{\alpha\beta}\right.$$

$$\left.-g_{\alpha\beta}^{(+)}(t-t')\,B_{\alpha\beta}\right]+e^{-i\omega_{mn}(t-t')}\left[g_{\alpha\beta}^{(-)}(t-t')\,C_{\alpha\beta}-g_{\alpha\beta}^{(+)}(t'-t)\,B_{\alpha\beta}\right]\}, \tag{57}$$

where $g_{\alpha\beta}^{(\pm)}(\tau)$ are the ordered correlation functions of the fluctuating electromagnetic field (8),

$$A_{\alpha\beta}\equiv\sum_{\sigma\sigma'}(-1)^{\sigma+\sigma'}\langle\alpha|1-\sigma\rangle\langle\beta|1-\sigma'\rangle\text{tr}\left\{\hat{T}_Q^{K\dagger}(J_mJ_m)\hat{T}_\sigma^1(J_mJ_n)\hat{T}_{\sigma'}^1(J_nJ_m)\hat{T}_{Q'}^{K'}(JJ')\right\}, \tag{58}$$

$$B_{\alpha\beta}\equiv\sum_{\sigma\sigma'}(-1)^{\sigma+\sigma'}\langle\alpha|1-\sigma\rangle\langle\beta|1-\sigma'\rangle\text{tr}\left\{\hat{T}_\sigma^1(J_nJ_m)\hat{T}_Q^{K\dagger}(J_mJ_m)\hat{T}_{\sigma'}^1(J_mJ_n)\hat{T}_{Q'}^{K'}(JJ')\right\}, \tag{59}$$

$$C_{\alpha\beta}\equiv\sum_{\sigma\sigma'}(-1)^{\sigma+\sigma'}\langle\alpha|1-\sigma\rangle\langle\beta|1-\sigma'\rangle\text{tr}\left\{\hat{T}_\sigma^1(J_mJ_n)\hat{T}_{\sigma'}^1(J_nJ_m)\hat{T}_Q^{K\dagger}(J_mJ_m)\hat{T}_{Q'}^{K'}(JJ')\right\}. \tag{60}$$

In the definitions (58) – (60) symbols $\langle\alpha|1-\sigma\rangle$ and $\langle\beta|1-\sigma'\rangle$ are transformation matrices from the circular components to the Descartes's one, that are inverse of that given by (30), and symbol $\text{tr}\{\cdots\}$ from now on stands for trace over atomic variables. Note that the linear on $\hat{V}(t)$ term in (56) vanishes in our case because of the average fluctuated field is zero at the thermal equilibrium: $\text{tr}_{th}\{\hat{E}_\alpha\}\equiv\langle\hat{E}_\alpha\rangle=0$.

It should be noted that ratio of $|g_{\alpha\beta}^{(+)}(t-t')|$ to $|g_{\alpha\beta}^{(-)}(t-t')|$ is proportional to the mean number of photons in the thermal equilibrium, $\langle n_{ph}\rangle\sim kT/\hbar\omega_{mn}\ll 1$. Therefore terms that proportional to $g_{\alpha\beta}^{(+)}(t-t')$ can be ignored in (57).

3.4 Master equation for multipole components in Markov-type approximation

Fluctuating field correlation functions $g_{\alpha\beta}^{(\pm)}(t-t')$ are nonzero only for the sufficiently small time difference $|\tau|\equiv|t-t'|$ comparable with the typical field correlation time τ_c. We will assume following (Loisell, 1973) that this correlation time is much less then typical variation times of the atomic multipole components. Thus, in the case of free space the lifetime of the atomic excited state much more than $\tau_c \approx 1/\omega_{mn}$. So, we can replace $\rho^{K'Q'}(J'J;t')$ by $\rho^{K'Q'}(J'J;t)$ and to take it out of the integral in (57). It is so-called Markov-type approximation.

It is also important to note that $\hat{T}^K_{Q'}(JJ')$ incoming in (58) and (60) are nonzero only if $J = J' = J_m$ because of its definition (32) and invariance of the trace under a cyclic permutation of the operators.

Taking into account assumptions mentioned above, property (8), and by making the change of variable $\tau \equiv t - t'$ in integration, we can represent (57) as

$$\frac{d\rho^{KQ}(J_m J_m; t)}{dt} = -\frac{d_{mn}d_{nm}}{3\hbar^2} \sum_{K'Q'} \sum_{\alpha\beta} \left[I^*_{\beta\alpha}(\omega_{mn}) A_{\alpha\beta} + I_{\alpha\beta}(\omega_{mn}) C_{\alpha\beta} \right] \rho^{K'Q'}(J_m J_m; t), \quad (61)$$

where

$$I_{\alpha\beta}(\omega_{mn}) \equiv \int\limits_0^\infty d\tau g^{(-)}_{\alpha\beta}(\tau) e^{-i\omega_{mn}\tau}. \quad (62)$$

In (62) we extended upper limit from t to ∞ because of $g^{(\pm)}_{\alpha\beta}(\tau)$ is in fact zero at $\tau \gg \tau_c$. The error of this replacement is negligible in Markov-type approximation.

Now we will show that integral (62) is expressed in terms of retarded Green function $G_{\alpha\beta}(\vec{r},\vec{r}';\omega_{mn})$. To prove that, let as consider Fourier transform $g^{(-)}_{\alpha\beta}(-\omega_{mn})$ of the function $g^{(-)}_{\alpha\beta}(\tau)$ defined by (12):

$$g^{(-)}_{\alpha\beta}(-\omega_{mn}) = \int\limits_{-\infty}^\infty d\tau g^{(-)}_{\alpha\beta}(\tau) e^{-i\omega_{mn}\tau} \quad (63)$$

Let us split this integral into two parts

$$g^{(-)}_{\alpha\beta}(-\omega_{mn}) = \int\limits_{-\infty}^0 d\tau g^{(-)}_{\alpha\beta}(\tau) e^{-i\omega_{mn}\tau} + \int\limits_0^\infty d\tau g^{(-)}_{\alpha\beta}(\tau) e^{-i\omega_{mn}\tau}. \quad (64)$$

Making the change of variable in integration $\tau \to -\tau$ in the first integral and utilizing relation (8), we can rewrite (64) as

$$g^{(-)}_{\alpha\beta}(-\omega_{mn}) = \int\limits_0^\infty d\tau \left(g^{(-)}_{\beta\alpha}(\tau) \right)^* e^{i\omega_{mn}\tau} + \int\limits_0^\infty d\tau g^{(-)}_{\alpha\beta}(\tau) e^{-i\omega_{mn}\tau}. \quad (65)$$

The second integral in (65) is just equal to $I_{\alpha\beta}(\omega_{mn})$, and the first one to its complex conjugation. So, (65) can be rewritten as follows

$$g^{(-)}_{\alpha\beta}(-\omega_{mn}) = I_{\alpha\beta}(\omega_{mn}) + I^*_{\beta\alpha}(\omega_{mn}). \quad (66)$$

Now comparing right-hand sides of (66) and (27), we obtain desired relation

$$I_{\alpha\beta}\left(\omega_{mn}\right) = -i\hbar\frac{1}{2}\left[1 + \coth\left(\frac{\hbar\omega_{mn}}{2kT}\right)\right]G_{\alpha\beta}\left(\vec{r},\vec{r}';\omega_{mn}\right) \tag{67}$$

It is yet mentioned after (37) that multipole components $\rho^{KQ}\left(J'J\right)$ transform under rotations contravariant to $\hat{T}_Q^K(J'J)$. It is convenient to introduce co-variant multipole components $\rho_Q^K\left(J'J\right)$ by convention

$$\rho_Q^K\left(J'J\right) \equiv (-1)^{J-J'-Q}\rho^{K-Q}\left(JJ'\right) = \left[\rho^{KQ}\left(J'J\right)\right]^*. \tag{68}$$

In these notations, making use of (67) and explicitly calculating traces in (58) and (60), one can finally represent (61) as follows [2]

$$\frac{d\rho_Q^K\left(t\right)}{dt} = -\gamma_0 \sum_{K'Q'} \Gamma_{QQ'}^{KK'}\rho_{Q'}^{K'}\left(t\right), \tag{69}$$

where

$$\gamma_0 = W_0 = \frac{4}{3}\frac{|d_{mn}|^2}{\hbar\left(2J_m+1\right)}\left(\frac{\omega_{mn}}{c}\right)^3 \tag{70}$$

is radiation decay rate of the excited degenerate state of the atom in vacuum, dimensionless relaxation tensor $\Gamma_{QQ'}^{KK'}$ can be represented as follows:

$$\Gamma_{QQ'}^{KK'} = \gamma_{QQ'}^{KK'} + i\Delta_{QQ'}^{KK'}, \tag{71}$$

where $\gamma_{QQ'}^{KK'}$ and $\Delta_{QQ'}^{KK'}$ are in general complex.

Geometrical part of $\gamma_{QQ'}^{KK'}$ and $\Delta_{QQ'}^{KK'}$ is represented by Clebsch-Gordan coefficient and dynamical one is proportional to retarded Green function:

$$\gamma_{QQ'}^{KK'} = \sum_{LM}\langle K'Q'LM|KQ\rangle\overline{G}_M^L\left(KK'L\right)\gamma(KK'L, J_mJ_n), \tag{72}$$

$$\Delta_{QQ'}^{KK'} = \sum_{LM}\langle K'Q'LM|KQ\rangle\widetilde{G}_M^L\left(KK'L\right)\gamma(KK'L, J_mJ_n), \tag{73}$$

where scalar coefficient $\gamma(KK'L, J_mJ_n)$ and irreducible tensors $\overline{G}_M^L\left(KK'L\right)$ and $\widetilde{G}_M^L\left(KK'L\right)$ are

$$\gamma(KK'L, J_mJ_n) = (-1)^{K+J_n-J_m}\frac{3}{2}\left(2J_m+1\right)\sqrt{\left(2K'+1\right)\left(2L+1\right)} \tag{74}$$

$$\times \left\{\begin{matrix} K & K' & L \\ J_m & J_m & J_m \end{matrix}\right\}\left\{\begin{matrix} 1 & 1 & L \\ J_m & J_m & J_n \end{matrix}\right\},$$

$$\overline{G}_M^L\left(KK'L\right) = \sum_{\alpha\beta\sigma\sigma'}\langle 1\sigma 1\sigma'|LM\rangle\langle 1\sigma|\alpha\rangle\langle 1\sigma'|\beta\rangle\overline{G}_{\alpha\beta}\left(KK'L\right), \tag{75}$$

$$\widetilde{G}_M^L\left(KK'L\right) = \sum_{\alpha\beta\sigma\sigma'}\langle 1\sigma 1\sigma'|LM\rangle\langle 1\sigma|\alpha\rangle\langle 1\sigma'|\beta\rangle\widetilde{G}_{\alpha\beta}\left(KK'L\right), \tag{76}$$

---

[2] hereinafter for simplicity we omit the dependence of ρ_Q^K on J_m: $\rho_Q^K\left(t\right) \equiv \rho_Q^K\left(J_mJ_m; t\right)$

and

$$\overline{G}_{\alpha\beta}\left(KK'L\right) = \frac{1}{2}\left[G''_{\beta\alpha}(\omega_{mn}) + (-1)^{K+K'-L}G''_{\alpha\beta}(\omega_{mn})\right] / \left(\frac{\omega_{mn}}{c}\right)^3,$$ (77)

$$\widetilde{G}_{\alpha\beta}\left(KK'L\right) = \frac{1}{2}\left[G'_{\beta\alpha}(\omega_{mn}) - (-1)^{K+K'-L}G'_{\alpha\beta}(\omega_{mn})\right] / \left(\frac{\omega_{mn}}{c}\right)^3.$$ (78)

Symbol G' and G'' in (77) and (78) denotes real and imaginary part of G, respectively, and symbols $\langle 1\sigma|\alpha\rangle$ and $\langle 1\sigma'|\beta\rangle$ are transformation matrices from the Descartes's components to the circu lar one, that given by (30).

3.4.1 Relaxation matrix symmetry

Note that $\overline{G}_{\alpha\beta}\left(KK'L\right)$ and $\widetilde{G}_{\alpha\beta}\left(KK'L\right)$, and consequently, $\overline{G}^L_M\left(KK'L\right)$ and $\widetilde{G}^L_M\left(KK'L\right)$, are symmetrical with respect to K and K'. As for the scalar $\gamma(KK'L, J_mJ_n)$, it changes upon permutation of K and K' as follows

$$\gamma(KK'L, J_mJ_n) = (-1)^{K-K'}\sqrt{\frac{2K'+1}{2K+1}}\gamma(K'KL, J_mJ_n)$$ (79)

because of invariance of $6-j$ symbol as regard to permutation of its columns.

Although tensor $G_{\alpha\beta}$ in general has no symmetry with respect to permutation of subscripts, tensors $\overline{G}_{\alpha\beta}\left(KK'L\right)$ and $\widetilde{G}_{\alpha\beta}\left(KK'L\right)$ have one, as one can see from (77) and (78),

$$\overline{G}_{\alpha\beta}\left(KK'L\right) = (-1)^{K+K'-L}\overline{G}_{\beta\alpha}\left(KK'L\right),$$ (80)

$$\widetilde{G}_{\alpha\beta}\left(KK'L\right) = -(-1)^{K+K'-L}\widetilde{G}_{\beta\alpha}\left(KK'L\right).$$ (81)

Irreducible tensors $\overline{G}^L_M\left(KK'L\right)$ and $\widetilde{G}^L_M\left(KK'L\right)$ in general are complex. Using relation $\langle 1\sigma|\alpha\rangle^* = (-1)^\sigma\langle 1-\sigma|\alpha\rangle$ and Clebsch-Gordan coefficients symmetry, one can show that

$$\left[\overline{G}^L_M\left(KK'L\right)\right]^* = (-1)^{L+M}\overline{G}^L_{-M}\left(KK'L\right),$$ (82)

$$\left[\widetilde{G}^L_M\left(KK'L\right)\right]^* = (-1)^{L+M}\widetilde{G}^L_{-M}\left(KK'L\right).$$ (83)

This relations allow to find the following symmetry of the relaxation matrix components

$$\left[\gamma^{KK'}_{QQ'}\right]^* = (-1)^{K'-K+Q-Q'}\gamma^{KK'}_{-Q-Q'},$$ (84)

$$\left[\Delta^{KK'}_{QQ'}\right]^* = (-1)^{K'-K+Q-Q'}\Delta^{KK'}_{-Q-Q'}.$$ (85)

On the other hand, from hermiticity of density matrix and equation (69) it is easy to obtain

$$\left[\Gamma^{KK'}_{QQ'}\right]^* = (-1)^{Q-Q'}\Gamma^{KK'}_{-Q-Q'},$$ (86)

that can be rewrite in terms of $\gamma^{KK'}_{QQ'}$ and $\Delta^{KK'}_{QQ'}$ as follows

$$\left[\gamma^{KK'}_{QQ'}\right]^* = (-1)^{Q-Q'}\gamma^{KK'}_{-Q-Q'},$$ (87)

$$\left[\Delta^{KK'}_{QQ'}\right]^* = -(-1)^{Q-Q'}\Delta^{KK'}_{-Q-Q'}.$$ (88)

Comparing (84) and (87) shows that $\gamma_{QQ'}^{KK'}$ is different from zero only for even $K + K'$. Similarly, comparing (85) and (88) shows that $\Delta_{QQ'}^{KK'}$ is different from zero only for odd $K + K'$.

These properties can be find more straightforward from symmetries (80) and (81) and definitions (75) and (76) that yield

$$\overline{G}_M^L (KK'L) = (-1)^{K+K'} \overline{G}_M^L (KK'L),$$ (89)

$$\widetilde{G}_M^L (KK'L) = (-1)^{K+K'+1} \widetilde{G}_M^L (KK'L).$$ (90)

Taking into account these properties that we can reformulate as $K + K'$ is even for $\overline{G}_{\alpha\beta} (KK'L)$ and odd for $\widetilde{G}_{\alpha\beta} (KK'L)$, one can see from (80) and (81) that part of $G_{\alpha\beta}$ which is symmetrical with respect to permutation of subscripts makes a contribution to $\overline{G}_{\alpha\beta} (KK'L)$ and to $\widetilde{G}_{\alpha\beta} (KK'L)$, and hence to $\Gamma_{QQ'}^{KK'}$, only when L is even. As for antisymmetrical part of $G_{\alpha\beta}$, it contributes to $\Gamma_{QQ'}^{KK'}$ only when L is odd.

When tensor $G_{\alpha\beta}$ is symmetrical (i.e., no external magnetic field), the form of tensor $\overline{G}_{\alpha\beta} (KK'L)$ as well of tensor $\widetilde{G}_{\alpha\beta} (KK'L)$ is simplified

$$\overline{G}_{\alpha\beta} (KK'L) = \delta_{L,2l}\delta_{K+K',2n} \left(\frac{c}{\omega_{mn}} \right)^3 G_{\alpha\beta}'',$$ (91)

$$\widetilde{G}_{\alpha\beta} (KK'L) = \delta_{L,2l}\delta_{K+K',2n+1} \left(\frac{c}{\omega_{mn}} \right)^3 G_{\alpha\beta}',$$ (92)

where n and l are integer. As a consequence, $\overline{G}_M^L (KK'L)$ and $\widetilde{G}_M^L (KK'L)$ are also simplified

$$\overline{G}_M^L (KK'L) = \delta_{L,2l}\delta_{K+K',2n} \left(\frac{c}{\omega_{mn}} \right)^3 [G'']_M^L,$$ (93)

$$\widetilde{G}_M^L (KK'L) = \delta_{L,2l}\delta_{K+K',2n+1} \left(\frac{c}{\omega_{mn}} \right)^3 [G']_M^L.$$ (94)

As stated above (see Eqs. (42) -(44)), in this case there are only four nonzero components of G_M^L in the proper coordinate system.

There is additional symmetry of the relaxation tensor $\Gamma_{QQ'}^{KK'}$ in the case. Using the fact that $\overline{G}_M^L (KK'L)$ and $\widetilde{G}_M^L (KK'L)$ are symmetrical with respect to K and K', evenness of L, relation (79) and also Clebsch-Gordan coefficient symmetry $\langle K'Q'LM|KQ \rangle = (-1)^{L+M} \sqrt{\frac{2K+1}{2K'+1}} \langle K - QLM|K' - Q' \rangle$, one can obtain

$$\Gamma_{QQ'}^{KK'} = (-1)^{K-K'+Q-Q'} \Gamma_{-Q'-Q}^{K'K}$$ (95)

that we can rewrite using (86) as follows

$$\Gamma_{QQ'}^{KK'} = (-1)^{K-K'} \left[\Gamma_{Q'Q}^{K'K} \right]^*.$$ (96)

This is just the symmetry of $\Gamma_{QQ'}^{KK'}$ relative to time reversal (Omont, 1977) that is natural in the absence of magnetic field.

In case of the atomic surroundings is axial symmetrical in addition, there are only two nonzero components of G_M^L in the proper coordinate system, G_0^0 and G_0^2. Therefor, only irreducible tensors $\overline{G}_0^L(KK'L)$ and $\widetilde{G}_0^L(KK'L)$ are nonzero and real in the system (see relations (82) and (83), and (93) and (94)). Consequently, only $\gamma_{QQ}^{KK'}$ and $\Delta_{QQ}^{KK'}$ are also nonzero and real (see relations (72 and (73)), hence,

$$\Gamma_{QQ'}^{KK'} = \delta_{Q,Q'}\Gamma_{QQ}^{KK'}. \tag{97}$$

So, in this case $\Gamma_{QQ}^{KK'}$ is real for even $K + K'$, imaginary for odd $K + K'$ and

$$\Gamma_{00}^{KK'} = 0 \tag{98}$$

for odd $K + K'$ because of the following Clebsch-Gordan coefficient symmetry

$$\langle K'0L0|K0\rangle = (-1)^{K'+L-K}\langle K'0L0|K0\rangle. \tag{99}$$

4. Deexcitation dynamics

Deexcitation of upper level is given by (69) with $K = Q = 0$

$$\frac{d\rho_0^0(t)}{dt} = -\gamma_0 \sum_{K'Q'} \Gamma_{0Q'}^{0K'}\rho_{Q'}^{K'}(t). \tag{100}$$

Hereinafter we suppose that there is no external magnetic field. In this case $\gamma_0\Gamma_{0Q'}^{0K'} = (-1)^{Q'}\gamma_{-Q'}^{K'}/\sqrt{2J_m+1}$, where $\gamma_{-Q'}^{K'}$ is defined by (40), multiplier $(-1)^{Q'}$ transforms covariant component $\rho_{Q'}^{K'}$ into contravariant one $\rho^{K'-Q'}$ and denominator $\sqrt{2J_m+1}$ reflect the fact that the right-hand side of (100) is variation in time of ρ_0^0, not of population that is $\sqrt{2J_m+1}\rho_0^0$ as in (39). To obtain temporal variation of the deexcitation, it is necessary to solve consistent differential equations, involving along with Eq. (100) also differential equations for $\rho_{Q'}^{K'}(t)$, incoming in its right-hand side.

Let us restrict themselves to the case of axial symmetrical atomic surroundings.

As it mentioned above, this case include half-space boundary and spherical particle. From (97), (98), (100), and also (99), it is follows that consistent differential equations, describing deexcitation dynamics in the proper coordinate system, include only multipole components with even K and $Q = 0$. The number of such components is $[J_m] + 1$ because of $0 \leqslant K \leqslant 2J_m$ as noted above (symbol $[J_m]$ here and further denotes the integer part of J_m). As the relevant $\Gamma_{00}^{KK'}$ are real in our case, from (96) we obtain that they are symmetrical relative to K and K'

$$\Gamma_{00}^{KK'} = \Gamma_{00}^{K'K}. \tag{101}$$

Hence, the number of different relevant components $\Gamma_{00}^{KK'}$ is $([J_m] + 1) \times ([J_m] + 2)/2$.

As is known , the general solution of $[J_m] + 1$ consistent linear homogeneous differential equations is given by a linear combination of $[J_m] + 1$ their eigen vectors, each of them varies in time exponentially with its own rate. The rates are eigen values of the consistent equations. The number of the eigen values is also in general equal to $[J_m] + 1$. So, the atomic deexcitation is also usually expressed as a linear sum of $[J_m] + 1$ exponentials.

In fact, the eigenvalues are relaxation rates of populations of magnetic sublevels $|J_m \pm M\rangle$ in the case under consideration. Indeed, relevant multipole components ρ_0^K incoming in the consistent differential equations, describing deexcitation dynamics, are linear combination of the populations of the sublevels $|J_m M\rangle$ (see (36)). In addition, the sublevels $|J_m M\rangle$ and $|J_m - M\rangle$ are transformed one into another (with the sign $(-1)^P$, where P is parity of the level m) under reflection in any plane through the symmetry axis (Landau & Lifshitz, 1977). Consequently, the relaxation rates of these sublevels are equal. So, the number of different relaxation rates is $[J_m] + 1$ as stated above with respect to the eigenvalues.

4.1 Deexcitation dynamics in the case of $J_m = 1$, $J_n = 0$

Let us consider in more detail the case when the angular momentums are $J_m = 1$ and $J_n = 0$. In the case under study, deexcitation dynamics is described by only two equations

$$\frac{d\rho_0^0(t)}{dt} = -\gamma_0 \left[\Gamma_{00}^{00}\rho_0^0(t) + \Gamma_{00}^{02}\rho_0^2(t) \right], \tag{102}$$

$$\frac{d\rho_0^2(t)}{dt} = -\gamma_0 \left[\Gamma_{00}^{02}\rho_0^0(t) + \Gamma_{00}^{22}\rho_0^2(t) \right]. \tag{103}$$

The eigen values $\gamma_\pm$ of the consistent equations are

$$\gamma_\pm = \gamma_0 \left[\Gamma_+ \pm \Gamma \right], \tag{104}$$

and fundamental solution matrix are

$$S(t) = \frac{1}{2} \begin{pmatrix} \left(1 - \dfrac{\Gamma_-}{\Gamma}\right) e^{-\gamma_- t} + \left(1 + \dfrac{\Gamma_-}{\Gamma}\right) e^{-\gamma_+ t} & \dfrac{\Gamma_{00}^{02}}{\Gamma} \left(-e^{-\gamma_- t} + e^{-\gamma_+ t}\right) \\ \dfrac{\Gamma_{00}^{02}}{\Gamma} \left(-e^{-\gamma_- t} + e^{-\gamma_+ t}\right) & \left(1 + \dfrac{\Gamma_-}{\Gamma}\right) e^{-\gamma_- t} + \left(1 - \dfrac{\Gamma_-}{\Gamma}\right) e^{-\gamma_+ t} \end{pmatrix}, \tag{105}$$

where dimensionless $\Gamma_\pm$ and Γ are defined as

$$\Gamma_\pm = \frac{1}{2} \left(\Gamma_{00}^{00} \pm \Gamma_{00}^{22} \right), \tag{106}$$

$$\Gamma = \sqrt{(\Gamma_-)^2 + (\Gamma_{00}^{02})^2}. \tag{107}$$

Specific solution column of the consistent equations (102)-(103), corresponding to initial conditions given by column $c = \mathrm{col}\left(\rho_0^0(0), \rho_0^2(0)\right)$, is obtained by multiplication of fundamental solution matrix on the right by column c.

It is known that the excited atomic states that are produced by the absorption of anisotropic resonance light are strongly polarized (Alexandrov et al., 1993; Happer, 1972; Omont, 1977). This atomic polarization results from the directionality or polarization of the light beam. So, immediately after excitation there are nonzero both $\rho_0^0(0)$ and $\rho_0^2(0)$.

However, let us consider the simplest case of isotropic excitation, when there is only population $\rho_0^0(0)$ on the upper level at the instant after excitation. So, the solution column in the case is given by

$$
\begin{pmatrix} \rho_0^0(t), \\ \rho_0^2(t), \end{pmatrix} = \frac{1}{2} \begin{pmatrix} \left(1 - \frac{\Gamma_-}{\Gamma}\right) e^{-\gamma_- t} + \left(1 + \frac{\Gamma_-}{\Gamma}\right) e^{-\gamma_+ t} \\ \frac{\Gamma_{00}^{02}}{\Gamma} \left(-e^{-\gamma_- t} + e^{-\gamma_+ t}\right) \end{pmatrix} \rho_0^0(0) . \tag{108}
$$

In the case under consideration that is $J_m = 1$, $J_n = 0$, dimensionless relaxation matrix elements are following: $\Gamma_{00}^{00} = (1/2)(c/\omega_{mn})^3 \mathrm{tr}(\mathbf{G}'')$, $\Gamma_{00}^{02} = -(\sqrt{2}/2)(c/\omega_{mn})^3 (G_{ZZ}'' - G_{XX}'')$, $\Gamma_{00}^{22} = \Gamma_{00}^{00} - \Gamma_{00}^{02}/\sqrt{2}$. So, relevant dimensionless $\Gamma_\pm$ and Γ are

$$
\Gamma_+ = \frac{3}{4} \left(\frac{c}{\omega_{mn}}\right)^3 (G_{ZZ}'' + G_{XX}''), \tag{109}
$$

$$
\Gamma_- = -\frac{1}{4} \left(\frac{c}{\omega_{mn}}\right)^3 (G_{ZZ}'' - G_{XX}''), \tag{110}
$$

$$
\Gamma = \frac{3}{4} \left(\frac{c}{\omega_{mn}}\right)^3 (G_{ZZ}'' - G_{XX}''). \tag{111}
$$

Substituting (109)-(111) into (105), we obtain

$$
S(t) = \frac{1}{3} \begin{pmatrix} [2e^{-\gamma_- t} + e^{-\gamma_+ t}] & \sqrt{2}[e^{-\gamma_- t} - e^{-\gamma_+ t}] \\ \sqrt{2}[e^{-\gamma_- t} - e^{-\gamma_+ t}] & [e^{-\gamma_- t} + 2e^{-\gamma_+ t}] \end{pmatrix} . \tag{112}
$$

Eigen values $\gamma_\pm$ in the case are

$$
\gamma_+ = \frac{3}{2} \left(\frac{c}{\omega_{mn}}\right)^3 G_{ZZ}'', \tag{113}
$$

$$
\gamma_- = \frac{3}{2} \left(\frac{c}{\omega_{mn}}\right)^3 G_{XX}''. \tag{114}
$$

In the case under consideration (i.e., $J_m = 1$, $J_n = 0$) it is possible such excitation conditions that upper level deexcitation is pure exponential. Such cases only three.

In the first case the atom is excited by light with linear polarization that is collinear to the symmetry axis. Such light excites only one upper sublevel with angular momentum projection on the symmetry axis $J_{mZ} = 0$. In this case the initial conditions column is given by

$$
c_0 \equiv \begin{pmatrix} \rho_0^0(0) \\ \rho_0^2(0) \end{pmatrix} = \frac{1}{\sqrt{3}} \begin{pmatrix} 1 \\ -\sqrt{2} \end{pmatrix} \rho_{00}(0),
$$

where $\rho_{00}(0)$ is population of the sublevel mentioned above. If we multiply fundamental solution matrix (112) on the right by column c_0, we get the variation in the time of the population and alignment of the upper level:

$$
\begin{pmatrix} \rho_0^0(t) \\ \rho_0^2(t) \end{pmatrix} = c_0 e^{-\gamma_+ t}
$$

In the second case the atom is excited by circular polarized light that propagates along symmetry axis. Now the only upper sublevel with angular momentum projection on the symmetry axis $J_{mZ} = +1$ (or $J_{mZ} = -1$ for the opposite circular polarization) is excited. Initial conditions column in the case is given by

$$c_1 \equiv \begin{pmatrix} \rho_0^0(0) \\ \rho_0^2(0) \end{pmatrix} = \frac{1}{\sqrt{3}} \begin{pmatrix} 1 \\ 1/\sqrt{2} \end{pmatrix} \rho_{11}(0), \tag{115}$$

where $\rho_{11}(0)$ is population of the exited sublevel. The solution corresponding to this column is

$$\begin{pmatrix} \rho_0^0(t) \\ \rho_0^2(t) \end{pmatrix} = c_1 e^{-\gamma - t}. \tag{116}$$

Lastly, in the third case the atom is excited by light with linear polarization that is orthogonal to the symmetry axis. It has been known that such polarization can be represented by the sum of the opposite circular polarization with the same amplitude, rotating in the plane that is orthogonal to the symmetry axis. This case is reduced to the previous one because of only two upper sublevels with angular momentum projection on the symmetry axis $J_{mZ} = \pm 1$ are excited independently with equal probability, and hence $\rho_{11}(0) = \rho_{-1-1}(0)$. The rates of decay of the both excited sublevels into the only low state are equal due to axial symmetry. Deexcitation dynamics in the case also given by (116).

These three exceptional cases of simple exponetial deexcitation can be physically interpreted as follows. In every case the excited state transforms to the only low state by means of one channel. The decay itself is induced by the optical transition oscillating dipole that arises due to interaction of the excited atom with the electric field quantum oscillations. Both the direction of the dipole oscillation and the direction of the exciting light polarization are the same due to the one and the same channel of excitation and deexcitation (see Fig. 1).

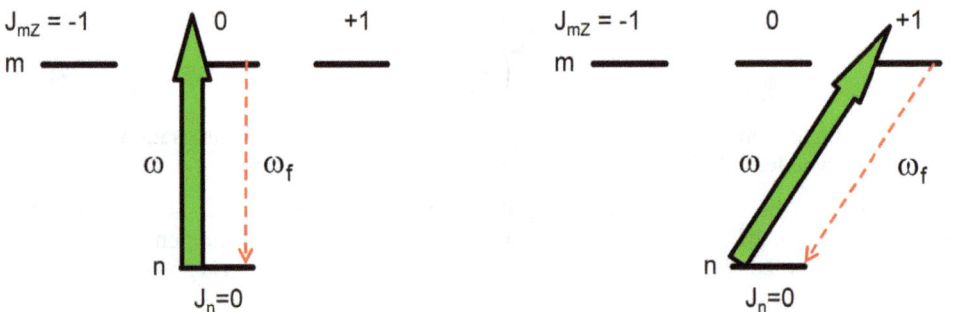

(a) Exciting light is linear polarized along (or transversely to) the symmetry axis passing through the atom and body; Z – axis is along (or transversely to) this axis

(b) Exciting light is circular polarized and propagates along the symmetry axis that is Z – axis

Fig. 1. Exceptional polarizations of the exciting light that led to the pure exponential decay of the excited atomic state (ω and ω_f are frequencies of the exciting light and fluorescence respectively)

Precisely owing to this fact, experimental results of the measurement of the decay of the fluorescence signal (Amos & Barnes, 1997; Chance et al., 1978; Drexhage et al., 1968; Fort & Grésillon, 2008; Kreiter et al., 2002; Snoeks et al., 1995; Vallée et al., 2001) are in good agreement with the simple model of the classic scattering dipole, in spite of the fact that fluorescence is the two-step process, rather than scattering.

It should be noted that consistent equations (102)-(103) describe deexcitation dynamics also in the case $J_m = 1$, $J_n = 1$, or $J_m = 1$, $J_n = 2$, and also in the case $J_m = 3/2$, $J_n = 1/2$, and either $J_m = 3/2$, $J_n = 3/2$, or $J_m = 3/2$, $J_n = 5/2$. Of course, specific values of the dimensionless Γ_{00}^{00}, Γ_{00}^{02}, and Γ_{00}^{22} in these cases differ from considered above.

It should be pointed out too that in the case $J_m = 3/2$ and $J_n = 1/2$ there is the only exciting light polarization, namely linear polarization along symmetry axis, that leds to the pure exponential decay of the excited state because of the relaxation rate equality of the excited sublevels ($JmZ = \pm 1/2$) due to the axial symmetry.

5. Conclusions

In the chapter we have proposed a general approach to the problem of deexcitation of a degenerate two-level atom near (inside) a body. On the basis of the approach the master equation for density matrix in the polarization moments representation was obtained.

We have shown that relaxation dynamics of a polarization moment is described in general by a consistent linear equations for all $2J_m + 1$ polarization moments of the excited level, where J_m is the total momentum of the level. We have expressed relaxation matrix elements of the consistent linear equations in terms of the field response tensor that can be found as the electric field of the classic oscillating unit dipole situated near the body.

We have found symmetry of the relaxation matrix.

An additional relaxation matrix symmetry is recognized in the case when there is no external quasistatic magnetic field, and as a result, the field response tensor is symmetrical one. Therefore, the tensor may be diagonalized. We have shown that relaxation matrix depends only on the trace of the field response tensor, on the difference between the most principal value of the diagonal response tensor and the half-sum of two others, and also on the difference between these two.

Axial symmetric atomic surroundings gives rise to one more additional symmetry of the relaxation matrix. In this case it depends only on the trace of the field response tensor and on the difference between its two principal values.

We have shown that deexcitation dynamics of the degenerate two-level atom in the conditions under consideration represents multiexponential decay. In the case of the axial symmetric atomic surroundings, the number of the exponential is equal to $[J_m] + 1$, where $[J_m]$ is the integer part of J_m. So, the simple exponential decay of the atomic excitation is possible only in two cases, namely, when $J_m = 0$ or $J_m = 1/2$. We have shown that simple exponential decay of the atomic excitation is also possible in the case of $J_m = 1$, $J_n = 0$ and on special polarizations of exciting light, namely on the linear polarization that is collinear or orthogonal to the axial symmetry axis, and on the circular polarizations rotating in the plane that is orthogonal to the symmetry axis. In this exceptional cases both the excitation and decay of the corresponding upper states follow the one and the same respective channel. Simple exponential decay of the atomic excitation is possible too in the case $J_m = 3/2$ and $J_n = 1/2$ when exciting light polarization is linear oriented along symmetry axis.

Our analysis have carried out in the absence of hyperfine structure on the combine energy levels. However, it can be easily expanded straightforward on general case by expanding quantum states irreducible basis of the total momentum, including both the total electronic momentum and the nuclear spin, into the direct product of states irreducible bases of the the nuclear spin and the electronic momentum. Just the late basis is involved into the electromagnetic interaction in the course of the allowed optical transition.

We have considered situation when degenerate two-level atom is situated in the vicinity of a body. Nevertheless, it is clear from the consideration that our treatment is more general and results obtained are true for an atom embedded in any anisotropic medium.

6. References

Agarwal, G. S. (1975a). Quantum electrodynamics in the presence of dielectrics and conductors. I. electromagnetic-field response functions and black-body fluctuations in finite geometries, *Phys. Rev. A* 11(1): 230–242.

Agarwal, G. S. (1975b). Quantum electrodynamics in the presence of dielectrics and conductors. IV. general theory for spontaneous emission in finite geometries, *Phys. Rev. A* 12: 1475–1497. URL: *http://link.aps.org/doi/10.1103/PhysRevA.12.1475*

Alexandrov, E. B., Chaika, M. P. & Khvostenko, G. I. (1993). *Interference of atomic states*, Springer-Verlag, Berlin, New York.

Allen, L. & Eberly, J. H. (1975). *Optical Resonance and Two-Level Atoms*, John Wiley and Sons, New York—London—Sydney—Toronto.

Amos, R. M. & Barnes, W. L. (1997). Modification of the spontaneous emission rate of Eu^{3+} ions close to a thin metal mirror, *Phys. Rev. B* 55: 7249–7254.
URL: *http://link.aps.org/doi/10.1103/PhysRevB.55.7249*

Arnoldus, H. F. & George, T. F. (1987). Quantum theory of atomic fluorescence near a metal surface, *J. Chem. Phys.* 87(8): 4263–4272. URL: *http://dx.doi.org/10.1063/1.452884*

Arnoldus, H. F. & George, T. F. (1988a). Spontaneous decay and atomic fluorescence near a metal surface or an absorbing dielectric, *Phys. Rev. A* 37: 761–769.
URL: *http://link.aps.org/doi/10.1103/PhysRevA.37.761*

Arnoldus, H. F. & George, T. F. (1988b). Symmetries of spontaneous decay for atoms near any surface, *Surface Science* 205(3): 617–636.
URL: *http://www.sciencedirect.com/science/article/pii/0039602888903056*

Barash, S. (1988). *Van der Waals Forces*, Nauka, Moscow. [in Russian].

Barnes, W. L. (1998). Fluorescence near interfaces: The role of photonic mode density, *Journal of Modern Optics* 45(4): 661–699.
URL: *http://www.tandfonline.com/doi/abs/10.1080/09500349808230614*

Berestetskii, V., Lifshitz, E. & Pitaevskii, L. (2008). *Quantum Electrodynamics*, Butterworth-Heinemann, Oxford.

Bernard, W. & Callen, H. B. (1959). Irreversible thermodynamics of nonlinear processes and noise in driven systems, *Rev. Mod. Phys.* 31(4): 1017–1044.

Biedenharn, L. C. & Louck, J. D. (1981). *Angular Momentum in Quantum Physics. Theory and Application*, Addison-Wesley, Massachusetts.

Blum, K. (1996). *Density Matrix Theory and Applications*, 2nd edn, Plenum Press, New York.

Callen, H. B., Barasch, M. L. & Jackson, J. L. (1952). Statistical mechanics of irreversibility, *Phys. Rev.* 88: 1382–1386. URL: *http://link.aps.org/doi/10.1103/PhysRev.88.1382*

Callen, H. B. & Welton, T. A. (1951). Irreversibility and generalized noise, *Phys. Rev.* 83: 34–40.
URL: *http://link.aps.org/doi/10.1103/PhysRev.83.34*

Chance, R. R., Prock, A. & Silbey, R. (1978). Molecular fluorescence and energy transfer near interfaces, *in* I. Prigogine & S. A. Rice (eds), *Advances in Chemical Physics*, Vol. 37, Wiley, New York, pp. 1–65.

Chew, H. (1987). Transition rates of atoms near spherical surfaces, *The Journal of Chemical Physics* 87(2): 1355–1360. URL: *http://link.aip.org/link/?JCP/87/1355/1*

Chew, H. (1988). Radiation and lifetimes of atoms inside dielectric particles, *Phys. Rev. A* 8(7): 3410–3416.

Das, P. & Metiu, H. (1985). Enhancement of molecular fluorescence and photochemistry by small metal particles, *J. Phys. Chem.* 89(22): 4680–4687.

Drexhage, K., Kuhn, H. & Schafer, F. (1968). Variation of fluorescence decay time of a molecule in front of a mirror, *Berichte Der Bunsen-Gesellschaft Fur Physikalische Chemie* 72(2): 329.

Dung, H. T., Knöll, L. & Welsch, D.-G. (2000). Spontaneous decay in the presence of dispersing and absorbing bodies: general theory and application to a spherical cavity, *Phys. Rev. A* 62: 053804. URL: *http://link.aps.org/doi/10.1103/PhysRevA.62.053804*

Fano, U. & Racah, G. (1959). *Irreducible Tensorial Sets*, Academic Press, New York.

Ford, G. W., & Weber, W. H. (1984). Electromagnetic interactions of molecules with metal surfaces, *Phys. Rep.* 113: 195.

Fort, E. & Grésillon, S. (2008). Surface enhanced fluorescence, *Journal of Physics D: Applied Physics* 41(1): 013001.
URL: *http://stacks.iop.org/0022-3727/41/i=1/a=013001*

Garrett, S. H., Wasey, J. A. E. & Barnes, W. L. (2004). Determining the orientation of the emissive dipole moment associated with dye molecules in microcavity structures, *Journal of Modern Optics* 51(15): 2287–2295.

Gersten, J. & Nitzan, A. (1981). Spectroscopic properties of molecules interacting with small dielectric particles, *The Journal of Chemical Physics* 75(3): 1139–1152.
URL: *http://link.aip.org/link/?JCP/75/1139/1*

Happer, W. (1972). Optical pumping, *Rev. Mod. Phys.* 44: 169–249.
URL: *http://link.aps.org/doi/10.1103/RevModPhys.44.169*

Hellen, E. H. & Axelrod, D. (1987). Fluorescence emission at dielectric and metal-film interfaces, *J. Opt. Soc. Am. B* 4(3): 337–350.
URL: *http://josab.osa.org/abstract.cfm?URI=josab-4-3-337*

Il'inskii, Y. A. & Keldysh, L. V. (1994). *Electromagnetic Response of Material Media*, Plenum Press, New York.

Klimov, V. V., Ducloy, M. & Letokhov, V. S. (1996). Spontaneous emission rate and level shift of an atom inside a dielectric microsphere, *J. Modern Opt.* 43(3): 549–563.

Klimov, V. V., Ducloy, M. & Letokhov, V. S. (2001). Spontaneous emission of an atom in the presence of nanobodies, *Quantum Electron.* 31(7): 569–586.

Klimov, V. V., Ducloy, M., Letokhov, V. S. & Lebedev, P. N. (1996). Radiative frequency shift and linewidth of an atom dipole in the vicinity of a dielectric microsphere, *J. Modern Opt.* 43(11): 2251–2268.

Klyshko, D. (2011). *Physical Foundations of Quantum Electronics*, World Scientific Publishing Company.

Kreiter, M., Prummer, M., Hecht, B. & Wild, U. P. (2002). Orientation dependence of fluorescence lifetimes near an interface, *J. Chem. Phys* 117(20): 9430–9433.
URL: *http://dx.doi.org/10.1063/1.1515732*

Landau, L. D. & Lifshitz, E. M. (1977). *Quantum Mechanics: Non-Relativistic Theory*, third edn, Pergamon Press, Ltd., New York.

Landau, L. D. & Lifshitz, E. M. (1980). *Statistical Physics*, part 1, third edn, Pergamon Press, Ltd., New York.

Lifshitz, E. M. & Pitaevskii, L. (1980). *Statistical physics : pt. 2: theory of the condensed state*, Vol. 9 of *Course of Theoretical Physics*, third edn, Pergamon Press, Ltd., New York.

Loisell, W. H. (1973). *Quantum Statistical Properties of Radiation*, Wiley, New York.

Lukosz, W. & Kunz, R. E. (1977). Light emission by magnetic and electric dipoles close to a plane interface. i. total radiated power, *J. Opt. Soc. Am.* 67(12): 1607–1615.
URL: *http://www.opticsinfobase.org/abstract.cfm?URI=josa-67-12-1607*

Macklin, J. J., Trautman, J. K., Harris, T. D. & Brus, L. E. (1996). Imaging and time-resolved spectroscopy of single molecules at an interface, *Science* 272(5259): 255–258.
URL: *http://www.sciencemag.org/content/272/5259/255.abstract*

Milonni, P. & Knight, P. (1973). Spontaneous emission between mirrors, *Optics Communications* 9(2): 119 – 122.
URL: *http://www.sciencedirect.com/science/article/pii/0030401873902393*

Nikolaev, G. N. (2006). Effective transfer of light energy to a nanoparticle by means of a resonance atomic lens, *JETP* 102(3): 394–405.

Omont, A. (1977). *Irreducible components of the density matrix: application to optical pumping*, 5 in *Progress in quantum electronics*, Pergamon Press, pp. 69–138 .
URL: *http://books.google.com/books?id=TmrRQgAACAAJ*

Purcell, E. M. (1946). Spontaneous emission probabilities at radio frequencies, *Phys. Rev.* 69(11-12): 681.

Ruppin, R. (1982). Decay of an excited molecule near a small metal sphere, *J. Chem. Phys.* 76(4): 1681.

Snoeks, E., Lagendijk, A. & Polman, A. (1995). Measuring and modifying the spontaneous emission rate of erbium near an interface, *Phys. Rev. Lett.* 74: 2459–2462.
URL: *http://link.aps.org/doi/10.1103/PhysRevLett.74.2459*

Sobelman, I. I. (1972). *Introduction to the theory of atomic spectra*, Vol. 40 of *International series of monographs in natural philosophy*, Pergamon Press, Oxford, New York.

Steiner, M., Schleifenbaum, F., Stupperich, C., Failla, A. V., Hartschuh, A. & Meixner, A. J. (2005). Microcavity-controlled single-molecule fluorescence, *Chem. Phys. Chem.* 6: 2190 – 2196.

Vallée, R., Tomczak, N., Gersen, H., van Dijk, E., García-Parajó, M., Vancso, G. & van Hulst, N. (2001). On the role of electromagnetic boundary conditions in single molecule fluorescence lifetime studies of dyes embedded in thin films, *Chemical Physics Letters* 348(3–4): 161–167.
URL: *http://www.sciencedirect.com/science/article/pii/S0009261401011198*

Varshalovich, D. A., Moskalev, A. N. & Khersonskii, V. K. (1988). *Quantum Theory of Angular Momentum*, World Scientific Pub Co Inc.

Wylie, J. M. & Sipe, J. E. (1984). Quantum electrodynamics near an interface, *Phys. Rev. A* 30(3): 1185–1193.

Wylie, J. M. & Sipe, J. E. (1985). Quantum electrodynamics near an interface. II, *Phys. Rev. A* 32(4): 2030–2043.

Yeung, M. S. & Gustafson, T. K. (1996). Spontaneous emission near an absorbing dielectric surface, *Phys. Rev. A* 54: 5227–5242.
URL: *http://link.aps.org/doi/10.1103/PhysRevA.54.5227*

Permissions

The contributors of this book come from diverse backgrounds, making this book a truly international effort. This book will bring forth new frontiers with its revolutionizing research information and detailed analysis of the nascent developments around the world.

We would like to thank Dr. Abbass A. Hashim, for lending his expertise to make the book truly unique. He has played a crucial role in the development of this book. Without his invaluable contribution this book wouldn't have been possible. He has made vital efforts to compile up to date information on the varied aspects of this subject to make this book a valuable addition to the collection of many professionals and students.

This book was conceptualized with the vision of imparting up-to-date information and advanced data in this field. To ensure the same, a matchless editorial board was set up. Every individual on the board went through rigorous rounds of assessment to prove their worth. After which they invested a large part of their time researching and compiling the most relevant data for our readers. Conferences and sessions were held from time to time between the editorial board and the contributing authors to present the data in the most comprehensible form. The editorial team has worked tirelessly to provide valuable and valid information to help people across the globe.

Every chapter published in this book has been scrutinized by our experts. Their significance has been extensively debated. The topics covered herein carry significant findings which will fuel the growth of the discipline. They may even be implemented as practical applications or may be referred to as a beginning point for another development. Chapters in this book were first published by InTech; hereby published with permission under the Creative Commons Attribution License or equivalent.

The editorial board has been involved in producing this book since its inception. They have spent rigorous hours researching and exploring the diverse topics which have resulted in the successful publishing of this book. They have passed on their knowledge of decades through this book. To expedite this challenging task, the publisher supported the team at every step. A small team of assistant editors was also appointed to further simplify the editing procedure and attain best results for the readers.

Our editorial team has been hand-picked from every corner of the world. Their multi-ethnicity adds dynamic inputs to the discussions which result in innovative outcomes. These outcomes are then further discussed with the researchers and contributors who give their valuable feedback and opinion regarding the same. The feedback is then collaborated with the researches and they are edited in a comprehensive manner to aid the understanding of the subject.

Apart from the editorial board, the designing team has also invested a significant amount of their time in understanding the subject and creating the most relevant covers. They scrutinized every image to scout for the most suitable representation of the subject and create an appropriate cover for the book.

The publishing team has been involved in this book since its early stages. They were actively engaged in every process, be it collecting the data, connecting with the contributors or procuring relevant information. The team has been an ardent support to the editorial, designing and production team. Their endless efforts to recruit the best for this project, has resulted in the accomplishment of this book. They are a veteran in the field of academics and their pool of knowledge is as vast as their experience in printing. Their expertise and guidance has proved useful at every step. Their uncompromising quality standards have made this book an exceptional effort. Their encouragement from time to time has been an inspiration for everyone.

The publisher and the editorial board hope that this book will prove to be a valuable piece of knowledge for researchers, students, practitioners and scholars across the globe.

List of Contributors

Sebastian Mackowski
Optics of Hybrid Nanostructures Group, Institute of Physics, Nicolaus Copernicus University, Torun, Poland

Mitsunori Yada and Yuko Inoue
Saga University, Japan

Motonari Adachi and Fumio Uchida
Fuji Chemical Co., Ltd., 1-35-1 Deyashikinishi-Machi, Hirakata, Japan

Keizo Nakagawa
Department of Advanced Materials, Institute of Technology and Science, The University of Tokushima, Minami-josanjima, Tokushima, Japan

Yusuke Murata
Toyo Tanso Co., Ltd., 5-7-12 Takeshima, Nishiyodogawa-ku, Osaka, Japan

Masahiro Kishida
Graduate School of Engineering, Kyushu University, 744 Motooka, Nishi-ku, Fukuoka, Japan

Masahiko Hiro
5Hitachi Chemical Co., Ltd., 2-1-1 Nishishinjuku, Shinjuku, Tokyo, Japan

Kenzo Susa
Trial Corporation, 2-195 Asahi, Kitamoto, Japan

Jun Adachi
National Instituite of Biomedical Innovation, 7-6-8 Asagi Saito, Ibaraki, Japan

Jinting Jiu
The Institute of Scientific and Industrial Research (ISIR), Osaka University, 8-1 Mihogaoka, Ibaraki, Japan

N. Venkatathri
Department of Chemistry, National Institute of Technology, Andhra Pradesh, India

Kazutaka Hirakawa
Faculty of Engineering, Shizuoka University, Japan

Jeffrey Yue, Xuchuan Jiang, Yusuf Valentino Kaneti and Aibing Yu
School of Materials Science and Engineering, University of New South Wales, Sydney, Australia

Udit Surya Mohanty, S. Y. Chen and Kwang-Lung Lin
Department of Materials Science and Engineering, National Cheng Kung University, Tainan, R.O.C, Taiwan

Shengyong Wu
Medical Imaging Institute of Tianjin, China

Margarita Sanchez-Dominguez
Centro de Investigación en Materiales Avanzados, S. C. (CIMAV), Unidad Monterrey, GENES-Group of Embedded Nanomaterials for Energy Scavenging, Mexico

Carolina Aubery and Conxita Solans
Instituto de Química Avanzada de Cataluña, Consejo Superior de Investigaciones Científicas (IQAC-CSIC), CIBER en Biotecnología, Biomateriales y Nanomedicina (CIBER-BBN), Spain

Hidemi Shigekawa and Shoji Yoshida
University of Tsukuba, Japan

Masamichi Yoshimura
Toyota Technological Institute, Japan

Yutaka Mera
University of Tokyo, Japan

Akihiko Matsuyama
Department of Bioscience and Bioinformatics, Kyusyu Institute of Technology, Japan

Sergey K. Filippov, Jiri Panek and Petr Stepanek
Institute of Macromolecular Chemistry, Academy of Sciences of the Czech Republic, Czech Republic

Braulio García-Cámara, Francisco González and Fernando Moreno
Grupo de Óptica, Departamento de Física Aplicada, Universidad de Cantabria, Spain

Raquel Gómez-Medina and Juan José Sáenz
Departamento de Física de la Materia Condensada, Universidad Autónoma de Madrid, Spain

Manuel Nieto-Vesperinas
Instituto de Ciencias de los Materiales de Madrid, CSIC, Spain

Gennady Nikolaev
Institute of Automation and Electrometry, Siberian Branch, Russian Academy of Sciences, Novosibirsk, Russia
Novosibirsk State University, Novosibirsk, Russia